Seeds of Woody Plants
in North America

Seeds of Woody Plants in North America

Revised and Enlarged Edition

James A. Young and Cheryl G. Young

DIOSCORIDES PRESS
Theodore R. Dudley, Ph.D., General Editor
Portland, Oregon

ISBN 0-931146-21-6
Printed in the United States of America

DISOCORIDES PRESS
9999 S.W. Wilshire, Suite 124
Portland, Oregon 97225

Library of Congress Cataloging-in-Publication Data

Young, James A. (James Albert), 1937-
    Seeds of woody plants in north America / James A. Young and Cheryl
G. Young. -- Rev. and enl. ed.
        p.   cm. -- (Biosystematics, floristic & phylogeny series ; v.
4)
    Rev. ed. of: Seeds of woody plants in the United States. 1974.
    Includes bibliographical references.
    ISBN 0-931146-21-6
    1. Tree crops--Seeds--Encyclopedias. 2. Woody plants--Seeds-
-Encyclopedias. 3. Tree crops--North America--Seeds--Encyclopedias.
4. Woody plants--North America--Seeds--Encyclopedias. I. Young,
Cheryl G. II. Young, James A. (James Albert), 1937- Seeds of
woody plants in the United States. III. Title. IV. Series.
SB170.Y68 1992
634.9'562--dc20   91-21710
                                                CIP

CONTENTS

PREFACE

Woody Plant Seed Manual was first published by the United States Department of Agriculture (USDA) in 1948 as Miscellaneous Publication 654. It was revised in 1974 as *Seeds of Woody Plants in the United States* and published by the Forest Service, USDA, as Agriculture Handbook 450. This revision covered the literature through 1968 and up to 1972, depending on the genera. Tremendous strides have been made in our knowledge of the propagation of woody plant species in the nearly two decades since the Handbook was last revised.

In the present volume we have more than doubled the number of genera included in the earlier Handbook—from 188 to 386 genera—and have added over 1000 new literature citations. This immense growth in the knowledge base reflects changing attitudes toward woody plant species. The original manual (1948) dealt with commercial timber species and woody species useful for shelterbelt plantings, largely on the Great Plains. This coverage was enlarged in the first revision (1974) to include many woody species useful as food and cover for wildlife. We have built on the basic interest in propagation of woody plants and have addressed more recent uses of woody plant material such as the use of native plant species in environmental plantings, renewed interest in plant material from Asia (especially China), and a general awareness of the importance of tropical forest species in the balance of the world's environment.

We have made an effort to cover the worldwide literature on the propagation of woody plant species from seeds. Some of the genera covered are adapted to limited areas of the United States, but they offer examples of the regeneration systems of different families of plants and expose the reader to the problems of forest regeneration in other environments. Discussion of these genera may be useful to researchers and plant propagators working in diverse environments.

Growth in the literature on germination of seeds of woody plants has not been distributed equally among genera. For commercially important genera, such as *Pinus*, there are hundreds of new literature citations. Some genera, such as *Acer* and *Fraxinus*, have served as

biological models for basic studies of the biochemical nature of deep-seated embryo dormancy. Important genera such as *Ulmus*, where seed dormancy is not a major problem, have not received a lot of attention from researchers. Seeds of many native plant species have been collected, threshed, stored, and germinated for the first time within the last decade.

This tremendous growth in the literature has forced us to streamline our presentation. Literature citations in Handbook 450 are not repeated in this volume, except for classic citations, such as the extensive writings of C. E. Heit, and early manuals, such as the one by Mirov and Kraebel (1939). We have attempted to streamline and standardize the presentation of tabular material. This revision drops much of the taxonomic information presented in Handbook 450. The user will need to use reference keys and manuals to identify plant material before using this book.

We have adopted the standard nomenclature of the Association of Official Seed Analysis (cited as ASOA 1985) for germination standards for all species where such standards have been published. We have made extensive use of the *Handbook of Seed Technology for Genebanks*, volume 2, *Compendium of Specific Germination Information and Test Recommendations*, written by R. H. Ellis, T. D. Hong, and E. H. Roberts of the University of Reading and published by the International Board for Plant Genetic Resources, Rome (cited as Genebank Handbook 1985).

This volume is not intended to serve as a manual on the vegetative propagation of woody plant species, but we have greatly expanded the coverage and introduced the concept of tissue culture techniques for woody plant propagation. This material was provided to introduce readers to sources of detailed information on these techniques.

We wish to acknowledge the expert editing of Dr. Elizabeth C. Dudley. Richard Abel conceived the idea of revising Handbook 450. Appreciation is expressed to the staff of the Life Science Library of the University of Nevada, Reno, for their assistance with reference material. Special thanks to Shannon Camille Juarez and Edward Estipona for many hours of assistance.

INTRODUCTION

As noted in the Preface, this is not a taxonomic manual. Plants must be identified before this volume is useful.

Plants are listed alphabetically by genus. Please note that some names have been changed from Handbook 450 to reflect correct modern usage. All measurements are given in metric units. Substrata for germination testing follow the standards of the Association of Official Seed Analysis:

B = between blotters
TB = top of blotters
T = paper toweling, either folded towel test or rolled towel test in horizontal or vertical position
S = sand or soil
TS = top of sand or soil
P = covered petri dish with (a) two layers of blotter, (b) three thicknesses of filter paper, or (c) top of sand or soil
C = creped cellulose paper
TC = top of creped cellulose paper

Appendix A provides an alphabetical listing of common names with their corresponding scientific names.

Appendix B gives a general indication of seed germination tendencies associated with specific genera and botanical families. More than 70% of the genera composing the legume family have seeds that require scarification compared to 13% for all genera. In the rose family, over 70% of the genera have species with seeds that require prechilling.

It also provides some statistics for those interested in the propagation of woody plants from seeds. Over 40% of the genera contain species whose seeds germinate without any pretreatment, provided the seeds have been properly harvested, threshed, and stored before being placed on a suitable substrate with moisture and adequate incubation temperatures. In contrast, only 2% of the genera contain species whose seeds are dormant and whose dormancy cannot be broken by conventional treatments.

The two major pregermination treatments are prechilling and scarification. Approximately 25% of the genera require prechilling to enhance or condition germination; 13% require seed coat or fruit pericarp removal or breaking by mechanical, chemical, or thermal means (scarification); and 6% require light to enhance or condition germination.

Roughly 10% of all the genera in this Appendix have species whose seeds are characterized by complex double dormancy. The most common form of this dormancy requires warm stratification followed by prechilling to condition germination. Seeds of these species often do not naturally germinate until two seasons after they mature. A few genera have more than two types of dormancy.

In evaluating these parameters it is important to remember that many types of seed dormancy are interrelated. Seeds may require either prechilling or light for germination, or a hard seed coat may prevent the leaching of an inhibitor from an embryo, but the inhibitor can be overcome by prechilling. In all genera there are exceptions to generalizations about types of germination systems. Despite these restrictions, Appendix B may provide valuable information about dormant seeds of a woody plant not covered in this volume.

OLEACEAE — OLIVE FAMILY

Abeliophyllum Nakai — Korean abelialeaf

GROWTH HABIT, OCCURRENCE, AND USE *Abeliophyllum* is a small shrub of rounded outline with multiple stems that are often straggly (Dirr 1983). At times it appears ragged-looking and disheveled. It is native to central Korea and was first introduced to the United States in 1924.

FLOWERING AND FRUITING The perfect white flowers are faintly tinged with pink and borne in axillary racemes. They appear in early spring.

PROPAGATION Freshly harvested seed should be planted immediately without pretreatment. Stored seeds may be dormant. Softwood cuttings root easily.

PINACEAE — PINE FAMILY

Abies Mill. — Fir

GROWTH HABIT, OCCURRENCE, AND USE *Abies* is a Northern Hemisphere genus of about 40 species found in North and Central America, Asia, Europe, and North Africa (Franklin 1974). Besides the native species, several species from Europe and Asia are grown in North America.

The typical appearance of *Abies* species is that of the classic evergreen. They are characterized by relatively narrow, often spirelike crowns, and distinct whorls of branches. Throughout much of the world, their height at maturity averages 15 to 30 m except near timberline, where they may be reduced to shrub height. Mature fir species native to western United States grow from 30 to 60 m in height on average sites.

Fir species occur from sea level to timberline, but a majority are found at middle to high elevations in mountainous areas and attain maximal development on these relatively cool, moist sites. Fir species are also found as components of boreal forests. A few species are found as components of low-elevation, temperate forests.

As long as there was an abundance of old-growth pine and Douglas fir in western North America, the lumber of fir species was generally not preferred in commerce because it was softer, weaker, and more perishable. On many mixed conifer sites species of pine were selectively logged and fir species left prior to World War II (Young et al. 1967). Old-growth trees of several species of fir are subject to various forms of heart rot which, in advanced stages, make them very dangerous to fall. Loggers avoided these low economic value, dangerous trees, accentuating the conversion of pine woodlands to *Abies* sites. Under pristine conditions in these pine woodlands, reproduction of firs was limited by frequent ground fires. As logging operations in the West exhausted old-growth stands of more desirable species and climbed to higher elevations in the mountains or went back to previously logged pine woodlands, the industry's utilization of fir species grew by default.

Exceptions to the general low quality of fir lumber have always been red (*A. magnifica*) and noble fir (*A. procera*) which produce strong, more durable wood. Late in the 19th century red fir was known as butterbox fir. The clear lumber it produced was preferred for butter and cheese boxes as fir wood did not influence the flavor of dairy products. Large quantities of red and noble fir are exported to Japan where the soft, white wood is popular in building construction.

Abies species are valuable sources of pulpwood. Balsam (*A. balsamea*) and Pacific silver (*A. amabilis*) fir are mainstays of the pulpwood industry in northeastern and northwestern United States, respectively. All native firs, except the rare *A. bracteata*, are utilized as pulpwood whenever supplies are adjacent to local markets.

Several species of fir are characteristic of subalpine landscapes. In such environments their form, texture, and color add to some of North America's most scenic areas. Fir species contribute to plant communities in critical watersheds where they are important factors in water yield and quality.

Increasing interest in growing native plants in gardens has heightened interest in culturing species of *Abies*. Where adapted, fir species make beautiful, but rather slow growing, lawn trees. Most native and some exotic *Abies* species are grown for Christmas trees and command premium prices.

During the last 2 decades the restocking of commercial fir sites for continued lumber production has become commonplace in contrast to the previous treatment of firs as secondary or weed species. Tree nurseries have made considerable strides in production techniques for transplanting stock.

GEOGRAPHIC RACES Foresters and botanists have long realized that where the ranges of certain species of fir overlapped there was considerable variability expressed. Likewise, for species which occupied a wide range of habitats over vast geographic areas, such as *A. grandis* and *A. concolor*, it was obvious that geographic races existed.

There have been 3 fairly recent monographic revisions of *Abies* (Franco 1950, Gaussen 1964, Lui 1971). The classification system of Lui (1971) has apparently been most widely accepted. He used subgenus *Pseudotorreya*, with a single section containing the species *bracteata*, and a second subgenus *Abies*, divided into 14 sections. Three subsections all contain continuously variable forms: *Grandis* contains *amabilis*, *concolor*, and *grandis*; *Nobilis* contains *magnifica* and *procera*; and *Balsameae* contains *balsamea*, *fraseri*, and *lasiocarpa*.

Recent studies by Critchfield (1988) offer evidence that crossing within sections is possible for *Abies*, but crossing between sections is difficult or impossible.

Franklin (1974) suggested that geographic origin is more important than specific or varietal name in selecting planting stock of firs in sections where the taxonomy is confused and hybridization probable. Use of local seed for planting stock is obviously the safest practice, even if self-limiting, until patterns of variation are better understood.

FLOWERING AND FRUITING The unisexual strobili of *Abies* are typically borne high in the crown. Female strobili are found on uppermost branches, where they occur singly or in small groups on the upper side of the previous year's twig growth. Male strobili cluster densely along the undersides of one-year-old twigs and generally occur lower in the crown than female strobili, although both male and female strobili are occasionally found on the same branchlet. Female strobili quickly elongate upward following bud burst, and in early developmental stages bracts are conspicuous (Fig. 1). Male strobili enlarge following bud burst but take on elongated, tassel form only during and after pollen shedding (Fig. 2).

At maturity, *Abies* cones are 7.5 to 25 cm long and typically cylindric or ovoid-shaped (Fig. 3). In some species, bracts are overgrown by cone scales early in development; in others, bracts remain conspicuous, nearly covering the entire surface of the cone in noble fir. Each scale bears 2 seeds at its base, although scales near the tip and base of the cone typically lack fertile seed. Ripening and seed dispersal take place the same fall and involve separation of cone scales and seed, leaving only the spikelike cone axis on the tree. Separation of scales and seed from the axis may be relatively active in species where scales are distorted during drying; in species where scales are not distorted, separation is passive, requiring branch movement or other disturbance for dislodgment. Wind is the chief agent for dispersal of seed. The winged seeds of red fir are often blown from the Sierra Nevada into the valleys on the margin of the Carson Desert, in western Nevada. Although most seed is usually disseminated in the fall, seedfall of some firs may continue well into winter (Table 1).

A mature seed has a large wing and is typically ovoid or oblong in shape (Figs. 4, 5). The rather soft seed coat is a shade of brown, tan, or, rarely, cream in color and contains resin vesicles. The number, character, and placement of the vesicles vary with species. Most of the seed is occupied by a fleshy endosperm and a well-developed embryo which may extend nearly the length of the endosperm (Fig. 6). The cotyledons, which vary from 3 to 14, are well differentiated.

Seed bearing of most firs typically begins at 20 to 30

Figure 1. *Abies procera*, noble fir: female strobili at receptive stage, ×1 (Franklin 1974).

Figure 2. *Abies procera*, noble fir: male strobili at time of pollen shedding, ×1 (Franklin 1974).

A. amabilis
Pacific silver fir

A. balsamea
balsam fir

A. concolor
white fir

A. fraseri
Fraser fir

A. grandis
grand fir

A. lasiocarpa var. *lasiocarpa*
subalpine fir

A. magnifica var. *shastensis*
Shasta red fir

A. procera
noble fir

Figure 3. *Abies*: mature female cones, ×0.5 (Franklin 1974).

years, with larger crops generally occurring at 2- to 4-year intervals (Table 1). There is evidence of a 3-year cycle for good crops of several northwestern firs, with climatic conditions determining whether the potential bumper crop is realized. Other species seem to have a 2-year periodicity of cone production.

Many agents reduce cone yields. Typically, a large amount of mature *Abies* cones is empty. In one example, red fir cones were found to contain 65 to 93% empty seed. Keen (1968) noted 72 to 75% sterile or empty seed in white and red fir, respectively. Extensive collections of unprocessed seed of noble, Shasta red, subalpine, Pacific silver, and grand fir have revealed similar high percentages of empty seed. The factors responsible have not been identified, but lack of pollination or genetic irregularities have been suggested. Collection of immature seed and improper cone and seed processing may also contribute to the nonviable fraction in commercial seed lots.

Seed production can also be reduced by adverse weather conditions, squirrels, and birds. Female strobili may be entirely or partially aborted up to 6 to 8 weeks after bud burst by late spring frosts. Pollen dispersal can be reduced by adverse weather. Fir cones are a preferred source of food for squirrels in some locations. Large quantities of cones are cut and cached; such cutting may also reduce future cone crops.

Cone and seed insects may significantly reduce seed yields and occasionally may totally destroy seed crops. Seed chalcids (*Megastigmus*) are common and may be abundant enough to have a major impact. Cone moths (e.g., *Barbara colfaxiana siskiyouana* and *Dioryctria abietella*) and cone maggots (*Earomyia*) cause the most conspicuous damage. Cone and scale midges may reduce seed yields.

COLLECTION OF FRUIT Fir cones are usually collected by hand from standing or recently felled trees, or from squirrel-cut cones on the ground or in caches. In the western United States extensive collections are made in 40- to 70-year-old stands which have developed in old burns; trees on these sites are often open-grown and easy to climb. Collection of squirrel-cut cones is easier, and there is no evidence that seed collected in this way is inferior. Observations in the Pacific Northwest suggest that squirrels do not begin to cut cones until the seeds have matured. The cones are typically cached in cool, moist microsites conducive to afterripening. The period available for cone collection is only about a month between fruit ripening and the beginning of seed dispersal (Table 1). Since fir cones disintegrate, collection is not possible after dispersal begins.

Germination of fir seeds increases almost up to the time of dissemination. Cone and seed maturity indices have been sought for noble and grand fir to identify the earliest possible collection date (Table 1). Proposed indices are based on crude fat content of the seed, cone specific gravity, loosening of seeds from cone scales, and color change of seed wings.

Artificial ripening studies have shown that germinability of seed can be improved by storing cones which are collected early under cool, moist conditions for several weeks. Storage in moist peat has proven deleterious.

EXTRACTION AND STORAGE OF SEED Seeds of most species of fir undergo 2 stages of ripening. The first involves movement of materials from the cone scale to the seed. The second involves metabolic changes in the seed itself. For this reason, seed should not be extracted from cones immediately after collection, particularly

Table 1. *Abies:* phenology of flowering and fruiting, and characteristics of mature trees (Franklin 1974).

Species	Flowering dates	Fruit ripening dates	Seed dispersal dates	Mature tree height (m)	Seed-bearing Age (years)	Seed-bearing Interval (years)
A. alba	May–June	Sept.–Oct.	Sept.–Oct.	40	25–30	2–3
A. amabilis	Apr.–June	Aug.–Sept.	Aug.–Sept.	60	30	4–8
A. balsamea	May	Aug.–Sept.	Oct.	18	15	3–5
A. bracteata	Apr.–May	Aug.	Sept.	30	–	3–5
A. concolor	May–June	Sept.–Oct.	Sept.–Oct.	55	40	2–4
A. firma	Apr.–May	Oct.	Oct.–Nov.	40	–	4–6
A. fraseri	May–June	Sept.–Oct.	Sept.–Nov.	22	15	3
A. grandis	Mar.–June	Aug.	Aug.–Sept.	62	20	3
A. homolepis	May–June	Sept.	Sept	28	–	5–7
A. lasiocarpa						
var. arizonica	June	Sept.–Oct.	Sept.–Oct.	34	50	2–3
var. lasiocarpa	June–July	Aug.	Sept.	34	20	2–4
A. magnifica						
var. magnifica	May–June	Aug.	Sept.–Oct.	49	35–45	2–3
var. shastensis	May–June	Aug.	Sept.–Oct.	49	30–40	2–3
A. mariesii	June	Sept.	Sept.–Oct.	22	–	5–7
A. nordmanniana	May	Sept.–Oct.	–	54	30–40	2–3
A. procera	June–July	Sept.	Oct.	71	12–15	3–6
A. sachalinensis	May–June	Sept.–Oct.	Oct.	31	–	2–4
A. veitchii	June	Sept.–Oct.	Oct.	25	30	5–6

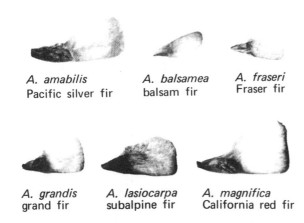

A. amabilis
Pacific silver fir

A. balsamea
balsam fir

A. fraseri
Fraser fir

A. grandis
grand fir

A. lasiocarpa
subalpine fir

A. magnifica
California red fir

Figure 4. *Abies*: seeds, ×1 (Franklin 1974).

not from early collected cones. Immediate extraction can result in seed of low viability.

In practice, sacked fir cones are usually stored for several weeks or months in drying sheds. Sacks of cones should not be stacked to maintain good air circulation and thus prevent heating and mold growth. Processing of fir cones is similar to processing of cones for other conifers (Table 2). Cones are kiln-dried at 30 to 38°C or air-dried for one to 3 weeks or more at 22 to 30°C. The partially or wholly disintegrated cones are tumbled, shaken, and screened to separate the seed. The seed is dewinged by hand or mechanically. Wings and other impurities are removed in one or more stages with air screens. Cone and seed yields and weight of seed are variable among species and collections (Table 2).

Abies seed is relatively fragile and can be damaged easily, especially during dewinging. Viability losses in

storage may actually be due to processing damage. Hand dewinging is recommended for seeds of balsam fir.

In the United States, seed is generally stored in unsealed drums, cans, or plastic bags at or near −15°C (Table 2). A low seed moisture content (9 to 12%) is best. Seed with high moisture content stores poorly. Fir seeds can be stored at low temperatures for 5 years or more without significant loss in viability.

Methods using gas-liquid chromatography analysis of terpenoids in the seed coat of *Abies* have been developed (Zavarin et al. 1976). This methodology has potential for distinguishing seed provenances.

GERMINATION We have not actively worked with *Abies* species except to grow a few common species from seed. Based on his extensive experience in the Pacific Northwest, Dr. Franklin did one of the more comprehensive efforts in the last revision of Handbook 450. There are 4 aspects to the knowledge we collectively term "germination ecology." These are (1) nursery practices, where the main concern is how to grow the plant from seed as economically as possible; (2) seed technology, where the concern is with testing and marketing seed of standard quality; (3) plant physiology, where the mechanisms of dormancy and germination are the concerns; and (4) field ecology, where the concern is how germination functions for the species in question in field environments. Ideally, understanding of these 4 factors has progressed in an interacting environment with free exchange of background knowledge from each specialty.

Laboratory studies of the germination of seed of *Abies* species illustrate 4 factors about the basic physiology of seed: (1) the embryo can be dissected from the seed coat and made to germinate without

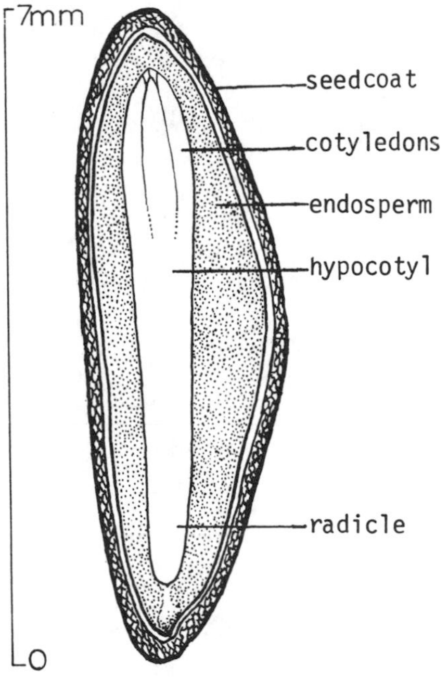

7mm

seedcoat

cotyledons

endosperm

hypocotyl

radicle

0

Figure 5. *Abies magnifica* var. *magnifica*, California red fir: longitudinal section through a seed, ×12 (Franklin 1974).

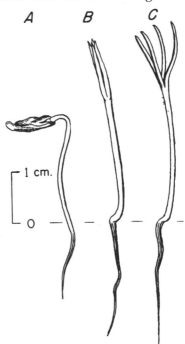

A   B   C

1 cm.

0

Figure 6. *Abies balsamea*, balsam fir: seedling development at 2, 5, and 7 days after germination (Franklin 1974).

Table 2. *Abies:* cone drying schedules, seed storage conditions, and cleaned seed weight (Franklin 1974, Browse 1979, and ASOA 1985).

Species	Cone processing schedule			Seed storage conditions		Cone wt./ 35 liters (kg)	Seeds/100 kg cones (kg)	Seeds/cone	Seeds/gram
	Air-drying time (days)	Kiln drying		Seed moisture (%)	Temperature (°C)				
		Time (hr)	Temperature (°C)						
A. alba	–	–	–	–	–	12.7	2.5	–	23
A. amabilis	60–80	6–14	30	–	–20	–	–	400	25
A. balsamea	20–30	0	–	9	–	15.9	–	134	130
A. cephalonica	–	–	–	–	–	–	–	–	18
A. cilicica	–	–	–	–	–	–	–	–	10
A. concolor	7–14	0	–	10	10	14.5	1.4	–	35
A. firma	14	0	–	–	10	–	–	–	24
A. fraseri	30–45	0	–	12	–12	–	–	–	125
A. grandis	60–180	6–14	30	–	–	–	–	115	50
A. homolepis	14	–	48	–	–	21.4	–	78	65
A. kawakamii	–	–	–	–	–	–	–	–	54
A. koreana	–	–	–	–	–	–	–	–	200
A. lasiocarpa									
var. arizonica	–	–	–	–	–	–	–	–	49
var. lasiocarpa	–	–	–	–	–	–	–	–	76
A. magnifica									
var. magnifica	8	0	–	–	–20	12.3	1.8	–	14
var. shastensis	–	–	–	–	–	–	–	–	16
A. mariesii	14	–	48	–	5	13.2	–	59	51
A. nordmanniana	–	–	–	–	–	15.9	5.5	–	16
A. pindrow	–	–	–	–	–	–	–	–	25
A. pinsapo	–	–	–	–	–	–	–	–	25
A. procera	60–180	6–14	30	–	–	–	–	500	30
A. sachalinensis	14	–	48	–	5	–	–	–	97
A. veitchii	–	–	–	–	–	–	–	–	130

dormancy problems, suggesting that dormancy is associated with the covering of the embryo; (2) light is necessary for germination in most species; (3) prechilling seed for 2 to 4 weeks enhances germination and (4) reduces or in some species replaces the light requirement for germination. The interplay of these factors is clearly shown in the standards for laboratory testing of fir seed (Table 3). The detailed studies of *A. fraseri* conducted by Adkins et al. (1984) and Blazich and Hinesley (1984) also clearly show these relationships. The detailed review table presented by Franklin (1974) illustrates how foresters, often working without lighted germinators, approached the germination problems of fir seed from the prechilling (stratification) aspect (Table 4).

One of the more recent and interesting studies of the prechilling approach to breaking dormancy of fir seeds is the work of Edwards (1982). Seed was prechilled and dried back to 35% moisture content, then stored for as long as 12 months. In some cases the storage period further enhanced germination. Using a variation of this procedure Tanaka and Edwards (1985) improved emergence of fir seedlings 10 to 19% over conventionally prechilled seed. Hall and Olson (1986) used similar techniques on seed of noble and Pacific silver fir and stored the treated seed for as long as 360 days without significant loss in viability. However, results were not consistent among seed lots. One of the problems with germinating seed of *Abies* species in the laboratory has always been infection of plates by fungal

species, especially *Rhizoctonia solani.* Heit and Natti (1969) used 70 to 75% a.i. dust of pentachloronitrobenzene to control fungal infection on fir seed during germination tests without influencing germination. Edwards and Sutherland (1979) used hydrogen peroxide solutions to reduce fungal spores on the surface of fir seed. The treatments did not result in enhanced germination of grand or amabilis fir seed.

The transfer of laboratory germination studies to field seedbeds seems to have several gaps and conflicts in the case of fir species seed. Many species occupy habitats with prolonged snow cover during the winter—red fir in the Sierra Nevada, for example, experiences some of the greatest annual snowfall accumulations known. It has been known for some time that deep snow cover provides a prechilling environment for fir seed. Is there sufficient light transmission in snow to influence the light requirement of fir seed? Students in the botany department at the University of California at Davis have conducted a series of outstanding studies on the ecology of red fir (e.g., Ustin et al. 1984). These studies have concentrated on the importance of light in the natural regeneration of this species beneath old-growth canopies. The emphasis in this work has been on moisture relations and light for seedling photosynthesis. Is there an influence of light flecks through canopy openings on germination through light quality, quantity, incubation temperature, and the interaction of all three?

Table 3. *Abies:* laboratory germination methods (compiled from Heit 1976a, ASOA 1985, and Handbook for Genebanks 1985).

Species	Substrata	Temperature (°C)	Duration (days)	Additional directions
A. alba	TB	20/30	28	Prechill 21 days at 3–5°C.
A. amabilis	P	15/25	28	Light; prechill 21 days at 0–5°C.
A. balsamea	TB, P	20/30	21	Light; prechill 28 days at 0–5°C.
A. bornmuelleriana	TB	20/30	18	Light; prechill 21 days at 3–5°C.
A. Borisii var. regis	TB	20/30	18	Light; prechill 21 days at 3–5°C.
A. cephalonica	TB	29/30	28	Light; prechill 21 days at 3–5°C.
A. cilicica	TB	20/30	28	Light; prechill 21 days at 3–5°C.
A. concolor	TB, P	20/30	28	Light; many lots complete in 14–21 days. May require prechilling 21 days at 3–5°C.
A. Ernestii	TB	20/30	21	Light; prechill 21 days at 3–5°C.
A. firma	TB	20/30	18	Light; prechill 21 days at 3–5°C.
A. fraseri	TB	20/30	21	Light; prechill 28 days at 3–5°C.
A. grandis	TB, P	20/30	28	Light; prechill 28 days at 3–5°C. Vermiculite can be used for substrate.
A. homolepis	TB, P	20/30	21	Light; prechill 21 days at 3–5°C.
A. Koreana	TB	20/30	21	Light; prechill 21 days at 3–5°C.
A. lasiocarpa	TB	20/30	21	Light.
A. lasiocarpa var. Arizonica	TB	20/30	28	Light.
A. magnifica	TB, P	20/30	21	Prechill 28 days at 3–5°C.
A. mariesii	TB	20/30	28	Light.
A. mayriana	TB	20/30	21	Light; prechill 21 days at 3–5°C.
A. nordmanniana	TB	20/30	18	Light; prechill 21 days at 3–5°C.
A. numidica	TB	20/30	28	Prechill 21 days at 3–5°C.
A. pinsapo	TB	20/30	28	Prechill 21 days at 3–5°C.
A. procera	TB, P	20/30	28	Light; prechill 14 days at 3–5°C. Vermiculite recommended as substrate. For dark germination prechill 21 days.
A. sachalinensis	TB	20/30	28	Prechill 21 days at 3–5°C.
A. spectabilis	TB	20/30	28	Light.
A. veitchii	TB	20/30	28	Prechill 21 days at 3–5°C.

Results in greenhouse experiments suggest that soil microbiological factors may influence germination of seed of fir species. Soils from different habitat types that were similar in physical characteristics produced differential initiation and duration of germination (Minore 1986). Past emphasis has been on the temperature of fir seed germination, but recent studies have taken the more difficult approach of studying moisture relations during germination (Thomas and Wein 1985, Jobidon 1986, Singh et al. 1986). There is evidence for seed of balsam fir that moisture must remain constantly available during the germination process in contrast to the seed of other types of conifers that can survive drying and rewetting cycles and still germinate.

There has been considerable concern in northeastern United States and Canada that acid rain would influence regeneration of forest species. Evidence to date indicates that seed of balsam fir would not be influenced in germination by current levels of acid rain so far as they affect soil water solutions (Scherbatskay et al. 1987).

NONGERMINATIVE TEST OF SEED QUALITY With a large percentage of fir seed in a given lot potentially empty, a number of tests have been developed to judge seed quality. The simplest are a cutting test or an X-ray analysis to determine percentage of filled seed. Hydrogen peroxide has been used, either to speed standard germination tests or as a key component of special tests. Tetrazolium tests have also been used, and results often correlate with seedling emergence

Table 4. *Abies:* nursery practices (Franklin 1974).

Species	Prechilling (days)	Season	Density (m²)	Depth (cm)
A. alba	0	Fall	340	1.9
A. amabilis	28	Spring	340	2.0
A. balsamea	0	Fall	700	0.6
A. bracteata	0	Fall	–	0.6
A. concolor	28	Spring	370	1.9
	0	Fall	340	1.9
A. firma	45	Spring	–	–
A. fraseri	0	Fall	230	0.6
A. grandis	0	Spring	375	2.0
	35	Spring	315	–
A. homolepis	45	Spring	–	–
A. lasiocarpa	0	Fall	–	–
A. magnifica	42	Spring	370	0.6
A. nordmanniana	60	Spring	525	2.5
A. procera	35	Spring	375	0.6
A. sachalinensis	45	Spring	–	–

obtained in nursery sowing. In comparison with standard germination tests, quick tests generally overestimate viablity of fir seed.

NURSERY AND FIELD PRACTICE In general, nursery practices follow local conditions and traditions (Table 4). Spring sowing of prechilled seed is the standard practice for most western North American and Japanese species of fir. *Abies alba, A. balsamea,* and *A. fraseri* are normally fall-sown without prechilling. In some European nurseries species such as noble and white fir are also sown in the fall.

Mulches of sawdust or straw are used by some nurseries to protect seedlings during the first winter. Fir seedlings have slow initial growth, and stock is usually outplanted as 2- to 3-year-old seedlings or 3- to 4-year-old transplants.

LEGUMINOSAE — LEGUME FAMILY

Acacia Mill.

GROWTH HABIT, OCCURRENCE, AND USE Acacia is a distinct genus of 450 to 500 species dispersed throughout the tropics and, to some extent, the temperate regions (Bailey 1951). The greatest development of the group is in Australia where nearly 300 species are found (Menninger 1962). About 70 species are found in subtropical portions of the United States (Whitesell 1974a). Certain species of *Acacia* are among the most beautiful flowering trees. However, some are hardy only in the nearly subtropical portions of Florida, Texas, and California. Several introductions of *Acacia* to Hawaii have become naturalized and are considered pests. Species that spread by root suckers can take over areas where they are adapted.

Several species, especially *A. senegal* and *A. arabica,* are commonly used for the production of gum arabic. Acacias are valuable for many other purposes: collectively they yield lumber, face veneer, furniture wood, fuel wood, tannin, resin, medicine, fiber, perfume, and dye stock. Some are hosts of valuable species of lac insects. The browse and fruit (legume) of *Acacia* species are rich sources of digestible protein for many herbivores. Some species have been proposed as fodder and fuel wood plants for semiarid and arid tropical areas of the world where tree and shrub resources have been overutilized (Borr and Atkinson 1970, Patil and Pathak 1977). Their relatively rapid growth, fixation of nitrogen, and production of valuable browse and seedpods make acacias popular species for reforestation projects.

As is the case with many legume species, *Acacia* seed is parasitized by insects during maturation of the legume. In many parts of the world seed beetles (*Bruchidae*) infest the seed (Halevy 1974). Insect larvae will render it nonviable if the insects are allowed to complete their life cycle. Large herbivores are attracted to maturing legumes of *Acacia* species in East Africa, and they consume large quantities of the fruit. Digestive fluids in the rumen selectively kill larvae of the bruchid beetle without reducing the seed viability (Lamprey et al. 1974). Digestive juices may also influence seedcoat. Domestic livestock consume significant quantities of fallen *A. tortilis* seedpods in arid

pastoral ecosystems of northwest Kenya (Coughenour and Detling 1986). These seeds are subsequently defecated in corrals where dense populations of this species arise. Periodic wildfires that burn through *Acacia* stands also selectively influence the bruchid beetle whose larvae are killed by lower temperatures than the minimal lethal temperature for seed viability (Sabiiti and Wein 1987). Again, the fire is a double-action factor: it not only kills the insect larvae, but also may influence subsequent germination through altering the seedcoat.

FLOWERING AND FRUITING *Acacia* flowers are perfect or polygamous, yellow in many species, but white in some. Because the genus is so large there is considerable variation in flower arrangement; frequently flowers are displayed in condensed, cylindrical or globular spikes. The fruit is a 2-valved dehiscent (two sections of a typical legume pod twist open and disperse seeds while still on plant) or an indehiscent legume that matures in late summer.

As would be expected the seed of *Acacia* occurs in a variety of shapes, sizes, and colors (Figs. 1, 2). An important feature of the seed is the hilum or scar where the placenta stalk connected it to the mother plant. This scar is important because it contains the strophiole which is implicated in the breaking of hard seededness in these species (Hanna 1984). The palisade cells of the strophiole in *Acacia* seed are much shorter than elsewhere in the seed coat, and the vascular bundle most closely approaches the surface. Depending on the species, the seed contains endosperm in thin to thick layers or lacks it entirely (Gunn 1984). Some species have seed with an aril on the hilum, which may aid dispersal. However, Tran (1981) considered the aril to influence germination. Acacias begin to bear seed between 2 and 4 years of age. Most species produce abundant seed crops every year.

COLLECTION, EXTRACTION, AND STORAGE OF SEED The collection method obviously depends on the species involved and the type of legume, whether indehiscent or dehiscent. Pods of indehiscent types can be picked from the ground beneath the trees. Seedpods of black wattle acacia (*A. mearnsii*) were swept from the ground

Figure 1. *Acacia*: seeds, ×4 (Whitesell 1974a).

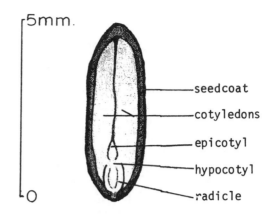

Figure 2. *Acacia melanoxylon,* blackwood acacia: longitudinal section through a seed, ×10 (Whitesell 1974a).

in seed orchards where the ground was cleared and prepared before seed maturity (Poggenpool 1978). Dehiscent pods must be picked from trees before they open. Whitesell (1974a) offered information on seed dispersal and yield for 4 species grown in California and Hawaii (Table 1).

Seeds of *Acacia* that are free of insect larvae can be stored for long periods. They should be thoroughly dried, stored in sealed containers, and protected from excessive heat (Cavanaugh 1987). Seeds stored in closed containers have remained viable for 50 years or more. Seeds of koa (*A. koa*), lying on the ground under trees, have been known to retain viability for as long as 25 years. Seed of black wattle acacia were stored for 17 years in open bottles without significant reduction in viability (Goo et al. 1979).

GERMINATION Seed of many species of *Acacia* is not dormant at maturity, but as further postmaturity drying occurs it becomes dormant (Cavanaugh 1987). Seed of *A. auriculiformis* has been shown to have about 60% germination at full maturity (Pukittayacamee and Hellum 1988). At 89 days postmaturity, germination increased to 77%, but by 110 days postmaturity, germination decreased to 51%. With other species in more arid environments the development of postmaturity dormancy with drying is more pronounced. Seed of *A. suaveolens* was 96% dormant after maturity, from a population that was 93% viable. Auld (1986a) suggested that techniques for breaking seed dormancy of *Acacia* could be placed in 5 groups:

(1) Immersion in hot or boiling water
(2) Acid scarification (usually concentrated H_2SO_4)
(3) Dry heat
(4) Mechanical scarification
(5) Exposure to microwave heat

Auld compiled the results of one or more of these five germination enhancement tests on 51 species of *Acacia*. Germination of untreated seed ranged from 0 to about 45% depending on species and collection techniques. Application of one or more of the 5 suggested enhancement treatments resulted in germination from 75 to 100%, again dependent on species and collection. For example, in different studies reported in the literature as cited by Auld (1986a), seed of *A. longifolia* was reported to have the following percentage germination: (1) control 0%, (2) boiling water 78%, (3) dry heat 100%, and (4) microwave 75 to 95%. No citation was given for acid scarification.

Acid scarification is usually conducted with concentrated sulfuric at the rate of 1.2 liters of acid per kilogram of seed at 25 to 27°C (Cavanaugh 1987). Experiments varying the ratio of acid to *Acacia* seed have been conducted. For *A. planifrons*, it was determined that the optimum ratio of sulfuric acid to seed was 160 ml per kilogram. *Acacia* seed that has been dried after scarification is regularly stored in Australia for as long as 6 months. Sniezko and Guaze (1987) stored acid-scarified seed in Africa for a year. Care needs to be taken so that heating does not occur during the acid

Table 1. *Acacia:* phenology of flowering and fruiting, pod size, and cleaned seed weight (Whitesell 1974a).

Species	Location	Flowering dates	Fruit ripening dates	Seed dispersal dates	Pod size (cm)	Seeds/gram
A. decurrens	California	Feb.–Mar.	–	–	10	70
A. koa	Hawaii	Jan.–July	June–July	Nov.	11	11
A. mearnsii	California	June	June–Oct.	Oct.	6	74
A. melanoxylon	California	Feb.–June	July–Nov.	Dec.	8	66
	Hawaii	May–June	–	–	–	–

scarification or the seed will be subject to fungal infection when planted.

Hot water appears to be the most effective treatment for breaking the hard seeded dormancy of *Acacia* seed in Australia, while acid scarification gives the best results with African species (Doran et al. 1983, Sniezko and Guaze 1987). An exception to this is the comprehensive information available from South Africa where filing seeds, dropping them in hot water, then soaking them is the most recomended treatment to enhance germination (Table 2) (Carr 1976).

The Australian hot water or moist heat treatment for *Acacia* seed is usually given as dropping the seed into boiling water and then removing the water from the heat source. Cavanaugh (1987) compiled the following information from the published literature to illustrate the variablity and outright lack of information for some species:

Species	Treatment
A. acminata	5 seconds maximum at 100°C
A. argyrodendron	None if fresh seed
A. cambagei	None if fresh seed
A. harpophylla	None if fresh seed
A. oncinocarpa	Unknown
A. peucer	Unknown
A. suaveolens	200 to 600 seconds at 80°C
A. sylvestris	Short periods up to 80°C
A. terminalis	30 seconds maximum at 100°C, or 100 to 600 seconds at 80°C

After treatment with hot water to break dormancy, the seeds of some species are very subject to fungal infection.

Dry heat has been used to scarify *Acacia* seed, but the maximum temperature and duration of treatment are specific for each species. The heat and duration requirements may vary among seed lots or among seeds of the same lot if moisture content is variable (Cavanaugh 1987).

Mechanical scarification of *Acacia* seed, except for hand treatment of small lots of seed, is difficult to accomplish. Most commercial scarifiers are built for small-seeded legume species that have thinner seed-coat than those of *Acacia*. The seedcoats of many species of *Acacia* is so tough that mechanical scarification can only be accomplished with excessive damage to the seed viability (Doran et al. 1983).

For dormant seed of an exotic species of *Acacia*, Auld (1986b) provides an excellent introduction to the extensive Australian literature and a summary of the response of many species to germination enhancement treatments. Microwave treatment to enhance germination of hard-seeded species is an interesting new aspect to seed physiology (Tran 1979). A domestic microwave oven that operates at a frequency of 2450 MHz is suitable for scarifying seed. Microwave heat has advantages over conventional sources of rapid heating and very uniform distribution of heat, even with bulk seed lots. Treatment duration varies from 30 seconds to 4 minutes.

Table 2. *Acacia:* species of southern Africa (Carr 1976).

Species	Seed parasitization	Germination[1]
Group 1: Thorns in pairs, inflorescence capitate.		
A. davyi	Some	F,B,S,
A. erioloba	Minimal	F,B,S,
A. erioloba	High	File two
A. exuvialis	Moderate	F,B,S
A. gerrardii	Moderate	F,B,S
A. grandicornuta	Moderate	F,B,S
A. haematoxylon	Moderate	F,B,S
A. hermanii	Minimal	hot water
A. karroo	Moderate	F,B,S
A. kirkii	High	F,B,S
A. nilotica ssp kraussiana	Low	F,B,S
A. permixta	Low	F,B,S
A. redacta	?	?
A. rehmanniana	Moderate	F,B,S
A. robusta	?	F,B,S
A. rogersii	None	F,S
A. sieberana	High	F,B,S
A. stuhlmannii	Low	File two
A. swazica	Moderate	F,B,S
A. tenuispina	Minimal	F,B,S
A. xanthoploea	High	F,B,S
Group 2: Straight thorns in pairs, inflorescence spicate.		
A. albida	Modcrate	F,B,S
Group 3: Straight and hooked thorns in pairs, inflorescence capitate.		
A. leuderitzii	Minimal	F,B,S
A. reficiens	Minimal	F,B,S
A. stolonifera	Minimal	F,B,S
A. tortilis ssp. heteracantha	Variable	F,B,S
Group 4: Hooked thorns in pairs, inflorescence spicate.		
A. burkei	Heavy	File two
A. caffra	Minimal	B,S
A. eobynsiana	Heavy	B,S
A. erubescens	Minimal	S
A. fleckii	Minimal	F,S
A. galpinii	Variable	F,B,S
A. hereroensis	High	F,B,S
A. mellifera ssp. detinens	Minimal	B,S
A. montis-usti	Minimal	F,S
A. nigrescens	Moderate	F,B,S
A. polyacantha	Moderate	F,B,S
A. welwitschii ssp. delagoensis	Heavy	F,B,S
Group 5: Hooked thorns in threes, inflorescence spicate.		
A. senegal ssp. leiorhachis	Minimal	F,B,S
Group 6: Prickles scattered along stems, inflorescence capitate.		
A. brevispica ssp. dregeana	High	B,S
A. kraussiana	Moderate	F,B,S
A. schweinfurthii var. schweinfurthii	Minimal	F,B,S
Group 7: Prickles scattered along stems, inflorescence spicate.		
A. ataxacantha	?	B,S

[1]F = file tests, B = drop into boiling water and remove from heat, S = allow to soak in cooling water for 24 hours before sowing, File two = file two sides of testa until cotyledons exposed.

Outside Australia, germination enhancement studies for *Acacia* species have been reported from several areas. In south Texas, Everitt (1983a, 1983b) reported that seed of *A. schaffneri* and *A. rigidula* required acid scarification for 30 to 45 minutes for germination, while that of *A. berlandieri* germinated without pretreatment. Vora (1989) suggested that the duration of acid scarification for seed of *A. schaffneri* collected in south Texas should be increased to 120 minutes, and the same treatment was necessary for seed of *A. smallii*.

In South Africa, Jeffery et al. (1988) reported on the influence of dry heat on the germination of a number of native and exotic species of *Acacia*. Golden wattle (*A. longifolia*) is an invasive alien in South Africa. McDowell and Mool (1981) compared its germination ecology to that of the native legume, *Virgilia oroboides*, with the idea of manipulating seedbed ecology to favor the native species. In Kenya, Murugi (1987) compared presowing treatments for native *Acacia* species that are important in forestry. For the hard seed of the native Kenyan tree *Acacia xanthoploea*, he found nicking the seed coat by hand to be the most efficient method of breaking seed dormancy. Al-Kinany (1981) studied not only the germination enhancement effects of presowing treatments on golden wattle, but he also measured subsequent seedling growth in relation to these treatments.

In an interesting study conducted in Sudan, Bebawi and Mohamed (1985) used different irrigation frequencies to selectively influence germination of species of *Acacia*. They related these results to the natural occurrence of the species relative to available soil moisture.

Whitesell (1974a) provided a complex table (Table 3) of the influence of pregermination treatments on seed of *Acacia* species adapted to California and Hawaii. As is the case with all studies of the germination of *Acacia* seeds, the results were dependent on species and collections of seeds within species.

In data compiled by Morrison (1987) it was shown that for mature seeds of *A. suaveolens* heavier than 20 mg, shown by a tetrazolium test to be more than 98% viable, less than 1% would germinate spontaneously on release from the fruit. The remaining seeds entered the soil seed bank. For seeds of six species of *Acacia* grown in Sudan, it was determined that 19% of the bulked populations of all species had seed germination without pretreatment (Bebawi and Mohamed 1985). In an unusual treatment for seeds of a tropical species, Zwaaw (1978) first treated seeds of black wood acacia for 3 minutes in 90°C water and then prechilled them for 3 to 6 weeks before planting. This combination of hot water and prechilling produced 98% germination. Hot water treatment at 100°C resulted in loss of seed viability for this species. Pathak et al. (1980) suggested that seed of *A. tortilis* produced in humid climates was most likely to exhibit polymorphic forms with differing scarification requirements. This seems to be the only mention of seed polymorphism in *Acacia* species. However, variation in characteristics of *A. nilotica* seeds produced in different provenances has been noted (Mathur et al. 1984).

NURSERY AND FIELD PRACTICE Properly pretreated *Acacia* seed should be covered with one cm of soil. Sowing time is in spring for the warm temperate zone of the United States mainland, and year-round in Hawaii, except during dry periods. Blackwood acacia is preferably outplanted from small stumps (seedlings trimmed to 1.5 cm) lifted from the seedbed one year after planting, or as transplanted seedlings 20 to 25 cm high. The best survival for koa planted in Hawaii is obtained with potted seedlings.

Table 3. *Acacia:* pregermination treatments, germination test conditions and results (Whitesell 1974a).

Species	Pretreatment	Substrata	Temperature (°C)	Germination (%)
A. decurrens	–	–	–	74
A. koa	Hot water	Soil	–	18
A. mearnsii	–	Soil	15	72
A. melanoxylon	Hot water	P	20	70

COMPOSITAE — SUNFLOWER FAMILY

Acamptopappus Gray — Goldenhead

GROWTH HABIT, OCCURRENCE, AND USE Goldenhead is a low, many-branched shrub found on the margins of the Mojave Desert and the deserts of the Great Basin (Young and Young 1985). *Acamptopappus sphaerocephalus* is a rounded shrub, 0.2 to 0.9 m high, that often is densely twiggy. Flowers are arranged in heads located on the tips of the branches. The fruit is an achene that is densely villous.

GERMINATION Achenes are difficult to collect because of the scattered nature of goldenhead stands. They are very dormant and have limited germination at low incubation temperatures (Kay et al. 1977). Seeds of this genus rapidly lost viability in all forms of storage (Kay et al. 1988).

ARALIACEAE — GINSENG FAMILY

Acanthopanax Miq. — Aralia

GROWTH HABITAT, OCCURRENCE, AND USE   *Acanthopanax* is composed of a score or more shrubs or small trees that are native to east Asia and planted in the United States as ornamentals (Bailey 1951). Fiveleaf aralia (*A. sieboldianus*) is a large shrub or small tree to 3 m in height. This species is characterized by arching branches and sharp, weak spines on the petioles. The leaves are palmate with 5 to 7 thin leaflets. Henry's aralia (*A. henryi*) is a similar species.

FLOWERING AND FRUITING   Flowers are arranged in umbels on the previous year's wood. They are about 2.5 cm across, solitary on peduncles 5 to 7 cm long, and greenish white in color. The fruit is a lustrous black, 2- to 5-seeded, 0.6 cm diameter berry that ripens in September and October and should be collected at that time. Seed should be extracted from the pulpy mass of the berry.

GERMINATION   The seed of *Acanthopanax* species requires 6 months of warm and 3 months of cold pretreatment for germination (Dirr and Heuser 1987). Henry's aralia roots from cuttings taken in late June and treated with 3000 ppm IBA-talc. Fiveleaf aralia cuttings should be taken in mid-August and treated with 8000 ppm of IBA-talc.

ACERACEAE — MAPLE FAMILY

Acer L. — Maple

GROWTH HABIT, OCCURRENCE, AND USE   Maples are deciduous (rarely evergreen) trees comprising approximately 150 species in North America, Asia, Europe, and North Africa (Harlow and Harrar 1958). Some species are sources of valuable lumber and veneer, and one (sugar maple, *A. saccharinum*) is used for the production of maple sugar and syrup (Olson and Gabriel 1974). Many maples have ornamental value because of their handsome foliage, interesting crown shape, flowers, or fruits. Consequently, they are widely used in landscape plantings. Several maples provide food and shelter for wildlife as well as valuable watershed cover. In the far western United States, where red-colored autumn leaves are rare, some species of *Acer* provide brilliant reds in the fall. Olson and Gabriel (1974) considered maples to rarely be used in reforestation plantings for the production of timber, but plantings of sugar maple were made for syrup and sugar production.

FLOWERING AND FRUITING   Maples are either monoecious, dioecious, or polygamodioecious, with regular, perfect or imperfect flowers that appear with or before the leaves. The fruit is composed of 2 fused samaras, which eventually separate on shedding, leaving a small, persistent pedicel on the tree. The fruit varies widely in shape, length of wings, and angle of divergence of the fused samaras (Fig. 1). Each filled samara typically contains a single seed without endosperm (Fig. 2). The fruits of most maples ripen in the autumn, but two important species in the eastern United States (red, *A. urubrum*, and sugar maple) ripen fruits in the spring before leaf development is complete (Table 1). Maple seeds turn from green to rose or yellowish brown when ripe. Seed dispersal is by wind or water.

COLLECTION, EXTRACTION, AND STORAGE OF SEED   Most maples produce seed at an early age, and bear almost annually (Table 1). Abbott (1974) offered the following samara production figures for red maple:

Mature height (m)	DBH (m)	Seeds/tree
12.9	1.8	49,700
6.2	0.7	11,900
10.2	1.4	16,800
8.6	1.5	15,700
13.5	1.5	54,000
18.6	3.8	955,000

Maple seed may be picked from standing trees or collected by shaking or whipping trees and collecting samaras on sheets of canvas or plastic spread on the ground. Samaras may also be collected from trees recently felled in logging operations. Samaras from species like red and sugar maple or boxelder (*A. negundo*) can be gathered from lawns, pavements, or the surface of water in pools. After collection, leaves and other debris can be removed by hand, screening, or fanning.

Maple seed is not generally extracted from fruit (samara). Dewinging reduces weight and bulk for storage. The separation of empty from filled samaras of sugar maple is done by floating samaras in n-pentane. Removal of empty samaras improves seed handling, storage, and control of seedbed density. The number of seeds per gram varies considerably among species of maple (Table 1).

Freshly collected samaras have a moisture content of 30 to 160% of dry weight. A period of drying is necessary before seeds are stored. Moisture content of

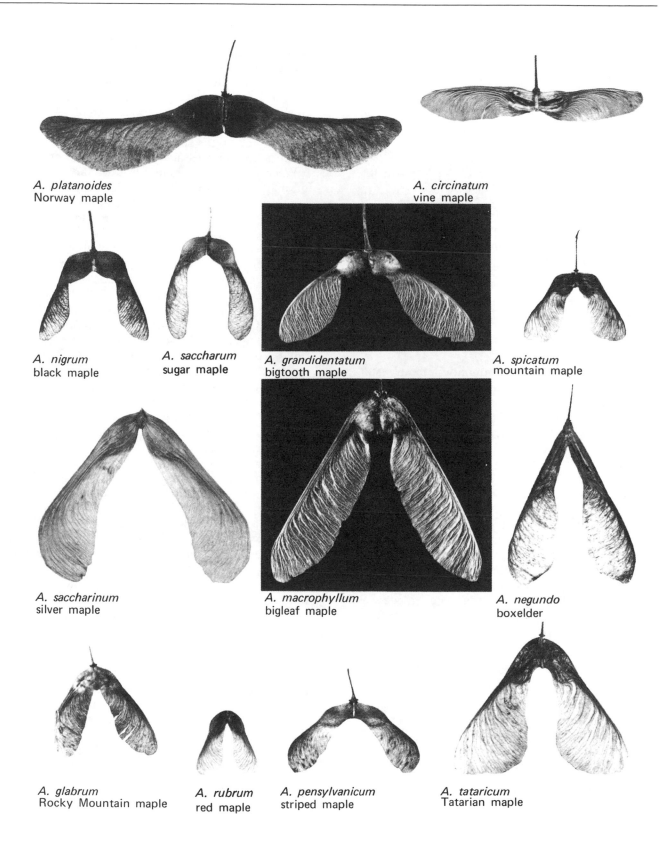

A. platanoides
Norway maple

A. circinatum
vine maple

A. nigrum
black maple

A. saccharum
sugar maple

A. grandidentatum
bigtooth maple

A. spicatum
mountain maple

A. saccharinum
silver maple

A. macrophyllum
bigleaf maple

A. negundo
boxelder

A. glabrum
Rocky Mountain maple

A. rubrum
red maple

A. pensylvanicum
striped maple

A. tataricum
Tatarian maple

Figure 1. *Acer*: samaras, ×1 (Olson and Gabriel 1974).

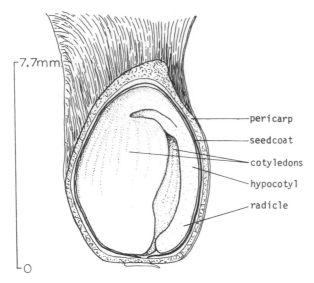

Figure 2. *Acer circinatum*, vine maple: longitudinal section of a seed showing bent embryo, ×7. On drying, seed shrinks leaving space between the seed coat and the pericarp (Olson and Gabriel 1974).

about 10 to 15% is recommended before storage. Dried seed of planetree maple (*A. pseudoplatanus*) loses its viability if moisture content is below 15%. Seed of bigleaf maple (*A. macrophyllum*) cannot be stored at room temperature or low temperatures, even for a short period of time. Seeds of most other species of maple can be stored for 1 or 2 years in sealed containers at 2 to 5°C. Seed of sugar maple has been kept for 54 months at 17% moisture at −10°C.

GERMINATION The seeds of many species of *Acer* are classified as being in deep dormancy. The occurrence of this deep dormancy has attracted world-class plant physiologists to study germination of some species of maple. Results obtained to date offer a glimpse of the complexity of biochemical pathways, but at the same time are fustrating because adequate procedures for preplanting treatments to break dormancy are not available for all species of maple.

It has long been obvious that the layers which form the samara or fruit that encloses the maple seed and the integument layers that constitute the actual seedcoat within the fruit played a dominant role in the dormancy of these seeds (e.g., Heit 1955). Nursery growers know the only way to break deep dormancy in most species of maple is prolonged stratification, either artificially or through fall planting in areas with cold winters. For some species a period of warm-moist stratification is necessary prior to prechilling.

Webb began investigating the stratification process for maple seed during the 1960s (e.g., Webb and Dumbroff 1969). He also collaborated with P. F. Waring of the University of Wales in a series of studies on the dormancy of maple seed. They suggested the seed covering could induce dormancy of the embryo by (1) restricting the flow of water or oxygen to the embryo, or through (2) mechanical restriction of embryo growth (Webb and Waring 1972). Both potential causes of dormancy were tested, and neither was identified as the cause of dormancy. Webb and Waring (1972) determined it required 40 to 60 days of prechilling to obtain germination of seed of planetree maple, but removal of the pericarp and testa reduced the prechilling requirement by two thirds. They suggested the testa restricted water flow from the embryo outward. This would slow the leaching or removal of germination inhibitors from the embryo. The exact nature of the embryo dormancy of *Acer* seed is still unknown, but change in the level of abscissic acid (Enu-Kwesi and Dumbroff 1980, Tilberg and Pinfield 1982) in relation to sequential changes in cytokinins is implicated as the mechanism (Webb and Waring 1972, Stadon et al. 1972, Tillberg and Pinfield 1981, Pinfield and Stobart 1982). Pozdova (1985) working in the USSR determined the influence of prechilling at 0 to 3°C over time on the cytokinin level in seed of *A. tataricum*. Cytokinins rose over time with prechilling and reached a maximun after 1 month. Interruption of prechilling reduced the cytokinin level and induced a secondary dormancy.

Table 1. *Acer*: phenology of flowering and fruiting, characteristics of mature trees, and cleaned seed weight (Olson and Gabriel 1974).

Species	Flowering dates	Fruit ripening dates	Seed dispersal dates	Mature tree height (m)	Year first cultivated	Seed-bearing Age (years)	Seed-bearing Interval (years)	Seeds/gram
A. campestre	Apr.–June	Aug.–Oct.	Oct.–Feb.	20.0	–	10	1	12.8
A. circinatum	Mar.–June	Sept.–Oct.	Oct.–Nov.	6.2	1826	–	1–2	10.0
A. ginnala	Apr.–June	Aug.–Sept.	Oct.–Nov.	6.2	1860	5	1	37.4
A. glabrum	Apr.–June	Aug.–Oct.	Sept.–Feb.	9.2	1882	–	1–3	29.6
A. grandidentatum	Apr.–May	Aug.–Sept.	Sept.–Dec.	9.2	1882	–	–	14.0
A. macrophyllum	Apr.–May	Sept.–Oct.	Oct.–Jan.	30.8	1812	10	1	7.2
A. negundo	Mar.–May	Aug.–Oct.	Sept.–Mar.	23.1	1688	–	1	29.5
A. pensylvanicum	May–June	Sept.–Oct.	Oct.–Nov.	10.8	1775	–	–	24.5
A. platanoides	Apr.–June	Sept.–Oct.	Oct.–Nov.	30.8	–	–	1	6.3
A. pseudoplatanus	Apr.–June	Aug.–Oct.	Sept.–Nov.	30.8	–	–	1	11.3
A. rubrum	Mar.–May	Aug.–Oct.	Apr.–June	27.7	1656	4	1	50.4
A. saccharinum	Mar.–May	Sept.–Oct.	Oct.–Dec.	27.7	1725	11	1	3.9
A. saccharum	Feb.–May	Aug.–Oct.	Apr.–July	30.8	–	30	3–7	15.5
A. spicatum	May–June	Sept.–Oct.	Oct.–Dec.	9.2	1750	–	–	48.7
A. tataricum	May–June	Aug.–Sept.	Sept.–Nov.	10.8	1759	–	–	25.0

Janerette (1977) conducted basic studies of the nature of germination of seeds of sugar maple at North Carolina State University. She bridged the gap between basic and applied plant physiology with studies of methods of stimulating germination of these seeds for germination testing (Janerette 1978, 1979a). She (1979a) also determined the prechilling requirement for sugar maple seeds could be greatly reduced by soaking them in cold water before starting the prechilling. Soaking for up to 14 days was benefical in shortening the prechilling requirement both in terms of the time for 50% of total germination and the length of the total duration of post prechilling germination.

Despite all the basic research on dormancy of *Acer* seed the practical nursery grower still has problems. Ideally the grower wants 80% germination in 2 weeks following planting of prechilled seeds. In practice, when working with *Acer* seeds such as sugar maple, the grower is lucky to get 30% germination (Carl 1983).

A study of sugar maple seeds from 32 production provenances showed (a) dry weight of seed and percentage of filled fruit varied genetically among and within provenances, (b) dry weight of seed and latitude of provenance were positively correlated—higher latitudes and colder temperatures yielded larger seeds—and (c) percentage of filled fruit was not related to dry weight of seed and was negatively related to latitude (Gabriel 1978). Changes in the anatomy, morphology, and stored food reserves of prechilled seeds of sugar maple were followed from a strongly dormant state through germination (Shih et al. 1985). No morphological or anatomical changes were observed before the first maturation of phloem elements on the 17th day of prechilling. Mature xylem elements were first observed on day 37 and first germination was noted on day 38. Changes in amounts of protein, lipid, and starch were not observed during prechilling, but mobilization of stored reserves was clearly evident with emergence of the radicles. These events appear closely associated with previously described peaks and patterns in growth regulator activity. Seeds of the 2 spring-fruiting species, red and silver (*A. saccharinum*) maple, require no pregermination treatment (Tables 2, 3). The Association of Offical Seed Analysis lumps all *Acer* species in a blanket germination standard. Internationally specific tests are proposed for the seeds of major species.

Table 2. *Acer:* germination test conditions and results for prechilled seeds (adapted from Olson and Gabriel 1974).

Species	Substrata	Temperature (°C)	Germination (%)
A. campestre	S	–	84
A. circinatum	C	20/30	19
A. ginnala	C	20/30	52
A. glabrum	S	15	0
A. grandidentatum	S	20/30	30
A. macrophyllum	S	20/30	90
A. negundo	S, Peat	–	96
A. pensylvanicum	S, Peat	–	2
A. platanoides	S, Humus	5/10	81
A. pseudoplatanus	SH	–	71
A. rubrum	S	20/30	91
A. saccharinum	TB	30	97
A. saccharum	S	5	95
A. spicatum	S	–	34
A. tataricum	S, Peat	–	75

Normally seed of red maple is considered germinable without pretreatment. However, it has been noted that several sources of red maple collected in Ontario were dormant and required prechilling for germination to occur (Wang and Hadden 1978). Maximum germination occurred when seed was prechilled for 6 to 8 weeks and then incubated at 20/30°C with an 8-hour photoperiod. Seed of red maple collected in northwest Ontario averaged 44% germination without prechilling if incubated at 5/15°C (Farmer and Goelz 1984). Germination was much lower at higher incubation temperatures. If the seed was prechilled 90 days, germination was not influenced by high incubation temperatures.

Boxelder seeds have variable prechilling require-

Table 3. *Acer:* laboratory germination methods (compiled from ASOA 1985 and Handbook for Genebanks 1985).

Species	Substrata	Temperature (°C)	Duration (days)	Additional directions
A. palmatum	TB,S	20	21	Prechill 120 days at 1–5°C, remove pericarp
A. platanoides	TB,S	20	21	Prechill 60 days at 1–5°C, remove pericarp
A. pseudoplatanus	TB,S	20	21	Prechill 60 days at 1–5°C, remove pericarp
A. rubrum	TB,S	20	21	
A. saccharinum	TB,S	20	21	
A. saccharium	TB,S	20	21	Prechill 60 days at 1–5°C, remove pericarp
Acer spp.[1]	TB	18–22	14	Excise embryo, or remove pericarp and prechill 60 days at 3–5°C. Alternative method: warm stratification 60 days at 25°C and prechill 28–168 days at 1–5°C.

[1]Data from ASOA standards.

ments to break dormancy depending on origin (Williams and Winstead 1972). Seeds from northern sources have shorter prechilling requirements and germinate at lower incubation temperatures. Seeds from southern sources, generally south of 37° north latitude, require longer periods of prechilling and germinate at warmer incubation temperatures.

A group of maples that has not received a lot of attention in the past is the *Trifolinata* section of *Acer*. Samaras of the trifoliate maples, especially *A. griseum*, *A. maximowiczianum*, and *A. triflorum*, have a ligneous pericarp which delays germination for several years (Fordham 1969). Stimart (1981) succeeded in germinating excised embryos of these species in 21 days in a lighted incubator on wick culture moistened with a solution of 10mg/liter of gibberellic acid. Excised embryos placed in the dark or intact seeds in any treatment failed to germinate.

NURSERY AND FIELD PRACTICE It is preferable to sow the seeds of most maples in the fall in mulched beds. Stratified seeds can be sown in the spring, but results are variable and often unsatisfactory. For example, seeds of sugar maple sown in the fall had nearly 100% emergence from fall seeding (Carl and Yawney 1977). Spring-sown seeds that had been stratified for 60, 40, or 30 days had emergence of 89, 75, and 31%, respectively. Seeds of red and silver maple should be sown in the spring soon after collection, but in some cases emergence will be delayed until the next spring.

Maple seed is usually sown 0.5 to 1 cm deep, broadcast or with a drill. Seedbed densities from 150 to 1500/m² have been recommended. However, densities at the lower end of this range appear most satisfactory for the production of vigorous plants. Shade is recommended during the period of seedling emergence and establishment (Fig. 3). Sometimes seedlings of maples are large enough to plant as 1-0 stock, but

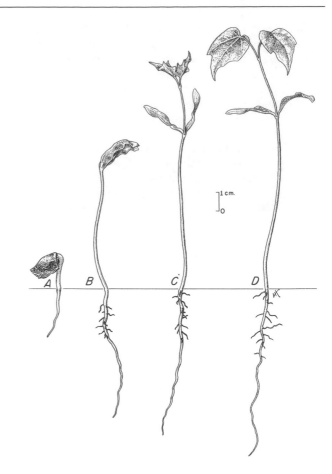

Figure 3. *Acer platanoides*, Norway maple: seedling development at 1, 3, 7, and 19 days after germination, ×0.5 (Olson and Gabriel 1974).

frequently 2-0 or even 2-2 stock is needed to ensure satisfactory results. In general, the larger the stock, the better the survival.

MYRTACEAE — MYRTLE FAMILY

Acmena DC. — Lillypilly

GROWTH HABIT, OCCURRENCE, AND USE Lillypilly (*Acmena smithii*) is a tall, graceful tree, often with branches, with a distinct weeping habit. It reaches about 8 m in height (Macoboy 1982), but is often listed in the literature as reaching over 30 m. It originated in eastern Australia, but is grown in South Africa, California, and Hawaii, among other places. Leaves are small and shiny; the fruit color is quite variable—from white to mauve—and 1 cm in diameter. *Acmena* is closely related to *Eugenia* and *Syzgium*.

FLOWERING AND FRUITING The white blossoms have no petals, but do have a mass of white stamens, and are attractive. Flowering occurs in early summer. The special attraction of the lillypilly is a profusion of berries in late summer.

PROPAGATION Lillypilly is progagated from seed or cuttings.

ACTINIDIACEAE — ACTINIDIA FAMILY

Actinidia Lindl.

GROWTH HABIT, OCCURRENCE, AND USE *Actinidia* is a group of vines that produce fragrant flowers and edible fruits. The group consists of about 20 species native from eastern Asia to the Himalayas (Bailey 1951). Bower actinidia (*A. arguta*) is extremely hardy, while some of the other species such as kiwi (*A. chinensis*) are only hardy to −8 or −12°C. Leaves of these species are alternate, entire or serrate.

FLOWERING AND FRUITING Flowers are polygamo-dioecious or dioecious, so male and female plants are needed. Flowers occur in axillary cymes or solitary. The fruit is a many-celled berry with numerous seeds imbedded in the pulp.

GERMINATION The seeds of most species of actinidia require some form of prechilling to induce germination (Table 1).

NURSERY AND FIELD PRACTICE The actinidia species are often propagated by cuttings (Table 1).

Tissue culture has proven easy with the actinidia species, and bower and kolomikta actinidia have been proliferated in culture at the Arnold Arboretum (Dirr and Heuser 1987). These species have been used as model systems to study cytokinin biochemistry.

Table 1. *Actinidia:* pregermination treatments and rooting techniques (Dirr and Heuser 1987).

Species	Pregermination treatment	Rooting technique
A. arguta	Fall plant or prechill 3 months.	Softwood/greenwood cuttings root readily.
A. chinensis	Seeds extracted fresh and planted will germinate. Store fruits at 5°C until ready to plant. Dried seeds develop dormancy that requires 3–4 weeks of prechilling to break.	Softwood cuttings produce variable results. Cultivars show variable propensities to root.
A. kolomikta	Seed is doubly dormant and requires 3 months of warm stratification and 3 months of prechilling. Alternative treatment involves 3 months of prechilling and enrichment of substrate with 500 ppm of GA_3 and 50 ppm of kinetin.	Not particularly easy to root. June cuttings treated with 8000 ppm IBA rooted 40%.
A. polygama	Fall planting or 3 months of prechilling is required.	Softwood cuttings root easily.

BOMBACACEAE — BOMBAX FAMILY

Adansonia L. — Baobab

GROWTH HABIT, OCCURRENCE, AND USE *Adansonia* is a group of remarkable trees native to dry areas of Australia, South Africa, and Malagasy (Madagascar) (Macoboy 1982). These trees only reach 13 m in height, but their trunks may reach 10 m in diameter. Named for the French botanist Adanson, these trees live for amazing lengths of time (some African specimens have been estimated to be 5000 years old) and often are the only surviving woody vegetation for miles around during times of drought. The trunk consists of light, fleshy wood that is mostly hollow chambers for storing water.

During drought periods the leaves are used for fodder by livestock. The fruits are used to make a refreshing, acid beverage, and the seeds are used to obtain a medicinal oil.

The trees are deciduous and in a dry year may not produce leaves. The leaves are preceded by 10 cm white flowers with reflexed petals. These open on long, hanging stems and are followed by velvety, cylindrical fruits up to 40 cm in length. The appearance of these fruits has led to the common name dead-rat tree.

LEGUMINOSAE — LEGUME FAMILY

Adenanthera L. — Bead tree

GROWTH HABIT, OCCURRENCE, AND USE The genus *Adenanthera* consists of 3 or 4 species of trees belonging to the legume family and related to the genus *Mimosa* (Bailey 1951). The most widely cultivated species is red sandlewood or peacock flower (*A. pavonina*), a tree that reaches 15 m in height and is native to tropical Asia and Africa. Branches are unarmed and leaves are bipinnate. Flowers are borne in spikelike racemes that reach 15 cm in length. They are white and yellow in the same cluster. The legume is 20 cm long and becomes quite twisted. The bright red seeds are known as Circassian beads and used for necklaces. The species is occasionally planted in the warmer parts of the United States.

GERMINATION Seeds of *A. pavonina* were soaked in 0, 20, 40, 60, or 80% sulfuric acid for 20, 40, or 60 minutes (Ahmed et al. 1986). Germination of control seeds was 2.5%. All treatments increased germination. The highest germination was 83% obtained with 60% sulfuric acid for 20 minutes. Freshly collected seed apparently does not have as hard a seedcoat as dried seeds, for Ng and Sanah (1979) reported that fresh seeds germinated slowly without pretreatment over an 11-week period.

ROSACEAE — ROSE FAMILY

Adenostoma H. & A. — Chamise

GROWTH HABIT, OCCURRENCE, AND USE *Adenostoma*, a genus of unarmed evergreen shrubs with resinous herbage, consists of 2 species found in chaparral vegetation of California (Young and Young 1985). Chamise (*A. fasciculatum*) dominates vast areas of chaparral, where it constitutes an extreme wildfire hazard. Red shank (*A. sparsifolium*) is another evergreen species found in chaparral communities.

The fruit of chamise is an achene enclosed in an indurate flower tube.

GERMINATION The achenes have hard seedcoats that require acid scarification for germination. An alternate method involves sowing the seed in soil in flats and burning pine needles on the soil surface (Emery 1964).

THEACEAE — TEA FAMILY

Adinandra Jack.

GROWTH HABIT, OCCURRENCE, AND USE *Adinandra* is a genus of about 80 species native to eastern and southeastern Asia, Taiwan, and Malaysia (Keng 1977). *Adinandra acuminata* is a small tree to 20 m in height. The green-brown bark flakes in small, soft pieces, and the inner bark is pink. The narrowly elliptic leaves are papery thin to leathery.

FLOWERING AND FRUITING The white flowers are borne on stalks 1 cm long. The fruit, a globose berry, is subtended by a calyx that contains few seeds.

GERMINATION Ng (1977a) considered seed of this species to have rapid germination, with more than 70% germinating in a 3- to 8-week period.

HIPPOCASTANACEAE — HORSECHESTNUT FAMILY

Aesculus L. — Buckeye, horsechestnut

GROWTH HABIT, OCCURRENCE, AND USE The buckeyes, occurring in North America, southeastern Europe, and eastern and southeastern Asia, include about 25 species of deciduous trees and shrubs (Rehder 1940). They are cultivated for their dense shade and ornamental flowers, and the wood of some species is occasionally used for lumber and paper pulp. They also provide wildlife habitat. The shoots and seeds of some buckeyes are poisonous to livestock (Bailey 1951). Six species are native to the United States, and horsechestnut (*Aesculus hippocastanum*) was introduced to North America from Europe (Rudolf 1974a).

None of the 7 species of *Aesculus* is used extensively in reforestation, but all are used for environmental forestry planting. This is particularly true for horsechestnut, which has been widely planted as a shade tree in Europe and eastern United States, where it sometimes escapes from cultivation (Bailey 1951). Texas buckeye (*A. glabra*) and yellow buckeye (*A. octandra*) are sometimes planted in Europe and eastern United States. California buckeye (*A. californica*) is occasionally planted in Europe and California. Arkansas buckeye (*A. × bushii*) is a natural hybrid that occurs in Mississippi and Arkansas. At least 5 other hybrids are known in cultivation.

FLOWERING AND FRUITING The flowers of *Aesculus* are irregular, white, red, or pale yellow in color, and are borne in showy clusters that appear after the leaves. Only those flowers near the base of the branches of the cluster are perfect and fertile; the others are staminate.

The fruit is a somewhat spiny or smooth, leathery, round or pear-shaped capsule with 3 cells (Fig. 1), each of which may bear a single seed. Sometimes only 1 cell develops; the remnants of the abortive cells and seeds remain clearly visible at maturity. When only 1 cell develops, the large seed is round to flat-shaped (Fig. 1). The ripe seeds are dark chocolate to chestnut-brown in color, smooth and shining, and have a large, light-colored hilum resembling the pupil of an eye. They contain no endosperm, the cotyledons being very thick and fleshy (Fig. 2). When ripe in the fall, the capsules split and release the seeds (Table 1).

COLLECTION, EXTRACTION, AND STORAGE OF SEED The fruits may be collected by picking or shaking them from the trees as soon as the capsules turn yellowish and began to split open, or by gathering them from the ground soon after they have fallen. They should be dried for a short time at room temperature to free the seeds from any parts of the capsule that may still adhere to them, but great care must be taken not to dry them too long. If the seeds are dried excessively the seedcoat becomes dull and wrinkled and seeds lose viability. Fresh horsechestnut seeds have a moisture content of 49%, while those slightly dried have a moisture content of 38%, based on fresh-weight. Fresh seed of red buckeye (*A. pavia*) was reported to have a moisture content of 56% on a fresh weight basis. The seeds should be sown at once or stratified for spring sowing.

Initial viability of fresh seeds of horsechestnut can be maintained for 6 months when they are stored in plastic bags at 5°C. This storage condition is the same as cold-moist stratification (prechilling) because of the high moisture content of fresh seed. When seeds of horsechestnut were stored at −2°C in sealed packages without added moisture for 13 months, germination dropped from 80 to 60% after 15 months, germination was only 25%. Data on cleaned seeds per kilo for the buckeye species are given in Table 1.

PREGERMINATION TREATMENT Seeds of Ohio buckeye (*A. glabra*), yellow buckeye, horsechestnut, and painted

Table 1. *Aesculus*: phenology of flowering and fruiting, characteristics of mature trees, cleaned seed weight, and germination test conditions and results (Rudolf 1974a).

Species	Flowering dates	Fruit ripening dates	Seed dispersal dates	Mature tree height (m)	Flower color	Seed-bearing Age (years)	Seed-bearing Interval (years)	Seeds/kg	Prechilling (days)	Substrata	Temperature (°C)	Germination (%)
A. californica	Apr.–Sept.	Sept.–Oct.	Nov.–Dec.	12.7	Whitish rose	5	1–2	28	0	Sand	20	56
A. glabra var. *arguta*	May	May–June	Sept.–Oct.	11.0	Yellow-green	8	1+	88	120	Sand	15/25	76
var. *glabra*	Mar.–Apr.	Sept.–Oct.	Sept.–Oct.	21.5	Yellow-green	8	—	123	120	Sand	20	59
A. hippocastanum	Apr.–June	Sept.–Oct.	Sept.–Oct.	24.6	Reddish-white	—	—	64	120	Sand	20/30	89
A. octandra	Apr.–June	Sept.	Sept.	22.7	Yellow	—	—	61	120	Sand	20/30	76
A. pavia	Mar.–June	Sept.–Oct.	Sept.–Nov.	8.6	Red	—	—	116	0	Sand	20/30	70
A. sylvatica	Apr.–May	Sept.–Oct.	July–Aug.	20.0	Pale yellow	8	1+	88	90	Sand	—	78

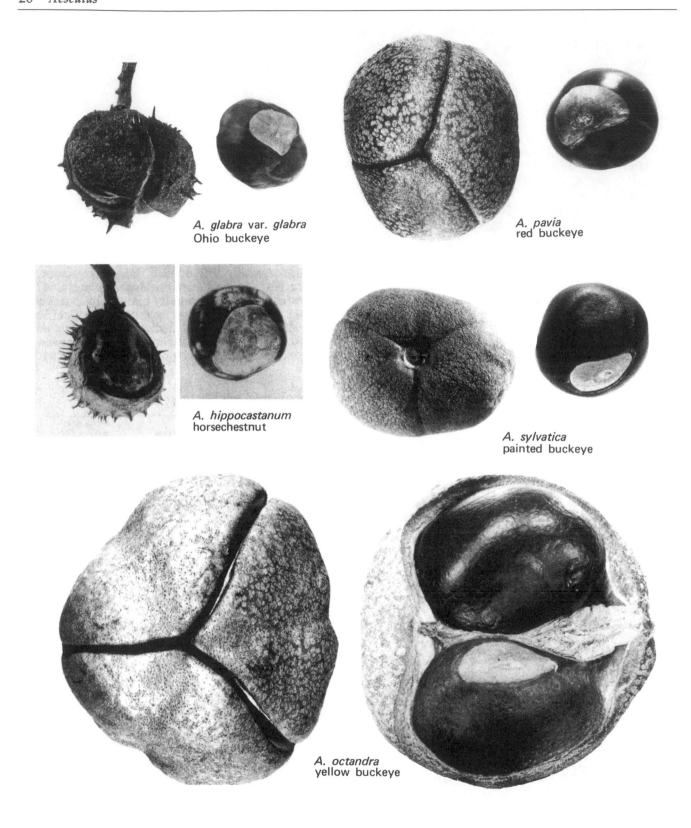

Figure 1. *Aesculus*: capsule and seed, ×1 (Rudolf 1974a).

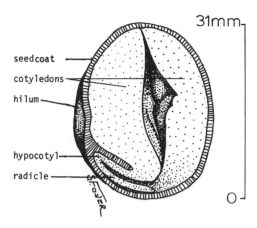

Figure 2. *Aesculus glabra* var. *glabra*, Ohio buckeye: longitudinal section through seed, ×1.5 (Rudolf 1974a)

buckeye (*A. sylvatica*) require prechilling to break dormancy. Rudolf (1974a) suggested that prechilling be done in moist sand or sand-peat mixtures at 5°C for about 120 days, or by storage in sealed containers at 2°C for 100 days or longer. The latter treatment would not normally constitute prechilling in the traditional cool moist stratification method, except for the high moisture content of these seeds. Fresh seeds of California buckeye and red buckeye (*A. pavia*) germinate without pretreatment.

GERMINATION TEST Prechilled buckeye seeds have been germinated in sand or on wet paper at alternating diurnal temperatures of 20/30°C (Table 1). The standard germination method given by the AOSA (1985) for seeds of red buckeye is substrate, top of creped cellulose paper; temperature 20/30°C; and duration, 28 days. Rudolf (1974a) offerred a non-prechilling procedure for seed of horsechestnut that he credited to the International Rules for Seed Testing. This procedure involved soaking them for 48 hours, and cutting off one-third of the seed at the scar end without removing the seedcoat. The portion of the seed with the scar was sown in flats of sand and incubated at 20/30°C for 21 days.

NURSERY AND FIELD PRACTICE Under natural conditions seeds of most *Aesculus* species germinate in the early spring. In the Mediterranean climate of California, the California buckeye germinates in the fall with the first effective rain after the summer drought. In nursery production buckeye seeds are sown in the fall as soon after collection as possible to prevent drying. They are sown 5 cm apart in rows spaced at 15 cm intervals, and covered with 2.5 to 5 cm of soil. Beds should not be over watered because the seeds rot readily. Normally, 1-0 stock is large enough for field planting. Seedling development is shown in Fig. 3.

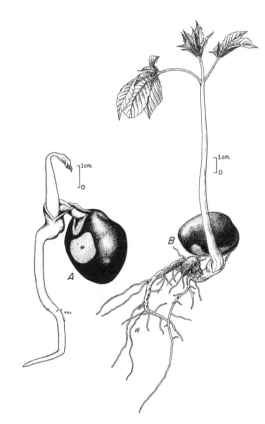

Figure 3. *Aesculus octandra*, yellow buckeye: seedling development at 2 and 4 days after germination (Rudolf 1974a).

ARAUCARIACEAE — ARAUCARIA FAMILY

Agathis Salisb. — Kauri or Daminar pine

GROWTH HABIT, OCCURRENCE, AND USE *Agathis* is a group of evergreen, resinous, coniferous trees found from the Philippines to Australia and New Zealand. The genus consists of about 20 species belonging to the same family as *Araucaria*. The leaves of these species are opposite or alternate, somewhat 2-ranked, flat, broad, and leathery in texture. The staminate catkins are cylindrical and borne axillary. The pistillate cones are ovoid to globose. The numerous cone scales are obovate. For the Queensland kauri (*A. robusta*), it takes 16 months

from pollination to cone maturity (Willan 1985). Seeds are winged on 1 side only and not united with the scales. Queensland kauri reaches 50 m in height.

Old-growth kauri trees provide valuable timber. The trees contain a resin that is used to prepare varnish. The fossil resin, called copal, is mined in New Zealand similarly to amber.

COLLECTION, EXTRACTION, AND STORAGE OF SEED Larvae of *Agathipaga*, a moth genus, may destroy over 50% of the seeds of several species of *Agathis* in Queensland and the western Pacific Islands (Willan 1985). Evidence indicates the seeds of most *Agathis* species are orthodox and can be dried and stored. Seeds of *A. australis* have been dried to 6% moisture content and stored at 5°C without loss of viability for 6 years. The same seeds stored at higher moisture content and temperature have lost viability in 14 months.

Cone size varies among species (Macoboy 1982):

Species	Cone diameter (cm)
A. australis	7.5
A. moeri	12.5
A. robusta	12.0
A. vitrensis	6.5

Cones of Queensland kauri borne in New Zealand contain up to 80% sound seed (Dakin and McClure 1975). Seed can be stored in plastic bags at 2 to 5°C for a few months.

GERMINATION Seeds of Queensland kauri sown in compost-soil mixture germinated after 4 days (Dakin and McClure 1975). Seedling percentage was 90% of the seeds sown.

Barton (1978) determined the seeds of Queensland kauri have optimum germination at 25°C, while some germination occurred at temperatures from 10.5 through 36°C. *Agathis australis* can be propagated by vegetative cuttings (White and Lovell 1984).

MELIACEAE — SENTAL FAMILY

Aglaia Lour.

GROWTH HABIT, OCCURRENCE, AND USE *Aglaia* is a widely distributed shrub of the Malaysian lowland forest that is poorly understood and not often classified to species, despite the estimate of 250 species in the genus. The leaves are pinnate with a terminal leaflet.

FLOWERING AND FRUITING The flowers are minute and round, white, pink, or yellow in color, often with a citronella scent (Corner 1952). The fruit, which does not open, are rather small, with a brightly colored thin rind and one to 3 cavities, each filled with a large seed that is surrounded by transparent white, yellow, or orange pulp.

Becker and Wong (1985) studied seed dispersal beneath an *Aglaia* species in the forest of Malaysia. Black hornbills fed on the fruit and apparently were long-distance dispersal agents. Squirrels rapidly cleaned up fruits under the trees, but the most vigorous seedlings were found some distance from the parent plants. Juvenile seedling growth was quite slow.

SIMAROUBACEAE — AILANTHUS FAMILY

Ailanthus altissima (Mill.) Swingle

GROWTH HABIT, OCCURRENCE, AND USE Native to China, *Ailanthus* is a short to medium-tall deciduous tree of value chiefly for shade and other environmental purposes, particularly in cities where soils are poor and the atmosphere smoggy. It is sometimes planted for shelterbelts, for game food and cover, and rarely, as in New Zealand, for timber (Little 1974). Ailanthus, or tree of heaven as it is often known, was introduced into cultivation in Europe in 1751 and brought to North America in 1784. It has become naturalized in many parts of the United States, from Massachusetts to southern Ontario, Iowa, and Kansas, and south to Texas and Florida, as well as from the southern Rocky Mountains to the Pacific Coast. In some localities ailanthus is so well established that it appears to be part of the native flora.

FLOWERING AND FRUITING Commercial seed consists of the one-celled, one-seeded, oblong, thin, spirally twisted samaras. These are 2.5 to 3 cm long, light reddish brown in color, and bear the seed at about the middle (Fig. 1). Flowers open in mid-April to July. Seeds ripen in large crowded clusters in September to October, continuing to the following spring (Illick and Brouse 1926). Ailanthus is a prolific seeder: trees 15- to 20-years-old bear considerable quantities. Seeds have no endosperm (Fig. 2).

COLLECTION, EXTRACTION, AND STORAGE OF SEED Ailanthus fruits are picked from standing trees by hand or flailed or stripped onto canvas at any time during the late fall and early winter. After collection the fruits should be spread out to dry to reach moisture equilibrium. Little (1974) suggested running the seeds through a macerator before cleaning. If it is really necessary to remove the seed from the fruit, a properly adjusted hammer mill or rubber belt thresher would seem a more desirable choice. Small quantities of fruits can be

Figure 1. *Ailanthus altissima*, ailanthus: samara, ×2 (Little 1974).

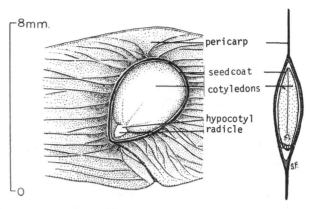

Figure 2. *Ailanthus altissima*, ailanthus: longitudinal section through a samara, ×6 (Little 1974).

threshed in a cloth bag by rubbing or trampling. An air screen can be used to separate the seeds from the papery portions of the fruit. Little (1974) indicated that 45 kg of fruit yield from 16 to 40 kg of seed. No explanation of this wide range in seed yield was given. The fruits number about 30 to 36 per gram, and the seeds 29 to 43 per gram.

Ailanthus seed should be dried and stored at 2 to 5°C in sealed containers. However, bulk storage in sacks without temperature control has been successful.

PREGERMINATION TREATMENT Little (1974) suggested that seeds of ailanthus require prechilling for 60 days for germination. The germination standards of the Association of Official Seed Analysis (1985) suggest incubation on top of germination paper (TB) at 20/30°C for 21 days. There may be seed collections of this species that require prechilling.

GERMINATION TEST We have already given the standard test. International rules for testing seeds of this species have, in the past, suggested removing the pericarp from the seed before testing. Published germination figures vary widely.

NURSERY AND FIELD PRACTICE Little (1974) offered the contradictory advice of stratifying the seeds over winter and planting the pretreated seeds in the spring. No reason was given for not planting in the fall. He also suggested maximum plant establishment is only 15 to 20% of the seed sown.

Ailanthus is a vigorously sprouting species that rapidly forms clumps. It is considered a pest in many areas because of this sprouting characteristic. It is also suspected of producing allelopathic influences on other plants.

LARDIZABALACEAE — LARDIZABALA FAMILY

Akebia Dcne.

GROWTH HABITAT, OCCURRENCE, AND USE Akebias are not well known in the United States. They are attractive, glabrous, twining shrubs from China and Japan that have potential as ornamental species and in conservation plantings. They make excellent cover for fences, trellises, and banks. Leaves appear early in the spring and hold to late November. *Akebia* species have been cultivated since 1845.

FLOWERING AND FRUITING The polygamomonocious flowers appear with the leaves. Pistillate flowers are chocolate-purple, the staminate flowers rose-purple.

Both are borne in the same pendant, axillary raceme. The fruits are 5 to 10 cm long, sausagelike pods of purple-violet color. They should be collected in October, and the seeds removed by hand and dried. Unfortunately, the fruits seldom set under cultivation.

GERMINATION Seeds of *Akebia* × *pentaphylla* require 1 month prechilling for maximum germination (Dirr and Heuser 1987). The seeds of fiveleaf akebia (*A. quinata*) and three-leaflet akebia (*A. trifoliata*) require 2 to 3 months of prechilling for germination.

LEGUMINOSAE — LEGUME FAMILY

Albizia Durazz.

GROWTH HABIT, OCCURRENCE, AND USE *Albizia* species are deciduous trees and shrubs bearing bipinnate leaves with small and numerous leaflets (Wick and Walters 1974). There are about 50 species in tropical and subtropical Asia, Africa, and Australia (Bailey 1951). One species occurs in Mexico. *Albizia falcata* is native to the Molucca Islands and New Guinea and is widely naturalized in Malaysia (Yap and Wong 1983). One of the better species of *Albizia* is silktree (*A. julibrissin*), a tree that grows to 12 m in height and has a large spreading crown. It has 10 to 25 pinnate leaves with 40 to 60 leaflets. Silktree is native from Iran to central China and was first introduced to cultivation in Europe in 1745.

Several species of *Albizia* produce valuable timber. Siris (*A. lebbek*) is a moderate- or large-sized deciduous tree, extensively planted in India in gardens, along roadsides, and in coffee and tea plantations as a cover crop (Babeley et al. 1986). It is widely planted in Queensland, Australia, as a shade tree, but is being investigated as a fodder tree for livestock production (Prinsen 1986). The heartwood is used for high-quality furniture and plywood production.

FLOWERING AND FRUITING Flowers are yellowish, white, or pink, in globose heads or cylindrical spikes that are axillary or paniculately arranged at the ends of the branches. The fruit is a large, strap-shaped, nonseptate pod without pulp (Fig. 1).

Albizia falcata begins flowering at 3 years and has 2 flowering periods annually—April and August.

COLLECTION, EXTRACTION, AND STORAGE OF SEED Pods of *A. falcata* ripen 2 months after flowering. The seeds, which are quite small, dehisce from the pods while the pods are still attached to the tree, making it necessary to collect pods on the tree just before full seed ripeness is reached. Pods are collected by cutting branches because the trees are difficult to climb. The green pods can be dried on screen trays in sheds. The small seeds have a funicle attached to the hilum. They are easily separated from the pods after they dry. The seeds number 50,000 to 62,000 per kilogram. Average annual seed yield per tree is 427 g at Sandakan, Sabah (Bowen and Eusebio 1983). Typical seed quality for *A. falcataria* follows:

Sound seed	53–74%
Insect-damaged seed	25–27
Fungal-damaged seed	8–17
Aborted seed	0–1

Air-dried seeds of *A. falcata* are put in plastic bags for storage at 4 to 8°C. Seeds stored in this manner have maintained viability for 18 months (Yap and Wong 1983). Seeds are shown in Figures 1 and 2.

GERMINATION Seeds of *Albizia* species have hard seed coats typical of many legumes. Without pretreatment germination is slow and germination capacity very low. Seeds of *A. falcata*, *A. chinensis*, and *A. richardiana* had germination enhanced by presoaking them in concentrated sulfuric acid. Seeds of siris can be scarified by soaking in concentrated sulfuric acid for 45 minutes, but filing to produce mechanical scarification of the seed coat provides more rapid germination and more vigorous seedlings (Babeley et al. 1986). In Australia

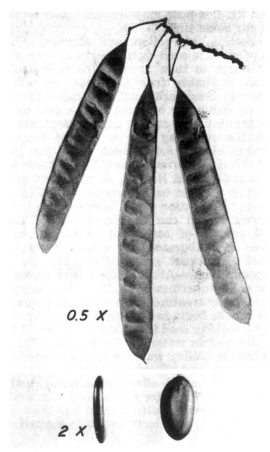

Figure 1. *Albizia jublibrissin*, silktree: longitudinal section through a seed, ×5 (Wick and Walters 1974).

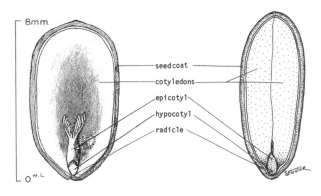

Figure 2. *Albizia julibrissin*, silktree: pods, ×0.5 and seed ×2 (Wick and Walters 1974).

the recommended pregermination treatment for seeds of siris is soaking them in cold water for 24 hours, then dropping them in boiling water, and allowing them to cool for an additional 24 hours (Prinsen 1986).

Seeds of *A. facataria* are pretreated by combining 3 parts boiling water and one part seeds, then allowing the seeds to cool for 2 days in the soak water (Yap and Wong 1983). Bowen and Eusebio (1983) used a more detailed approach for seeds of the same species: Seeds were dropped in water of the following temperatures for 30 seconds and subsequent germination recorded:

Water temperature (°C)	Germination (%)
30	15
40	16
50	16
60	25
70	38
80	49
90	57
100	51

The authors considered there was seed mortality at the 100°C temperature.

NURSERY AND FIELD PRACTICE In Australia, siris is propagated by stub transplants: seedlings are grown on nursery beds and then lifted, and the tops trimmed back to a 5 cm stub, and the roots pruned to 25 cm (Prinsen 1986). Survival of these transplants is virtually 100% in the field. The stub transplants grow to 5 m in 4 years. Seedling development is shown in Figure 3.

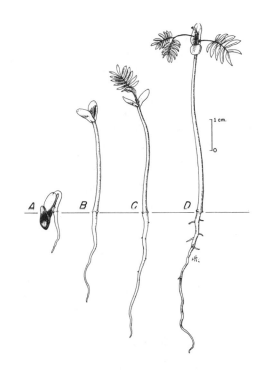

Figure 3. *Albizia julibrissin*, silktree: seedling development at 1, 3, 5, and 8 days (Wick and Walters 1974).

EUPHORBIACEAE — SPURGE FAMILY

Aleurites J. R. & G. Forst. — Candlenut

GROWTH HABIT, OCCURRENCE, AND USE The striking candlenut tree (*A. moluccana*) is found on hillside forests of the Pacific Islands and southeast Asia, where its pale, mealy foliage stands out from darker tropical vegetation (Macoboy 1982). It is one of the great domesticated trees of the world, with a thousand uses, and has been adopted as the official tree emblem of Hawaii, where it was probably imported by Polynesian ancestors. The genus contains about 4 species native to the same area as the candlenut.

The tung-oil tree (*A. fordii*) is the source of tung oil, which is used as a drying agent in paint (Bailey 1951). Candlenut wood was used by Polynesians to build canoes. A gum extracted from the sap was used to strengthen tapa cloth. The sap of candlenut is typical for the spurge family in being milky white.

Candlenut may reach 20 m in height and is densely clothed with hand-sized, 3- or 5-lobed leaves that are pale green with a rusty fuzz underneath. The tiny flowers, borne in panicles several times a year, are followed by clusters of 5-cm nuts that resemble European walnuts.

COLLECTION, EXTRACTION, AND STORAGE OF SEED In the Philippines the fleshy fruits of *Aleurites* species are placed in barrels with water. After a day or two the pulp becomes soft. The fruits are then mashed carefully with a tamper, without crushing the seeds. Water is added so the pulp floats free from the seeds which sink to the bottom (Willan 1985).

GERMINATION The seedcoats or pericarp of *Aleurites* fruits are hard and require scarification. The nuts are spread on the ground and covered with 3 cm of dry *Imperata* grass which is set on fire. As soon as the the grass is burned, the nuts are placed in cold water. It is believed the thermal shock causes the seedcoats to crack (Willan 1985). In controlled experiments Eakle and Garica (1977) used sulfuric, hydrochloric, and nitric acids and hot water in scarification treatments, but did not enhance the germination of candlenut seeds. *Aleurites* species can be propagated from cuttings.

BETULACEAE — BIRCH FAMILY

Alnus Hill. — Alder

GROWTH HABIT, OCCURRENCE, AND USE *Alnus* includes about 30 species native to the Northern Hemisphere and the Andes of South America (Bailey 1951). Their most common native habitats are high mountains, swamps, and bottomlands along streams (Schopmeyer 1974a). Alders are among the first species to become established naturally on many denuded areas. Seedlings have been planted successfully for reforestation of spoil banks and soil fertility is improved through fixation of atmospheric nitrogen. Alders are planted for wildlife food and cover and for ornamental use. In the Pacific Northwest, red alder (*A. rubra*) is harvested for pulpwood and as a hardwood for furniture manufacturing.

FLOWERING AND FRUITING Clusters of male and female catkins occur on the same tree in late winter or spring (Table 1). Strobiles (catkins) of most species are 1 to 1.5 cm long when mature (Fig. 1), but those of Nepal alder (*A. nepalensis*) and red alder are larger, having lengths of 1.2 to 2.4 cm. Flowering occurs in abundance by the time trees are a decade old, and good crops occur at least once every 4 years (Table 1). In a recent study Farmer et al. (1985) estimated that 60% of the catkins of green alder (*A. viridis* ssp. *crispa*) were fertile and produced some seed. They also estimated the annual seed fall for this species in Ontario, Canada, as 0.14 to 2.4 million seeds per hectare.

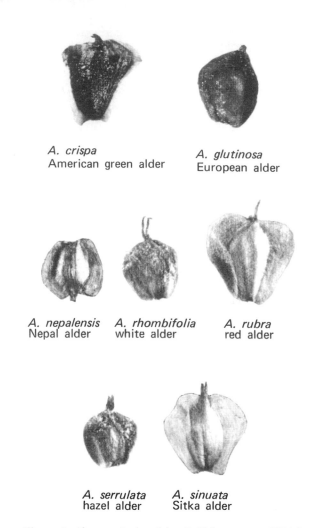

A. crispa
American green alder

A. glutinosa
European alder

A. nepalensis
Nepal alder

A. rhombifolia
white alder

A. rubra
red alder

A. serrulata
hazel alder

A. sinuata
Sitka alder

Figure 2. *Alnus*: nuts (seeds), ×8 (Schopmeyer 1974a).

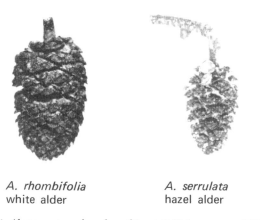

A. rhombifolia
white alder

A. serrulata
hazel alder

Figure 1. *Alnus*: mature female catkins, ×2 (Schopmeyer 1974a).

Seeds are small nuts borne in pairs on the bracts of the strobiles. When released they are dispersed by wind, and in some species by water. Seeds of European alder (*A. glutinosa*) have remained viable after floating for 12 months in still water. The nuts of red and Sitka alder (*A. sinuata*) have broad wings about as wide as the body of the nut. In other species the wings are reduced to a narrow border (Fig. 2). Seeds contain no endosperm (Fig. 3).

COLLECTION, EXTRACTION, AND STORAGE OF SEED Strobiles may be collected from standing or recently felled trees when the bracts (scales) start to separate on the ear-

liest strobiles. They will open after being exposed in drying racks in a well-ventilated room for several weeks at ambient air temperature. They can be opened in a shorter time by drying them in a kiln at 25 to 35°C. Most seeds fall out of the strobiles during the drying process. The remainder, if needed, may be extracted by tumbling or shaking.

Air screening seeds of European alder has produced purities as high as 90%. Quality, however, usually is low because only a small proportion of the empty seeds can be separated. Soundness in most seed lots has been between 30 and 70% (Table 1). The number of seeds per gram ranges from 600 to 1500. Except for seed of American green alder (*A. crispa*), higher numbers of seeds per gram may indicate low quality. Numbers ranging from 1760 to 4400 seeds per gram have been found for samples of Nepal, red, and thinleaf alder (*A. tenuifolia*), but less than 5% of the seeds in these samples were full. Such low per-

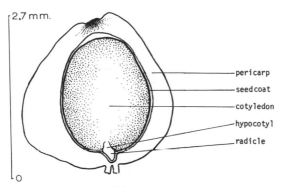

Figure 3. *Alnus*: (A) *A. glutinosa*, European alder: seedling development at 1 and 7 days germination; (B) *A. tenuifolia*, thinleaf alder: two older seedlings (Schopmeyer 1974a).

centages of good seed are common in sparse seed crops.

Air-dried seeds can be stored in sealed containers at 2 to 5°C. Under these conditions, viability has been maintained for 2 years for seeds of European alder and for 10 years for speckled alder (*A. rugosa*).

PREGERMINATION TREATMENT Germination capacity of fresh seeds of white (*A. rhombifolia*) and thinleaf alder was equally good for prechilled and non-prechilled seed. Seeds of European alder and European speckled alder (*A. incana*) also germinated promptly without prechilling; but dried seed, at a moisture content of 8 to 9%, was dormant (Table 1). Germination of dried seeds, after prechilling for 180 days at 5°C, was higher than that of fresh seed. Maximum germination, however, was obtained only when prechilling was followed by a 3-day period at −20°C.

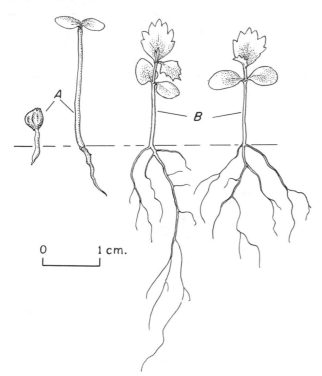

Figure 4. *Alnus rubra*, red alder: longitudinal section through nut, ×6 (Schopmeyer 1974a).

Table 1. *Alnus*: phenology of flowering and fruiting, characteristics of mature trees, cleaned seed weight and soundness, and germination test conditions and results for prechilled seeds (Schopmeyer 1974a).

Species	Flowering dates	Fruit ripening dates	Seed dispersal dates	Mature tree height (m)	Year first cultivated	Seed-bearing Age (years)	Seed-bearing Interval (years)	Seeds/gram	Seed soundness (%)	Prechilling (days)	Temperature (°C)	Germination (%)
A. crispa	Spring	Aug.–Oct.	When ripe	3.1	1782	–	–	2800	42–90	60	20/30	28
A. glutinosa												
Fresh	Mar.–May	Fall	Overwinter	31.4	1866	–	–	700	–	0	25	29
Dried	Mar.–May	Fall	Overwinter	31.4	1866	–	–	700	–	180	25	34
A. incana												
Fresh	Feb.–Apr.	–	–	20.0	–	>25	1–4	1470	51	0	25	28
Dried	Feb.–Apr.	–	–	20.0	–	>25	1–4	1470	51	0	25	28
A. nepalensis	–	Winter	Winter	30.0	1916	10	–	1460	–	–	–	–
A. rhombifolia	Mar.	Fall	–	24.0	1885	–	–	1480	–	0	20/30	59
A. rubra	Spring	Aug.–Oct.	Fall–winter	22.7	1884	> 10	4	660	71	0	15/25	56
A. rugosa	Mar.–May	Fall	–	8.0	–	–	–	–	30–60	–	–	–
A. serrulata	Spring	Fall	–	8.0	1769	–	–	–	–	–	18/25	36
A. sinuata	–	–	–	12.3	1903	–	–	–	–	–	–	–
A. tenuifolia	Spring	Aug.–Sept.	–	9.2	1880	–	–	1490	–	0	20/30	4

GERMINATION There has been considerable research on the germination of seeds of red alder. Radwin and DeBell (1981) obtained germination of 59 to 87% without pretreatment. Elliott and Taylor (1981) also found dormancy was not significant in untreated seed, but that certain trees produced seeds that required pre-chilling. The tree-to-tree differences in germination of red alder seeds had previously been noted by Kenady (1978). Bormann (1983) determined seeds of red alder responded to red and far-red light in a reversible fashion, implicating the role of phytochrome in their germination process. Germination was strongly inhibited under conditions of low irradiance similar to that which would occur with seed burial or under a broadleaf canopy. For germination testing, both constant temperatures and diurnally alternating temperatures have been used. Light has been recommended for germinating seeds of alder, but seeds of most species germinate in the dark (see red alder above).

NURSERY AND FIELD PRACTICE Spring sowing is preferred by nursery growers in Pennsylvania, Washington, and California, but fall sowing is preferred in New York. Sowing depths of 0.25 to 0.5 cm have been used for European and red alder. In California, seed of red alder has been diluted with vermiculite and drilled 1 cm deep. Seed of Nepal alder has been mixed with sand and spread over nursery beds to give densities of 250 seedlings per square meter. Out-planting usually is done with 1-0 stock. Germination is epigeal (Fig. 4). Berry and Torrey (1985) provide details of how to inoculate seedlings of alder with a suspension of *Fankia* (Actinomycetales) to insure they will be able to fix nitrogen.

Direct seeding in the field has been done successfully with speckled and European alder. In Pennsylvania speckled alder was established by broadcasting on previously disked areas and on sod. Seed was collected in the fall and sown the following February and March. In England, European alder was spot seeded on a shallow blanket bog, and each seed spot fertilized with phosphate.

ROSACEAE — ROSE FAMILY

Amelanchier Medic — Serviceberry

GROWTH HABIT, OCCURRENCE, AND USE Serviceberries include about 25 species of small deciduous trees and shrubs native to North America, Europe, and Asia. Most species provide browse and edible fruits for domestic livestock and wildlife and many have attractive flowers (Brinkman 1974a). Saskatoon serviceberry (*Amelanchier alnifolia*) and downy serviceberry (*A. aborea*) have been used to a limited extent for shelter and wildlife plantings, but other species also should be considered for these environmental uses. Geographic races of *Amelanchier* undoubtedly occur in widely distributed species such as Saskatoon and downy serviceberry. Several natural hybrids are known (Cruise 1964).

FLOWERING AND FRUITING The perfect white flowers appear in terminal clusters early in the spring, before

A. florida
Pacific serviceberry

A. laevis
Allegheny serviceberry

A. alnifolia
saskatoon serviceberry

A. florida
Pacific serviceberry

Figure 1. *Amelanchier*: pomes, ×2 (Brinkman 1974a).

Figure 2. *Amelanchier*: seeds, ×6 (Brinkman 1974a).

the leaves in some species (Table 1). Fruits are berry-like pomes (Fig. 1) that turn dark purple or black when they ripen (Table 1). Each fruit contains from 4 to 10 small seeds, although some of these are usually abortive. Fertile seeds are dark brown with a leathery seedcoat (Fig. 2) and with the embryo filling the seed cavity (Fig. 3). Fruits usually are eaten by birds or animals as soon as they ripen.

COLLECTION OF FRUIT To minimize losses to wildlife, fruits must be picked from the plants as soon as possible after ripening. Unless the seeds are to be extracted promptly, the fruits should be spread out in thin layers to dry. Loss of viability will result if the fruits are allowed to overheat.

EXTRACTION AND STORAGE OF SEED Seed extraction is usually accomplished by macerating the fruits in water and washing them over screens. This removes most of the pulp. After drying and rubbing through the screens, the seeds and remaining debris can be run through an air screen to remove small, aborted seeds and bits of fruit. Seed yield and weight vary con-

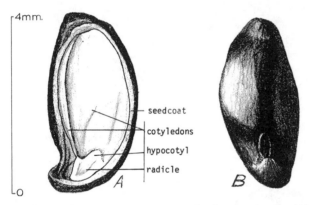

Figure 3. *Amelanchier sanguinea*, roundleaf serviceberry: (A) longitudinal section through seed, and (B) exterior view; both at ×12 (Brinkman 1974a).

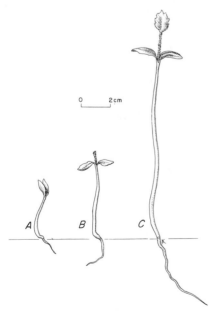

Figure 4. *Amelanchier* spp: seedling development at 3, 5, and 7 days after germination (Brinkman 1974a).

Table 1. *Amelanchier*: phenology of flowering and fruiting, characteristics of mature trees, cleaned seed weight, pregermination treatments, and germination test conditions and results (Heit 1971, Brinkman 1974a, Dirr 1987).

Species	Flowering dates	Fruit ripening dates	Mature tree height (m)	Year first cultivated	Ripe fruit color	Seeds/ gram	Warm Stratification (days)	Prechilling (days)	Substrata	Temperature (°C)	Germination (%)
A. alnifolia	May–June	July–August	6.1	1826	Blue-purple	180	0	90	Sand	—	60–70
A. arborea	Mar.–June	June–August	18.9	1623	Red-purple	176	120	120	Sand	—	95
A. canadensis	Mar.–April	May–June	7.7	1641	Black	211	0	120	Sand	—	98
A. florida	April	August	12.3	1826	Purplish black	118	0	90–120	Sand	20/30	54
A. laevis	Mar.–June	June–August	9.2	1870	Dark purple	—	0	120	K	20/30	—
A. sanguinea	May–June	July–Sept.	3.1	1824	Dark purple	185	0	30–90	P	20/30	10
							—	60+	—	20	60–70
								—		—	—

siderably (Table 1). Serviceberry seeds can be stored in sealed containers at 5°C.

PREGERMINATION TREATMENT Embryos of all species show dormancy that can be overcome by prechilling (Crocker and Barton 1931). The seedcoat of some species may also retard germination. Scarification of Allegheny serviceberry (*A. laevis*) in concentrated sulfuric acid followed by prechilling improved germination. One always wonders about such treatments because acid scarification may influence the basic metabolism of the seeds as well as physically remove at least a portion of the seed coverings. The necessary time period for prechilling varies, but most species require 2 to 6 months. More recent studies have not significantly advanced knowledge over that which was known in 1974 for the serviceberry species. Weber et al. (1982) found 3 to 5 months of prechilling at 1 to 3°C was necessary for germination of seeds of downy serviceberry. In a very interesting study Robinson (1986) suggested that either mechanical scarification or inges-

tion by cedar waxwing birds enhanced germination of *Amelanchier* seeds.

GERMINATION TEST Germination of stratified seed can be tested in sand or sand-peat mixtures. A constant temperature of 21°C or alternating temperatures of 20/30°C have been equally successful. Light does not appear to enhance germination (Table 1). Germination is epigeal (Fig. 4). Under natural conditions germination apparently begins in early spring during snowmelt.

NURSERY AND FIELD PRACTICE Serviceberry seed may be sown in the fall or prechilled seed can be sown in the spring. Many seeds do not germinate until the second spring. They should be sown as soon as possible after collection and beds kept mulched until germination begins the following spring. Seed should be sown in drill rows at the rate of 80 to 85 sound seeds per linear foot and covered with 0.5 cm of soil. For Saskatoon serviceberry, half-shade is beneficial the first year.

MYRTACEAE — MYRTLE FAMILY

Amomyrtus luma (Md.) D. Legrend & Kausel

GROWTH HABIT, OCCURRENCE, AND USE *Amomyrtus luma* is a tree which grows to 20 m in height and is native to Chile. Ramirez et al. (1980) investigated its germina-

tion and photoperiod. Seed stores with less loss of viability if the pericarp is not removed.

LEGUMINOSAE — LEGUME FAMILY

Amorpha L. — False indigo

GROWTH HABIT, OCCURRENCE, AND USE In North America, the false indigos include about 15 closely related species of deciduous shrubs or subshrubs (Brinkman 1974b). Some species die back to the ground every year. The most important species provide wildlife food and cover. Because of their handsome foliage and flowers, some species also are suitable for environmental planting. Indigobush amorpha (*A. fruticosa*) is the tallest species (Table 1) and has been grown for wildlife plantings. At least four varieties of indigobush amorpha are recognized, and some of these appear to be geographic races.

FLOWERING AND FRUITING The irregular perfect flowers of false indigo are blue to violet-purple in color and are borne in the spring or summer (Table 1). The fruit is a short, indehiscent, somewhat curved and gland-dotted pod containing 1 (rarely 2) small glossy seeds (Figs. 1, 2). Commercial seed usually consist of the dried seed pods. Good seed crops of California amorpha (*A. californica*) are borne every 2 years, and similar frequencies probably are typical of the other species. Disper-

| A. fruticosa | A. canescens |
| indigobush amorpha | leadplant amorpha |

Figure 1. *Amorpha*: (A) pod and seed of *A. fruticosa*, indigobush amorpha; (B) seed of *A. cunescens*, leadplant amorpha; all ×5 (Brinkman 1874b).

sal, mostly by animals, occurs in the fall.

COLLECTION, EXTRACTION, AND STORAGE OF SEED The ripe pods can be striped from the branches and should be spread out in thin layers for a few days to permit superficial drying.

Extraction of the seeds is not necessary, as the pods are largely one-seeded and do not interfere with germination. The seeds can be extracted by threshing the

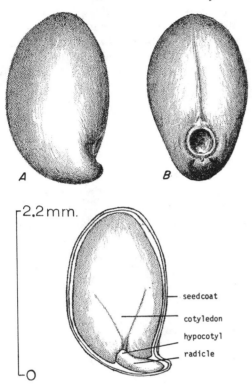

Figure 2. *Amorpha canescens*, leadplant amorpha: exterior views of seed and embryo, ×20 (Brinkman 1974b).

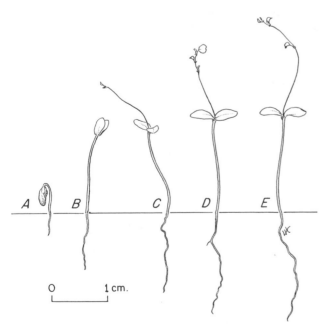

Figure 3. *Amorpha canescens*, leadplant amorpha: seedling development at 1, 2, 8, 20, and 52 days after germination (Brinkman 1974b).

Table 1. *Amorpha*: phenology of flowering and fruiting, characteristics of mature trees, ripe fruit and cleaned seed weight, and germination test conditions and results (Brinkman 1974b).

Species	Flowering dates	Fruit ripening dates	Seed dispersal dates	Mature tree height (m)	Year first cultivated	Ripe fruit/kg	Seeds/gram	Light (hr)	Substrata	Temperature (°C)	Germination (%)
A. californica	May–July	July–Sept.	Aug.–Sept.	2.8	—	—	84	0	Soil	—	42
A. canescens	June–Aug.	Aug.–Sept.	Fall	1.0	1833	211,000	650	8	Sand	20/30	28
A. fruticosa	May–June	Aug.	—	5.5	1724	114,000	170	8	Sand	20/30	63
A. nana	May–July	July	July	1.0	1811	132,000	—	0	Sand	20/30	70

pods by hand with a flail or by running them through a properly adjusted hammer mill. Little is known about seed yield or storage (Table 1). Seeds of leadplant amorpha (*A. canescens*) were stored for 22 months at 5°C, followed by 16 months at room temperature with little loss of viability. Seed of indigobush amorpha has retained viability 3 to 5 years at room temperature.

PREGERMINATION TREATMENT No treatment is necessary for fresh seed of California amorpha (Mirov and Kraebel 1939) and probably not for fall-sown seed of leadplant amorpha.

GERMINATION Stored seeds of all species have impermeable seedcoats and show a high percentage of dormant seeds. Germination of some seed lots has been improved by soaking the seed in hot water for about 10 minutes. Prechilling has been used in preparation for spring sowing in nursery beds. Germination test conditions reported in the literature (Table 1) are so uncontrolled that germination work needs to be started with a basic determination of the nature of dormancy. Germination is epigeal (Fig. 3).

NURSERY AND FIELD PRACTICE Seed is usually sown in the fall. This probably should be the practice until germination characteristics and storage-germination relations are worked out. The pods are sown in rows and covered with a thin layer of soil. The various species of amorpha can also be propagated from cuttings, layers, or suckers (Bailey 1951).

VITACEAE — GRAPE FAMILY

Ampelopsis Michx.

GROWTH HABITAT, OCCURRENCE, AND USE The genus *Ampelopsis* consists of about 25 species of mostly tendril-bearing, ornamental vines native to North America and Asia (Bailey 1951). Several of the important species are listed below.

Species	Year first cultivated	Distribution
A. humulifolia	1868	Northern China
A. Bodinieri	1907	Central China
A. cordata	1796	Eastern United States
A. brevipedunculata	1870	East Asia
A. Delavayana	1900	Central China
A. aconitifolia	1868	Northern China
A. japonica	1867	Japan, northern China
A. arborea	—	Eastern United States to Mexico
A. megalophylla	1900	Central China
A. mumulifolia	—	Northern China

Technically this genus differs from *Vitis* by the absence of shedding bark, by cymose rather than paniculate inflorescences, and by bisexual flowers and spreading petals that fall separately. Its leaves are alternate, simple or compound, with long petioles.

FLOWERING AND FRUITING The flowers are rather inconspicuous, but the fruits are stunning. The fruit is a berry that ranges from yellow to pale lilac and porcelain blue. The small seeds need to be cleaned from the pulp.

GERMINATION The seeds of the most commonly cultivated species require prechilling (Dirr and Heuser 1987): Prechilling for 3 months is the ideal treatment for *A. brevipedunculata* and *A. humulifolia*. Seeds of *A. aconitifolia* collected in September, cleaned, and sown immediately, germinated 40%, but the same seeds prechilled for 1 and 7 months, germinated 60%, respectively.

NURSERY AND FIELD PRACTICE Untreated summer cuttings of porcelain ampelopsis (*A. brevipedunculata*) rooted 90% in 30 days (Dirr and Heuser 1987).

ANACARDIACEAE — CASHEW FAMILY

Anacardium L. — Cashew

GROWTH HABIT, OCCURRENCE, AND USE The genus *Anacardium* consists of about 8 species of trees and shrubs native to the American tropics (Bailey 1951). Cashew (*A. occidentalis*) is widely cultivated in the tropics for its fruit. It is a spreading evergreen tree that reaches 10 m in height. The stems have a milky juice. The flowers are yellowish pink and about 0.8 cm across. The fruit is a kidney- or heart-shaped nut that is edible when roasted. The enlarged receptacle is fleshy, bright yellow or red in color, and is also edible.

GERMINATION Slow imbibition by dry intact seeds is the main cause of delayed germination in cashew (Subbaiah 1982/1983). The problem is greatest with large seeds. Presoaking for 1 or 2 days, or removal of the waxy layer of the pericarp by treatment with chloroform or acetone promotes imbibition and reduces the time required for germination. Light and gibberellin have also been reported to enhance germination. Saplaco and Revilla (1973) reported cashew seedling emergence was enhanced by placing the seeds on their side in the seed bed.

ERICACEAE — HEATH FAMILY

Andromeda L. — Bog rosemary

GROWTH HABIT, OCCURRENCE, AND USE *Andromeda* consists of evergreen low shrubs of cold regions in the Northern Hemisphere (Bailey 1951). The leaves are entire with short petioles. The flowers are pinkish and borne in nodding terminal umbels. The fruit is a capsule that dehisces from 5 valves and usually contains many seeds. Bog rosemary (*A. polifolia*) is a low shrub from a creeping rootstock, and is widely distributed in northern Europe, Asia, and North America. This species has been cultivated since 1768. Downy rosemary (*A. glaucophylla*) is similar to bog rosemary except the leaves are white-tomentulose beneath. It has been cultivated since 1879.

GERMINATION Downy rosemary seeds will germinate without pretreatment (Dirr and Heuser 1987). Best results occur by seeding in milled sphagnum without covering seeds. Flats can be placed under mist for the germination period. Bog rosemary seeds sown in the spring have sporadic germination, apparently indicative of a prechilling requirement. Winter cuttings of both species can be rooted fairly easily.

MYRTACEAE — MYRTLE FAMILY

Angophora Cav. — Apple gum

GROWTH HABIT, OCCURENCE, AND USES Apple gums are a small Australian genus resembling the eucalyptus. They have achieved world popularity in other dry areas such as California and South Africa, but are native to the fast-draining sandstone of the east coast of Australia. These trees have elegant orange or pinkish bark, which peels unevenly from the trunk. The apple gums have two leaf forms—pale green, heart-shaped juvenile foliage and long drooping adult leaves up to 12.5 cm long. The two species that are commonly planted are smoothbarked apple gum (*Angophora costata*) and dwarf apple gum (*A. cordifolia*).

FLOWERING AND FRUITING The cream-colored flowers are largely mass of stamens, but unlike eucalyptus they also have small petals. The fruits are like peanuts, but ribbed.

GERMINATION Seeds germinate without pretreatment.

DIPTEROCARPACEAE — DIPTEROCARP FAMILY

Anisoptera Korth.

GROWTH HABIT, OCCURENCE, AND USE *Anisoptera* is a genus of about 14 species distributed from Burma to New Guinea (Corner 1952). It is similar to *Dipterocarpus*, but with fissured bark and small stipules that do not leave ringlike scars.

FLOWERING AND FRUITING The inflorescence is many-flowered. The fruit is an edible nut fused with the calyx.

GERMINATION Seeds of *Anisoptera laevis* had 79% germination after 4 weeks of testing in a study conducted in Malaysia (Ng and Sanah 1979).

BIGNONIACEAE BIGNONIA FAMILY

Anisostichus Bur. — Cross vine

GROWTH HABITAT, OCCURRENCE, AND USE Cross vine is a woody, tendril-climbing, evergreen vine native to the southeastern United States. It climbs to 20 m and is used as an ornamental because of the profusion and beauty of its flowers. The red-orange flower consists of a funnelform-campanulate corolla that is strongly 2-lipped. The fruit is a capsule 10 to 17 cm long. The single species *Anisostichus capreolata* has been cultivated since 1653.

GERMINATION The seeds germinate without pretreatment (Dirr and Heuser 1987). Softwood roots readily in June and July.

ANNONACECE — ANNONA FAMILY

Annona L. — Custard apple

GROWTH HABIT, OCCURRENCE, AND USE The genus *Annona* consists of 50 species of shrubs and trees, mostly native to tropical America. Several species yield important edible fruits. Pond apple (*A. glabra*) is native to southern Florida. Sugar apple (*A. squamosa*) is a small branching tree to 6.5 m. It is widely grown in tropical America where the fruits are used in sherbets and for jellies. Common custard apple (*A. reticulata*) is a small tree to 8 m in height that is native to tropical America.

FLOWERING AND FRUITING Flowers are super-axillary, often opposite the leaves, and solitary or in clusters. The calyx usually is tubular and 3-parted. The petals are 6 in number in 2 series. The fruit is a syncarp that is large and fleshy and formed from the fusion of the pistil and receptacle.

GERMINATION These trees are propagated by seed, and seed dormancy can be a problem. Seeds of *A. muricata*, *A. reticulata*, and *A. squamosa* are reported not to be dormant, but still require 3 months to germinate. Seeds of *A. cherimula*, *A. crassiflora*, and *A. diversifolia* exhibit considerable dormancy and require several months of prechilling before they will germinate. Despite the seedcoat of *Annona* species, which is thick and heavily lignified, seed imbibes water rapidly. In *A. squamosa* the tiny embryo is embedded in a large endosperm, and it continues to develop after being shed from the tree. Gibberellin applied at 350 ppm may be effective in enhancing germination.

RUBIACEAE — MADDER FAMILY

Anthocephalus A. Rich.

GROWTH HABIT, OCCURRENCE, AND USE *Anthocephalus* is a deciduous group of trees reaching 30 m in height, with outstanding limbs, drooping slightly at tips. The leaves are large, 20 by 15 cm, and more or less ovate. Flowers are borne singly at the end of short side twigs and are yellowish white. It is a common tree in the forest of southern Asia.

GERMINATION Seeds of *A. chinensis* are sensitive to light quality. When exposed to continuous light or dark, they have low germination; apparently they are phytochrome sensitive and require red light for germination. Enrichment of the germination substrate with potassium nitrate or gibberellin will not overcome the light requirement.

EUPHORBIACEAE — SPURGE FAMILY

Aporusa Blume — Crescent tree

GROWTH HABIT, OCCURRENCE, AND USE *Aporusa* is a genus of evergreen shrubs and small trees, few of which exceed 16 m in height. They are widely distributed as understory shrubs in the forests of Malaysia. The crescent tree (*A. benthamina*) is one of the most striking small forest trees in Malaysia.

FLOWERING AND FRUITING The fruit is a small, round or oblong capsule with a thin, often yellow, red, or purple rind. It splits from the base to the apex into 2 to 4 thin, bony segments exposing the seeds, which are covered with a thin, yellow, orange, or red pulp.

GERMINATION Seeds of *A. arborea* had 87% germination after 10 weeks testing in a study conducted in Malaysia (Ng and Sanah 1979).

THYMELAEACEAE — DAPHNE FAMILY

Aquilaria Lam. — Malayan eaglewood

GROWTH HABIT, OCCURRENCE, AND USE *Aquilaria* is a genus of medium-sized trees reaching 30 m in height. The trunk of older trees is strongly fluted at the base. The leaves are rather small, elliptic in outline, and with a long tip. The flowers are pale greenish or yellowish white and fragrant. The fruit is 2.5 cm long and 2 cm wide, flattened, egg-shaped, and green in color. The seeds dangle out of the fruit on strings attached to one end of each seed at maturity. The seeds are 0.6 cm long, pear-shaped, and orange-brown in color. *Aquilaria maloccensis* is found throughout the Malayan peninsula in lowland forest.

The wood of these species is normally soft, pale, and odorless. In certain old trees that are dying, hard and fragrant wood is found in isolated pieces. These sections of dark wood are chipped out, cleaned, and sold as eaglewood or aloewood (Whitmore 1972). *Aquilaria agallocha* (eaglewood) is grown in Assam for commer-

cial agarwood production. The trees are naturally infected with a fungus, *Cytosphaera mangiferae*. This infected wood produces fragrant fumes (incense) when burnt.

GERMINATION Seeds of *Aquilaria agallocha* were sown in the fruit to avoid damage to the seeds in one of the few reported experiments with seeds of this genus (Bhaskar 1984). Only about 4% emergence was obtained, but seedling survival was no problem. The fruit numbered 1200 to 1600 per kg, and each contained 1 or 2 seeds.

ARALIACEAE — GINSENG FAMILY

Aralia L.

GROWTH HABIT, OCCURRENCE, AND USE *Aralia* consists of about 20 species of deciduous trees, shrubs, and herbs native to Australia, Asia, and North America (Bailey 1951). The trees and shrubs are spiny and the herbaceous species are either spiny or smooth stemmed (Blum 1974). They are used in environmental plantings and some of the woody species are widely planted as oddities because of clublike branches. Some North American species have potential value for planting (Table 1).

The *Aralia* species enjoy considerable academic interest from botanists because of (a) their unusual growth forms (White 1984) and (b) the asynchrony in flowering between the sexes (Barrett and Thomson 1982, Baua et al. 1982, Thomson et al. 1982, Flanagan and Moser 1985).

FLOWERING AND FRUITING The flowers of *Aralia* are polygamous, white or green, and occur in umbels or panicles. Flowering occurs from May to September depending on the species; fruit matures in late summer or fall. The fruit is a small berrylike drupe containing 2 to 5 compressed, crustaceous, light reddish brown nutlets that are round, oblong, or egg-shaped. Each nutlet contains one compressed, light brown seed with a thin coat that adheres closely to the fleshy endosperm (Figs. 1, 2). The nutlet is the seed of commerce for *Aralia*.

COLLECTION, EXTRACTION, AND STORAGE OF SEED *Aralia* fruits may be collected when they begin to fall from the plants in autumn (Table 1). The seeds are ripe when the endocarps of the nutlets become hard and brittle;

this ripening may occur somewhat later than the ripening of the pulp. The fruits should be run through a macerator, with water, immediately after collection. This will prevent fermentation and enable the pulp and empty seed to float off or be removed by screening.

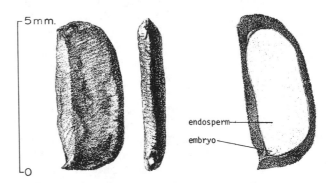

Figure 1. *Aralia spinosa*, devil's walkingstick: nutlets (seeds); ×3 (Blum 1974).

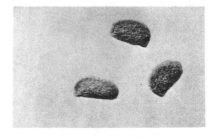

Figure 2. *Aralia nudicaulis*, wild sarsaparilla: exterior view of nutlets in two planes and longitudinal section, ×8 (Blum 1974).

Table 1. *Aralia*: phenology of flowering and fruiting, growth habit, characteristics of mature trees, and cleaned seed weight (Blum 1974).

Species	Flowering dates	Fruit ripening dates	Seed dispersal dates	Growth habit	Mature tree height (m)	Year first cultivated	Seeds/gram
A. hispida	June–July	Aug.–Sept.	Aug.–Sept.	Shrub	0.3–1.0	1788	220–225
A. nudicaulis	May–June	Aug.	Aug.	Shrub	0.1–1.1	1731	185–240
A. spinosa	July–Aug.	Aug.–Sept.	Sept.–Oct.	Tree	7–8	1688	230–345

Purity of seed processed in this manner is high (90%), but soundness in some lots may only be from 30 to 60%. The cleaned seeds are quite small (Table 1). Cleaned seed that has reached moisture equilibrium should be stored in sealed containers at low temperatures.

PREGERMINATION TREATMENT *Aralia* seeds have dormant embryos, and some species, notably bristly aralia (*A. hispida*), appear to have hard seedcoats (Heit 1967c). Scarification followed by prechilling may be required to obtain germinatiopn of these seeds. Dormancy of

seeds of devil's walkingstick (*A. spinosa*) can be satisfied by prechilling only. There is some evidence that a period of afterripening at warm temperatures before prechilling is beneficial in breaking the dormancy of *Aralia* seeds.

NURSERY AND FIELD PRACTICE Considering the limited knowledge of the germination ecology of *Aralia* species all that can be recommended in terms of nursery practice is to acid scarify seeds for 30 to 40 minutes and then broadcast the seeds in the fall so they will receive natural stratification over the winter.

ARAUCARIACEAE — ARAUCARIA FAMILY

Araucaria Benth.

GROWTH HABIT, OCCURRENCE, AND USE The araucarias, consisting of 15 species, are evergreen coniferous trees generally confined to the Southern Hemisphere. They are found in South America, Australia, New Guinea, New Caledonia, New Hebrides, and Norfolk Island under tropical, subtropical, and temperate climates (Record and Hess 1943, Dallimore and Jackson 1967).

The araucarias are noted for their long, straight, clear boles, and symmetrical crowns; many are useful for timber, and some are cultivated as ornamental trees (Street 1962).

Several species have been introduced to California, Florida, and Hawaii (Table 1). Araucaria species are generally found on sites ranging from sea level to 2150 m, with 125 to 250 cm of rainfall, and well-drained soils. Columnar araucaria (*Araucaria columnaris*) and Norfolk Island pine (*A. heterophylla*) have been widely planted in Hawaii (Walters 1974a). The botanical identity of these two species is often confused.

FLOWERING AND FRUITING Araucarias generally begin to flower and fruit between the age of 15 to 20 years. Male

and female flowers are typically found on different parts of the same tree. Male flowers usually appear at the base of the crown in young trees and female flowers at the top. As the tree grows older, the male and female flowers come closer to each other. Bisexual flowers are also found. After pollination the female flowers develop slowly; the cones mature in about 2 years (Ntima 1968). The mature cones are ovoid to almost spherical, ranging in size from 10 by 5 cm for hoop pine (*A. cunninghamii*) to 30 by 20 cm for bunya-bunya (*A. bidwillii*).

Upon maturing, cones turn from green to brown. They disintegrate on the tree or fall to the ground and disintegrate. The brown seeds are kite-shaped and generally fall within the periphery of the crown (Figs. 1, 2). The times of flowering, of seed development and dispersal, and seed crop intervals for important species of araucaria are given in Table 1.

COLLECTION, EXTRACTION, AND STORAGE OF SEED Collection of cones should begin when the first trace of brownness is observed on the cone. The second-year cones are generally picked from felled trees (Ntima

Table 1. *Araucaria:* phenology of flowering and fruiting, characteristics of mature trees, distribution, and cleaned seed weight (Walters 1974a).

Species	Flowering dates	Seed ripening dates	Seed dispersal dates	Seed-bearing interval (years)	Mature tree height (m)	Occurrence Native	Occurrence United States	Seeds/ gram
A. augustifolia	–	Apr.–May	May–August	Annually	25–37	Brazil Argentina Paraguay	Hawaii	0.10
A. bidwillii	Sept.–Oct.	Jan.–Feb.	Jan.–Feb.	1–2	30–43	Queensland	California Florida	0.07
A. columnaris	Dec.–Jan.	Dec.–Feb.	Dec.–Feb.	3–4	62	New Caledonia	Hawaii Florida	2.20
A. cunninghamii Early flowering	Dec.–Jan.	Dec.	Dec.	4–5	62	New Guinea	Florida California Hawaii	4.40
Late flowering	April–May	–	–	–	62	New Guinea	Florida California Hawaii	4.40
A. heterophylla	Sept.	April	April–May	3–4	62	Norfolk Is.	Hawaii	0.56

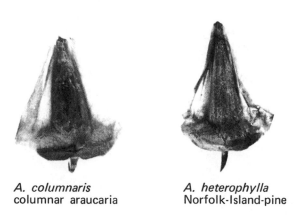

A. *columnaris*
columnar araucaria

A. *heterophylla*
Norfolk-Island-pine

Figure 1. *Araucaria*: seed, ×1 (Walters 1974a).

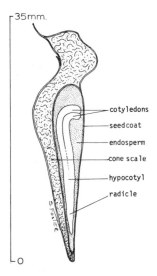

Figure 2. *Araucaria heterophylla*, Norfolk Island pine: longitudinal section through a seed, ×2 (Walters 1974a).

1968). Collection must be timed correctly to get the highest proportion of mature and fertile seed. A method for timing cone maturity is to pick a cone and measure the time it takes to disintegrate; ripe cones spontaneously disentegrate within 7 days (Walters 1974a). Collected cones are spread on shelves in single layers for drying and should be turned daily. They normally will began to disintegrate within a few days. Those which fail to disintegrate within 10 days are discarded, being considered too immature. The average seed weight ranges from 12.9 g for bunya-bunya to 0.2 g for hoop pine (Table 1) (Walters 1974a). Ethanol and pentane mixtures have been used to separate filled from empty seeds of hoop pine.

Araucaria seeds have short viabilities under atmospheric conditions and should be sown within a month of collection (Ntima 1968). Seed storage for these species has recently attracted a lot of research. To be stored for any period of time, seeds of the araucarias must be partially dried. The various species are highly sensitive as to how much drying the seeds will withstand and still remain viable. The following breakdown of safe drying levels is based on Tompsett (1982, 1983, 1984) and Scowcroft (1988): For *A. araucana, A. angustifolia, A. hunstenii,* and *A. bidwillii,* the minimum acceptable moisture level is 25 to 40%; for *A. columnaris, A. rulei, A. nemorosa,* and *A. scopulorum,* it is 12%; and for *A. cunninghamii* it is 2%. At the proper moisture level seeds can be stored for some time at 5°C. They should be sown as soon as they are removed from cold storage because they rapidly lose viability.

Pritchard and Prendergast (1986) conducted a complex study on the influence of drying on the subsequent development of excised embryos of *A. huntsteinii.* Aseptically excised embryos reacted different from nonsurface sterilized embryos. Apparently this work has considerable significance to the *in vitro* conservation of recalcitrant seed tissue, but how to translate the findings to practical seed technology is not clearly apparent.

GERMINATION No pregermination treatments are needed for araucaria seeds (Walters 1974a). Under suitable moisture and temperature conditions (20 to 30°C), germination may began about 10 days after sowing. Germination is delayed by cooler temperatures, sometimes taking 50 days (Ntima 1968). Seed quality varies from year to year; if sufficient pollen is available, seed quality is usually good. Aquila and Ferreira (1984) suggested that seed scarification improved the germination of seeds of parana pine (*A. angustifolia*). They also found that germination was faster with seeds stored for 4 months in polyethylene bags, but total germination was lower.

NURSERY AND FIELD PRACTICE Araucaria can be grown under high or low shade. For both types of shade, seeds are sown during spring. Seeds should be treated with a fungicide to prevent damping-off. With high shade, the seeds of all species except bunya-bunya are sown in flat-bottomed drills about 1 cm deep and covered with softwood sawdust. If hardwood sawdust is used it must be treated with a fungicide.

Bunya-bunya seeds are sown in drills, but are handled differently than other species. The drills (furrows) for bunya-bunya are 7.5 to 10 cm deep. A few months after sowing, "tubers" (*fusiform radicle*) are formed. The seed beds are redug, and the tubers are collected. The tubers are either planted directly into soil-filled tubes or stored until they are needed. Exposure to sunlight before planting breaks tuber dormancy, and the plants begin to grow. Almost every seed produces a tuber, and all tubers develop into plants (Walters 1974a). With low shade the seed is broadcast sown on well-prepared beds and covered with 1.8 cm of sawdust. The aim of both types of sowing is to have a stocking rate of 120 to 160 plants per m^2.

Newly sown beds are given full overhead shade for several days. Best shoot development occurs when the beds are give 75% shade for the first several months and 50% shade for the next 3 months (except for hoop pine). Shading should be removed gradually to give full sunlight 2 weeks before potting. Full light is not admitted until nearly 1 year after sowing hoop pine. When 75% of the seedlings are 15 to 22 cm tall, they are lifted and potted. These operations need to be done carefully to avoid root damage. Potting is done about 5 months before field planting. Potted seedlings should be given full shade. The shade should be removed to full sunlight a month before transplanting to the field. Seedlings are generally outplanted when 2 years old (Walters 1974a).

With these complicated nursery practices one wonders what the natural regeneration might be. For nursery production of seedlings of these species in temperate climate, it is well to remember that the seedlings of most species are very susceptible to frost.

ERICACEAE — HEATH FAMILY

Arbutus menziesii Pursh — Pacific madrone

GROWTH HABIT, OCCURRENCE, AND USE Pacific madrone is one of approximately 14 species of *Arbutus* that are trees and shrubs native to the Mediterranean region, the Canary Islands, and North America (Bailey 1951). It is an evergreen tree varying in height from 8 to 40 m and occuring in coastal regions from southwestern British Columbia to southern California (Roy 1974a). The wood has been used for flooring, cabinet work, small turnery, charcoal, and fuel wood. The bark has been used to tan leather. The foliage is moderately browsed by wildlife. The attractive berries are eaten by birds, especially band-tail pigeons and quail. The fruits have narcotic properties and the leaves are astringent (Roy 1974a). Pacific madrone is one of the most strikingly beautiful trees of the Pacific Northwest. Despite being cultivated since 1827 it is rarely used in environmental plantings. A relatively slow rate of growth and difficulty in propagating may explain the lack of use of this handsome tree.

FLOWERING AND FRUITING The bisexual flowers are about 0.8 cm long, globular or urn-shaped, and are borne in dense racemes of which several are supported on a thick and stiff main stem about 12.5 to 15 cm long. The calyx is free of the ovary and is 5-parted, nearly to the base. The corolla is white with 5 lobes that are spreading or recurved and much shorter than the swollen tube. The 10 stamens are shorter than the corolla. The anthers are short and have two slender, dorsal awns. The superior ovary is glandular and roughened, and terminates in a columnar style that protrudes from the corolla to expose an obscurely 5-lobed stigma.

The madrone fruit is a berry 0.8 to 1.25 cm in diameter, bright red or orange-red when ripe, and with a thin, rough, granular skin (Fig. 1). The generic name was derived from "arboise," a Celtic word for rough fruit. The fruit has a dry, mealy flesh, is generally 5-celled, and contains about 20 hard, compressed seeds (Fig. 2). Flowering occurs from March to June and fruit ripens in September and October. Fruit remains on the tree until December. Minimum seed-bearing age is 3 to 5 years and abundant seed crops occur nearly every year.

COLLECTION, EXTRACTION, AND STORAGE OF SEED Berries of Pacific madrone may be collected from standing trees from October to December. They can be dried at room temperature or the seeds can be separated from the pulp directly after harvesting (Mirov and Kraebel 1939). For separation the fruits should be soaked in water to soften the pulp. They can be run through a macerator and the seeds separated by flotation. The seeds should be thoroughly dried before storage.

The dried fruits or seeds can be stored at room temperature for 1 or 2 years. Longer storage should be in sealed containers at 3 to 5°C (Mirov and Kraebel 1939). Fresh madrone fruits weigh from 0.4 to 0.7 g.

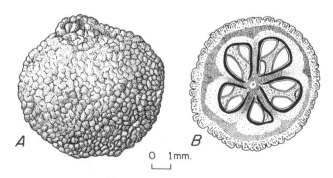

Figure 1. *Arbutus menziesii*, Pacific madrone: (A) exterior view of fruit, ×5; and (B) transverse section of fruit showing its 5 carpels, ×5 (Roy 1974a).

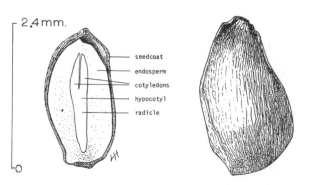

Figure 2. *Arbutus menziesii*, Pacific madrone: longitudinal section through a seed (left) and exterior view of seed (right), ×16 (Roy 1974a).

Dried fruits weigh about 0.2 g each. The small seeds number 400 to 700 per gram.

GERMINATION The fleshy layer should be removed from the seeds before they are prechilled. Prechilling at 2 to 5°C is recommended for 60 days. Prechilled seeds have a germination capacity as high as 90% in 38 days incubation at 22°C on blotter paper.

NURSERY AND FIELD PRACTICE Pacific madrone has been propagated by starting seeds in flats and transplanting the seedlings to individual containers. It has also been propagated vegetatively by grafting, layering, and rooting of cuttings.

The litter of Pacific madrone has been shown to have an allelopathic influence on seedlings of Douglas fir in the Pacific Northwest (Tinnin and Kirkpatrick 1985). Pacific madrone seedlings are unlikely to survive in undisturbed conifer stands because of pathogens, invertebrates, drought, and litter fall (Tappeiner et al. 1986). Resprouting of madrone on the forest floor is controlled by the level of light that reaches the ground under the forest canopy (Minor 1986a). On a worldwide basis the only other recent mention of *Arbutus* in the literature came from Israel where Karschow (1975) studied the seedling morphology and germination of *A. andrachene*.

LORANTHACEAE — MISTLETOE FAMILY

Arceuthobium Bieb. — Dwarf mistletoe

GROWTH HABIT, OCCURRENCE, AND USE Dwarf mistletoe plants are yellow or brown, leafless, fragile-jointed, and parasitic on coniferous trees. Dwarf mistletoe species reduce growth of heavily infested trees, cause increased mortality, reduce cone and seed yield, and predispose trees to other diseases.

The genus *Arceuthobium* comprises about 41 taxa, 33 in the New World and 8 in the Old World (Hawksworth and Wiens 1984). The taxonomy of the genus has only recently been studied in detail and revisions are expected to continue.

FLOWERING AND FRUITING Flowers are solitary or several from the same axil, and unisexual. Male flowers are usually 3-merous, and compressed. The fruit is a round locule opening by a circular slip. Pistillate flowers are ovoid and compressed, with 2 teeth. The fruit is a fleshy berry mounted on a recurved pedicel. The seeds are viscid and adhere upon landing. Members of the genus *Arceuthobium* produce seeds that are forcibly discharged when mature (Baker and French 1986). Seeds are usually discharged less than 30 m (Mathiasen 1988).

GERMINATION Seed dispersal for most species occurs in September. The fruits discharge in the morning as temperatures increase and relative humidity decreases. The activities of birds and small mammals may also release seeds. Germination for most species of dwarf mistletoe that have been studied does not begin until the spring following dispersal. In a study of *A. americanum* in Colorado, Hawksworth (1965) determined that germination did not start until the following May, but eventually 93% of the seeds germinated. Similar patterns of germination were found for seeds of the same species in Alberta, but germination only reached 10 to 50%.

The seeds of various species of *Arceuthobium* vary greatly in their prechilling requirements (Beckman and Roth 1968). Seeds of *A. pusillum* do not require prechilling for germination while seeds of other species require as long as 100 days (Scharpf and Parmeter 1962, Scharpf 1970, Knutson 1971).

Seeds of *Arceuthobium* are sometimes stored for use in experiments. Cold temperatures and high humidity provide the most desirable storage conditions (Livingston and Blanchette 1986).

ERICACEAE — HEATH FAMILY

Arctostaphylos Adans. — Manzanita

GROWTH HABIT, OCCURRENCE, AND USE *Arctostaphylos* includes about 50 species of evergreen shrubs, and occasionally small trees. The manzanita species are characterized by very crooked branches and bark that is dark red or chocolate-colored, smooth, and polished. Manzanita species are native to North and Central America, with one species, bearberry (*A. uva-ursi*), circumpolar. The name of the genus is derived from the Greek *arktos* (bear) and *staphule* (grape), in reference to bears eating the fruits. It has been passed on from the first version of the *Woody Plant Seed Handbook* that the fruits of manzanita are used as preserves and that fruit drinks and the leaves have medicinal value (Berg 1974). However, the importance of these species to wildland management lies in their long-term persistence in chaparral and woodland situations that are repeatedly burned. Many species of manzanita sprout from massive woody crowns after being burned. The seeds of some, perhaps most, species germinate after wildfires or after slash disposal fires done in site preparation for forest reforestation plantings. Manzanita fields characterize a type of environment in the Sierra Nevada and trans-Cascade Mountain areas of the far West.

The manzanitas are an extremely diverse, and one would expect, evolving group of plants. They range in stature from the nearly prostrate bearberry to the 3 m tall bigberry manzanita (*A. glauca*). The intricate form of the trunks and branches and their striking stem colors that contrast with green to gray leaves make these drought-tolerant species useful in environmental plantings (Everett 1964). Inability to germinate seeds for direct planting has limited the use of manzanitas for such plantings.

FLOWERING AND FRUITING   Small white or pink perfect flowers bloom in the early winter in the Mediterranean parts of the Pacific Coast and in the spring in the interior. The fruit is a fleshy to mealy drupe, 0.6 to 1.5 cm in diameter, which varies in color from red to brown to dark brown, depending on the species. Consult local floras for flowering times and fruit characteristics for a particular species. The fruits ripen in the early summer in Mediterranean climates and in the fall in the interior and mountainous areas. The fruits consist of 4 to 10 stony seeds that may be separate or variously coalesced or united into a single solid stone (Figs. 1, 2). In the case of greenleaf manzanita (*A. patula*) the woody seeds are polymorphic with a wide range in seed size and consequently in the thickness of the woody material covering the seed.

COLLECTION, EXTRACTION, AND STORAGE OF SEED   Fruit can be picked from the shrubs or picked up off the ground. When the crop is heavy, fruit can be collected quite rapidly. Bearberry fruit is bright red when ripe; Eastwood manzanita (*A. glandulosa*) fruit is red brown; greenleaf manzanita fruit is brown to black; and bigberry manzanita fruit is light to dark brown. The outer fleshy portion of the fruit can be separated by maceration and flotation, or by rubbing dried fruits through a screen and using an air screen. Seed counts per gram range from 2.4 for the large stones of bigberry to 128 for bearberry.

GERMINATION   Seeds of manzanita have hard seedcoats and embryo dormancy. Pregermination treatments that have been used on bearberry seeds include immersion in sulfuric acid followed by both warm stratification and prechilling. Duration of acid scarification ranged from 3 to 6 hours, with 60 to 120 days warm stratification at 20/30°C followed by 60 to 90 days at 5°C and final incubation in sand, peat, or sand-peat mixtures at 20/30°C. After all this preparation roughly 50% germination was attained. It has been suggested that the type of substrate used for warm stratification is important. Limited testing of bigberry and Eastwood manzanita seeds at the Rancho Santa Ana Botanical Garden in California produced only slight germination after prolonged periods of acid scarification. Acid scarification of greenleaf manzanita seeds for 4 hours followed by 120 days prechilling has produced 20% germination when the pretreated seeds

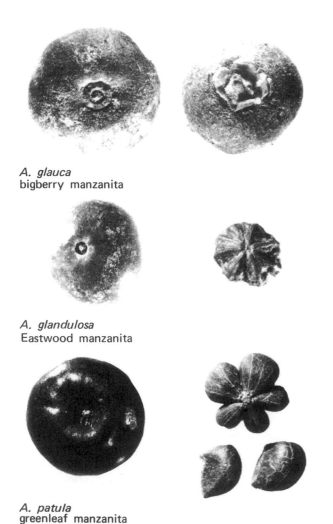

A. glauca
bigberry manzanita

A. glandulosa
Eastwood manzanita

A. patula
greenleaf manzanita

Figure 1. *Arctostaphylos*: fruits (drupes) and seeds (stones), ×3 (Berg 1974).

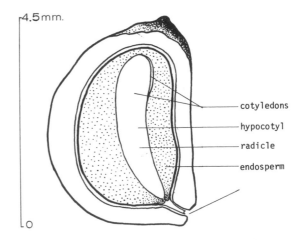

4.5 mm.

cotyledons

hypocotyl

radicle

endosperm

Figure 2. *Arctostaphylos uva-ursi*, bearberry: longitudinal section through a seed, ×12 (Berg 1974).

were incubated at 20/30°C. We have had similar results with seeds of greenleaf manzanita. Dissected embryos are dormant and require prechilling for at least 40 days for appreciable germination to occur.

At the base of each manzanita nutlet or seed there is a channel through which the radicle is forced during germination. Moisture apparently enters through this channel to initiate germination. The channel is plugged at seed maturity with a hard material that is slightly softer than the woody covering of the nutlet. Acid scarification dissolves this plug or softens the material so that germination can proceed. The polymorphic nature of the seeds makes the optimum duration of the scarification process difficult to determine. This is complicated by some seeds coalescing together and being more resistant to acid than single seeds. Acid scarification of sufficient duration to induce some germination undoubtedly kills some seeds and fails to clear the blocked channel on others in the same seed lot.

Another approach to the germination of seed of manzanita involves burning 7 to 10 cm of pine needles on a flat that has been sown with manzanita nutlets. The suggestion is that manzanita germinates in nature after wildfires. We have found that results with this method are highly variable depending on the moisture content of the planting soil and the depth of planting. Dry soil, shallow planting (0.5 cm), and leaving the flats outside over winter after burning give the best results for greenleaf manzanita.

In the one study of manzanita that has been reported recently Carlson and Sharp (1975) acid scarified seeds of greenleaf and pine-mat (*A. nevadensis*)

manzanita in a complex process of dumping and reapplying acid. The results were poor establishment of both species.

NURSERY AND FIELD PRACTICE Again there is relatively little experience to report from the literature. Acid scarified manzanita seeds planted in the fall in mulched beds produce limited success. Work with these species since the 1930s has been only simple variations of this basic system (Fig. 3).

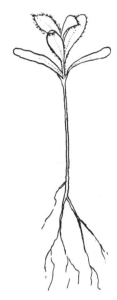

Figure 3. *Arctostaphylos patula*, greenleaf manzanita: seedling at 1 month, actual size (Berg 1974).

MYRSINACEAE — MYRSINE FAMILY

Ardisia japonica (Thunb.) Bl. — Japanese ardisia

GROWTH HABIT, OCCURRENCE, AND USE The ardisia species are a group of about 200 plants native mainly to tropical and subtropical regions. Japanese ardisia (*Ardisia japonica*) is an evergreen shrub that reaches 0.4 m in height. The lustrous leaves are crowded at the end of stems.

FLOWERING AND FRUITING The white flowers are about 1.2 cm across and arranged in 2- to 6-flowered panicles. The fruit is a red drupe about 6 mm across.

GERMINATION The seeds germinate without pretreatment (Dirr and Heuser 1987).

ARISTOLOCHIACEAE — BIRTHWORT FAMILY

Aristolochia L. — Dutchman's pipe

GROWTH HABITAT, OCCURENCE, AND USE *Aristolochia* is a genus of herbs and woody plants, the latter often climbing, distributed in warm regions, and most numerous in South America. Dutchman's pipe (*A. durior*) is a woody climber to 3 m. Leaves of this species are nearly glabrous and 15 to 35 cm broad when mature. Dutchman's pipe is found in eastern North America from Minnesota south to Kansas.

FLOWERING AND FRUITING The flowers are U-shaped, inflated above the ovary, and greenish yellow in color with 3 lobes and a spreading, flat, brown-purple limb 2.5 cm across. The fruit is a capsule.

GERMINATION Excellent germination of Dutchman's pipe seed has been obtained after 3 months of prechilling (Dirr and Heuser 1987). The plants can also be increased by dividing or rooting of cuttings.

ROSACEAE — ROSE FAMILY

Aronia Med. — Chokeberry

GROWTH HABIT, OCCURRENCE, AND USE The chokeberries discussed here are three closely related species of deciduous shrubs (Gill and Pogge 1974a). Black chokeberry (*A. melanocarpa*) is only 0.5 to 1 m tall. The red (*A. arbutifolia*) and purple (*A. prunifolia*) are medium sized, one to 1.5 m tall. Red and purple chokeberry are practically identical ecologically and the only satisfactory way to distinguish between them is by the color of their ripe fruit. Both species have pubescence on younger branches, leaf stems, and lower leaf surfaces. In contrast, black chokeberry is smooth, or has a few scattered hairs, on these parts. The combined ranges of the three species include most of the eastern United States and adjacent Canada. All are moderately tolerant to shading and prefer moist, usually acidic soils. The most likely habitats are bogs and swamps, low woods, clearings, and damp barrens. However, each species will tolerate drier conditions, and black chokeberry is better adapted than the others to growth in drier thickets or clearings on bluffs or cliffs. All are valuable as food for wildlife in fall and winter. Their handsome foliage, flowers, and fruits also make them attractive as ornamentals, but none has been cultivated extensively.

Figure 1. *Aronia arbutifolia*, red chokeberry: leaf and cluster of fruits (pomes), ×1 (Gill and Pogge 1974a).

FLOWERING AND FRUITING The white bisexual flowers bloom for 2 to 3 months during March to July, the local flowering period depending on latitude and elevation. Fruit ripening dates are similarly dependent and range from August to November. Fruit drop from the plants begins shortly after ripening and may continue through the winter and spring. The fruits are rather dry, berrylike pomes (Fig. 1) containing 1–5 seeds (Fig. 2), some of which may be abortive. Natural seed dispersal is chiefly by animals. Black chokeberry fruits shrivel soon after ripening and most of them drop. Purple chokeberries shrivel at the beginning of winter whereas pomes of the red-fruited species remain plump and bright into the winter. Red chokeberry may yield fruit at 2 years of age and produces good seed crops almost every year. Black chokeberry yields a good crop about every second year.

COLLECTION, EXTRACTION, AND STORAGE OF SEED If loss to birds is a hazard, fruits should be handpicked as soon as they are ripe. Otherwise, they should be picked within a month or so. The delay should be least with black chokeberries and can be longest with the red-fruited species.

Commercial seed usually consist of the dried pomes or "dried berries" as typically listed in seed catalogues. Apparently it is not general practice, seeds can be separated from the fruits by maceration and flotation. A blender can be used for maceration of small lots of seed. Seeds should be dried before storage. The dried fruits weigh about 16 per gram and dried, cleaned seeds run 550 per gram.

PREGERMINATION TREATMENT Chokeberry seeds have an internal dormancy that has been broken by prechilling in a moist medium at 2 to 5°C. Seeds of red chokeberry have been prechilled 90 days; black chokeberry seeds may require longer prechilling, while the seeds of purple chokeberry require only 60 days.

GERMINATION TEST Germination tests of prechilled seeds have been conducted with substrata of soil, sand, and peat for 30 days at 20/30°C or a constant 20°C. Germination started in 8 days and was virtually complete in 20 to 30 days (Crocker and Barton 1931).

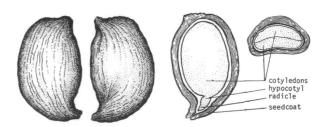

Figure 2. *Aronia melanocarpa*, black chokeberry: exterior views of seed, longitudinal section and transverse section; all at ×16 (Gill and Pogge 1974a).

Germination of seed that is not prechilled is quite low. Germination is epigeal.

NURSERY AND FIELD PRACTICE In some nurseries the dried fruits have been soaked in water for a few days, mashed, and the whole mass stratified until spring. The recommended sowing depth is 1.8 cm. Germination begins a few days after sowing. As a general rule of thumb, 454 g of cleaned seeds yields about 10,000 usable plants. Outplanting may be done with 2-year-old seedlings.

COMPOSITAE — SUNFLOWER FAMILY

Artemisia L. — Sagebrush

GROWTH HABIT, OCCURENCE, AND USE In his study of section *Tridentatae* of *Artemisia*, Beetle (1960) offered the following characterization of the woody sagebrush species:

> In western North America, from southern Canada to northern Mexico, a group of closely related, often hybridizing, sagebrushes comprise an endemic section of the worldwide genus *Artemisia* L. They occur in varying amounts over about 422,000 square miles (10.8 million hectares) in the 11 western states.

Sagebrush species were the dominant of approximately 10% of the original landscape of what became the adjacent 48 states. Sagebrush comes in many forms and sizes from tiny pigmy sagebrush (*A. pymeae*) to plants of basin big sagebrush (*A. tridentata* ssp. *tridentata*) that, growing as phreatophytes may reach 2 m in height. Normally the sagebrush species are grouped into big sagebrush species which mature from 0.5 to 1 m tall and the low or dwarf sagebrush species which mature under 0.5 m in height.

The classic monograph on *Artemisia* was completed by Hall and Clements (1923). The most recent and comprehensive revision of the *Tridentatae* portion of *Artemisia* has been done by E. Durrant McArthur and associates (McArthur et al. 1981). Leila Shultz slightly modified McArthurs classification to include 13 taxons in the subgenus (Shultz 1984). Most of the landscape-characterizing, woody species of sagebrush belong to the subgenus *Tridentatae*, but there are important species outside that group such as bud sage (*A. spinescens*).

FLOWERS AND FRUITING The inflorescence of sagebrush plants range from spikelike (spiciform), through racemelike, to paniculate. For basin big sagebrush the individual flower heads are often produced profusely with one to 3 in a subsessile cluster at each node on the flower stalk. The involucre is composed of bracts 3 to 4 mm long and 2 mm wide with 15 bracts in a 4 to 5 imbricate series. The flowers are perfect with trumpet-shaped corollas about 2 to 3 mm long with 5 teeth. The achenes are 0.6 to 0.75 mm wide and 1.2 to 1.7 mm long (Figs. 1, 2). Seeds of big sagebrush range from 3,000 to 5,000 per gram (Young and Evans 1980, ASOA 1989):

Species	Seeds/gram
A. ludoviciana	8900
A. nova	4500
A. tridentata	
ssp. *tridentata*	3500–4000
ssp. *vaseyana*	3800–4900
ssp. *wyomingensis*	3700–3800

ARTEMISIA

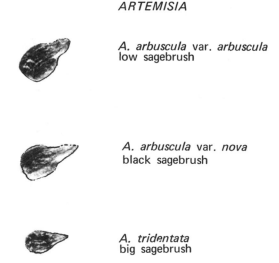

A. *arbuscula* var. *arbuscula*
low sagebrush

A. *arbuscula* var. *nova*
black sagebrush

A. *tridentata*
big sagebrush

Figure 1. *Artemisia*: achenes (cleaned seed), ×8 (Deitschman 1974).

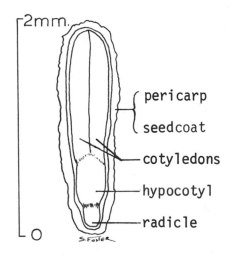

Figure 2. *Artemisia nova*, black sagebrush: longitudinal section through an achene, ×30 (Deitschman 1974).

The members of the subgenus are apparently all fall bloomers that set and mature seeds relatively late in the fall or early winter. The range of sagebrush species is so great there is considerable variations among species and with geographic location within species. Consult local floras for more precise information on phenology.

COLLECTION, EXTRACTION, AND STORAGE OF SEED Determination of seed ripeness is difficult with sagebrush species because of the small seed size. When seed can be easily removed from the heads by shaking and is too hard to easily be crushed with a thumbnail, it is ready to harvest. Because seed matures so late in the season, it often has to be harvested quickly after maturity to avoid losses and storm damage. We harvest sagebrush seed by clipping the seed stalks and bagging the material in paper bags for air drying. Deitschman (1974) recommended beating the seed stalks over containers or use of vacuum harvestors. A mature stand of big sagebrush will produce 50 million seeds per ha, but this is only a yield of 12.5 kg per ha and it is difficult to recover through harvesting 10% of this seed production.

Most methods of harvesting seeds of sagebrush produce a mass of thrashy material from seed stalks and flower parts with the very small seeds lost in the mixture. This fluffy material is often sold as sagebrush "seed". It is bulky to transport and difficult to meter when seeding. It is also difficult to determine seed quality from this material. Seeds of sagebrush can be threshed by rubbing the the inflorescence through a screen and the seed separated from the chaff by using an air screen and pneumatic seed blower (Young and Young 1985).

GERMINATION Big sagebrush was long considered a plant species that easily regenerated. Recent interest in restoring the natural vegetation on mine spoils have shown that big sagebrush is not an ease species to establish by artifical seeding. The very small seeds are adapted for germination on or very near the surface of seedbeds (Jacobson and Welch 1987). Proper seed placement in artifical seeding is difficult. McDonough and Harniss, working with seeds of big sagebrush collected in eastern Idaho found in a series of studies that seeds of mountain big sagebrush (*A. tridentata* ssp. *vaseyana*) have afterripening requirements that must be satisfied before the seeds will germinate

(McDonough and Harniss 1974a, 1974b; Harniss and McDonough 1976). These seeds responded to prechilling. They also had relatively low total germination for big sagebrush seeds collected from native stands. Germination standards exist for two or three woody or semiwoody species and several perennial herbaceous species (ASOA 1985):

Species	Substrate	Temperature (°C)	Duration (days)	Additional directions
A. tridentata	P	15/20	21	Dormant lots may require 14 days prechilling.
A. nova	P	15 or 20	21	—
A. ludoviciana	P	15/25	14	—
A. absinthium	P	20/30	21	—
A. dracunculus	P	20/30	21	—
A. maritima	P	20/30	21	—
A. vulgaris	P	20/30	21	—

We have just completed a 5 year study of germination of seeds of basin and mountain big sagebrush, both from native stands and reciporal common gardens. For collections from the western Great Basin no afterripening or other forms of dormancy were found for either subspecies. Germination capacity of at least 80% was common over a wide range of incubation temperatures. Alternating temperatures of 15/20°C or 15/25°C most frequently produced optimum germination.

NURSERY AND FIELD PRACTICE For direct drilling seeds of sagebrush species should be mixed with an extender such as rice hulls or vermiculite for better control of seed metering. The seedbed should be well prepared and firm. The seed placement should be at the bottom of a shallow drill.

Big sagebrush plants arc the only species of temperate desert that can be started in flats, in the fall, transplanted to pots or tubes and out planted the next spring. This will only work if the plants are transplanted to weed free conditions for the first season. Most species must be held over the summer in containers and transplanted very early the next spring. Sagebrush plants in spaced plantings grow very rapidly and will produce significant amounts of seed the second year.

MORACEAE MULBERRY FAMILY

Artocarpus Frost. — Breadfruit

GROWTH HABIT, OCCURRENCE, AND USE *Artocarpus altilis*, breadfruit, is a large tropical relative of the mulberry. The genus consist of about 50 milky-juiced trees of the Asiatic tropics and Polynesia. Breadfruit was dispersed in Polynesia by human migration (Macoboy 1982). It is

a striking broad tree to 15 m in height with heavy, profuse foliage. The deeply lobed leaves are often considered the most beautiful of any tree and may reach 1 m in length. Polynesians used the leaves for roofing, clothing, and food wraps for the oven. The peeled fruit can be baked, boiled, or picked when unripe and will keep if buried in pits for years. When ripe it can be pounded into a paste and used for a desert.

FLOWERING AND FRUITING The pistillate flowers are borne in dense heads, with the perianth tubular and immersed in the fleshy rachis. The fruit is a syncarp, containing the large seeds or may be seedless. If seeds occur, they are imbedded in the flesh.

PROPAGATION Breadfruit is propagated from shoots which arise spontaneously from the roots. There is no mention of propagation from seeds.

ANNONACEAE — ANNONA OR CUSTARD-APPLE FAMILY

Asimina Adans. — Pawpaw

GROWTH HABIT, OCCURRENCE, AND USE The genus *Asimina* is a group of 8 or 9 shrubs or small trees native to North America (Bonner and Halls 1974a). The more or less edible fruit provides food for wildlife (Bailey 1951). Pawpaw (*A. triloba*) is a small tree to 5 m high found from Ontario and Michigan to Florida and Texas.

FLOWERING AND FRUITING Pawpaw flowers are solitary, perfect, and greenish purple (Bonner and Halls 1974a). They are very primitive for woody angiosperms, and the reproductive biology of these primitive plants creates considerable interest among botanists (Zimmerman 1960, Norman and Clayton 1986). The flowers appear in the spring during March to May about the same time as the leaves. Pollination apparently is a problem and limits fruit set (Willson and Schemske 1980).

The fruits are fleshy berries that contain several dark brown, shiny seeds (Fig. 1). Pawpaw fruits are about 5 to 17 cm long; those of small flower pawpaw (*A. parviflora*) are only 1.9 to 5 cm long. Both fruits are greenish yellow before maturity and turn brown to black as they ripen in August and September. The fleshy part of the fruit is edible, but there appear to be different fruit types (Bonner and Halls 1974a). Those with white flesh are barely edible, while others are larger and have a yellowish or orange flesh with a much better flavor. The seeds themselves are oblong, rounded, flat, and bony (Fig. 2).

A. parviflora
smallflower pawpaw

A. triloba
pawpaw

Figure 1. *Asimina*: fruits and seeds, ×1 (Bonner and Halls 1974a).

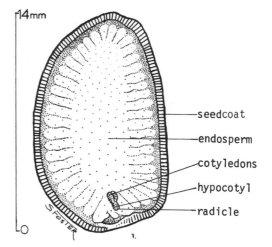

Figure 2. *Asimina parviflora*, small flower pawpaw: longitudinal section through a seed, ×4 (Bonner and Halls 1974a).

COLLECTION, EXTRACTION, AND STORAGE OF SEED Pawpaw fruits should be picked or shaken from the trees as soon as the flesh is soft. The seeds may be extracted by macerating the fruits in water and floating off the pulp, but the entire fruit may be sown (Bonner and Halls

1974a). Seed yield, purity, and soundness are as follows:

	Small flower	Pawpaw
Cleaned seeds/kg fruit	—	168
Cleaned seeds/kg	2860	1533
Purity (%)	98	100
Sound seeds (%)	94	96

GERMINATION Seed germination is very slow because the seeds have dormant embryos, and the seedcoats are slowly permeable. Prechilling for 60 days at 5°C resulted in germinative capacities of 50, 62, and 82% for three samples of pawpaw seeds (Bonner and Halls 1974a). Prechilling for 100 days has been recommended, but germination still may be slow and irregular. Fall sowing of untreated seeds does not improve results. No specific test conditions have been reported, but alternating temperatures of 20/30°C on a moist medium have been satisfactory.

NURSERY AND FIELD PRACTICE Pawpaw seeds may be sown in the fall without pretreatment, or prechilled and sown in the spring. They should be covered with 1.5 cm of nursery soil. Some shade is helpful in germinating seedlings (Fig. 3). Another method is to plant fresh seeds, before they dry, in pots of sand, and then to keep them in a cool cellar or similar place. As the seeds sprout, they can be picked out and transplanted into nursery beds. Pawpaw can also be propagated by layering and root cuttings. There are many horticultural varieties.

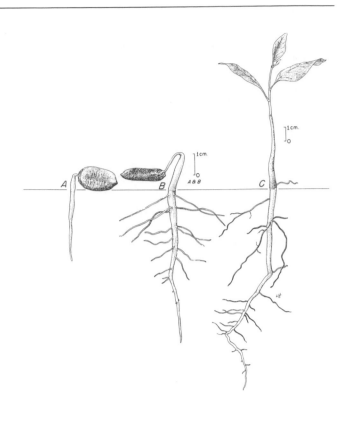

Figure 3. *Asimina triloba*, pawpaw: seedling development at 2, 9, and about 20 days germination (Bonner and Halls 1974a).

CHENOPODIACEAE — GOOSEFOOT FAMILY

Atriplex L. — Saltbush

GROWTH HABIT, OCCURRENCE, AND USE *Atriplex* is a genus of herbaceous and woody species that usually have grayish to whitish herbage that is scurfy with inflated hairs (Foiles 1974). The genus is essentially cosmopolitan with over 100 species. One of the unique aspects of some species is that sodium, usually a plant poison, is a stimulant and may even be essential for their nutrition (McArthur and Sanderson 1983). *Atriplex* species play dominant roles in the vegetation of areas where soluble salts accumulate in the soil. Saltbushes are important browse species in semiarid to arid range lands in Australia, North and South America, and central Asia. *Atriplex* species also are important species in brackish marsh situations.

Atriplex canescens, fourwing saltbush, is one of the few shrubs that can be established on wild lands in western North America through direct seeding. Shadscale (*A. confertifolia*) is a landscape characterizing species in the more arid portions of the intermountain area of western United States. Shadscale apparently evolved rapidly to occupy habitat exposed by the post-Pleistocene drying of pluvial lakes in the Great Basin (Stutz 1983). Many salt-bush species occur as complex polypoid series.

Some of the variable forms of Nuttall saltbush (*A. nuttallii*) are hardly woody and barely more than 10 cm tall at maturity. Plants of fourwing saltbush are extremely rapid growers and under favorable conditions reach 1 m in 2 growing seasons. The diploid gigas form of fourwing that occurs as a relict species on sand dunes in the Great Basin is especially rapid growing (Stutz et al. 1975). Quailbush (*A. lentiformis*) may reach 2 m in height.

FLOWERING AND FRUITING Saltbush may be monoecious or dioecious. The flowers are small and green in axillary clusters or glomerules, or in panicled spikes. Staminate flowers have a 3- to 5-parted calyx. Pistillate flowers consist of a naked pistil enclosed between a pair of appressed, foliaceous bracts that enlarge in the fruit, may be partly united, and are more or less expanded, thickened and/or equipped with appendages.

The fruit is a utricle, usually with a membranous pericarp (Figs. 1, 2). The seed within the fruit is flattened and surrounded by scanty endosperm. The fruit is dispersed by wind and animal activities. It is very

ATRIPLEX

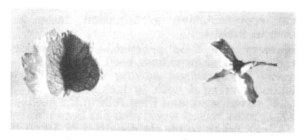

A. canescens
fourwing saltbush

A. confertifolia
shadscale saltbush

A. nuttallii
Nuttall saltbush

Figure 1. *Atriplex*: fruits (utricles), ×2 (Foiles 1974).

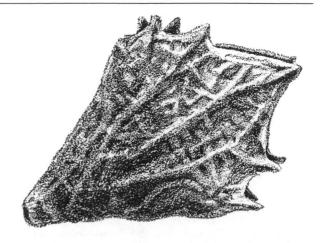

Figure 2. *Atriplex semibaccata*, trailing saltbush: utricle, ×16 (Foiles 1974).

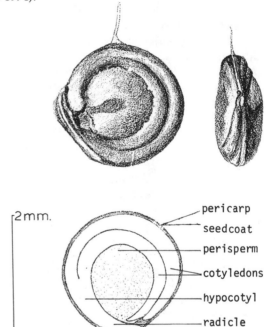

pericarp
seedcoat
perisperm
cotyledons
hypocotyl
radicle

2mm.

Figure 3. *Atriplex semibaccata*, trailing saltbush: exterior views in two planes of achenes removed from their utricles, and longitudinal section through an achene, ×16 (Foiles 1974).

important in the diet of livestock wintering on desert ranges. Fruits of the more spinescent species are licked from the ground beneath the shrubs.

COLLECTION, EXTRACTION, AND STORAGE OF SEED The mature utricle is commonly referred to as seed in the industry. The seeds can be collected by hand, mechanical stripping, or vacuum harvesting, depending on the species. Shadscale seeds often stay on the plant until the next spring. The fruits are often dewinged by a hammer mill, which reduces the material to be transported after cleaning by 50%. Seed of trailing saltbush (*A. semibaccata*) (Fig. 3) can be threshed from the fruit and separated with an air screen. Air-dried seeds (6 to 8% moisture) can be stored in warehouses in bags for prolonged periods without loss of viability (Table 1).

GERMINATION The saltbushes are a complex and diverse group and it is difficult to generalize about germination processes. However, they largely share the problems associated with seeds that do not dehisce from complex, lignified fruits. The bracteoles, photosynthetic in most species, often eventually harden and may become a barrier to moisture imbibition, gas exchange, and leaching of inhibitors from the embryo or they may become a physical restraint to germination (Warren and Kay 1983). Burton et al. (1983) to enzymatically degraded the bracts of Gardner salt-

bush (*A. gardneri*) to enhance germination. Delignification of the cell walls was necessary to enhance enzymatic degradation, but germination was not enhanced.

As previously mentioned, excessive soluble salts in the soil often are a part of the habitat of saltbushes. Many species apparently take up salts to balance osmotic levels across cell membranes to use water supplies with reduced osmotic potentials. Several species of *Atriplex* appear to concentrate excess salts in the flower bracteoles. These salts may interfere with subsequent germination. The presence of salts in the bracteoles is often compensated for by washing, or washing and wringing the fruits as a pregermination treatment (Ansely and Abernethy 1983, Young et al. 1980).

Table 1. *Atriplex:* seed sales and storage information (Plummer 1983).

Species/Cultivars	Annual sales (tons)	Storage life (years)
A. canescens 'Marana', 'Ricon', and 'Wytana'	22.5	10
A. confertifolia	12.5	10
A. corrugata	1.0	3
A. cuneata	0.5	8
A. gardneri	4.0	5
A. lentiformis 'Casa'	0.5	3
A. tridentata	0.5	4

The seeds of many species of saltbush have embryo dormancy in addition to the problems in germination resulting from the bracts. Prechilling has been used to overcome one form of embryo dormancy. Binet (1965, 1966) determined that germination of coastal species of *Atriplex* was enhanced by prechilling for 30 days at 5°C. Ward (1967) found that dormancy of fat hen (*A. hasta = A. triangularis*) seeds was removed by prechilling for 14 days at 4°C.

Several species of saltbush have lower seed germination at constant versus alternating incubation temperatures. For some species, otherwise dormant seeds can be induced to germinate with alternating incubation temperatures (Ignaciuk and Lee 1980, Young et al. 1980). Alternating temperatures interact with germination at lowered osmotic potentials. *Atriplex glabriuseula* and *A. laciniata* are less sensitive to increased salinity concentrations at optimal alternating temperature regimes (Ignaciuk and Lee 1980). In contrast, Sankary and Barbour (1972) found that total germination of seeds of cattle saltbush (*A. polycarpa*) was similar at alternating and constant temperatures. Sharma (1976) demonstrated that the optimum conditions for germination of *A. vesicaria* and *A. nummularia* shifted to slightly higher temperatures at low osmotic potentials, but the effects of lowering water potential were not as marked at temperatures above or below the optimum. Similar results were obtained with seeds of coastal species (Binet 1965, 1966).

Beadle (1952) noticed that seeds of saltbushes are polymorphic, but thought the different sized seeds were of no significance in germination characteristics of the species. Koller (1957) noted that flat seeds of *A. dimorphostegia* germinated early in the season while germination of humped seeds was delayed. Uchiyama (1981) found that the germination ability of heavier seeds of this species of saltbush was much higher (93%) than that of lighter seeds (10%). Khan and Ungar (1984a) found that large seeds of fat hen had enhanced seedling growth at low salinity levels when incubated at warm day and cool night temperatures (5/25°C, 12/12 hr).

Seeds of fat hen exhibit a very pronounced morphological and physiological polymorphism (Khan and Ungar 1984b). Seed size varied from 1 to 2.8 mm and was a predictor of the likelihood of successful establishment through its effect on germination and seedling vigor. Seedling dry weight was positively correlated with seed size. The degree of salt tolerance increased progressively with increasing seed size. Small and large seeds of fat hen contain different kinds and relative amounts of phenolic compounds that cause dormancy (Khan and Ungar 1986). Small seeds had gentistic, salicylic, syringic, and chlorogenic acids and catechol; medium seeds had gentistic and salicylic acids, catechol, and protocatechol; and large seeds had gentisic, caffeic, and 2-hydroxy-5 methoxy benzoic acids and protocatechol. The presence of endogenous inhibitors, salicylic, syringic, and chlorogenic acids, and catechol in small seeds, but not in large, may account for dormancy in these seeds. Dormancy in the small fat hen seeds was reversible by application of gibberellic acid and kinetin.

The bottom line on *Atriplex* germination is to watch for seed polymorphism: dormant seeds may need washing, prechilling, and growth regulators to enhance germination.

NURSERY AND FIELD PRACTICE It is fairly easy to obtain stands of fourwing saltbush by direct seeding. Care must be taken to match seed sources with the climate of the site being seeded. Fourwing saltbush seeds collected in warm desert situations may not be winter hardy when planted in temperate desert environments. Seeding rates of 20 to 40 kg/ha have been used. Often the saltbush seed is drill box mixed with grass seeds.

Transplanting of fourwing saltbush seedlings from pots or tubes to the field can be done with a high level of success. Seedlings (Fig. 4) can be started in flats, transplanted to pots or other containers with sufficient soil volume so they can be held over in lath houses until the following winter, and then transplanted to the field in early spring.

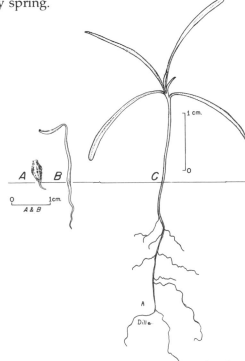

Figure 4. *Atriplex canescens*, fourwing saltbush: seedling development at 1 and 2 days after germination and at a later stage (Foiles 1974).

CORNACEAE — DOGWOOD FAMILY

Aucuba **Thunb.**

GROWTH HABIT, OCCURRENCE, AND USE *Aucuba* is a small genus of plants native to central and eastern Asia (Bailey 1951). The aucubas are dioecious, evergreen shrubs with stout branches and opposite leaves.
FLOWERING AND FRUITING The flowers are small and borne in terminal panicles. Japanese aucuba (*A. japonica*) bears a bright red, one-seeded drupe.
GERMINATION The seeds are initially dormant and require at least 3 months of prechilling to germinate (Dirr and Heuser 1987).

OXALIDACEAE — WOOD-SORREL FAMILY

Averrhoa **L.** — Five-corner fruit

GROWTH HABIT, OCCURRENCE, AND USE The genus *Averrhoa* consists of 2 evergreen trees from tropical Asia introduced to tropical America and south Florida for production of edible fruits. Leaves are alternate, odd-pinnate, and somewhat sensitive.
FLOWERING AND FRUITING The flowers are very small, fragrant, and borne in small clusters, sometimes on naked branches or the trunk or even the underside of leaves. The fruits appear to be carved from orange wax. Inside each 5-parted fruit is a mass of watery pulp, which tastes like a cross between apricot and passion fruit. Depending on species and cultivar the fruits contain a few to many seeds.
GERMINATION Trees are propagated by cuttings. Little is known about seed germination.

VERBENACEAE — VERBANA FAMILY

Avicennia **L.** — Mangrove

GROWTH HABIT, OCCURRENCE, AND USE *Avicennia* is a small genus that occurs on the shores of tropical America and from the Malay Peninsula, China, and Australia to the Philippine Islands. One species, white mangrove (*A. marina*) grows in South Africa. Honey mangrove (*A. nitida*) has very fragrant flowers and grows from Florida and Texas to tropical America. The members of the genus are trees and shrubs with simple, opposite leaves that are stalked and untoothed.
FLOWERING AND FRUITING The inflorescence consists of short, dense cymes. The flowers are bisexual. The fruit is a flattened capsule. The usually single seed germinates while the fruit is still on the tree.
Avicennia species are pioneering mangroves that colonize the fringes of lagoons. The seedlings are about 2.5 cm long, cone-shaped, and easily borne and scattered by tidal waters. After seedlings fall from the tree the cotyledons spread and develop.
GERMINATION Recalcitrant propagules of *A. marina* were stored under different humidities to achieve both rapid and slow drying (Farrent et al. 1984). Results indicate germination starts as soon as the seeds are shed (note previous indication that seeds germinated on mother plant) and any storage rapidly reduced subsequent seedling survival. Propagules dried rapidly retained viability longer than ones dried slowly. The viability depended on the rate of drying rather than the absolute moisture content reached.

EUPHORBIACEAE — SPURGE FAMILY

Baccaurea Muell. Arg.

GROWTH HABIT, OCCURRENCE, AND USE *Baccaurea* is a genus of evergreen trees from the lowland jungles of southern Asia and Australia. Trees are characterized by very thin bark and spirally arranged, simple leaves. The leaves are borne on a long stalk with a distinct swelling or knee at the tip.

FLOWERING AND FRUITING The flowers are tiny and yellowish-green in color. The fruit is a medium-to-large berry. Most species bear edible fruit. The berry contains one to 6 large oblong seeds, each surrounded with juicy or creamy pulp in a transparent skin.

GERMINATION Seeds of *B. pyriformis* had 96% germination after 3 weeks testing in a study conducted in Malaysia (Ng and Sanah 1979). Germination test conditions suggested by Genebank Handbook (1985) are substrate, sand; incubation temperature, 25 to 20°C; duration, 35 days; and additional instructions, continuous light.

COMPOSITAE — SUNFLOWER FAMILY

Baccharis L.

GROWTH HABIT, OCCURRENCE, AND USE The genus *Baccharis* is composed of about 250 species of deciduous or evergreen shrubs and herbs native to North and South America (Olson 1974a). Some species are used as ornamentals, some for erosion control, and some for medicinal purposes. Groundselbush (*B. halimifolia*) is a serious, introduced weed in forestry plantations in Australia (Panetta 1972). It is native to the Eastern Seaboard and Gulf states of the United States (Bailey 1951). *Baccharis* plants are generally considered poor browse species and some species are poisonous to livestock. Dwarf baccharis (*B. pilularis* var. *pilularis*) has special use as a fire protection species in southern California.

The baccharis species vary considerably in size. Dwarf baccharis is scarcely 0.25 m high. Coyote bush (*B. pilularis* var. *consanguinea*) is an erect, rounded shrub, one to 4 m high, that is found in the foothills of California and western Oregon. Mule fat (*B. viminea*) is a willowlike shrub that reaches 4 m in height and is found along dry water courses in the southwestern United States. Narrowleaf baccharis (*B. angustifolia*) is an evergreen shrub native to the southeastern United States. As ornamental species the baccharis are hardy as far north as Boston and are quite salt tolerant. Most species will grow in soils ranging from pure sand to heavy clay (Dirr and Heuser 1987).

FLOWERING AND FRUITING The white or yellowish male and female flowers, borne on separate plants, are in heads that occur in clusters (Olson 1974a). The female flowers develop into compressed, usually 10-ribbed achenes, tipped by a pappus of bristly hairs 1 cm long or less (Figs. 1, 2). The achenes are dispersed by wind soon after ripening. Seed crops are borne annually.

COLLECTION, EXTRACTION, AND STORAGE OF SEED The ripe fruits of *Baccharis* can be collected by hand or by brushing them into containers or ground cloths. The achenes should be allowed to reach moisture

Figure 1. *Baccharis angustifolia*, narrowleaf baccharis: top, achene with pappus, ×2; bottom, with pappus removed, ×20 (Olson 1974a).

equilibrium in paper bags in a dry storage area before final storage. They may be rubbed through a fine screen to remove the pappus for ease in handling. Olson (1974a) provided the following information on achene and cleaned seed weight:

Species	Achenes/gram	Cleaned seeds/gram
B. angustifolia	—	5,000
B. pilularis	180	
var. *consanguinea*	—	8,300
var. *pilularis*	—	18,000
B. viminea	110	11,000

The differences in numbers between achenes and cleaned seeds per gram seem wrong. Perhaps the achene weights reflect field-collected material that contained flower parts and trash in addition to achenes. Cleaned seeds of *Baccharis* can be stored at 2 to 5°C in sealed containers.

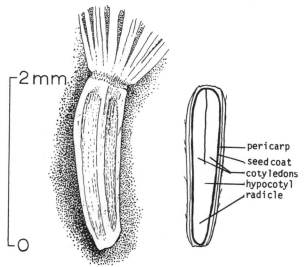

Figure 2. *Baccharis viminea*, mule fat baccharis: left achene with pappus; right, longitudinal section through an achene, ×22 (Olson 1974a).

GERMINATION There is considerable variation between what Olson (1974e) suggested for native species in the United States and more recent information available from detailed studies in Australia. Olson (1974e) suggested that *Baccharis* species germinated without pretreatment:

Species	Substrate	Temperature (°C)	Germination (%)
B. angustifolia	C	15	21
B. pilularis			
var. *pilularis*	—	20/30	92
var. *consanguinea*	P	25	54
B. viminea	—	20/30	82

Panetta (1972) determined that freshly harvested achenes of groundsel-bush lacked innate dormancy, but that fluctuations in temperature and light were necessary to elicit a full germination response. Under conditions of intermittent light, a fluctuation amplitude of less than 1°C was sufficient to promote germination to 50% of its maximal value. Germination was minimal at constant temperatures in the absence of light, but the introduction of 7.5°C temperature fluctuations induced it in about 25% of the seeds tested. This work suggests that individuals interested in propagating the other native *Baccharis* species should try the influence of fluctuating temperature and light on seed germination.

NURSERY AND FIELD PRACTICE Seeds may be sown in the fall or early spring in flats or seedbeds using sandy soil mixtures, or one of the perlite or sphagnum moss seeding mixtures. Seeds usually germinate within 7 to 15 days. Plants large enough for 10 cm pots can be taken from outside beds within 4 months. A germinating seedling of Coyote bush is shown in Fig. 3.

Figure 3. *Baccharis pilularis* var. *consanguinea*, coyote bush: seedling development 60 days after germination (Olson 1974a).

BARRINGTONIACEAE — BARRINGTONIA FAMILY

Barringtonia J. R. & G. Frost. — Fish poison tree

GROWTH HABIT, OCCURRENCE, AND USE Scattered about beach areas of the Indian and Pacific oceans is a broad, handsome tree species that appears superfically to resemble the American magnolia. It has the same buttressed trunk and leathery, glossy leaves. The leaves of the fish poison tree (*B. asiatica*) fall irregularly throughout the year and become beautifully colored. *Barringtonia* flowers are brilliant puffballs of fragrant stamens up to 15 cm in diameter. They appear at dusk and are gone by dawn. Although generally found by the ocean, *Barringtonia* species grow equally well inland, and make fine garden specimen trees in warm climates. *Barringonia acutangula*, with spectacular clusters of red flowers, and *B. racemosa*, with white flowers, are also grown as ornamentals.

The most remarkable thing about the tree is its heart-shaped, 4-sided fruits that develop after the flowers. These consist of a corky, fibrous husk containing 1 seed. They are buoyant and will float for long distances. Native fishers all over the Pacific use the fruits as net floats. They have discovered that the fruits sprinkled in a lagoon will stun fish and bring them quickly to the surface.

PROPAGATION *Barringtonia* species are propagated from seed, and need a subtropical climate. They are popular trees in Hawaii.

ACANTHACEAE — ACANTHUS FAMILY

Beloperone californica Nees — Choparosa

GROWTH HABIT, OCCURRENCE, AND USE *Beloperone californica* is a shrub found along water courses in the creosote bush scrub desert of southwestern United States and adjacent Mexico. It has potential as a native garden species.

The fruit is a capsule. The seeds germinate without pretreatment (Young and Young 1985).

TILIACEAE — LINDEN FAMILY

Belotia A. Rich.

GROWTH HABIT, OCCURRENCE, AND USE *Belotia campbellii* is a fast growing pioneer tree from the tropical forest of southern Mexico.

GERMINATION Seeds of this species are initially dormant due to endogenous dormancy, but germination will occur after storage (Vazquez-Yanes 1981).

BERBERIDACEAE — BARBERRY FAMILY

Berberis L. — Barberry

GROWTH HABIT, OCCURRENCE, AND USE The barberries include about 280 species of evergreen or deciduous, spiny or unarmed shrubs (rarely small trees) native to Asia, Europe, North Africa, and North, Central, and South America (Rudolf 1974b). Some authorities place about 90 species (evergreen, unarmed plants) in a separate genus, *Mahonia*. Because of their handsome foliage and often attractive flowers and fruits, many of the barberries are grown for ornamental purposes. The barberries are also of value for wildlife plantings and erosion control. The majority are susceptible to the black stem rust of wheat, which restricts their use. Species having resistance to black stem rust of wheat are Oregon grape (*B. aquifolium*), Korean barberry (*B. koreana*), Japanese barberry (*B. thunbergii*), Cascade barberry (*B. nervosa*), and creeping barberry (*B. repens*).

FLOWERING AND FRUITING The perfect yellow flowers are borne in the spring in clusters, in spikes, or individually, depending on the species (Rehder 1940). The fruit (Fig. 1) is a berry with one to several seeds (Figs. 2, 3). Good fruit crops are borne almost annually and ripen in the summer and fall (Table 1). Seeds are dispersed by birds and mammals. Color of ripe fruit and other characteristics vary with species (Table 1).

Figure 1. *Berberis nervosa*, Cascade barberry: a spike of berries, ×1 (Rudolf 1974b).

Figure 2. *Berberis aquifolium*, Oregon grape: seeds, ×10 (Rudolf 1974b).

Table 1. *Berberis*: phenology of fruiting and flowering, characteristics of mature trees, cleaned seed weight, and germination test conditions and results (Rudolf 1974b).

Species	Flowering dates	Fruit ripening dates	Year first cultivated	Mature tree height (m)	Growth habit	Ripe fruit color	Seeds/gram	Prechilling (days)	Substrata	Temperature (°C)	Germination (%)
B. aquifolium	Apr.–May	July–Aug.	1823	3.1	Evergreen	Blue	73	90	Sand	20/30	25
B. fremontii	May–June	Sept.–Oct.	1895	4.6	Evergreen	Black	93	0	–	–	85
B. haematocarpus	Spring	Aug.–Sept.	1916	3.1	Evergreen	Red	227	–	–	–	–
B. koreana	May–June	Sept.–Oct.	1905	1.8	Deciduous	Red	84	60	Sand	15	88
B. nervosa	Apr.–May	Aug.	1822	1.8	Evergreen	Blue	51	–	–	–	77
B. nevinii	Mar.–May	July–Aug.	–	1.8	Evergreen	Red	126	90	Soil	–	77
B. repens	May–June	June–July	1822	2.5	Evergreen	Purple	127	196	P	22	74
B. thunbergii	Spring	May–Sept.	1864	2.5	Deciduous	Red	64	90	P	15/25	90
B. vulgaris	Apr.–June	Sept.–Oct.	–	2.5	Deciduous	Purple	84	40	P	15/25	91

Table 2. *Berberis:* standard germination procedures (ASOA 1985, Genebank Handbook 1985).

Species	Substrata	Temperature (°C)	Duration (days)	Additional directions	Source
Berberis species	–	–	–	Prechill at 1–5°C for 6–13 weeks	Genebank Handbook
B. thunbergii	P	18/22	10–14	Excise embryo	ASOA
B. thunbergii	TB	18/22	10–14	Excise embryo	Genebank Handbook
B. vulgaris	P	18/22	10–14	Excise embryo	ASOA
B. vulgaris	TB	18/22	10–14	Excise embryo	Genebank Handbook
Mahonia aquifolium	–	–	–	Prechill at 1–5°C for 6–13 weeks	Genebank Handbook

COLLECTION, EXTRACTION, AND STORAGE OF SEED   Ripe barberry fruits may be picked using gloves, or they may be flailed into receptacles spread beneath the bushes. They may then be run through a macerator with water and the pulp separated by flotation. The seeds should be dried and either sown immediately or stored in sealed containers at temperatures slightly above freezing. Seeds of Japanese and creeping barberry have been stored for 4 years at 2 to 5°C in sealed containers. Viability of seeds of Fremont barberry (*B. fremontii*) and creeping barberry was maintained for 5 years in sealed containers in an unheated shed. Seed yield are shown in Table 1.

PREGERMINATION TREATMENT   Seeds of Fremont and red barberry (*B. haematocarpus*) germinate without pretreatment. Seeds of the other species have internal dormancy and require prechilling to stimulate germination (Table 1). Standard germination procedures have been developed for certain species (Table 2).

NURSERY AND FIELD PRACTICE   Whole berries or, preferably, cleaned seed may be sown in the fall, or prechilled and sown in the spring. Injury from damping-off is more likely if fruits are used. Fall-sown beds should be mulched before germination begins. The seeds should be covered with 0.5 to 1 cm of soil plus 5 cm of sand. Germination is epigeal (Fig. 4). For European barberry (*B. vulgaris*) 22% of the seeds sown produced plants. Barberry species are important commercial nursery species and are propagated by seed, grafting, and cuttings (Table 3).

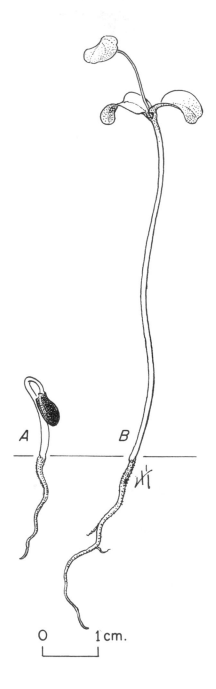

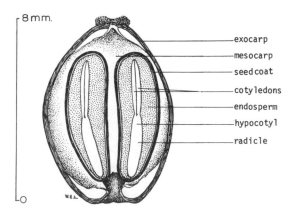

Figure 3. *Berberis thunbergii,* Japanese barberry: longitudinal section through two seeds in a berry, ×6 (Rudolf 1974b).

Figure 4. *Berberis thunbergii,* Japanese barberry: seedling development at 1 and 16 days after germination (Rudolf 1974b).

Table 3. *Berberis:* propagation techniques (Dirr and Heuser 1987).

Species	Propagation techniques	Species	Propagation techniques
B. buxifolia	Seeds: fall plant or prechill 2–3 months. Cuttings: soft to green wood.	B. × mentorensis	Seeds: no information. Cuttings: root easily.
B. candidula	Seeds: fall plant or prechill 2–3 months. Cuttings: easy year-round except very new growth.	B. × ottawensis	Seeds: fall plant or prechill.
		B. sargentiana	Seeds: no information. Cuttings: treat November cuttings with 3000 ppm IBA-talc.
B. × chenaultii	Seeds: 50% germination without pretreatment, prechilling 2–3 months recommended.	B. × stenophylla	Seeds: fall plant or prechill 2–3 months.
B. circumserrata	Seeds: fall plant or prechill 2–3 months.	B. thunbergii	Seeds: 5% germination for untreated seeds; prechilled seeds consistently give 90% germination. Fall plant or prechill 1–2 months. Cuttings: treat July or August cuttings with 1000–5000 ppm IBA solution.
B. darwinii	Seeds: fall plant or prechill 2–3 months.		
B. gagnepainii	Seeds: no information. Cuttings: November cuttings rooted 100% without treatment.		
B. gilgiana	Seeds: fall plant or prechill 1 month. Cuttings: timing important for this deciduous species.	B. triacanthopora	Seeds: fall plant or prechill 2–3 months.
		B. vernae	Seeds: fall plant or prechill 3 months.
B. × gladwynensis	Seeds: apparently does not fruit. Cuttings: fall cuttings root well.	B. verruculosa	Seeds: no information. Cuttings: treat early spring or late fall cuttings with 5000 ppm IBA solution in a 10-second dip.
B. julianae	Seeds: fall plant or prechill 3 months. Cuttings: treat winter cuttings with 10,000 ppm IBA.		
B. koreana	Seeds: fall plant or prechill 2–3 months. Cuttings: treat mid-August cuttings with 8000 ppm IBA-talc.	B. vulgaris	Seeds: fall plant or prechill at least 40 days.

BETULACEAE — BIRCH FAMILY

Betula L. — Birch

GROWTH HABIT, OCCURRENCE, AND USE The birches consist of about 40 species of deciduous trees and shrubs occurring in the cooler parts of the Northern Hemisphere (Brinkman 1974c). Several species produce valuable lumber; others, because of their graceful habit, handsome foliage, and bark are useful for ornamental plantings. Nearly all species provide food and cover for wildlife, and some are important because they promptly regenerate on cutover and disturbed areas. This characteristic makes the birches important in the reclamation of coal strip mine spoils.

In the United States, yellow birch (*Betula alleghaniensis*) is the species most often used in forest plantings for timber production. In Scandinavian countries and the USSR white birch (*B. pendula*) and hairy birch (*B. pubescens*) are widely planted. Cultivars of these species and paper birch (*B. papyrifera*) are often planted as ornamental species.

FLOWERING AND FRUITING The flowers are monoecious and borne in catkins. Staminate catkins are formed in late summer or autumn, remain naked during the winter, and open after considerable elongation in the spring. The conelike pistillate catkins—terminal on short, spurlike lateral branches and with closely overlapping scales—appear with the leaves (Table 1). When they ripen in late summer or autumn, the fruits

Table 1. *Betula:* phenology of flowering and fruiting, characteristics of mature trees, and cleaned seed weight (Brinkman 1974c).

Species	Location	Flowering dates	Fruit ripening dates	Seed dispersal dates	Mature tree height (m)	Year first cultivated	Seed-bearing Age (years)	Seed-bearing Interval (years)	Seeds/ gram
B. alleghaniensis	Midrange	Apr.–May	Aug.–Oct.	Sept.–Spring	31.0	1800	40	2	990
B. davurica	Japan	May	Oct.	–	20.0	1883	–	2	1600
B. glandulosa	Midrange	June–Aug.	Aug.–Oct.	Sept.–Mar.	1.8	1880	–	–	8500
B. lenta	Midrange	Apr.–May	Aug.–Sept.	Sept.–Nov.	24.6	1759	40	1–2	1400
B. nigra	North	Apr.–May	May–June	May–June	31.0	1736	–	–	825
B. papyrifera	Midrange	Apr.–June	Aug.–Sept.	Aug.–Spring	21.5	1750	15	2	3040
B. pendula									
dewinged	USSR	Apr.–June	July–Aug.	July–Sept.	20.0	–	15	2–3	5300
winged	USSR	Apr.–June	July–Aug.	July–Sept.	20.0	–	15	2–3	1750
B. populifolia	Midrange	Apr.–May	Sept.–Oct.	Oct.+	12.3	1750	8	1	940
B. pubescens	Germany	May–June	Aug.–Sept.	Fall–Winter	20.0	1789	15	2–3	3800
B. pumila var. glandulifera	Midrange	May–June	Sept.–Oct.	Oct.–Mar.	3.1	1762	–	1–2	5300

become brown and woody and are erect or pendulous (Fig. 1). Each scale may bear a single small nut (seed) (Figs. 2, 3), which is oval in shape with two persistent stigmas at the apex. The seed turns from greenish tan to light brown or tan when mature. After seed-fall, the strobiles slowly disintegrate on the trees with the apices persisting on the branchlets. Seed dispersal is usually by wind, sometimes by water. Seed may be blown some distance over crusted snow. The production of paper birch seed has been estimated at 1.75 to 90 million seeds per hectare depending on the production year (Bjorkbom 1971). The Soviets put a lot of research effort into developing methods of predicting the seed yield and quality of birch seed (e.g., Kosnikov and Nikulin 1985).

COLLECTION OF SEED Birch seed is collected by picking or stripping the strobiles (while they are still green enough to hold together) from standing trees or shrubs or from trees recently felled in logging operations. Because ripe strobiles shatter readily, they are usually put directly into bags rather than allowed to fall on canvas with the attendant seed loss.

EXTRACTION AND STORAGE OF SEED Freshly collected strobiles usually are rather green and thus are subject to heating; they should be spread out to dry for several weeks until they began to disintegrate. They can be shattered by flailing and shaking and the seeds separated from most of the scales and debris by air

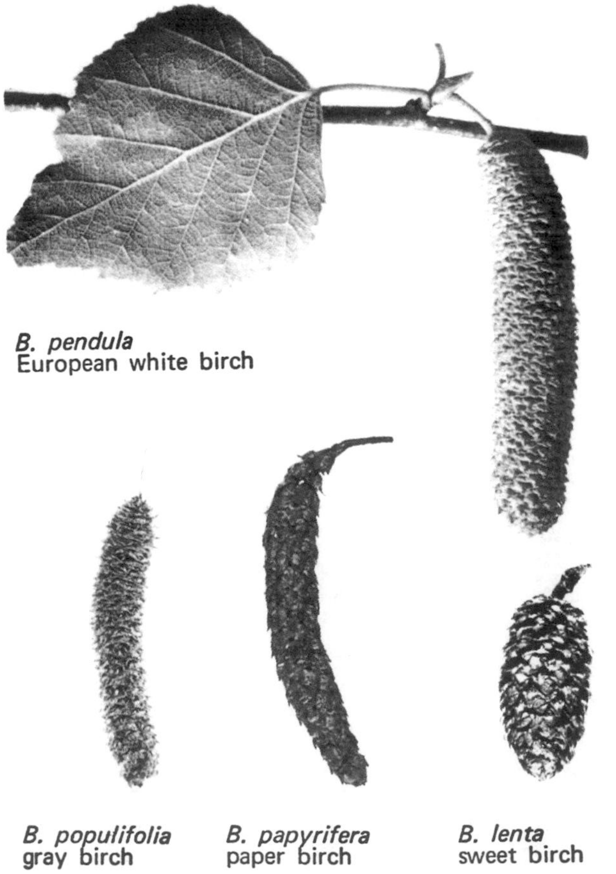

screening. Round-holed screens of the following sizes (sizes given in inches because the screens for air screens are not commonly available in metric sizes) have proven satisfactory for these species:

Species	Screen size (inch)
B. alleghaniensis	8/64
B. nigra	10/64
B. papyrifera	8/64
B. pendula	1/10
B. pubescens	1/10
B. pumila	6/64

Birch seeds are very small and light; the number per gram and yield per liter vary considerably among species (Table 1).

For paper birch, Bjorkbom and Marquis (1965) found that a higher proportion of the seed was viable in good years than in poor, and that the germination energy of sound seed also was greater in years of high seed production. The percentage of viable birch seed can be estimated by examining the seed under transmitted light.

Seed of most species of birch can be stored at one to 3% moisture content and temperatures of 2 to 5°C (Heit 1967a). Other test have shown that seeds of sweet birch (*B. lenta*), paper birch, and gray birch (*B. populifolia*) with one to 5% moisture content can be kept for 1.5 to 2 years at room temperature; if the moisture content is much higher, percent germination may drop even though the seed is stored at low temperatures.

GERMINATION Early tests for germination of birch seed suggested that prechilling was a requirement for germination. Later work indicated that seed was sensitive to light and supplying light during germination overcame prechilling requirements. Brinkman (1974c) suggested that germination of birch species with light always gave better results than prechilling and germination in the dark (Table 2). The germination capacities for many of the birch species reported in Table 2 appear to the authors to be rather low.

Recent research results may offer a better understanding of the germination ecology of birch species. John Bevington of Moravian College provided the results of a well-thought-out and designed study of the germination of seeds of paper birch (Bevington 1986). This study combined various aspects of the germination ecology with seed collections from different geographical areas. At incubation temperatures from 14 to 18°C seeds of paper birch from northern sources a had higher percentage germination than seeds from sources collected farther south. Northern sources of paper birch had seeds with thinner, more translucent pericarps than those from southern sources. They also had higher germination over a wider range of temperatures. At 15°C seeds had progressively higher germination as the photoperiod was increased. This was not a true photoperiod effect because seeds responded to the total amount of light rather than to the relative

B. pendula
European white birch

B. populifolia
gray birch

B. papyrifera
paper birch

B. lenta
sweet birch

Figure 1. *Betula*: ripe female strobiles, ×1 (Brinkman 1974c).

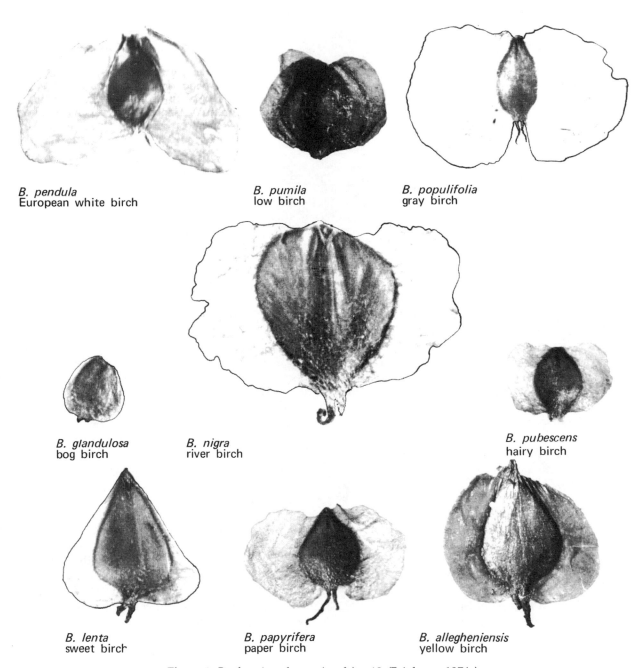

B. pendula
European white birch

B. pumila
low birch

B. populifolia
gray birch

B. glandulosa
bog birch

B. nigra
river birch

B. pubescens
hairy birch

B. lenta
sweet birch

B. papyrifera
paper birch

B. alleghaniensis
yellow birch

Figure 2. *Betula*: winged nuts (seeds), ×12 (Brinkman 1974c).

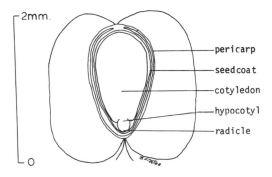

Figure 3. *Betula nigra*, river birch: longitudinal section through a
nut (seed) (Brinkman 1974c).

Table 2. *Betula:* germination test conditions and results (Brinkman 1974c).

Species	Prechilling (days)	Light (hours)	Substrata	Temperature (°C)	Germination (%)
B. alleghaniensis	60	8+	S	10/32	27
	0	8+	–	20/30	59
B. davurica	0	8+	–	20/30	18
B. glandulosa	Winter	–	S	10/30	24
	0	20	Perlite	20/25	3
B. lenta	70	8+	S	10/32	43
B. nigra	60	8+	S	20/30	34
	0	20	Perlite	20/25	73
B. papyrifera	75	8+	S	10/32	34
	0	8+	P	–	47
B. pendula	40	8+	–	–	30
	0	8+	S	20/30	36
B. populifera	90	8+	–	–	64
B. pubescens	60	8+	–	20/30	40
	0	20	–	15/25	87
B. pumila var. *glandulifera*	0	8+	P	18/25	31

length of light and dark periods. At temperatures near 15°C the effect may have ecological significance in preventing fall germination. Prechilling at 3°C promoted subsequent germination, with seeds of northern collections being more responsive than those from southern areas. Prechilling enhances paper birch seeds' sensitivity to light.

Working with seeds of *B. maximowicziana* in Japan, Nagata and Black (1977) determined that unchilled seeds required repeated exposure to red light to induce germination. Prechilled seeds required only an single exposure. The prechilling requirement could be replaced with enrichment of the germination substrate with GA3. Extreme diurnal temperature fluctuations could be substituted for light requirements with prechilled seeds.

Pratt (1986) investigated the germination of seeds of gray birch collected from trees growing on anthracite mine spoils in northern Pennsylvania. In the laboratory seeds of gray birch would germinate at constant incubation temperatures of 55°C, but germination declined rapidly under fluctuating temperature conditions. No germination occurred under fluctuating temperatures in excess of 30°C. This observation was related to the field germination period on these harsh mine spoil sites.

Working on the Baltic coast of Sweden, Granstron and Fries (1985) investigated the persistance of birch seeds in the soil seedbanks under stands of conifers. They found that 94% of the artificially established birch seed populations were lost in the first year. They interpreted this as indicating the birches were colonizing species of recently cut forest only if a recent seed source was available.

Methods for testing seeds of birch species for germination in the laboratory reflect the prechill versus light-during-germination aspects of their physiology (Table 3). The influence of GA3 enrichment and fluctuating temperatures on prechilled seeds should be investigated.

NURSERY AND FIELD PRACTICE Birch seed usually is sown soon after collection in the late summer or fall, but seed may be sown in the spring after prechilling for 4 to 8 weeks. Seed is broadcast and covered as lightly as possible; if the seedbed can be kept moist, the seed is sown without covering. Germination is epigeal (Fig. 4) and usually complete in 4 to 6 weeks after spring sowing. Birch seedlings require shade for 2 to 3 months during the summer. Tree percent is low; only 15 to 20% of European white birch and hairy birch seed will produce 1-0 seedlings. A seedling density of 265 to 475 per m² is desirable. Stock is usually planted as 1-0 or 2-0 seedlings. Birch species are very important in the commercial nursery business and are propagated by cuttings as well as by tissue culture techniques (Table 4).

Table 3. *Betula:* suggested methods for testing germination of birch species (ASOA 1985, Genebank Handbook 1985).

Species	Substrata	Temperature (°C)	Light	Duration (days)	Source
Betula sp.[1]	TB	20/30	Yes	21	Genebank Handbook
	P	20/30	Yes	21	ASOA
B. papyrifera	TB	20/30	–	21	Genebank Handbook
B. pendula	TB	20/30	–	21	Genebank Handbook
B. pubescens	TB	20/30	–	21	Genebank Handbook

[1]Alternative method: Prechill at 1–5°C for 4 weeks, plus light.

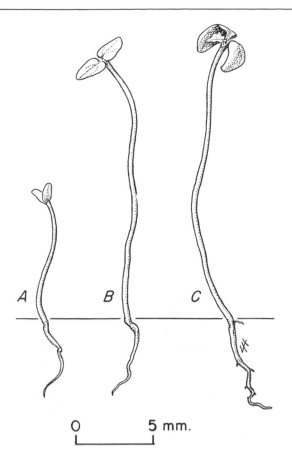

Figure 4. *Betula populifolia*, gray birch: seedling development at 1, 10, and 40 days germination (Brinkman 1974c).

Table 4. *Betula*: propagation techniques (Dirr and Heuser 1987).

Species	Propagation techniques
B. albo-sinensis	Seeds: fall plant or prechill 2–3 months plus light.
B. alleghaniensis	Seeds: fall plant or prechill 2–3 months plus light.
	Grafting: *B. pendula* suitable understock.
	Tissue culture: axillary bud explant.
B. ermanii	Seeds: no pretreatment required.
B. grossa	Seeds: fall plant or prechill 1–2 months plus light.
B. jacquemontii	Seeds: fall plant or prechill 1–2 months plus light.
	Cuttings: mid-August.
B. lenta	Seeds: no pretreatment with light; dark germination requires 1 month prechilling.
B. maximowiczine	Seeds: light required for germination.
	Cuttings: treat early July cuttings with 1% IBA.
B. nigra	Seeds: no pretreatment necessary if planted after spring maturity.
	Cuttings: treat June–July cuttings with 1000 ppm IBA.
B. papyrifera	Seeds: conflicting reports in literature.
	Cuttings: timing critical, base of cutting just becoming firm.
	Tissue culture: axillary bud explant.
B. pendula	Seeds: collect in fall and plant in peat-perlite in greenhouse at 18°C.
	Cuttings: treat late May and June cuttings with 8000 ppm IBA-talc.
	Tissue culture: cambial explants.
B. platyphylla var. *japonica*	Seeds: no pretreatment required or prechill 1–2 months.
var. *szechuanica*	Cuttings: timing critical.
	Tissue culture: stem tip and nodal explants.
B. populifolia	Seeds: no pretreatment necessary or prechill 1–2 months plus light.
	Cuttings: treat with 50 ppm IBA.

BOMBACACEAE — BAMBAX FAMILY

Bombax L. — Silk cotton tree

GROWTH HABIT, OCCURRENCE, AND USE From the tropical forests of Asia, South America, and Africa comes *Bombax*, a genus of splendid trees (Macoboy 1982). The scientific name refers to the filaments obtained from the bulky seedpods that are used as a substitute for kapok.

Red silk cotton tree (*Bombax malabaricum*) from southeast Asia is the most commonly seen species in gardens in Australia, Hawaii, and South Africa. The tree reaches 35 m or more in height, with a widely buttressed trunk at maturity. Its leaves are strikingly digitate and up to 50 cm across, consisting of 3 to 7 widely spaced leaflets.

Bombax ellipticum is a deciduous species from Mexico. Its compound leaves of 5 leaflets may be 45 cm in diameter. The leaves are copper red when young.

Bombax barrignon is a smaller ornamental species from tropical America.

FLOWERING AND FRUITING In the spring red silk cotton tree produces a stunning crop of flowers 17.5 cm in diameter. These appear at the ends of the branches shortly after the tree loses its foliage for a brief period in winter. The fruit of the *Bombax* is a pod.

GERMINATION The ornamental *Bombax* species reportedly establish easily from seed (Macoboy 1982). Seeds of *B. valetonii* germinate readily according to Ng (1980).

ANACARDIACEAE — CASHEW FAMILY

Bouea Meissn.

GROWTH HABIT, OCCURRENCE, AND USE *Bouea* is a genus of evergreen trees to 20 m, with very dense and bushy crowns. The bark is light grayish brown and finely fissured. *Bouea* is widely grown in southeastern Asia as a fruit tree.

FLOWERING AND FRUITING The flowers are very small, greenish yellow in color and borne in short, axillary panicles. The fruit is a mangolike structure with a fibrous, leathery stone. The cotyledons are bright purple when the stone is cut.

GERMINATION Seeds of *B. macrophylla* had 100% germination 3 weeks after planting in a study conducted in Malaysia, but only 5 seeds were used in the test (Ng and Sanah 1979).

STERCULIACEAE — STERCULIA FAMILY

Brachychiton Schot. et Endl. — Bottle tree

GROWTH HABIT, OCCURRENCE, AND USE The genus *Brachychiton* consist of about 11 species of Australian trees that are grown in warm climates (Bailey 1951). They are considered by some to be Australia's most beautiful trees. *Brachychiton* species are among the most variable of trees in size and shape of fruit and leaves, and color and size of the generally bell-shaped flowers (Macoboy 1982). *Brachychiton acerifdries* (illawarw flame), *B. bidwillii* (pink lacebark), and *B. discolor* (white lacebark) have a tall growth form, while *B. rupestre* (Queensland bottle tree) has a bloated trunk. *Brachychiton* species thrive in warm, dry climates such as those of California, South Africa, and the Mediterranean.

FLOWERING AND FRUITING The fruit is a woody follicle that does not dehisce until ripe. The seeds and pulp inside the fruit are hairy and often cohering.

GERMINATION The *Brachychiton* species are established by direct seeding and cuttings.

COMPOSITAE — SUNFLOWER FAMILY

Brickellia Ell. — Brickle bush

GROWTH HABIT, OCCURRENCE, AND USE *Brickellia* contains about 100 species native to North and South America. Brickle bush is from 0.5 to 1 m tall with many stems that have a tomentose covering.

California brickle bush (*B. californica*) is a shrub found in many different plant communities from California to Texas and New Mexico. This species has potential for native plant gardens in southwestern United States (Young and Young 1985).

FLOWERING AND FRUITING The flower heads are borne in small clusters at the ends of lateral branches.

GERMINATION The fruit is an achene that germinates without pretreatment (Emery 1964). Seeds of *Brickellia* species stored at room or low temperatures maintained viability for at least 7 years (Kay et al. 1988).

MORACEAE — MULBERRY FAMILY

Broussonetia L'Hér. ex Vent. — Paper mulberry

GROWTH HABIT, OCCURRENCE, AND USE The paper mulberry (*Broussonetia papyrifera*), so called because the bark was used for paper making in the Orient, is a tree reaching 16 m in height. It is native to China and Japan. Introduced into cultivation in the United States in 1750, it has become naturalized from New York to Florida. This small tree has wide-spreading branches that form a rounded canopy.

FLOWERING AND FRUITING The flowers are small—the staminate in nodding, cylindrical spikes, and the pistillate in dense, globular, short-stalked, small heads. The fruit is a collection of orange-red drupelets standing in the persistent perianths and bracts, forming a small globular syncarp.

GERMINATION AND PROPAGATION Seeds germinate without pretreatment (Dirr and Heuser 1987). Cuttings taken during July and August, from short shoots with a heel, root readily. Suckers taken in the spring will also root. Explants produce callus cells, but it is not known if shoots result.

TILIACEAE — LINDEN FAMILY

Brownlowia Roxb.

GROWTH HABIT, OCCURRENCE, AND USE The genus *Brownlowia* consists of about 17 species native to southeast Asia. The leaves are spirally arranged and the underside of the leaf is a dull brownish color due to minute scales. *Brownlowia* species are common in mangrove forests.The flowers are small and borne in panicles. The fruits consist of one to 5 separate bodies arranged in a rosette with each containing one large seed.

GERMINATION Seeds of *B. heleriana* stored 1 day before testing had 41% germination after 5 weeks in a study conducted in Malaysia (Ng and Sanah 1979).

SIMAROUBACEAE — QUASSIA FAMILY

Brucea J. F. Mill.

GROWTH HABIT, OCCURRENCE, AND USE *Brucea amarissima* is a lax evergreen shrub or spindly tree up to 6 m tall. The pinnate leaves are up to 50 cm long. The genus includes about 75 species native to Asia. These plants are related to *Ailanthus*. Their bitter fruits are used by the Chinese as medicine.

FLOWERING AND FRUITING The flowers are 2.5 cm wide and red except for a green ovary; they are borne in spikes. The fruit is a berry 0.6 cm long that is green at first and then turns purplish black. The seeds are dispersed by fruit bats.

GERMINATION Seeds of *Brucea javanica* had 36% germination after 39 weeks of testing. It required 32 weeks of testing to attain 50% of the final germination (Ng and Sanah 1979).

LOGANIACEAE — LOGANIA FAMILY

Buddleia L. — Butterfly Bush

GROWTH HABIT, OCCURRENCE, AND USE The genus *Buddleia* consists of about 100 species native to tropical and subtropical portions of Asia, the Americas, and South Africa (Bailey 1951). The butterfly bushes are grown outdoors in the southern United States and elsewhere in North America as greenhouse species.

FLOWERING AND FRUITING The fragrant flowers consist of a campanulate or funnelform corolla with 4 lobes. The fruit is a capsule that contains numerous minute seeds.

GERMINATION Seeds of the major species cultivated in the United States germinate without pretreatment (Dirr and Heuser 1987). Alternate leaf butterfly bush (*B. alternifolia*) can be propagated with cuttings taken in mid-August and treated with 8000 ppm of IBA-talc. Orange-eye butterfly bush (*B. davidii*) can be rooted very easily from summer cuttings. Meristem explants of this species produce shoots and subsequent virus-free plants (Dirr and Heuser 1987).

SAPOTACEAE — SAPOTE FAMILY

Bumelia langinosa (Michx.) Pers. — Gum bumelia

GROWTH HABIT, OCCURRENCE, AND USE Gum bumelia is a spiny shrub or small tree (up to 19 m) found from southern Georgia to southern Illinois and west to southern Kansas and southern Arizona; it also grows in northern Mexico. It is deciduous to the north and evergreen in the southern part of its range. Gum bumelia has value as wildlife food. It is also planted as an ornamental, and to some extent as a shelterbelt species. It is deeply tap rooted and extremely drought resistant.

FLOWERING AND FRUITING . The perfect white flowers are borne on small fascicles 0.5 to 1.25 cm across that open in June to July (Bonner and Schmidtling 1974). The fruit is a single-seeded drupe 0.75 to 2.5 cm long. It turns purplish black as it ripens in September and October and persists on the tree into winter. The single seed is 0.6 to 1.25 cm long, rounded, brownish, and shiny (Figs. 1, 2).

COLLECTION, EXTRACTION, AND STORAGE OF SEED Fruits should be picked as soon as they turn purplish black. The fleshy outer coat may be removed by careful maceration in water. Four samples from Texas and Oklahoma yielded 50–60 cleaned seeds/40 kg fruits, 13 cleaned seeds/g, 94% purity, and 88% sound seed (Bonner and Schmidtling 1974). Longevity of seed in storage apparently is still not known.

GERMINATION Gum bumelia seeds germinate slowly and may be influenced by the seedcoat and internal conditions. Scarification by soaking in concentrated sulfuric acid for 20 minutes, followed by 4 to 5 months prechilling at 2 to 5°C, has been recommended. Prechilling alone at 5°C has also been successful. Preliminary trials on each seed lot are necessary to determine if scarification is necessary. Germination can be tested in flats of sand or sand and peat at incubation temperatures of 20/30°C. Test duration of 60 to 90 days is required for prechilled seeds. Germinative capacity ranges from 40 to 50%.

NURSERY AND FIELD PRACTICE Seeds should be spaced at the rate of 80 per meter of drill row and covered lightly with soil. Outplanting at the age of 2 years is suggested.

Figure 1. *Bumelia langinosa,* gum bumelia: seed ×8 (Bonner and Schmidtling 1974).

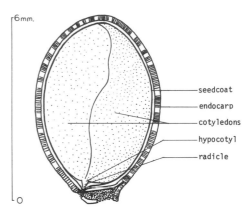

Figure 2. *Bumelia langinosa,* gum bumelia: longitudinal section through a seed ×8 (Bonner and Schmidtling 1974).

PALMAE — PALM FAMILY

Butia Becc.

GROWTH HABIT, OCCURRENCE, AND USE *Butia* is a genus of a half-dozen monoecious, pinnate palms of South America that are planted in California and Flordia as ornamental species. The trees are not very tall, usually under 5 m in height. The leaves are pinnate with a glaucous or bluish tint and 25 to 50 pinnae. The fruits are more or less soft drupes about 2.5 cm long.

Butia monosperma is a common species throughout India and Burma and is a useful species in rural development projects.

GERMINATION Germination is influenced by the incubation temperature (Sento 1972). Tetrazolium can be used as an approximate index of seed viability.

Large and small seeds of *B. monosperma* were identified as weighing 1.0 and 0.56 g, respectively (Pathak and Patil 1985). Soaking in ascorbic acid (100–1000 ppm) did not improve germination of large seeds, but did improve that of small seeds. Small seeds produced more roots compared to shoots than large seeds. Germination information for members of the genus is provided by the Genebank Handbook (1985):

Species	Germination techniques
B. bonnetii	Seeds not difficult to germinate. Can be tested in moist sand at incubation temperatures of 20/30°C for 300 days.
B. capitata	Very dormant. Untreated seeds require up to 960 days to germinate. Gibberellic acid at 100–2000 ppm enhances germination.
B. eriospata	Very dormant. Untreated seeds require up to 22 months to germinate.
B. yatay	Very dormant. When tested in moist sand at 20/30°C these seeds require from 2 to 20 months to germinate.

Carpenter (1987) investigated the germination of seeds of *B. capitata*. Fruit coloration was the key to subsequent seed germination. Seeds from dark yellow fruits had higher germination than seeds from greenish fruits. The seeds required 90 to 120 days afterripening before sowing.

BUXACEAE — BOX FAMILY

Buxus L. — Box

GROWTH HABIT, OCCURRENCE, AND USE The boxes consist of about 30 species of shrubs or small trees native to central Europe, the Mediterranean region, eastern Asia, the West Indies, and Central America (Rehder 1940). These species are widely grown as hedges and edging species. Their leaves are opposite, leathery, and entire.
FLOWERING AND FRUITING The flowers are apetalous and borne in terminal or axillary clusters. The fruit is a 3-celled capsule. The fruits and flowers are not showy.
GERMINATION *Buxus* species can be propagated by seed or cuttings (Dirr and Heuser 1987):

Species	Propagation technique
B. harlandii	Seeds: no information.
	Cuttings: June cuttings, treated with 3000 ppm of IBA, rooted 100%.
B. microphylla	Seeds: some germination without pretreatment, but prechilling 1 to 3 months is recommended.
	Cuttings: skilled nursery growers produce rooted cuttings at all seasons.
B. sempervirens	Seeds: require no pretreatment for 70+% germination, but prechilling speeds uniform germination.
	Cuttings: cuttings taken in mid-July root without treatment.

PALMAE — PALM FAMILY

Calamus L. — Rattan

GROWTH HABIT, OCCURRENCE, AND USE *Calamus* species provide the raw material for rattan, which is widely used for a variety of products including furniture. Formerly the source of rattans was native stands in southern Asia. Overutilization has largely exhausted the resource and there is considerable interest in establishing plantations (Generalo 1977, Manokaran 1978). In India *C. tenuis* is found in moist habitats in the sub-Himalayan region.
FLOWERING AND FRUITING The fruits of *C. tenuis* have an outer skin with a yellowish white shine. The fruit clusters attract ants and birds to their sour-sweet pulp. The fresh fruits number about 1100 per kg (Gulati and Sharma 1983).

GERMINATION Studies were conducted in Malaysia to determine the best germination substrate for *C. manan* and *C. caesius* seeds (Ahmad 1983). Emergence of seedlings of the latter was from 13 to 89%, but there was no consistent relationship with substrate. Emergence of seedlings of the former was from 0.2 to 87%, with the best emergence occurring from a sawdust substrate.

The best seedling emergence for *C. tenuis* was obtained by sowing whole fruits within 10 days of collection.

LEGUMINOSAE — LEGUME FAMILY

Calliandra Benth.

GROWTH HABIT, OCCURRENCE, AND USE *Calliandra* is a group of leguminous trees and shrubs consisting of about 120 species that are widely distributed in the tropics. These woody plants, hardy only in the southern United States, are not as widely known as *Albizia, Leucaena,* and *Acacia* species, but are similarly useful for fuel wood and fodder plantings. The plants are mostly unarmed with bipinnate leaves divided into numerous leaflets.
FLOWERING AND FRUITING Flowers are purplish or white in color and borne in globose heads or clusters. The corolla is small and obscured by the numerous, long,

silky, purple or white stamens. The pods are straight or nearly so, and usually compressed, with thickened edges. They dehisce elastically on the tree, indicating they must be collected before they are completely ripe.
GERMINATION Limited information indicates the seeds often have hard seedcoats and must be scarified for germination to occur. Seeds of *C. calothyrsus* have been successfully scarified by pouring 80°C water over them and allowing them to cool for 24 hours (Diagana 1985).

VERBENACEAE — VERBENA FAMILY

Callicarpa L. — Beauty bush

GROWTH HABIT, OCCURRENCE, AND USE The genus *Callicarpa* consists of about 40 species of tropical and subtropical environments in Asia, Australia, and North and Central America (Bailey 1951). They are widely planted as ornamental species. American beauty bush (*C. americana*) is a shrub or small tree up to 2 m in height. The leaves of this species have a covering or scurffy-stellate down and are glandular-dotted.
FLOWERING AND FRUITING Flowering commences in June with fruit showing traces of color in late August to September.

The flowers of *C. americana* are small, pink, bluish, or white in color, and borne in axillary clusters. The fruit is a small, subglobose blue, berrylike drupe. The rich metallic-colored fruit in autumn is the most attractive aspect of these species. All flowers occur on new growth so the shrubs can be cut to ground level in the winter.
GERMINATION Generally the small seeds germinate without pretreatment (Dirr and Heuser 1987):

Species	Propagation techniques
C. americana	Seeds: No pretreatment necessary. Nursery growers usually clean pulp from seeds before planting, but it is not necessary. Cuttings: Softwood cuttings taken from June to September and treated with 1000 ppm IBA-talc root readily.
C. bodinieri	Seeds: No pretreatment necessary; fresh seeds germinate readily. Cuttings: Treat late June cuttings with 5 to 10 ppm IBA in a 24-hour soak.
C. dichotoma	Seeds: No pretreatment necessary, but pre-chilling makes germination more uniform. Cuttings: Treat late June cuttings with 5 to 10 ppm IBA in a 24-hour soak.
C. japonica	Seeds: No pretreatment necessary, but pre-chilling makes germination more uniform. Cuttings: October cuttings can be rooted in sand. IBA treatment hastens rooting.

CUPRESSACEAE — CYPRESS FAMILY

Callitris Vent.

GROWTH HABIT, OCCURRENCE, AND USE White cypress (*Callitris glauca*) is a small- to medium-sized tree usually growing to 18 m tall, but occasionally reaching 30 m. The natural range of this species is from central Queensland to Victoria, Australia. Areas where it is native typically have from 5 to 15 days of frost and annual precipitation from 35 to 55 cm.

Mature cones are spherical, dark brown, 1 to 2 cm in diameter and consist of 3 large flexible and 3 alternating woody scales. The scales are slightly wrinkled on the outer surface and bear a small point on the tip. The seeds are light brown, winged, and 0.6 to 0.8 cm wide. The wood is widely used in eastern Australia.

GERMINATION Little detailed information is available on germination of *Callitris* species (Langkamp 1980):

Species	Temperature (°C)	Duration (days)	Viability (%)
C. canescens	—	25–44	
C. columellaris	30	8–50	23
C. drummondii	—	23–96	
C. intratropica	—	30	2
C. preissii	15	14–43	54
C. rhomboidea	—	28	50

According to Goor and Barney (1976), there are 10 seeds of *C. glauca* per gram, average germination is 50%, and seeds should be stored dry in sealed containers at low temperatures.

ERICACEAE — HEATH FAMILY

Calluna Salisb. — Heather

GROWTH HABIT, OCCURRENCE, AND USE *Calluna* is a monotypic genus native to Europe and Asia and adventive in North America. Over 600 cultivars of *C. vulgaris* are used as ornamentals. Heather is a low, evergreen shrub that blooms profusely in late summer.

FLOWERING AND FRUITING Closely related to the genus *Erica*, heather differs in its deeply 4-parted, longer, colored calyx that conceals the corolla. The flower is subtended by 4 bracts that resemble a calyx. The fruit is a dehiscent capsule with few seeds.

GERMINATION Heather seeds require no pretreatment for germination (Dirr and Heuser 1987). They should be seeded on milled peat, not covered, and place under mist or in a high humidity environment. Germination takes place in 3 to 5 weeks.

There are many different methods of rooting heather. Cuttings taken in August and treated with 1000 ppm IBA-solution root easily.

CUPRESSACEAE — CYPRESS FAMILY

Calocedrus decurrens (Torr.) Florin — Incense cedar

GROWTH HABIT, OCCURRENCE, AND USE *Calocedrus* is the North American representative of the related genus *Libocedrus* (in which it was formerly included) that forms a Pacific basin group of conifers. About 7 species of *Libocedrus* are native to portions of South America, New Zealand, New Caledonia, New Guinea, Taiwan, and southeastern China (Bailey 1951).

The range of incense cedar spans about 15° of latitude from the southern slopes of Mount Hood in Oregon, southward within and adjacent to the Cascade Range to the Siskiyou, Coastal, and Sierra Nevada ranges to Sierra de San Pedro in northwestern Mexico (Stein 1974a). It extends inland from the coastal fog belt to arid parts of central Oregon and western Nevada. In elevation, the distribution extends from 275 to 2030 m in the north to 920 to 2980 m in the south. Incense cedar grows on many kinds of soil including the harsh habitats provided by soils derived from serpentine.

The wood of incense cedar is variable in color, very durable, light, moderately soft, uniformly textured, easy to split and wittle, and finishes well. During the last 2 decades a considerable industry has developed for its utilization, especially for export products to Japan, such as pencil stock. One reason for the growth of this industry has been pricing policy in federal timber sales where incense cedar was considered virtually a waste product.

Cultivated since 1853, incense cedar is widely grown outside its natural range as an ornamental species. Several horticultural cultivars are recognized.

FLOWERING AND FRUITING Yellowish green staminate

flowers develop terminally on twigs as early as September, even before the current year's cones have opened. The staminate cones are 0.6 cm long and golden in color. Pollen shed is during late winter or early spring.

Pistillate cones, each containing up to 4 seeds, hang singly (Fig. 1), are scattered throughout the crown, and mature in 1 growing season. Cones become reddish brown at maturity. The winged seeds are about 2.5 cm long and 0.8 cm wide (Fig. 2). Although appearing to have only 1 wing, each seed actually has 2—a long, wide wing extending lengthwise beyond the seed on one side and a narrow, much shorter one barely emerging parallel to it on the opposite side. The wings are persistent and project past the narrow radicle end of the seeds rather than from the cotyledon end as in many conifers (Fig. 3). During germination, the radicle clearly emerges from the winged end of the seed (Fig. 4).

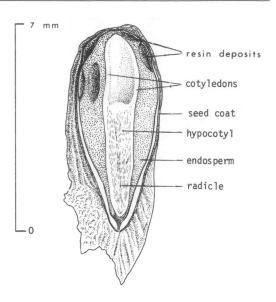

Figure 3. *Calocedrus decurrens*, incense cedar: longitudinal section through a seed, ×8 (Stein 1974a).

Figure 1. *Calocedrus decurrens*, incense cedar: cones hang singly from the branch tips, ×0.75 (Stein 1974b).

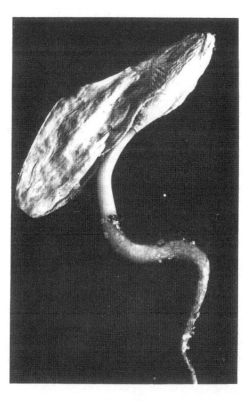

Figure 4. *Calocedrus decurrens*, incense cedar: germinating seed, × 3, with radicle and hypocotyl emerging from winged end (Stein 1974a).

Seed dispersal extends from September through November. Seed crops are often light and there is wide geographical variability in crop abundance. Even in light years occasional trees will have abundant cones. COLLECTION, EXTRACTION, AND STORAGE OF SEED Cones are generally handpicked from standing or felled trees. They can be stripped with a rakelike tool and should be handled in cloth bags.

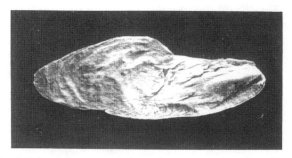

Figure 2. *Calocedrus decurrens*, incense cedar: seed with wing intact, ×3 (Stein 1974a).

Under warm, low humidity conditions cones will air-dry in about 3 to 7 days. Seeds separate from open cones. The wings should not be removed to avoid damage to the seeds.

One hundred liters of cones weighs 250 to 310 kg and yields from 2.2 to 6.6 kg of seed. A minimum of 14 and a maximum of 64 seeds per gram were found in 55 samples. ASOA (1985) standards list 32 seeds per gram for this species.

Incense cedar seeds do not keep well in ordinary dry storage, but high viability can be maintained in very cool, dry storage.

PREGERMINATION TREATMENT Incense cedar seeds require prechilling for germination. ASOA (1985) standards are as follows: substrata, the top of blotters or creped paper; incubation temperature, 20/30°C; and test duration, 28 days. Seeds should be prechilled 30 days before testing.

NURSERY AND FIELD PRACTICE Fall-sown seeds have higher and more uniform emergence. Seedlings must be protected from early spring frost. Spring-sown seeds have to be prechilled before planting.

The winged seeds are difficult to drill so they are usually broadcast. Planting depth is usually 0.6 cm. Germination is epigeal (Fig. 5). Seedling height reaches 5 to 10 cm the first season. Seedbed densities of 260 to 320 per m² are satisfactory. Generally 1-0 or 2-0 seedling stock is used for outplanting. Tree percentage ranges from 20 to 75%.

Cuttings are difficult to root. Hardwood cuttings taken in August may root. Mid-November cuttings treated with hormones and given a variety of propagation manipulations have been rooted (Dirr and Heuser 1987).

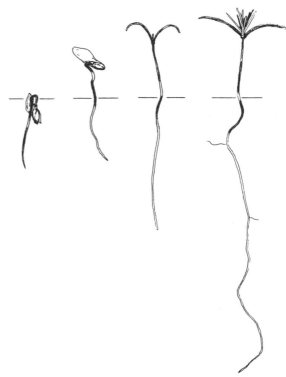

Figure 5. *Calocedrus decurrens*, incense cedar: seedling development 4, 7, 10, and 17 days after germination, ×0.5 (Stein 1974a).

GUTTIFERAE — GARCINIE FAMILY

Calophyllum L.

GROWTH HABIT, OCCURRENCE, AND USE The Garcinie family is almost exclusively tropical in distribution. No species of Guttiferae is indigenous to the United States, but species of *Calophyllum* extend from South and Central America into Mexico and the West Indies (Laurence 1951). Many species of *Calophyllum* develop characteristic boat-shaped fissures in their bark, and these fissures join and eventually form deep, longitudinally running ridges (Stevens 1974). Nearly all species of *Calophyllum* can be recognized by their leaves. *Calophyllum* species typically have cymose inflorescences. Some species are dioecious with a few hermaphroditic or polygamous flowers. The fruit is a drupe. The mesocarp contains fibers and is shrunken and often wrinkled when dried. In India the pathogen *Colletotrichum gloeosporioides* infects fruits of *C. inophyllum* (Wadia et al. 1984).

Calophyllum species contain many unique chemical compounds (Dharmaratne et al. 1984). *Calophyllum inophyllum* is considered the most beautiful of sea shore trees in the western Pacific (Menninger 1962).

GERMINATION Germination is known for only a few species. It takes one to 2 months (rarely up to 4 months). The radicle may break the wall of the stone to one side of the base as in *C. euryphyllum*, or push out a basal plug as in the case of *C. peekelii*. Seedling emergence is hypogeal (Stevens 1974).

Fruits of *C. brasiliense* var. *antillanum* sown without endocarps had 100% germination after 8 weeks, while fruits with cracked endocarps had 95% germination, and entire fruits had 75% germination (Zentsch and Diaz 1977). All groups reached 100% germination after 16 weeks. The endocarp restricted water uptake. Fruits stored below 0°C and 35% moisture content did not germinate. Transplanted seedlings of *C. brasiliense* were moved from a forest in Belize to beneath and beyond the canopy of *Byrsonima crassifolia* (Kellman 1985). Transplants only survived under the tree canopy.

CALYCANTHACEAE — CALYCANTHUS FAMILY

Calycanthus L. — Sweet-scented shrub

GROWTH HABIT, OCCURRENCE, AND USE The genus *Calycanthus* consists of 4 species of deciduous shrubs native to North America. These shrubs are often grown as ornamentals because of their sweet-scented flowers. Winter buds are small and naked. The flowers are large, and terminal on leafy branches. The sepals and petals are brownish purple; there are many stamens.

Carolina allspice (*C. floridus*) is a shrub with branches and leaf petioles pubescent. It reaches 3 m in height. The leaves are 12.5 cm long, dark green above and pale gray-green and densely pubescent below.

FLOWERING AND FRUITING The flowers of Carolina allspice are about 5 cm across, fragrant, and dark reddish brown. The fruits consist of many one-seeded achenes completely enclosed in a receptacle.

COLLECTION, EXTRACTION, AND STORAGE OF SEED Fruits can be collected in October or November in the north, and as early as August in Georgia. The urn-shaped receptacle should be collected when the color changes from green to brown. Seed collected early when the seedcoat can still be broken with a fingernail will germinate without pretreatment. If fruit collected when the seedcoat is hard, prechilling for 2 to 3 months will be required for germination (Dirr and Heuser 1987).

Cuttings of Carolina allspice taken in July and treated with 8000 ppm of IBA-talc readily rooted. Other species, such as (*C. fertillis*), will root from mid-July cuttings that are slightly wounded.

THEACEAE — TEA FAMILY

Camellia L.

GROWTH HABIT, OCCURRENCE, AND USE Camellias are associated with the culture of Japan, but 70% of the 90 or more known species have been found in China and the Indochina peninsula. They are woody plants that become small trees. Generally these plants have beautiful, glossy leaves. Their native habitat is mountainous, subtropical forest, where they grow in partial shade.

The flowers of most wild species of *Camellia* are neither large nor spectacular, often being 4 cm or less in diameter and plain white in color. Even smaller are the flowers of the most widely cultivated camellia, *C. sinensis*, which is the major source of tea. Most of the ornamental *Camellia* species descend from wild *C. japonica*, a rather scraggly looking tree 15 m height, found naturally in Japan, China, and Korea. Camellias are grown as ornamental species in coastal Australia, New Zealand, the southeastern United States, cis-montane California and Oregon, and in parts of western Europe warmed by the Gulf Stream.

The third most widely cultivated species is *C. sasanqua* a slender, densely-foliaged tree that grows to 5 m in height. This species is native to southern Japan and nearby islands, where it was originally cultivated for the oil extracted from its fruits.

The fourth most widely cultivated species is *C. reticulata*, found naturally in the forest of southern China at elevations of 2000 to 3000 m. It is an open-growing tree up to 16 m in height, with large, heavily veined leaves and pink flowers up to 9 cm in diameter in the wild.

FLOWERING AND FRUITING The camellia fruit is a loculicidal, woody capsule, with a persistent central axis. The few seeds are subglobose or angular in shape.

GERMINATION Propagation techniques for *Camellia* species follow (Dirr and Heuser 1987):

Species	Propagation technique
C. japonica	Seeds: Collect in fall before seedcoat hardens. Dry seed should be soaked in hot water for 24 hours. One nursery prechills seeds until radicle emerges.
	Cuttings: Very difficult to root; patience required. The best time to take cuttings is when wood turns color from green to brown.
	Grafting: pot grafting in January-February gives best results.
	Tissue culture: Plantlets have been produced from callus derived from excised cotyledons.
C. sasanqua	Seeds: Collect in fall before seedcoat hardens. Dry seed should be soaked in hot water for 24 hours. One nursery prechills seeds until radicle emerges.
	Cuttings: Treat mid-January cuttings with 3000 ppm IBA.
	Tissue culture: Fresh explants from juvenile material initiate shoots.
C. sinensis	Seeds: Prechill at 5 to 10°C for 3 weeks. It may be necessary to remove seedcoat.
	Cuttings: Treat July–August cuttings with 1000 ppm IBA-solution.

BIGNONIACEAE — TRUMPET CREEPER FAMILY

Campsis radicans (L.) Seem. — Common trumpet creeper

GROWTH HABIT, OCCURRENCE, AND USE Common trumpet creeper (*C. radicans*), a deciduous vine, is native from Texas and Florida north to Missouri, Pennsylvania, and New Jersey (Bonner 1974a). It has been introduced into New England. The vine is sometimes used in erosion control, but its greatest value is as a wildlife food.

FLOWERING AND FRUITING The large, orange-to-scarlet, perfect flowers are 5 to 9 cm long. They appear from May through September. The fruit is a 2-celled, flattened capsule about 5 to 15 cm long (Fig. 1) that matures from September to November. The small, flat, winged seeds (Figs. 1, 2) are dispersed chiefly by wind as the mature capsules split open on the vine in October through December. The capsules turn from green to gray-brown as they mature. Good seed crops are borne annually.

COLLECTION, EXTRACTION, AND STORAGE OF SEED Ripe capsules should be gathered when they turn grayish brown in the fall before splitting open. Seeds can be extracted by hand-flailing. One sample yielded 300 seeds per gram (Bonner 1974a), with 98% purity and 52% soundness. The longevity of common trumpet creeper seeds in storage is not known.

GERMINATION The seeds exhibit some embryo dormancy. Prechilling for 60 days at 5°C is recommended. Germination tests in sand have been run for 30 days at 20/30°C. Germination is epigeal (Fig. 3).

NURSERY AND FIELD PRACTICE Seedlings can be grown in nursery beds either from untreated seeds sown in the fall or from prechilled seeds sown in the spring. Cuttings are easily rooted from June to September from soft wood (Dirr and Heuser 1987).

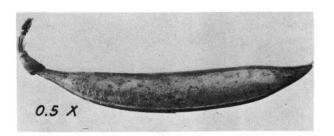

0.5 X

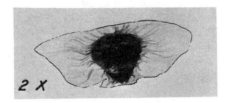

2 X

Figure 1. *Campsis radicans*, common trumpet creeper, fruit, ×0.5, and seed ×2 (Bonner 1974a).

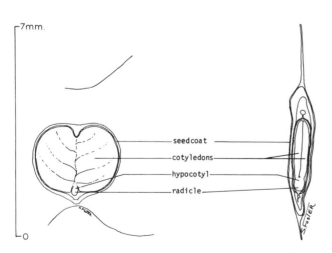

7mm.

seedcoat
cotyledons
hypocotyl
radicle

Figure 2. *Campsis radicans*, common trumpet creeper: longitudinal section through a seed, ×8 (Bonner 1974a).

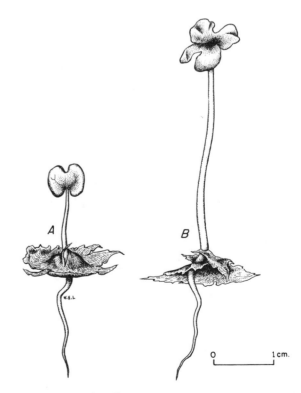

A   B

0   1 cm.

Figure 3. *Campsis radicans*, common trumpet creeper: seedling development at 1 and 9 days after germination (Bonner 1974a).

BURSERACEAE — TORCHWOOD FAMILY

Canarium L.

GROWTH HABIT, OCCURRENCE, AND USE The genus *Canarium* comprises about 150 species native to Africa, Indo-Malaysia, tropical Australia, and Polynesia. These species are large trees up to 30 m high with medium buttresses and gray bark. The leaves are large, up to 45 cm long, with 3 to 5 pairs of leaflets. Species in Java are used as fast-growing shade trees.

FLOWERING AND FRUITING The flowers are 5 cm in diameter, pale yellowish white in color, and fragrant. They are unisexual, with 3 sepals and 6 stamens in the male flower, and an ovary with 3 cavities in the female.

The fruit is faintly 3-angled, greenish at first and turning bluish black with maturity. The stone contains a single oily kernel valued as a food and a source of cooking oil.

GERMINATION Seeds of *C. megalanthum* had 95% germination after 4 weeks incubation in a test conducted in Malaysia (Ng and Sanah 1979).

LEGUMINOSAE — LEGUME FAMILY

Caragana arborescens Lam. — Siberian peashrub

GROWTH HABIT, OCCURRENCE, AND USE Siberian peashrub, sometimes called caragana or peatree, is one of the most hardy small deciduous trees or shrubs planted on the northern Great Plains (Dietz and Slabaugh 1974). Introduced into the United States in 1752, peashrub is native to Siberia and Manchuria and occurs from southern USSR to China. Botanical varieties include the dwarf form (*Caragana arborescens* var. *nana*) and Lorberg peashrub (*C. arborescens* var. *pendula*). Several other species of *Caragana* are adapted to North America, but are largely restricted to botanical gardens and plant material centers. These include Russian peashrub (*C. frutex*), Maximowicz peashrub (*C. maximowiciziana*), littleleaf peashrub (*C. microphylla*), pygmy peashrub (*C. pygmaea*), and Chinese peashrub (*C. sinica*). *Caragana* readily adapts to sandy, alkaline soil and open, unshaded sites on the northern Great Plains where it grows to heights of 2.5 m. It is extensively used for shrub buffer strips and windbreaks on farmlands and for hedges and outdoor screening in

Figure 1. *Caragana arborescens*, Siberian peashrub: fruit, ×2 (Dietz and Slabaugh 1974).

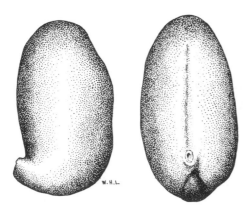

Figure 2. *Caragana arborescens*, Siberian peashrub: seeds, ×11 (Dietz and Slabaugh 1974).

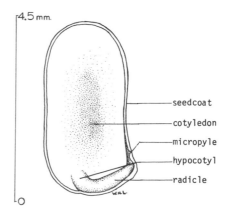

Figure 3. *Caragana aborescens*, Siberian peashrub: longitudinal section through a seed, ×11 (Dietz and Slabaugh 1974).

many towns and cities of the upper Midwest.

This genus has had limited use for wildlife and erosion plantings in the Great Lake states. It has shown promise for deer-range revegetation programs in the Black Hills of South Dakota. Siberian peashrub is one of the few exotic shrubs that has shown promise for use in riparian habitat restoration in the Great Basin.

FLOWERING AND FRUITING The yellow bisexual flowers appear from April to June. The fruit is a pod 2.5 to 5 cm long (Fig. 1), and contains approximately 6 reddish brown, oblong to spherical seeds about 2.5 to 4 mm in diameter (Figs. 2, 3). Fruits ripen to amber or brown from June to July. Seed dispersal is usually completed by mid-August in most areas on the Great Plains. Shrubs take about 3 to 5 years to reach commercial seed-bearing age, and good crops occur nearly every year.

COLLECTION, EXTRACTION, AND STORAGE OF SEED The optimum seed-collection period for Siberian peashrub is less than 2 weeks, usually in July or early August. Since the seeds are ready to collect as soon as the fruit ripens, the pods should be gathered from the shrub by hand as soon as they begin to open.

Pods should be spread out to dry in a protected area until they pop open. The seeds are then easily extracted by light sifting, beating, and air screening. The average number of cleaned seeds per gram is 41. The normal yield range is 28 to 44 kg of cleaned seed per 100 kg of fresh fruit.

Seeds have been successfully stored in sealed containers at low temperatures. Studies in Canada have shown that Siberian peashrub seeds remain viable for at least 5 years when stored dry at room temperature. Germination capacity of seeds stored 5 years was 93%, compared to 94% for seeds stored 1- and 2-year periods.

PREGERMINATION TREATMENT Germination has been enhanced by prechilling seeds in moist sand (5 to 10% moisture content) for 15 days at 5°C, or in vermiculite for 40 days. Prechilled seeds can be dried at 32°C for 3 hours without subsequent loss of germination.

GERMINATION TEST Germination tests can be conducted at 20/30°C on paper or sand for 14 days with prechilled seeds. Dropping the seeds in hot water (65°C) and allowing the water to cool for 24 hours has been reported to enhance germination (Dirr and Heuser 1987).

NURSERY AND FIELD PRACTICE Seed can be drilled or broadcast in late spring. Dry seed sown in the spring has been reported to germinate more promptly if soaked for 10 to 12 hours before sowing. Many nursery growers recommend 50 to 160 seeds per meter of row. Planting depths range fron 0.6 to 1.25 cm. It may be important to inoculate the seed with *Rhizobia*.

Siberian peashrub can be seeded on the Great Plains with a cover crop of oats during the last week of July or the beginning of August. The shrubs are large enough to lift the following fall. To be large enough to transplant, the seedlings should be 30 cm or more in height at the time of lifting.

Cuttings taken in late July root well without treatment. IBA will injure cuttings (Dirr and Heuser 1987). The more desirable horticultural varieties are top grafted on *C. arborescens* seedlings.

RHIZOPHORACEAE — MANGROVE FAMILY

Carallia Roxb.

GROWTH HABIT, OCCURRENCE, AND USE *Carallia* is a genus of about 20 species native to Madagascar, tropical Asia, and Australia. These evergreen trees reach 15 to 20 m in height. The genus is common in damp coastal areas in southeast Asia.

FLOWERING AND FRUITING The flowers are 0.6 cm in diameter, pale greenish in color, and borne on short stalks in clusters. The fruit is 0.6 cm in diameter and rounded in outline, translucent pink in color, and with persistent sepals at the top.

GERMINATION Seeds of *C. brachiata* that had been stored for 4 days had 68% germination over a 15-week period in a test conducted in Malaysia (Ng and Sanah 1979).

CARICACEAE — PAPAYA FAMILY

Carica L. — Papaya

GROWTH HABIT, OCCURRENCE, AND USE The genus *Carica* consists of about 25 species native to tropical and subtropical America (Bailey 1951). Papaya (*Carica papaya*) is widely planted in the tropics for its edible fruit. The papaya tree reaches 6 m in height with several erect stems bearing heads of leaves. The trunks are hollow.

FLOWERING AND FRUITING Male papayas bear their creamy-green flowers in dangling racemes, and the female flowers are borne in small, neat clusters. The massive melon-sized fruit is borne close to the trunk among the leaves. The black seeds are borne in the center of the fruit.

SAXIFRAGACEAE — SAXIFRAGE FAMILY

Carpenteria californica Torr.

GROWTH HABIT, OCCURRENCE, AND USE Carpenteria is an erect evergreen shrub, 1 to 2 m high. It is restricted to granite ridges and slopes of the Sierra Nevada foothills between 460 and 1230 m elevation between the San Joaquin and Kings rivers in Fresno County, California. It was introduced into cultivation in 1880 and is a valuable ornamental species (Neal 1974a).

FLOWERING AND FRUITING Flowers are large, white, perfect, and bloom from June to August. Fruits are leathery, beaked, conical, capsule 0.9 to 1.0 cm long.

COLLECTION, EXTRACTION, AND STORAGE OF SEED Fruits hold many small seeds in 5 to 7 cells (Fig. 1). Seeds may be collected from July to October. Cleaned seeds are 3300 to 4700 per gram.

GERMINATION Germination without pretreatment is excellent. This species suckers freely and can be easily propagated from cuttings.

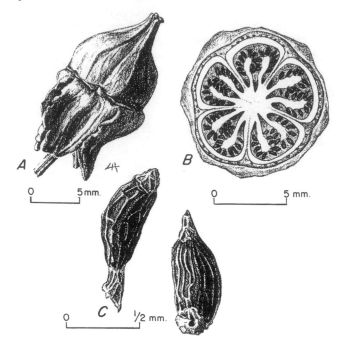

Figure 1. *Carpenteria californica,* carpenteria: (A) exterior view of fruit; (B) cross section of fruit; (C) exterior view of seeds in two planes (Neal 1974a).

BETULACEAE — BIRCH FAMILY

Carpinus L. — Hornbeam

GROWTH HABIT, OCCURRENCE, AND USE The hornbeams include about 26 species of deciduous trees and shrubs native to the Northern Hemisphere from Europe to eastern Asia, south to the Himalayas, and in North and Central America (Rudolf and Phipps 1974). European hornbeam (*Carpinus betulus*) and American hornbeam (*C. caroliniana*) have been planted for conservation purposes in the United Sates, and several species are used in environmental plantings. The hornbeams occur mainly as understory trees in rich, moist bottom lands, coves, and lower protected slopes. The wood of some species is used for specialty purposes, and all species produce wildlife food.

GEOGRAPHIC RACES Several varieties of European hornbeam are recognized, although none appears to be a geographic race. Because of the extensive range of this species, however, geographic races probably have developed (Rudolf and Phipps 1974). Botanists recognize a northern and a southern form of American hornbeam, which differ in bark, leaf, and bract characteristics (Vines 1960).

FLOWERING AND FRUITING Staminate and pistillate flowers appear in the spring on the same tree. Fruits are ribbed nutlets, each with an involucre (Figs. 1, 2), and are borne in clusters. They ripen from late summer to fall. Seeds are dispersed from fall to spring (Table 1); they are blown short distances and are distributed mainly by birds.

COLLECTION, EXTRACTION, AND STORAGE OF SEED The fruits are harvested while they are light greenish brown to brown—before they become dry in late fall—by flailing them onto ground cloths or by handpicking from the trees. They should perferably be picked slightly green, but fully developed in size when collected (Rudolf and Phipps 1974). In Poland, however, late fall (October to November) harvests of well-ripened fruits of European hornbeam have yielded seed of good germinative capacity (Suszka 1968).

After collection, the ripe fruits are spread out in thin layers in a cool, well-aerated room or shed; they should dry superficially and then be placed in a dewinging machine or beaten in bags to separate seeds from involucres. Chaff is removed by screening and air screening. Seed yield is about 50% of fruit weight (Rudolf and Phipps 1974).

Seeds may be partially dried and stored in sealed containers, then removed 100 to 120 days before planting for prechilling. An alternative method is to

Table 1. *Carpinus*: phenology of flowering and fruiting, characteristics of mature trees, cleaned seed weight, pregermination treatments, and germination test conditions and results (Rudolf and Phipps 1974).

Species	Flowering dates	Fruit ripening dates	Seed dispersal dates	Mature tree height (m)	Seed-bearing Age (years)	Seed-bearing Interval (years)	Seeds/gram	Warm pretreatment Temperature (°C)	Warm pretreatment Duration (days)	Prechilling Temperature (°C)	Prechilling Duration (days)	Prechilling Temperature (°C)	Prechilling Duration (days)	Germination
C. betulus	Apr.–May	Aug.–Nov.	Nov.–Spring	20	10–30	1–2	28	15	28	5	98	15	70	18–90
											–	–	5	180
C. caroliniana	Mar.–June	Aug.–Oct.	Nov.–Spring	12	15	3–5	66	15/30	60	5	60	12/28	60	1–5

prechill and then store. European hornbeam seeds dried to 10% moisture content were stored in sealed containers at 3°C for 14 months with no loss of viability.

GERMINATION Dormancy, apparently caused by conditions in the embryo and endosperm, may be overcome by prechilling (Table 1). Germination tests have been made in sand or sand plus peat mixtures (Table 1). Rudolf and Phipps (1974) suggested a new test was needed for seeds of American hornbeam. Considering the variability in results for seeds of European hornbeam (18 to 90%), both tests could stand improvement.

NURSERY AND FIELD PRACTICE The optimum seedbed is continuously moist, rich, loamy soil protected from extreme changes in temperature. The natural germination of seeds in the wild is often delayed until the second year after dispersal. For good germination in the first year, seeds should be collected while still green and sown immediately or prechilled over winter and sown the next spring. This prolonged prechilling over winter seems to be in conflict with the prechilling periods previously given. More modern comments on propagation techniques are shown in Table 2.

Table 2. *Carpinus*: propagation techniques (Dirr and Heuser 1987).

Species	Propagation techniques
C. betulus	Seeds: collect green in September, sow immediately or prechill 3–4 months. Soak dry seeds 6 hours, those that float are not viable. Cuttings: treat July cuttings with high concentrations of IBA. Cuttings require dormancy period after rooting.
C. caroliniana	Seeds: germination generally easier than with *C. betulus*. Green seeds planted immediately after collection had 24% emergence. Prechilling for 15–18 weeks increased emergence to 43–58%. Dry seed requires 2 months warm incubation before prechilling. Cuttings: no information.
C. cordata	Seeds: prechill 3–4 months or plant as collected. Cuttings: treat June cuttings with 3% IBA-talc.
C. japonica	Seeds: collect in mid-November and prechill 4 months. Cuttings: treat July cuttings with high concentrations of IBA. Cuttings require dormancy period after rooting.
C. laxiflora	Seeds: collect in mid-September and prechiull 3–4 months. Cuttings: treat July cuttings with high concentrations of IBA. Cuttings require dormancy period after rooting.
C. orientalis	Seeds: untreated seeds do not germinate. Prechill 4 months or warm incubate and then prechill. Cuttings: no information.
C. turczaninovii	Seeds: 2 months prechilling produced 60% germination. Cuttings: treat June cuttings with 3% IBA-talc.

Seeds should be sown at a density of 300 to 400 per m² and covered with 1.5 cm of soil. Fall-sown beds should be mulched with burlap, pine needles, or other material to protect from frost. Beds need to be shaded the first year. Field planting is normally done with 2-0 stock.

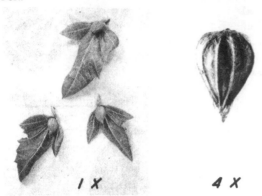

Figure 1. *Carpinus caroliniana*, American hornbeam: 3 nutlets with involucre, ×1; and a nutlet with involucre removed ×4 (Rudolf and Phipps 1974).

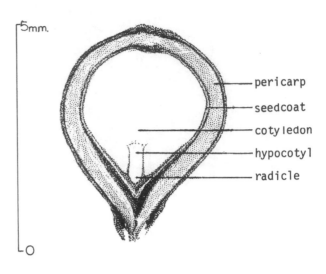

Figure 2. *Carpinus caroliniana*, American hornbeam: longitudinal section through a nutlet, ×12 (Rudolf and Phipps 1974).

JUGLANDACEAE — WALNUT FAMILY

Carya Nutt. — Hickory

GROWTH HABIT, OCCURRENCE, AND USE   There are about a dozen hickories native to the United States. Pecan (*Carya illinoensis*) is widely grown as a nut crop in the South and Southwest. To a lesser extent, shellbark hickory (*C. laciniosa*) and shagbark hickory (*C. ovata*) are also grown for nut production (Bonner and Maisenhelder 1974a).

FLOWERING AND FRUITING   Hickories are monoecious and flower in the spring (Table 1): The staminate flowers develop from axils of leaves of the previous season or from inner scales of the terminal buds at the base of the current growth. The pistillate flowers appear in short spikes on peduncles terminating in shoots of the current year. *Carya* fruits are ovoid, globose, or pear-shaped nuts enclosed in husks developed from the floral involucre (Fig. 1). Husks are green prior to maturity; they turn brown to brownish black as they ripen. They become dry at maturity and split away from the nut into 4 valves along sutures. The husks of mockernut (*C. tomentosa*), shellbark, pecan, shagbark, and nutmeg hickory (*C. myristicaeformis*) split at the base at maturity, usually releasing the nut. Husks of pignut (*C. glabra*), bitternut hickory (*C. cordiformis*), and water hickory (*C. aquatica*) split only in the middle, or slightly beyond, and generally cling to the nuts. The nut is 4-celled at the base and 2-celled at the top. The bulk of the edible embryonic plant is cotyledonary tissue (Fig. 2).

COLLECTION, EXTRACTION, AND STORAGE OF SEED   Nuts can be collected from the ground after natural seedfall or after shaking the trees or flailing the limbs. Persistent husks may be removed by hand, or by running the fruits through a corn sheller. Shagbark and shellbark hickory have been known to produce 425 to 570 and 570 to 850 liters of nuts respectively per tree. Good crops of all species are produced at intervals of one to 3 years (Table 1). Nuts can be stored 3 to 5 years in closed containers at 5°C and 90% relative humidity (Bonner and Maisenhelder 1974a). It is now generally considered, however, that this storage method is not correct. Pecan and shagbark hickory nuts should be stored at 3°C and 5% moisture (Bonner 1976). The nuts of *Carya* species are orthodox as far as storage is concerned, which means they can be stored at low moisture contents. Bonner and Turner (1980) developed a microwave oven procedure for rapid determinations of seed moisture content of hickory fruits.

PREGERMINATION TREATMENT   Hickories exhibit embryo dormancy, which can be overcome by prechilling in a moist medium at 2 to 5°C for 30 to 150 days (Table 1). Naked stratification in plastic bags is suitable for most species. Seeds in storage for a year or more may require only 30 to 60 days prechilling (see below for an update on prechilling). Pit stratification with about 0.5 m of compost, leaf, or soil cover to prevent freezing will suffice. Prior to the cold treatment, nuts should be soaked in cold water at room temperature for 2 to 4 days with 1 or 2 changes of water per day.

Pecan seed germination and dormancy have been recently reviewed by Dimalla and Van Staden (1977).

C. aquatica
water hickory

C. cordiformis
bitternut hickory

C. glabra
pignut hickory

C. myristicaeformis
nutmeg hickory

C. illinoensis
pecan

C. laciniosa
shellbark hickory

C. ovata
shagbark hickory

C. tomentosa
mockernut hickory

5   10   15 cm.

Figure 1. *Carya*: hickory nuts with husks attached and removed (×1). Size and shape of the nuts may vary greatly within each species and may differ from the examples shown here (Bonner and Maisenhelder 1974a).

Table 1. *Carya*: phenology of flowering and fruiting, characteristics of mature trees, fruit and cleaned seed weight, pregermination treatments, and germination test conditions and results (Bonner and Maisenhelder 1974a).

Species	Flowering dates	Fruit ripening dates	Seed dispersal dates	Mature tree height (m)	Year first cultivated	Seed-bearing Age (years)	Seed-bearing Interval (years)	Fruits/ liter	Seeds/ fruits	Seeds/kg	Prechilling (days)	Substrata	Temperature (°C)	Germination (%)
C. aquatica	Mar.–May	Sept.–Nov.	Oct.–Dec.	30.8	1800	20	1–2	–	–	360	90	S	20/30	92
C. cordiformis	Apr.–May	Sept.–Oct.	Sept.–Dec.	30.8	1689	30	3–5	–	160	340	90	S	20/30	55
C. glabra	Apr.–May	Sept.–Oct.	Sept.–Dec.	28.0	1750	30	1–2	–	165	440	120	S	20/25	60
C. illinoensis	Mar.–May	Sept.–Oct.	Sept.–Dec.	43.0	1766	10–20	1–2	12.5	–	220	90	S	20/30	50
C. laciniosa	Apr.–June	Sept.–Nov.	Sept.–Nov.	36.9	1800	40	1–2	–	138	66	120	S	20/30	–
C. myristicaeformis	Apr.–May	Sept.–Oct.	Sept.–Dec.	30.8	–	30	1–2	18.0	–	270	120	C	20/30	60
C. ovata	Apr.–June	Sept.–Oct.	Sept.–Dec.	30.8	1911	40	1–3	21.8	69	220	150	S	20/30	80
C. tomentosa	Apr.–May	Sept.–Oct.	Sept.–Dec.	30.8	1766	25	2–3	6.3	143	200	150	S	20/30	60

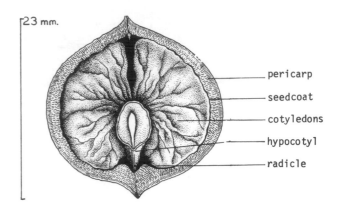

Figure 2. *Carya ovata*, shagbark hickory: longitudinal section through the embryo of a nut with husk removed, ×2 (Bonner and Maisenhelder 1974a).

Because of the economic importance of pecan as a nut crop, there has been considerable research on its seed maturation and germination (Wood 1984). Not only does prechilling at 3 to 5°C enhance germination of dormant seeds of the hickories, it also speeds and makes more uniform the germination of nondormant seeds (Table 2). Bonner (1976) recommended that fresh pecan seeds should be prechilled for 30 to 60 days. Fresh shagbark hickory nuts and those stored up to 2 years need 90 to 120 days of prechilling; prechilling can be reduced to 60 days for seeds stored over 2 years.

Table 2. *Carya*: laboratory germination methods for nondormant and dormant seeds (based on Genebank Handbook 1985 and ASOA 1985).

Species	Germination techniques
C. aquatica	Nondormant seeds: alternating temperatures 21/32°C, in soil. Dormant seeds: prechill 90–120 days at 0–3°C, then as above.
C. cordiformis	Nondormant seeds: alternating temperatures 20/30°C, in sand, peat, or soil. Dormant seeds: prechill 90–120 days at 0–7°C then as above.
C. glabra	Nondormant seeds: alternating temperatures 20/30°C, in sand, peat, or soil. Dormant seeds: prechill 90–120 days at 0–5°C, germinate at 21/27°C.
C. illinoensis	Nondormant seeds: alternating temperatures 20/30°C, on top of paper. Dormant seeds: prechill 30 days at 3°C, then as above.
C. laciniosa	Nondormant seeds: alternating temperatures 20/30°C, in sand, soil, or peat. Dormant seeds: prechill 90–120 days at 0–5°C, then as above.
C. myristicaeformis	Nondormant seeds: alternating temperatures 20/30°C, between paper. Dormant seeds: prechill 60–120 days at 0–5°C, then as above.
C. ovata	Nondormant seeds: alternating temperatures 20/30°C, between paper. Dormant seeds: prechill 90–150 days at 3–7°C, then as above.
C. tomentosa	Nondormant seeds: alternating temperatures 20/30°C, in sand, soil, or peat. Dormant seeds: no information.

There are other methods of speeding the germination process for pecan nuts. It has been shown that prechilling does not always produce germination for this species, even when total germination capacity is increased (Adams and Thielges 1978). Research in South Africa suggest that delay in germination of pecans is caused by mechanical restriction of the shell, which is easily overcome by increasing the incubation temperature or artificially heating germination beds to 30 to 35°C (van Staden et al. 1976). On the other hand, researchers attempting to grow northern cultivars of pecans in the South for use as rootstock have found it necessary to prechill the seeds (Madden and Tisdale 1975).

GERMINATION TEST Adequate germination tests can be made on prechilled nuts in flats of sand, peat, or soil, or on thick layers of moist kimpack, creped cellulose, or similar material, at diurnally alternating temperatures of 20/30°C (Table 1). Viability can be checked with a tetrazolium test. Modern rules for test germination of the nuts of hickory species are shown in Table 2.

NURSERY AND FIELD PRACTICE Either fall sowing with untreated seed or spring sowing with prechilled seed may be used. Excellent results with fall sowing have been reported for shagbark hickory, but good mulching is necessary. Drilling rows 20 to 30 cm apart and with planting 20 to 25 nuts per meter of row aare recommended; drilling depth should be 2 to 3.75 cm. Mulch should remain until germination is complete. Shading is generally not necessary, but shellbark hickory seedlings may profit from it. Protection from rodents may be required for fall sowing.

Apparently cuttings of *Carya* species are difficult to root (Dirr and Heuser 1987). Success with pecan has been minimal, and successful cuttings apparently often do not overwinter. Generally, juvenile wood produces more roots than mature wood when treated with a 2% IBA-solution.

Pecans are also difficult to graft. A bud patch is recommended. This can be accomplished in August in the field. Lateral bud stem section explants through tip culture have led to successful tissue culturing.

VERBENACEAE — VERBENA FAMILY

Caryopteris Bunge — Bluebeard

GROWTH HABIT, OCCURRENCE, AND USE About a half-dozen species of deciduous shrubs native to eastern Asia constitute the genus *Caryopteris*. These shrubs are widely planted as ornamental (Bailey 1951). They have short-petioled, opposite leaves.

FLOWERING AND FRUITING The flowers are borne in many-flowered cymes. Each flower consists of a 5-lobed corolla, with a short cylindrical tube and spreading limb. One segment of the corolla is larger and fringed. The fruit is a dry capsule; its 4 valves separate into 4 somewhat-winged nutlets.

GERMINATION Seeds of common bluebeard (*Caryopteris incana*) germinate without pretreatment (Dirr and Heuser 1987). The seeds should be collected in late summer or early fall, and dried under cover before being stored in sealed containers. Seeds should be planted in late spring.

Cuttings of bluemist shrub (*C.* × *clandonensis*), taken as soon as new growth is substantial and treated with 1000 ppm IBA-solution, root readily (Dirr and Heuser 1987).

PALMAE — PALM FAMILY

Caryota L.

GROWTH HABIT, OCCURRENCE, AND USE *Caryota* is a genus of tall palm trees that reach 18 m in height. The group is characterized by leaves with few, broad, bipinnate leaflets. The genus is native to India, southern Asia, and Australia.

FLOWERING AND FRUITING Flowers are usually in groups of 3—the central and lowest is female and the others male. The globose fruit contains 1 or 2 seeds.

GERMINATION Tukas (*C. mitis*) fruits had 45% germination without pretreatment in a test conducted in Malaysia (Ng and Sanah 1979). Seeds of fish tail or toddy palm (*C. urens*) are difficult to germinate and may require between 6 and 7 months before they will germinate either in greenhouse sowing or at 20/30°C in moist sand (Loomis 1958, 1961; Ishihata 1974; Bonker 1976).

LEGUMINOSAE — LEGUME FAMILY

Cassia L. — Senna

GROWTH HABIT, OCCURRENCE, AND USE The genus *Cassia* consists of about 400 species native to tropical or cool temperate regions. *Cassia beareana* is a tree native to tropical Africa that reaches 10 m in height. Golden-shower (*C. fistula*) also reaches 10 m in height; it is native to India and has become naturalized in the West Indies. Partridge pea (*C. fasciculata*) is a herbaceous species of *Cassia* found from New England to Florida and Texas. Wild senna (*C. marilandica*) is a species that only reaches 1 m in height; it is found in the eastern United States.

FLOWERING AND FRUITING Flowers are nearly regular, and solitary or borne in racemes or panicles. The calyx teeth are nearly equal in length and typically longer than the floral tube. The corolla consists of 5 spreading claw petals, nearly equal in length. The fruit is a pod (legume) that can be either sessile or stalked.

GERMINATION As frequently happens with seeds of legumes, the seeds of most *Cassia* species have hard seedcoats that require scarification. Sheikh (1980) compared acid scarification and hot water soaking for enhancing germination of seeds of *C. fistula*. Acid scarification for 30 minutes in concentrated H_2SO_4 was the most effective treatment. Khan et al. (1984) determined that the untreated seeds of *C. holosericea* had virtually no germination. Acid scarification increased germination to 56% and seeds mechanically scarified had 96% germination.

Goor and Barney (1976) reported that *C. siamea*, which is native to Malaysia and widely used in afforestation work, has 30,000 to 35,000 seeds per kg, and commonly has 75 to 80% germination after acid scarification for 20 minutes.

FAGACEAE — BEECH FAMILY

Castanea Mill. — Chestnut

GROWTH HABIT, OCCURRENCE, AND USE *Castanea* is a genus of small- to medium-sized trees with about 11 species found in southwestern and eastern Asia, southern Europe, north Africa, and the eastern United States (Sander 1974a). American chestnut (*C. dentata*) was formerly one of the most valuable timber species in the Appalachian region. Its nuts were an important wild-life food and were extensively marketed for human consumption as well. In the years since the chestnut blight was discovered in New York in 1904, this disease has spread throughout the range of the American chestnut and completely destroyed it as a commercial species. A limited amount of seed is still obtained from sprouts of blight-killed trees, but these sprouts seldom live long enough to produce many nuts. Propagation of the American chestnut is almost futile anywhere within its natural range, and although the search for resistant trees started in 1918 and continues, none has been found.

Japanese chestnut (*C. crenata*) and Chinese chestnut (*C. mollissima*), both highly resistant to the blight, and European chestnut (*C. sativa*) have been used in an attempt to replace American chestnut as a timber tree and source of nuts. Chinese chestnut is the most promising and has been widely planted throughout the eastern United States, mostly in orchards for nut production. The chestnuts are also useful ornamental trees in lawns and parks.

There are no named superior cultivars of the American chestnut. There are, however, at least four superior strains of Chinese chestnut: Abundance, Kuling, Meiling, and Nanking. These have been selected for the quality of their nuts and are extensively propagated for orchard purposes.

Much hybridization work has been done to find a blight-resistant hybrid with forest tree form and fast growth that will replace American chestnut. Some progress has been made and a few crosses have produced promising offspring.

FLOWERING AND FRUITING *Castanea* is monoecious, but individual trees are largely self-sterile and cross-pollination is necessary to ensure good seed crops. Two distinct types of flowers are borne on the present year's growth. Unisexual male catkins appear near the base of the flowering branch, while nearer the apex, bisexual catkins can be found. The pistillate flowers occur singly or in clusters of 2 to 3, near the base of the bisexual flowers. The rest of the catkin bears male flowers that reach full bloom 2 to 3 weeks after those on the unisexual catkin. The pistillate flowers reach full development sometime between these two periods of male flowering. Flowering begins in April or May in the southeastern United States and in June in the northeastern states.

The fruiting structure is a very spiny, globose involucre (bur) that encloses 1 to 3 true nuts (Fig. 1). The nuts begin to ripen in August or September in the southeastern United States, while in the Northeast fruit ripening begins in September or October. There is much variation in the size, color, and yield of nuts,

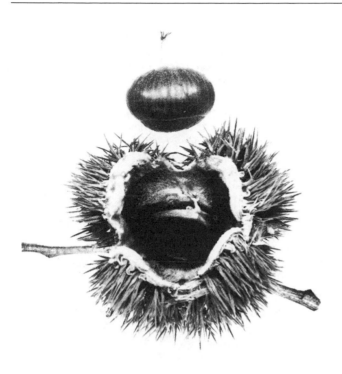

Figure 1. *Castanea dentata*, American chestnut: fruit (bur) and nut, ×1 (Sander 1974a).

even among trees grown from seed of an individual tree.

The nuts are generally somewhat flattened and range from light to dark brown; sometimes they are nearly black. The nuts of American chestnut are small, 1.25 to 2 cm wide and 2.5 cm long. The exotic species bear larger nuts that are 2 to 3 cm across. The embryo has large cotyledons and the seed contains no endosperm (Fig. 2).

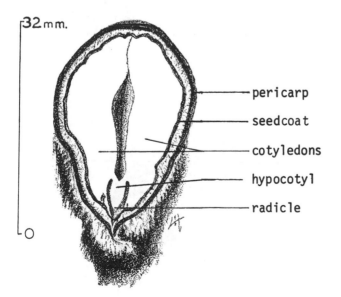

Figure 2. *Castanea dentata*, American chestnut: longitudinal section through a nut, ×1.8 (Sander 1974a).

COLLECTION, EXTRACTION, AND STORAGE OF SEED Chestnut fruits are perishable and must be harvested promptly when ripe. Harvesting should begin as soon as the burs begin to split open. Fallen nuts or burs may be gathered from the ground beneath the trees, or the burs may be picked from the trees. The mature nuts should be gathered at least every other day. This frequent collection is especially important if the weather is hot and dry. Within a week, nuts on the ground or in open burs on the tree may become dry and hard and lose their viability (Table 1).

Table 1. *Castanea:* mature tree characteristics and seed weight (Sander 1974a).

Species	Mature tree height (m)	Year first cultivated	Seeds/kg
C. crenata	10.8	1876	33
C. dentata	43.0	1800	288
C. mollissima	21.5	1853	135
C. sativa	21.5	1880	33

Because of this perishable nature, chestnuts must be sown or stored promptly after harvesting to maintain viability. Freshly harvested chestnuts should be cured by spreading them in thin layers in wire trays and keeping them out of direct sunlight in a well-ventilated place. Curing time will vary with the humidity; usually 1 to 7 days at 15 to 22°C will be adequate. Fresh nuts have been stored over winter in a moist medium at −1 to 3°C as a form of prechilling. Moisture and humidity during storage are critical. Moisture content of nuts should be about 40 to 45%, and the relative humidity in the storage container should be maintained at close to

Table 2. *Castanea:* propagation techniques for nurseries (Dirr and Heuser 1987).

Species	Propagation techniques
C. dentata	Seeds: prechill 3 months.
	Tissue culture: cambial tissue explants through callus have produced budlike structure.
C. mollissima	Seeds: germinate without pretreatment, but prechilling 1–3 months is optimum. Seed weevils may destroy embryo, hot water treatment (50°C) for 30 minutes will control.
	Cuttings:most attempts have failed, but juvenile material treated with growth regulators will rot.
	Grafting: difficult, but pot grafting January–February and dormant grafting in field have worked.
	Tissue culture: axillary shoot proliferation has been attained.
C. sativa	Seeds: no information.
	Cuttings: mature cuttings difficult, but juvenile cuttings can be rooted.
	Grafting: numerous approaches work—cleft-tongue grafting, or chip, patch, or shield budding.
	Tissue culture: lateral buds from seedlings develop axillary shoot. Cotyledonary tissue has also been successfully cultured.

70%. If the nuts get too dry, they become hard and lose their viability; if too much moisture is present they will mold or decay.

Although chestnuts have been stored for a year or more, it is generally not advisable to store them for more than 6 to 8 months, because of losses from decay. Nuts stored for longer periods may also begin to germinate while in storage.

PREGERMINATION TREATMENT Chestnut seed requires a period of prechilling to overcome seed dormancy before it will germinate. In normal practice, over-winter storage under cold, moist conditions will be more than adequate to overcome dormancy. If seeds are planted in the fall, no prechilling is necessary, but nuts should be kept in cold storage until planted.

GERMINATION TEST Prechilled nuts of American chestnut have been germinated in a moist medium at temperatures from 15 to 22°C. Germinative energy after 28 days was 92% and germination capacity after 42 days was 100%. Rules for testing chestnut seeds recommend presoaking the nuts for 48 hours, cutting off one-third of the seed at the end with the scar, removing the testa, and germinating the seed segments in moist sand at 20/30°C for a 21 days.

NURSERY AND FIELD PRACTICE Chestnuts may be planted in either the fall or spring. Fall planting should be done in September or October with maturee nuts that have been cured and kept in cold storage from the time they are harvested until they are planted. Nuts planted in the spring should be prechilled over winter and planted as early as the soil can be worked.

Nuts planted in fall or spring should be placed in beds 2.5 to 5 cm deep and spaced 7.5 to 10 cm apart in the nursery beds. Fall-planted beds should be mulched with 2.5 to 10 cm of coarse straw. The mulch should be removed in the spring when the nuts start to germinate. Rodents can remove many chestnuts (Khutortsov and Anikeeko 1981), and seedlings must be protected. Generally about 80% of the nuts sown produce seedlings. Seedlings can be transplanted after one year. Propagation techniques are presented in Table 2.

FAGACEAE — BEECH FAMILY

Castanopsis Spach — Chinkapin

GROWTH HABIT, OCCURRENCE, AND USE The genus *Castanopsis* includes about 30 species of evergreen shrubs and trees (Hubbard 1974a). Golden chinkapin (*C. chrysophylla*) and Sierra chinkpin (*C. sempervirens*) are native to the United States and found in the mountains of the Pacific Coast. The remaining species are native to the mountains of southern and eastern Asia.

Golden chinkapin reaches 15 to 46 m in height. *Castanopsis chrysophylla* var. *minor* is an evergreen shrub that reaches 1 to 3 m in height. It is found at higher elevations and in drier habitats than is the typical variety. In certain localities, plants are found varying from shrubs to large trees. Sierra chinkapin is a spreading shrub 0.3 to 2.5 m high at maturity. Several species of chinkapin have been cultivated since 1845 for ornamental uses, some species produce lumber, some are useful for erosion control, and most of them provide nuts which are food for wildlife. Both the shrubby chinkapins are considered low value browse species in the western United States. The Sierra chinkapin is widely used in environmental plantings in the Lake Tahoe basin.

FLOWERING AND FRUITING Chinkapin flowers are unisexual, with staminate and pistillate flowers on the same plant. The staminate flowers are borne in groups of 3 in the axils of bracts, forming densely flowered, erect, cylindrical catkins 0.4 to 7.5 cm long. One to three pistillate flowers are borne in an involucre, usually at the base of the staminate catkins or borne in short separate catkins. The fruit consists of 1 to 3 nuts (Figs. 1, 2) enclosed in a spiny bur. The nuts mature in the fall of the second year (Hubbard 1974a):

Species	Flowering dates	Fruit ripening dates	Seed dispersal dates
C. chrysophylla	June–Feb.	Aug.–Sept.	Fall
var. *minor*	June–Sept.	Sept.–Oct.	Fall
C. sempervirens	July–Aug.	Sept.–Oct.	Fall

Figure 1. *Castanopsis sempervirens*, Sierra chinkapin: nuts, ×2 (Hubbard 1974a).

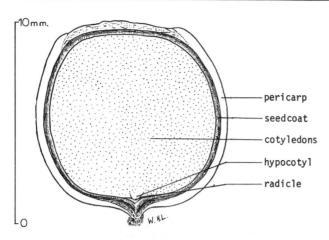

Figure 2. *Castanopsis chrysophylla,* golden chinkapin: longitudinal section through a nut, ×5 (Hubbard 1974a).

COLLECTION, EXTRACTION, AND STORAGE OF SEED Collectors should handpick burs in late summer or early fall, after they are ripe but before they open. The collected burs should be spread out to dry in the sun or in a warm room. After drying, they may be run through a fruit disintegrator (no description given) or shaker to separate the nuts (Hubbard 1974a). *Castanopsis chrysophylla* yiels 2100 seeds/kg, *C. chrysophylla* var. *minor* yields 1540 seeds/kg, and *C. sempervirens* yields 2640 seeds/kg.

When stored in sealed containers at 5°C, chinkapin seeds retain their viability well for at least 2 years, and probably longer. Viability of one sample of golden chinkapin seed stored in this manner dropped only from 50 to 44% in 5 years.

GERMINATION Germinative capacity of untreated seeds of golden chinkapin in three tests ranged from 14 to 53%. For Sierra chinkapin, germination was 30% in 25 days. Germination is hypogeal (Fig. 3) and usually takes place 16 to 24 days after sowing. Prechilling does not increase germination.

NURSERY AND FIELD PRACTICE Little information is available on nursery practices for chinkapins. Seed has been covered with 5 cm of soil. There are problems of survival after emergence.

Figure 3. *Castanopsis sempervirens,* Sierra chinkapin: seedlings at one month, actual size (Hubbard 1974a).

CASUARINACEAE— CASUARINA FAMILY

Casuarina L.

GROWTH HABITAT, OCCURRENCE, AND USE The genus *Casuarina* includes about 25 species native to Australia and the Pacific Islands. A few species are planted in southern California. Branchlets are very slender and equisetumlike, usually jointed, striated, and more or less deciduous. The leaves are represented by whorls of minute scales. The plants are monoecious or dioecious. The staminate flowers occur in slender spikes that terminate the green or gray branchlets. Pistillate flowers occur in short dense heads in the axils along the axis, and are without perianth. The female flower becomes a dry conelike body. The fruit is a thin, winged nutlet.

This is a monogeneric family of very strange appearing trees. Horsetailtree (*C. equisetifolia*) is a narrow, very tall tree (15 to 25 m), producing hardwood. It grows in sandy soils at sea level, and is rarely found above 1500 m (Goor and Barney 1976). Its natural range extends from Burma, India, Malaysia, and Indonesia into northern and eastern Australia.

This species is much planted in the tropics and is naturalized in southern Florida. River oak casuarina (*C. cunninghamiana*), native to Australia and New Caledonia, is grown in Florida, California, and Hawaii (Olson and Pettys 1974). Longleaf casuarina *C. glauca*) is also native to Australia and is grown in Hawaii.

COLLECTION, EXTRACTION AND STORAGE OF SEED The winged fruits of *Casuarina* aid in seed dispersal (Figs. 1, 2). Seeds are normally seeded planted with the wings intact. In India "cones" of horsetail tree are placed on trays in the sun, and a thin cloth is secured on top of the trays to keep seeds from being lost to the wind (Willan 1985). The trays are treated with 10% BHC powder to keep ants from removing seeds. In 3 days the seeds are ready to be removed from the fruiting structure. After that the fruiting structure breaks down and becomes mixed with the seeds. Good seed crops are produced virtually every year. The seeds are small, about 1500 per gram (Goor and Barney 1976). Seeds dried to 6 to 16% moisture and stored at −7 to 3°C retain viability for at least 2 years.

GERMINATION Seeds germinate without pretreatment. Germination of 70% is common (Goor and Barney 1976). Olson and Pettys (1974) suggest that light is required for germination:

Species	Light (hr.)	Temperature (°C)	Substrata	Germination (%)
C. cunninghamiana	16	20/30	K	55
	—	—	Sand	85
C. equisetifolia	16	20/30	Sand	34
C. glauca	16	20/30	Sand	57

The data they present suggest that for riveroak casuarina light is not necessary for germination.

NURSERY AND FIELD PRACTICE Horsetail tree regenerates from coppice stands as well as from seeds, and this is apparently true for the other species in the genus. In Hawaii, seeds of horsetail tree and longleaf casuarina are broadcast on beds in the spring and covered with 0.6 cm of soil. Recommended seeding densities are 200 to 300 per m². Seedlings may be lifted and planted when 4 to 6 months old. In South Africa, seedling yield per kilo of seed is 17,600 for riveroak casuarina.

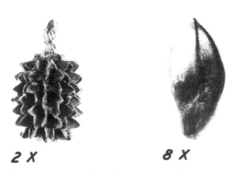

Figure 1. *Casuarina cunninghamiana*, river oak casuarina: (left) multiple fruit, ×2; (right) samara, ×8 (Olson and Pettys 1974).

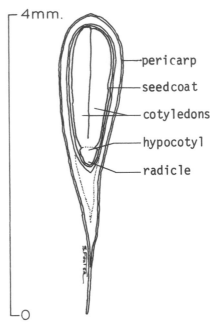

Figure 2. *Casuarina cunninghamiana*, river oak casuarina: longitudinal section through a samara, ×20 (Olson and Pettys 1974).

BIGNONIACEAE — TRUMPET CREEPER FAMILY

Catalpa Scop.

GROWTH HABIT, OCCURRENCE, AND USE The catalpas include about 10 species of deciduous, or rarely evergreen, trees native to North America, the West Indies, and eastern Asia (Rehder 1940). Southern catalpa (*Catalpa bignoniodes*) and northern catalpa (*C. speciosa*) are native to the United States and have been planted quite widely outside their native range, especially the northern catalpa (Bonner and Graney 1974). Mature trees attain heights of 9.2 to 19 m. Both species have been grown to some extent in Europe. Catalpas are used in shelterbelt and ornamental plantings and have minor value as timber trees, mainly for post and small poles.

FLOWERING AND FRUITING Attractive clusters of purplish, perfect flowers of both species are borne in May and June. Fruits of the two species ripen in October, and good crops are borne every 2 to 3 years beginning at about age 20. Mature fruits are round, brown, 2-celled capsules, 15 to 50 cm long (Fig. 1). In late winter or early spring the capsules split into halves to disperse the seeds. Each capsule contains numerous oblong, thin, papery, winged seeds 2.5 to 5 cm long and 0.6 cm wide (Fig. 2). Removal of the papery outer seedcoat reveals an embryo with flat, round cotyledons (Fig. 3).

COLLECTION, EXTRACTION, AND STORAGE OF SEED Fruits should be collected only after they have become brown and dry. When dry enough, the seeds can be extracted by light beating and shaking. Pods of northern catalpa collected in February and March had seeds of higher quality than those collected in the fall. Seeds of southern catalpa are slightly smaller than those of northern catalpa: *C. bignoniodes* yields 46 seeds/gram, *C. speciosa* yields 40 seeds/gram (Bonner and Graney 1974). Dry, cold storage is satisfactory for seeds of both species, at least over winter. Long-term storage indicates that seeds of southern catalpa keep for at least 2 years.

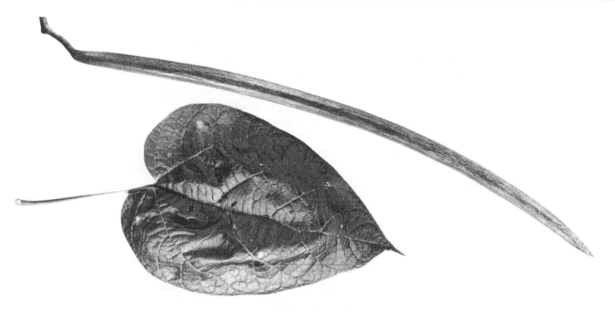

C. bignonioides
southern catalpa

Figure 1. *Catalpa bignoniodes*, southern catalpa: capsule and leaf, ×0.5 (Bonner and Graney 1974).

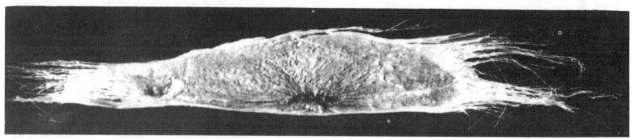

C. speciosa
northern catalpa

Figure 2. *Catalpa speciosa*, northern catalpa: seed, ×4 (Bonner and Graney 1974).

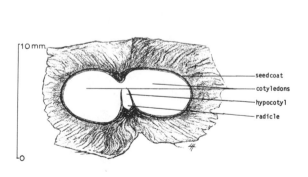

Figure 3. *Catalpa speciosa*, northern catalpa: longitudinal section through a seed, ×3 (Bonner and Graney 1974).

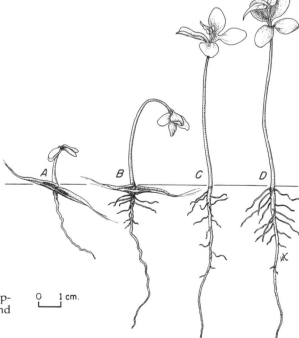

Figure 4. *Catalpa bignoniodes*, southern catalpa: seedling development at 1, 5, 8 and 20 days after germination (Bonner and Graney 1974).

GERMINATION Seeds of catalpa germinate promptly without pretreatment. Tests should be made on wet germination paper for 21 days at 20°C night and 30°C day temperatures; light is not necessary. Fosket and Briggs (1970) considered seeds of northern catalpa to be photosensitive in terms of both light quality and photoperiod. Other moist media are satisfactory. Germination capacities in excess of 90% have been obtained in about 12 days with good quality seed. Germination is epigeal, and the emerging 2-lobed cotyledons look like 4 leaves (Fig. 4).

NURSERY AND FIELD PRACTICE Catalpa seeds should be sown in late spring in drills at the rate of about 100 per meter and covered with 0.28 cm of soil. A pine needle mulch is recommended for southern catalpa. In Louisiana, this species starts germination about 12 days after March sowing and germination is about 80%. Nematodes, powdery mildews, and the catalpa sphinx may give trouble in the nursery. Catalpas are normally planted at 1-0 stock.

Cuttings of catalpa can be rooted if taken in late December and treated with 8000 ppm of IBA-talc. Summer softwoods can also be rooted (Dirr and Heuser 1987).

RHAMNACEAE — BUCKTHORN FAMILY

Ceanothus L.

GROWTH HABIT, OCCURRENCE, AND USE The genus *Ceanothus* consists of about 55 species of shrubs restricted to North America. Most are found along the Pacific Coast, and only 2 species are found east of the Mississippi River (Reed 1974). The ceanothus species are mainly evergreen or deciduous shrubs or small trees. Some species have been cultivated for many years. Many species are used as ornamentals, and many prostrate or semiprostrate forms as soil stabilizing cover, especially in California. The ceanothus species are important browse species for wildlife and domestic range animals and shelter for wildlife. Deerbush ceanothus (*C. integerrimus*) is one of the most important summer browse species in California. Many species bear root nodules and have been shown to be very important in increasing soil nitrogen in forest situations.

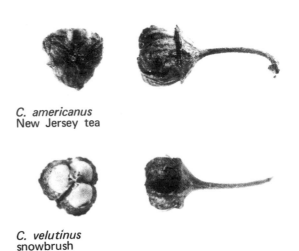

C. americanus
New Jersey tea

C. velutinus
snowbrush

Figure 1. *Ceanothus*: capsules ×4 (Reed 1974).

Table 1. *Ceanothus*: phenology of flowering and fruiting, characteristics of mature trees, cleaned seed weight, pregermination treatments and results (Reed 1974).

Species	Flowering dates	Fruit ripening dates	Mature tree height (m)	Year first cultivated	Seeds/ gram	Hot water treatment Temperature (°C)	Duration (days)	Prechill (days)	Germination (%)
C. americanus	May–July	Aug.–Oct.	0.3–1	1713	250	100	Cool	90	65
C. aboreus	Feb.–Aug.	May–Oct.	3–9	1911	100	90	Cool	0	90
C. cordulatus	May–June	July–Sept.	0.7–2	–	365	Hot	Cool	0	70
C. crassifolius	Jan.–June	Aug.–Sept.	1–3	1927	120	72	Cool	90	76
C. cuneatus	Mar.–June	May–June	1–5	1848	100	72	Cool	90	61
C. diversifolius	Spring	Apr.–June	1	1941	185	Hot	Cool	90	51
C. fendleri	Apr.–Oct.	Aug.–Dec.	0.5–1	1893	–	–	–	–	–
C. greggii	Mar.–Apr.	July	0.6–2	–	50	100	1 min	60	100
C. impressus	Feb.–Apr.	June	–	–	240	100	Cool	60	62
C. integerrimus	Apr.–Aug.	June–Aug.	1–5	1850	150	85	Cool	90	100
C. oliganthus	Feb.–Apr.	May–June	1–7.6	–	150	80	Cool	90	92
C. prostratus	Apr.–June	July	0.1–0.5	–	90	100	Cool	115	71
C. rigidus	Dec.–Apr.	May–June	1–2	1847	160	90	Cool	0	85
C. sanguineus	Apr.–June	June–July	1.3–3	1812	290	100	5 min	0	Good
C. sorediatus	Mar.–Apr.	May–July	1–6	–	–	100	5 min	60	100
C. thyrsiflorus	Jan.–June	Apr.–July	1–8	1837	–	72	Cool	90	83
C. velutinus	June–Aug.	July–Aug.	1–2	1853	200	90	Cool	30	82

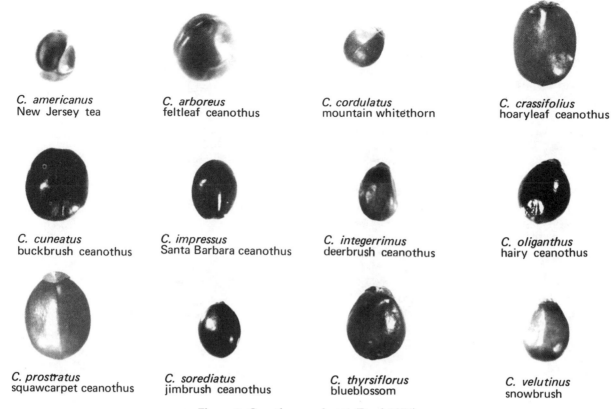

C. americanus New Jersey tea	*C. arboreus* feltleaf ceanothus	*C. cordulatus* mountain whitethorn	*C. crassifolius* hoaryleaf ceanothus
C. cuneatus buckbrush ceanothus	*C. impressus* Santa Barbara ceanothus	*C. integerrimus* deerbrush ceanothus	*C. oliganthus* hairy ceanothus
C. prostratus squawcarpet ceanothus	*C. sorediatus* jimbrush ceanothus	*C. thyrsiflorus* blueblossom	*C. velutinus* snowbrush

Figure 2. *Ceanothus*: seeds, ×6 (Reed 1974).

FLOWERING AND FRUITING Flowers are small, bisexual, regular, blue, white, purple, lavender, or pink, borne in racemes, panicles, or umbels. The 5 sepals are somewhat petallike, united at the base with a glandular disk in which the ovary is immersed. The 5 petals are distinct, hooded, and clawed; the 5 stamens with elongated filaments are opposite the petals. The ovary is 3-celled, 3-lobed with a short 3-cleft style. The fruit is drupaceous or viscid at first, but soon dries up into a 3-lobed capsule (Fig. 1), separating when ripe into 3 parts. Flowering and fruit ripening dates vary greatly with the individual species (Table 1). Seeds are smooth, and varied in size according to species (Figs. 2, 3; Table 1). Feltleaf ceanothus (*C. arboreus*) and hoaryleaf ceanothus (*C. crassifolius*) start producing seed in 1 and 5 years of age, respectively. Fendler (*C. fendleri*) and desert ceanothus (*C. greggii*) are reported to have good seed crops annually.

COLLECTION, EXTRACTION, AND STORAGE OF SEED Capsules of ceanothus species split at maturity and the seeds are ejected with considerable force (Evans et al. 1987). To collect seeds, it is best to tie bags around fruiting branches. Reed (1974) considered ripening seed capsules should not be cut and left on ground clothes covered with netting because seeds would not ripen correctly. This is the common practice of many commercial seed collectors. After drying and seed ejection, seeds can be cleaned with an air screen.

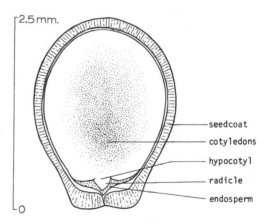

Figure 3. *Ceanothus americanus*, New Jersey tea: longitudinal section through a seed, ×20 (Reed 1974).

Seeds can be dried and stored in containers at low temperatures for prolonged periods. Viable seeds of snowbush ceanothus (*C. velutinus*) have been recovered from forest seedbanks in which the seeds are estimated to be 200 years old.

PREGERMINATION TREATMENT Ceanothus species have been shown to have both hard seedcoats and embryo dormancy. Hot water treatments soften hard seedcoats and prechilling generally solves embryo dormancy (Table 1). Many of the ceanothus species are fire successional species that dominate areas after wild-fires. Their germination ecology is apparently adapted to remaining dormant for long periods in the litter and soil seedbanks and rapidly germinating after wildfires. GERMINATION TEST Standard germination tests apparently do not exist for most species of ceanothus. Radwan and Crouch (1977) broke the dormancy of seeds of red stem ceanothus (*C. sanguineus*) with 90°C hot water or dry heat at 100°C followed by prechilling. Germination was further enhanced by enriching the substrate with 250 ppm of potassium nitrate. Germination trials in this study were conducted in petri dishes on germination paper.

NURSERY AND FIELD PRACTICE Fall planting to satisfy prechilling requirements has given the best results. Seeds should be pretreated with hot water to break the hard seedcoats. Seedlings are often transplanted to containers and grown for a season before being transplanted into the field (Fig. 4). Care should be taken to inoculate seedlings with nitrogen-fixing organisms. Inland ceanothus (*C. ovatus*) can easily be rooted from softwood cuttings (Dirr and Heuser 1987).

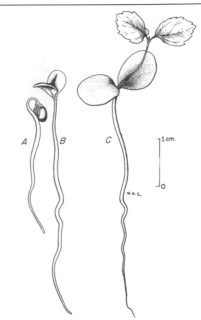

Figure 4. *Ceanothus americanus*, New Jersey tea: seedling development at 1, 5, and 15 days after germination (Reed 1974).

MORACEAE — MULBERRY FAMILY

Cecropia Loefl.

GROWTH HABIT, OCCURRENCE, AND USE The genus *Cecropia* has about 50 species and is native to the Central and South American tropics. The fruit is a drupe that attracts animals and birds. Sork (1984) observed a single tree of *Cecropia obtusifolia* produced 900,000 fruits that were used by 48 species of animals. This species is a common pioneer of felled areas in tropical forest (Vazquesz-Yanes 1979, 1980).

GERMINATION Seeds are spread by birds, bats, and primates and remain dormant in the soil until ground disturbance produces appropriate conditions for germination. Light is necessary for germination, and with proper light, germination occurs over a wide range of temperatures.

MELIACEAE — MAHOGANY FAMILY

Cedrela L. — Toon

GROWTH HABIT, OCCURRENCE, AND USE . The *Cedrela* species are tall trees with attractively colored wood. This genus consists of about 20 species from southeastern Asia to the Caribbean, Central and North America, and India (Macoboy 1982). *Toona* (Endl.) Roemer = *Toon* is the currently accepted genus for the Old World species. The *Cedrala* species strongly resemble *Ailanthus*, to which they are related. They often have heavily buttressed roots. The leaves are long and pinnate with many leaflets and generally are deciduous. In the case of toon (*C. sinensis*) the young foliage is a delicate pink, onion-flavored and edible. The foliage turns a magnificent gold in autumn. Toon is one of the world's most valuable timber species, and is also grown for its decorative spring foliage. The tiny flowers are in weeping panicles and are followed by small fruits. The fruit is a leathery or woody capsule that dehisces at the top by 5 valves. The seeds are winged.

West Indian cedar (*C. odorta*) is a tree that reaches 33 m in height. This tropical species has fragrant wood that is insect repellent and is used for cigar box construction.

GERMINATION No pretreatment is required for fresh seed that is sown without storage (Dirr and Heuser 1987). Stored seed may require prechilling. The Genebank Handbook (1985) suggests testing seeds on the top of blotter paper with an incubation temperature of 20/30°C for 28 days.

Studies of toon seeds collected from the Ivory Coast have shown that dry storage of seeds at very low temperatures did not reduce subsequent germination (Corbineau et al. 1985).

PINACEAE — PINE FAMILY

Cedrus Trew — True Cedar

GROWTH HABIT, OCCURRENCE, AND USE The true cedars, consisting of 4 closely related species of medium- to large-sized evergreen trees, are native to the Syrian Mountains, the Himalayas, the Atlas Mountains, and Cyprus (Rudolf 1974). The oily, sweet-scented wood is very durable, and is an important source of timber in the Himalayas and North Africa. In addition, an oil is obtained from the distillation of *Cedrus* wood. In South Africa deodar cedar (*C. deodara*) is used in shelter belts. The 4 species are widely planted, especially on the Pacific Coast of the United States.

GEOGRAPHIC RACES Stock of cedar of Lebanon (*C. libani*) grown from seed collected at the highest elevations where the species occurs in Asia Minor has proven hardy in Massachusetts, whereas stock from other sources must be grown farther south. The hardiest and best-formed trees are reported to grow in the Taurus Mountains of Turkey. Atlas cedar (*C. atlantica*) can be grown as far north as New York in sheltered positions, while deodar cedar (*C. deodara*) can be grown safely only in California and the southern United States. This statement is not entirely true, however, because Atlas and deodar cedar both grow in Reno, Nevada, under high elevation, low relative humidity, and −30°C temperature extremes.

FLOWERING AND FRUITING The male and female flowers of the true cedars are borne separately on the same trees. The male flowers, which bear pollen only, are in upright cylindrical catkins about 5 cm long. The female flowers, which develop into cones, are small, upright, ovoid bodies, greenish or purplish in color, about 1.25 to 2.5 cm long and 2.5 cm in diameter. Although pollination takes place in the summer or fall, the cones do not begin to grow until the following spring. They do not attain full development until the second or sometimes third year. The mature, barrel-shaped cones (Fig. 1) are 5 to 10 cm long, erect, brown, and resinous; they are borne on short, stout stalks and are characterized by numerous closely appressed, very broad scales, each containing two seeds (Table 1). The cones break up from fall to spring following maturity, leaving the central axis on the tree, as in *Abies*. The irregular, triangular mature seeds are rather soft and oily and have a membranous, broad wing several times larger than the seed itself (Figs. 2, 3). Seed production is variable among species (Table 1).

Figure 1. *Cedrus libani*, cedar of Lebanon: mature cone, ×1 (Rudolf 1974c).

Figure 2. *Cedrus libani*, cedar of Lebanon: seed with wings, ×2. (Rudolf 1974c).

Table 1. *Cedrus:* phenology of flowering and fruiting, characteristics of mature trees, and cone characteristics (Rudolf 1974c).

Species	Flowering dates	Cone ripening	Seed dispersal dates	Seeds/gram	Ripe cone color	Cone dimensions (cm)	Mature tree height (m)	Year first cultivated
C. atlantica	June–Fall	Fall	Fall–Spr.	14	Light br.	6 × 4.5	40	1840
C. brevifolia	–	–	–	–	Light br.	7 × 4	25	1879
C. deodars	Sept.–Oct.	Fall	Sept.–Dec.	8	Reddish br.	8 × 5	500	1831
C. libani	Summer–Fall	Fall	Fall–Spr.	12	Grayish br.	9 × 5	40	1638

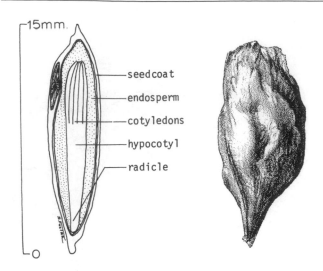

Figure 3. *Cedrus brevifolia*, Cyprian cedar: longitudinal section through a seed and exterior view of a dewinged seed, ×4 (Rudolf 1974c).

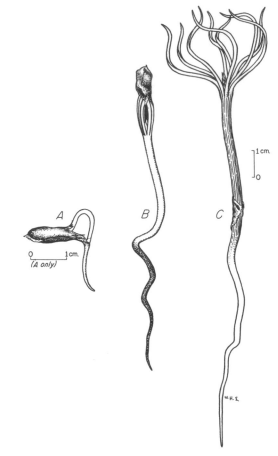

Figure 4. *Cedrus libani*, cedar of Lebanon: seedling development at 1, 4, and 8 days after germination (Rudolf 1974c).

Commercial seed bearing of deodar cedar begins from about 30 to 45 years of age, and good seed crops are borne every 3 years with light seed crops in the intervening years. The seed supply of this species is reduced at times by the activities of birds and the seed weevil (*Euzophera cedrella*).

COLLECTION, EXTRACTION, AND STORAGE OF SEED   Cones may be picked directly from the tree, or cone-bearing twigs may be cut from standing or felled trees just before ripening is complete. Small collections of seed may be made by raking up fallen cone scales beneath the trees (Toth 1980). One hundred liters of cones weighing 20 to 27 kg yield 6.6 kg of seed.

The cones of Atlas, Cyprian cedar (*C. brevifolia*), and cedar of Lebanon have been opened by soaking them in warm water for 48 hours. Cones of deodar cedar have been opened in the sun. After the scales are dry, they can be placed in a cone shaker to remove the seeds. These are then dewinged and cleaned with an air screen. Commercial seed of Atlas cedar has been found to be about 50 to 65% sound.

The seeds are oily and do not keep well under ordinary storage conditions. Entire cones of Atlas cedar, Cyprian cedar, and cedar of Lebanon have been stored over winter with satisfactory results. Cedrus seed has retained viability for 3 to 6 years when dried to a moisture content of less than 10%, placed in sealed containers, and held at temperatures of −1 to 3°C.

GERMINATION   The seeds exhibit little or no dormancy and will germinate without pretreatment. However, prechilling at 3 to 5°C for 14 days has been recommended to hasten results.

ASOA rules specify germination test of prechilled seeds on top of blotters for 21 days at 15°C. International Seed Testing (ISTA) rules specify diurnally alternating temperatures of 20/30°C. Light apparently is not needed. The substrate may be sand.

Table 2. *Cedrus*: propagation techniques (Dirr and Heuser 1987).

Species	Propagation techniques
C. atlantica	Seeds: no prechilling requirement, but nurseries get faster and more uniform germination by prechilling 1 month.
	Cuttings: difficult to root; late February cuttings treated with 8000 ppm IBA-talc give about 20% success.
	Grafting: cultivars of all true cedars are pot grafted in midwinter on seedlings of *C. deodara* using side veneer grafts.
C. deodara	Seeds: no prechilling requirement, but nurseries get faster and more uniform germination by prechilling 1 month. Damping-off can be a problem.
	Cuttings: rooted successfully using previous year's wood taken from October to December.
	Grafting: universal understock for all true cedars. In colder climates the cultivars 'Kashmir' and 'Shalimar' might be rooted and used as understocks.
C. libani	Seeds: no prechilling requirement, but nurseries get faster and more uniform germination by prechilling 1 month.
	Cuttings: difficult; best results with November cuttings in sand or peat without treatment.
	Grafting: universal understock for all true cedars.

Thapiyal and Gupta (1980) studied the dormancy of deodar cedar from 5 different sources. Germination without prechilling ranged from 16 to 69%. One week of prechilling significantly improved the germination of all the seed lots. Prechilling at 9°C was more effective than at 3°C.

NURSERY AND FIELD PRACTICE True cedar seed may be sown in the fall or in the spring, in drills 10 to 15 cm apart at a rate of 260 to 370 per m². In studies in India with deodar cedar seeds Chandra and Ram (1980)

found 10 mm to be the optimum planting depth. Some authorities suggest soaking the seeds for 2 to 3 hours before sowing (Struck and Whitcomb 1977). In northern areas, fall-sown beds should be mulched over winter and the beds covered with burlap on cold nights in the spring during germination. In India 58% of the deodar cedar seeds planted produced plants (Fig. 4).

Cedrus species are difficult to root from cuttings (Table 2) (Nicholson 1984).

BOMBACACEAE — BOMBAX FAMILY

Ceiba Medik.

GROWTH HABIT, OCCURRENCE, AND USE The genus *Ceiba* consists of about 12 species mostly in tropical America, but extending to Asia and Africa. These trees have tall, straight, often spiny trunks that may reach 50 m in height and 3 m in diameter. Kapok tree (*Ceiba pentandra*) has widely spaced branches arranged in irregular whorls around the trunk. The leaves are pinnate, with 5 to 7 spear-shaped leaflets about 15 cm long.

Whitish spring flowers are followed by elliptical capsules that contain the kapok fiber. The capsules may be 25 cm long. The kapok tree produces an industrial crop in southern Asia and there are references to the establishment of seed orchards for this species for use in forestry, but we did not locate any references for seed germination of *Ceiba* species. Willan (1985) points out the obvious: seeds are distributed by wind.

CELASTRACEAE — STAFF-TREE FAMILY

Celastrus scandens L. — American bittersweet

GROWTH HABIT, OCCURRENCE, AND USE American bittersweet is a deciduous climbing or twining shrub of eastern North America (Wendel 1974). It occurs in thickets, in stands of young trees, along fence rows, and along streams, usually in rich soil. It occurs naturally from southern Quebec; west to southern Manitoba; and south to Oklahoma and central Texas, Arkansas, Tennessee, northern Alabama, and western North Carolina. The plant is valuable for ornamental purposes and game food and cover. It was introduced into cultivation in 1736.

FLOWERING AND FRUITING The small greenish, polygamodioecious or dioecious flowers, open from May to June, and are borne in racemelike clusters at the ends of branches. The light reddish seed are about 0.6 cm long and are borne in flesh arils, two of which are usually found in each of the the 2 to 4 cells composing the fruit, a dehiscent capsule (Fig. 1). The yellow to orange capsules ripen from late August to October. They split open soon thereafter, exposing the seeds covered with showy red arils (Figs. 2, 3). Good seed crops are borne annually and may persist on the bushes throughout much of the winter.

COLLECTION, EXTRACTION, AND STORAGE OF SEEDS The ripe fruit should be collected as soon as the capsules sep-

Figure 1. *Celastrus scandens*, American bittersweet: fruiting branch, ×1 (Wendel 1974).

Figure 2. *Celastrus scandens,* American bittersweet: seed with aril removed, ×8 (Wendel 1974).

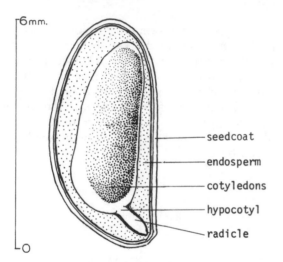

seedcoat

endosperm

cotyledons

hypocotyl

radicle

Figure 3. *Celastrus scandens,* American bittersweet: longitudinal section through a seed, ×10 (Wendel 1974).

arate, or from about mid-September to as long as they hang on the vines, but rarely later than December.

EXTRACTION AND STORAGE OF SEED Collected fruits should be spread out in shallow layers and allowed to dry for 2 to 3 weeks. The seeds are removed from the capsule by flailing or running them through a hammer mill with water. The seeds are allowed to dry for another and cleaned with an air screen. American bittersweet has about 4 to 8 seeds per fruit. The average numer of seeds per gram is 57.

The seed is usually sown in the fall as soon as it is collected. For longer storage the fleshy aril should be cleaned from the seed which should be stored in sealed containers at 3–4°C.

PREGERMINATION TREATMENT Seeds of American bittersweet require prechilling or fall sowing for adequate germination to occur.

GERMINATION TEST ASOA (1985) standards suggest that bittersweet seeds be tested on paper at incubation temperatures of 18 to 22°C for 10 to 14 days. For a quick check of viability the excised embryo technique is recommended.

NURSERY AND FIELD PRACTICE Seeds should be broadcast on a firm seedbed and covered with a mixture of sawdust and sand. The beds should be covered with shade cloth until germination occurs. Emergence usually occurs about 20 days after planting. Young seedlings are susceptible to damping-off (Fig. 4). About 6600 plants are obtained from each kilo of seed planted. For vegetative propagation techniques see Table 1.

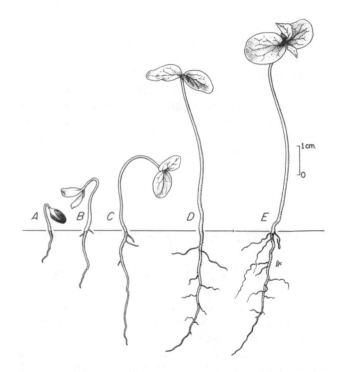

Figure 4. *Celastrus scandens,* American bittersweet: seedling development at 1, 2, 5, 10, and 39 days after germination (Wendel 1974).

Table 1. *Celastrus:* propagation techniques (Dirr and Heuser 1987).

Species	Propagation techniques
C. × *loeseneri*	Cuttings: use root pieces taken in mid-December.
C. *orbiculatus*	Seeds: 3 months prechilling required. Cuttings: treat early August cuttings with 8000 ppm IBA-talc.
C. *scandens*	Seeds: aril should be removed because it inhibits germination. Prechill 3 months. Cuttings: treat early August cuttings with 8000 ppm IBA-talc.

ULMACEAE — ELM FAMILY

Celtis L. — Hackberry

GROWTH HABIT, OCCURRENCE, AND USE The hackberries comprise a large, widespread genus that includes about 70 species of shrubs and trees in the Northern Hemisphere (Bonner 1974b).

FLOWERING AND FRUITING The small, greenish flowers appear in the spring as the new leaves emerge (Bonner 1974b):

Species	Flowering dates	Fruit ripening dates	Seed dispersal dates
C. laevigata	Apr. May	Sept.–Oct.	Oct.–Dec.
C. occidentalis	Apr.–May	Sept.–Oct.	Oct.–Winter
C. reticulata	Mar.–Apr.	Late fall	Fall–Winter

These species are polygamo-monoecious. Hackberry fruits are spherical drupes 0.6 to 1.25 cm in diameter with a thin pulp enclosing a single bony nutlet (Figs. 1, 2). Good seed crops are borne practically every year, and the fruits persist on the branches into winter (Bonner 1974b):

Species	Mature tree height (m)	Seed-bearing age (years)	Year first cultivated	Seeds/ gram
C. laevigata	25	15	1811	13
C. occidentalis	40	—	1656	9
C. reticulata	14	—	1890	11

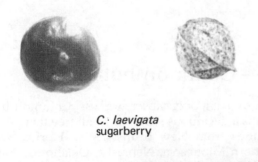

C. laevigata
sugarberry

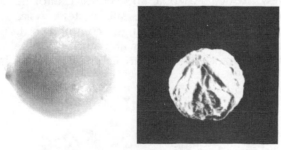

C. reticulata
netleaf hackberry

Figure 1. *Celtis*: (left) fruits and (right) seeds ×4 (Bonner 1974b).

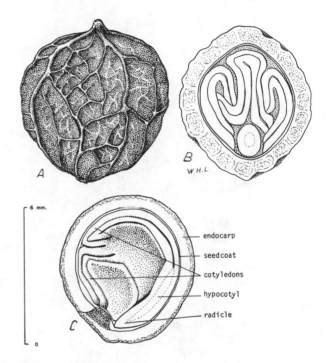

Figure 2. *Celtis occidentalis*, hackberry: A, exterior view of seed; B, longitudinal section, ×6 (Bonner 1974b).

COLLECTION, EXTRACTION, AND STORAGE OF SEED Mature fruits can be picked by hand from trees as late as midwinter. Collection is easier after the leaves have fallen. Limbs of sugarberry (*Celtis laevigata*) can be flailed to knock the fruits onto ground sheets spread under the trees. Unless the fruits are collected early in the season, they do not need drying.

Twigs and trash can be removed by air screening, and the fruits can be depulped by wet maceration. The last step is not essential, but it has been reported to aid germination. Seed yield runs about 110 to 175 seeds per kg of fruit for the North American species.

PREGERMINATION TREATMENT Hackberry seeds exhibit dormancy that requires prechilling (Bonner 1974b):

Species	Prechilling (day)	Temperature (°C)	Germination (%)
C. laevigata	90	20/30	55
C. occidentalis	90	20/30	47
C. reticulata	120	20/30	37

The Genebank Handbook (1985) recommends incubating seeds of *Celtis occidentalis* at 20/30°C, with light, after prechilling for 12 weeks at 3 to 5°C. Bonner (1974b) suggested that the optimum incubation temperature for sugarberry seeds was 20/30°C after 16 weeks prechilling. He considered a secondary dormancy to be introduced during testing. Excised embryo tests were not recommended because of the difficulty in removing the embryo without excessive damage. Fulbright et al. (1986) investigated the germination of spiny hackberry (*C. pallida*) seeds. They were highly dormant and required mechanical

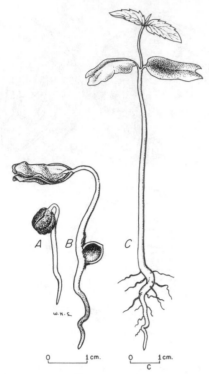

Figure 3. *Celtis laevigata*, sugarberry: seedling development at 1, 2, and 5 days after germination (Bonner 1974b).

scarification plus enrichment with GA_3 and 3 days moist heat (30°C) followed by 2 weeks prechilling to induce about 50% germination. Vora (1989) obtained very high emergence of spiny hackberry seedling simply by soaking them in water before planting. Lodhi and Rice (1971) considered sugarberry trees to be allelopathic to understory vegetation, which may be a factor in the natural regeneration of the species.

NURSERY AND FIELD PRACTICE Both fall sowing of untreated seeds and spring sowing of prechilled seeds have given satisfactory results. Seeds may be broadcast or drilled 20 to 25 cm apart and should be covered with 1.25 cm of firmed soil. Beds should be mulched with straw or leaves held in place with bird screen until germination starts. Germination is epigeal (Fig. 3). For additional propagation techniques see Table 1.

Table 1. *Celtis*: propagation techniques (Dirr and Heuser 1987).

Species	Propagation techniques
C. jessoensis	Seeds: prechill 3 months or fall plant. Cuttings: treat early August cuttings with 3000 ppm IBA-talc.
C. laevigata	Seeds: Fall plant or prechill 2–3 months. Cuttings: no published results. Grafting: chip budding in summer.
C. occidentalis	Seeds: prechill 2–3 months. Cuttings: take cuttings from summer soft wood at end of bud formation, treat with 5000–10,000 ppm IBA solution; root cuttings also possible. Grafting: not easy to bud, but chip budding in summer has been used.

RUBIACEAE — MADDER FAMILY

Cephalanthus occidentalis L. — Common buttonbush

Figure 1. *Cephalanthus occidentalis*, common buttonbush: fruiting head, ×1 (Bonner 1974c).

GROWTH HABIT, OCCURRENCE, AND USE Common buttonbush is a deciduous shrub or small tree that grows on wetlands from New Brunswick to Florida, west to southern Minnesota, Nebraska, Oklahoma, southern New Mexico, Arizona, and central California. It also occurs in Cuba, Mexico, and eastern Asia. In the southern part of its range, common buttonbush reaches heights of 3 to 6 m at maturity, but is shrubby in other areas. The seeds are eaten by many birds, and

Figure 2. *Cephalanthus occidentalis*, common buttonbush: single fruit, ×8 (Bonner 1974c).

the tree has some value as a honey plant. The species was cultivated as early as 1735 (Bonner 1974c).

FLOWERING AND FRUITING The perfect, creamy-white flowers are borne in clusters of globular heads and open from June to September. The fruiting heads (Fig. 1) become reddish brown as they ripen in September and October. Single fruits are 0.6 to 0.8 long (Fig. 2). Each fruit is composed of 2 or occasionally 3 or 4 single-seeded nutlets (Fig. 3) which separate from the base.

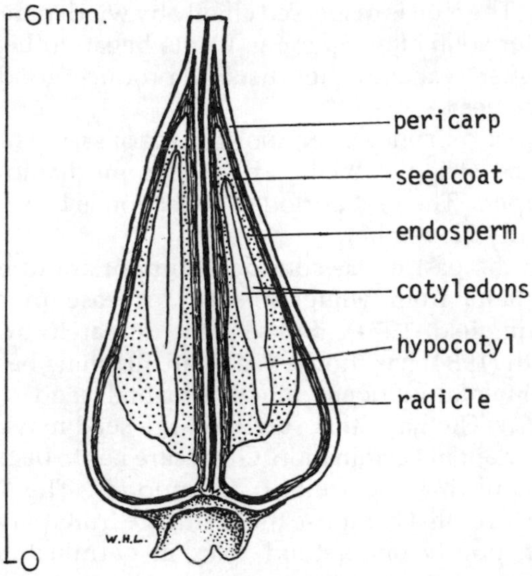

Figure 3. *Cephalanthus occidentalis*, common buttonbush: longitudinal section through the two nutlets of a single fruit, ×12 (Bonner 1974c).

COLLECTION, EXTRACTION, AND STORAGE OF SEED Collection can begin as soon as the fruiting heads turn reddish brown. Many heads disintegrate after they become ripe, but some persist through the winter

months. When the heads are dry, a light flailing will break them into separate fruits. There are about 300 individual fruits per gram.

GERMINATION Common buttonbush seeds germinate promptly without pretreatment. Germination is epigeal (Fig. 4). Seeds generally emerge about 10 to 14 days after planting. Cuttings taken in July to August and placed in sand or peat root without treatment (Dirr and Heuser 1987).

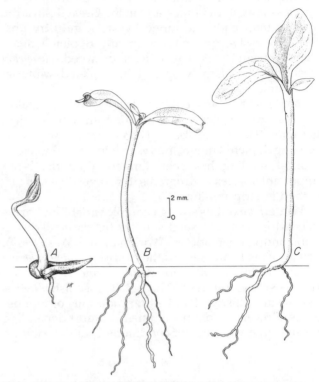

Figure 4. *Cephalanthus occidentalis*, common buttonbush: seedling development at 1, 23, and 40 days after germination (Bonner 1974c).

CEPHALOTAXACEAE — PLUM YEW FAMILY

Cephalotaxus Sieb. et Zucc. — Plum yew

GROWTH HABIT, OCCURRENCE, AND USE The genus *Cephalotaxus* consists of 5 coniferous trees and shrubs and is native to Asia. The species are similar to the yews of the genus *Taxus*. The leaves are linear and pointed, arranged in 2 rows. The leaf midrib is prominent with glaucous bands beneath. The staminate flowers are in clusters of 1 to 8. The pistillate flowers have several bracts, each of which bears 2 naked ovules. The fruit is drupelike, with the seed enclosed in a fleshy envelope.

Plum yew (*Cephalotaxus harringtonia*) reaches 10 m in height and has gray, fissured bark. The leaves are 2-ranked, with the individual needles 5 cm long. The

fruit is ovate, about 2.5 cm long and greenish purple in color. The specific name is in honor of the Earl of Harrington of Elvaston Castle near Derby, England; he first grew the tree in cultivation in 1830.

Chinese yew (*C. fortunii*) is a tree growing to 10 m with reddishbrown bark which peels in flakes. The branches of this species have pendulous ends. The fruit of the Chinese plum yew is 3 cm in diameter.

GERMINATION Seeds of *Cephalotaxus* species require 3 months of prechilling or fall planting. Rooting of cuttings is difficult and results are variable. Cuttings taken in the fall and treated with 5000 to 10,000 ppm of IBA-talc have rooted (Dirr and Heuser 1987).

CHENOPODIACEAE — GOOSEFOOT FAMILY

Ceratoides lanata (Pursh) J. T. Howell — Winterfat

GROWTH HABIT, OCCURRENCE, AND USE Winterfat is a drought-resistant, low shrub, that is native to dry, sandy to shallow clay loam soils from Saskatchewan and Manitoba, south to California, Texas, and Mexico (Springfield 1974). Winterfat is a very valuable browse species on western ranges, and in the Great Basin is the species that made the range livestock industry possible. Excessive grazing has greatly depleted many stands of winterfat. The poisonous weed, *Halogeton glomeratus*, has largely spread in depleted winterfat communities.

Winterfat has been cultivated since 1895 and makes an attractive species for planting in native plant gardens. There has been considerable interest in seeding depleted rangelands with winterfat. Success in artifical seeding has been limited. Winterfat is an important species in programs to revegetate lands disturbed in strip mining.

Winterfat exhibits strong ecotypic variability. Care is needed in selecting seed sources for particular locations and soil situations (Workman and West 1969). Tall growing forms such as the cultivar *Ceratoides lanata* Hatch are most suited for winter ranges.

FLOWERING AND FRUITING The small, greenish flowers usually are unisexual and either dioecious or monoecious. Flowers bloom from June to August depending on elevation and weather conditions. Fruits usually are ripe in October. They are borne on a compound spike (Fig. 1). The fruit is a one-seeded utricle consisting of a nutlet enclosed in 2 bracts, each bearing fluffy white hairs. The pubescent membranous pericarp is fused to the seedcoat. The curved embryo almost completely encircles a very thin layer of residual endosperm (Figs. 2, 3). The fruit is dispersed chiefly by wind in the fall or winter soon after ripening. Plants began to bear fruit the first year and they usually produce good crops every year.

COLLECTION, EXTRACTION, AND STORAGE OF SEED The seeds can be stripped by hand or with mechanical seed strippers. The best period of collection is late October to early November.

In the past it was considered important to remove the hairs from winterfat seed for ease in drilling (Springfield 1974). Subsequent research by Terry Booth (1984) has shown that clean threshing has a very detrimental influence on seed viability and seedling quality. The hairs that surround the seed may also be important in germination. Great care has to be taken in manipulating the fruits of chenopods. The various structures that compose these strange fruits play many, often poorly understood roles in germination. The number of cleaned seeds per gram ranges from 240 to 460. Seed viability decreases rapidly with storage, but is maintained longer with refrigerated storage in sealed containers.

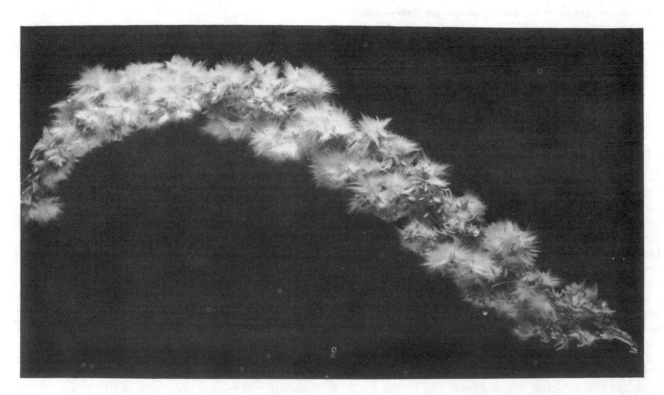

Figure 1. *Ceratoides lanata*, winterfat: fruiting spike, approximately ×1 (Springfield 1974).

Figure 2. *Ceratoides lanata*, winterfat: cleaned seed (nutlet), × 12 (Springfield 1974).

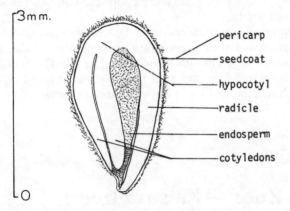

Figure 3. *Ceratoides lanata*, winterfat: longitudinal section through a nutlet, ×16 (Springfield 1974).

GERMINATION Seeds germinate naturally during cold or cool weather. Dettori et al. (1984) investigated the germination of winterfat seeds from native stands, seeds of the cultivar 'Hatch' and seeds of Eurasian winterfat (*C. latens*) at 55°C constant and alternating temperatures:

Species or cultivar	Mean germination (%)	Mean of optima temperature (%)
C. lanata	23	53
'Hatch'	35	73
C. latens	72	98

Germination was highly variable among sources within the same species of winterfat. Moyer and Lang (1976) found considerable variability in germination at low temperatures of different collections of winterfat seed. Allen et al. (1987) devoted an entire study to determining seed quality in winterfat seeds.

Seeds of winterfat require 2 to 3 months afterripening for optimum germination. Moisture or low temperature stress reduces germination.

NURSERY AND FIELD PRACTICE For range revegetation, fruits should be sown during cool weather, as high soil temperatures are detrimental to seedling establishment. Shallow seeding or seeding on the soil surface in shallow depressions may be the most desirable technique. Seedling emergence is very rapid. Plants can be container grown and transplanted.

LEGUMINOSAE — LEGUME FAMILY

Ceratonia siliqua L. — Carob

GROWTH HABIT, OCCURENCE, AND USE Carob is native to the eastern Mediterranean, from the southern coast of Asia Minor to Syria (Alexander and Sheppard 1974). This small, 2.6 to 15.4 m high, evergreen tree has long been cultivated as a forage crop on a wide variety of soils in Asia, Europe, and North Africa. It was introduced to the United States in 1854. Carob is adapted to the warm climates of southern Florida, the Gulf States, New Mexico, Arizona, and southern California. It is chiefly valuable in the United States as an ornamental tree, but has been used to some extent in environmental plantings. Carob pods, a rich source of protein, are a potential forage for livestock. Carob gum, which surrounds the endosperm, has been used in a variety of food products.

FLOWERING AND FRUITING The flowers, borne in small, lateral, red racemes, are polygamo-trioecious. They bloom from September to December in California depending on the variety. The fruit is a coriaceous, indehiscent pod 10 to 30 cm long, 0.6 to 2 cm thick, filled with a sweet, pulpy substance bearing 5 to 15 obovate, transverse, brown, bony seeds about 0.6 cm wide (Figs. 1, 2). Fruits ripen, turn dark brown, and begin to fall

Figure 1. *Ceratonia siliqua*, carob: seed, ×2 (Alexander and Sheppard 1974).

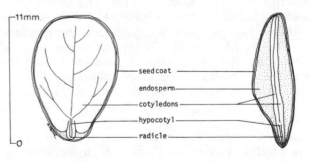

Figure 2. *Ceratonia siliqua*, carob: longitudinal section through a seed, ×3 (Alexander and Sheppard 1974).

from September to November depending on the variety and weather conditions. Plants bear fruit when 6 to 8 years old, and abundant crops occur every second year. Average annual yield of fruit per tree is 90 to 110 kg.

COLLECTION, EXTRACTION AND STORAGE OF SEED Fruits may be collected on the ground, or the ripe pods may be shaken from the tree onto ground cloths. Pods shaken from the tree should be allowed to remain on the ground for 2 to 3 days until completely dry. Because of their high sugar content, pods are likely to become moldy and quickly infested with a small scavenger worm (*Paramycelios transitella*) if wet weather occurs during the harvesting season. Since the worms infect the pods while they are still attached to the tree, it is advisable to limit collection to dry years. Drying the pods or fumigation should solve the insect problem.

Seeds are easily extracted after the pods have been air-dried a few days. If the pods are to be stored for a time before extraction, they should be fumigated. Cleaned seeds average 5 per gram. Seed yield from fruit runs about 11 to 31 kg of seed per 45 kg of fruits.

PREGERMINATION TREATMENT Seeds from freshly collected fruits will germinate without pretreatment. If the seeds are allowed to dry, over time they develop hard seedcoats. Acid scarification in concentrated H_2SO_4 for 1 hour breaks this dormancy. An alternative treatment is to drop the seeds into boiling water and allow the water to cool overnight.

GERMINATION TEST Germination tests have been conducted on moist vermiculite at 18°C for 34 days. Apparently the tests were conducted on fresh or scarified seeds.

NURSERY AND FIELD PRACTICE Scarified seeds should be sown in soil or vermiculite under partial shade. They can be sown in either the spring or fall. The seedlings develop a long tap root that is easily injured so they should be grown in appropriate containers.

CERCIDIPHYLLACEAE — KATSURA TREE FAMILY

Cercidiphyllum Sieb. et Zucc. — Katsura tree

GROWTH HABIT, OCCURRENCE, AND USE Katsura tree is a monotypic genus native to east Asia containing deciduous trees with short spurs on the branches. Leaves are opposite on the shoots, solitary on the spurs. Flowers are solitary, appearing before the leaves. *Cercidiphyllun japonicum* has a bushy-pyramidal habit with a maximum of 30 m height. Usually the trees have several trunks. Leaves are 5 to 12 cm long, glabrous, dark blue-green above, and glaucescent beneath. The leaves are purplish when young, and turn bright yellow or partly scarlet in autumn. The fruit is a pod 2 cm long.

COLLECTION, EXTRACTION, AND STORAGE OF SEED The seeds are small and winged. The fruits should be collected in late October or November before they start to split open (Dirr and Heuser 1987). They can be dried and the seeds removed by shaking. Stored in glass jars at 5 to 8 °C, they remain viable for at least 3 years.

GERMINATION Seeds of Katsura tree require no pretreatment and germinate in 7 to 14 days. Cuttings should be taken in mid-July and treated with 2% IBA. The weeping cultivar can be chip budded or cleft grafted on seedling understock during the summer months.

LEGUMINOSAE — LEGUME FAMILY

Cercis L. — Redbud

GROWTH HABIT, OCCURRENCE, AND USE The genus *Cercis* consists of about 7 species native to North America, southern Europe, and southwestern, central, and eastern Asia (Roy 1974b). These are deciduous small trees and shrubs with unarmed branchlets that lack terminal buds. The leaves are simple, alternate, roundish with a heart-shaped base, palmately veined with 3 to 9 prominent nerves, and with long, slender, terete petioles.

Eastern redbud (*C. canadensis*) is found on abandoned farms, cutover woodlands, and in forest understories. It is a small- to medium-sized tree, generally 7 to 11 m tall, but occasionally reaching 15 m. It has been cultivated since 1641 as an ornamental tree in the northeastern United States, and occasionally in western Europe. It is browsed by white-tailed deer, its seeds are eaten by birds, including bobwhite quail, and it is valuable as a honey plant.

Western redbud (*C. occidentalis*) is a tall, rounded or spreading shrub, usually with many long, erect stems clustered at the base. Generally it is not regarded as a tree, but occasionally in sheltered places it grows 2.5 to 6 m high with a single, smooth grayish trunk 5 to 8 cm in diameter. It grows on well-drained and slightly acid soils of foothills and flats, commonly in open woodlands and chaparral, and generally between elevations of 75 and 1000 m, but sometimes as high as 1380 m. It has been planted as an ornamental throughout the Pacific region.

Young leaves, twigs, sprouts, and pods are browsed by livestock and wildlife. The roots of western redbud were used by Indians in basketry. B. L. Kay, the longtime wildland seeding specialist at the University of California, considered redbud to be the most beautiful of the native shrubs. It was the only wildflower he would pick each spring for his wife. The long-lasting flowers are breathtaking as cut flowers, the new leaves are attractive and unique, and the seed pods are very unusual and attractive.

FLOWERING AND FRUITING Redbud flowers are brilliant pink to reddish purple and, rarely, white. They are bisexual and are borne on thin jointed pedicels in short lateral, umbellike fascicles from the old wood, thus covering the branches with a brilliant flame of color in early spring before the leaves appear.

The fruits are stalked, oblong or broad-linear, flat legumes (pods) with 2 thin, reticulate-veined valves (Fig. 1). They are straight on the upper edge where the suture is winged, curved on the lower edge, acute on the ends, tipped with the thickened remnants of the style and contain several seeds.

The seeds are small, compressed, obovate to rounded, brown, and hard. Their straight embryos are surrounded by endosperm (Figs. 1, 2). Seedcoats are light tan to dark brown and are composed of small thick-walled cells. Seeds of eastern redbud are about 0.6 cm long, and those of western redbud are somewhat larger (Roy 1974b). In both species some of the pods open on the tree in late autumn to release a few seeds, but many pods hang unopened on the tree during most of the winter.

COLLECTION, EXTRACTION, AND STORAGE OF SEED Collection of eastern redbud seeds generally can begin in late summer when the pods turn dark and the seeds are brown, and can continue through November. In Oklahoma, however, seeds collected in the late fall or winter are worthless because they are infested with insects. Under these circumstances, seed collection should be made as soon as the pods are ripe. Pods can be picked from trees or dropped onto ground cloths by flailing or shaking the branches. Pods should be placed in loosely woven bags or spread in a protected area to dry.

If redbud pods are only partially dry when collected, they should be spread to dry in the sunlight for several days. Seeds should be threshed from the pods and separated from the chaff by air screening. Eastern

redbud cleaned in this manner averaged 90% in purity and 85% in soundness. After the seed is thoroughly dried, it can be stored in sealed containers at 3 to 5°C. Cleaned seeds per gram and other yield data are as shown in Table 1 (Roy 1974b):

PREGERMINATION TREATMENT Seeds of redbud have hard seedcoats as well as embryo dormancy. Both prechilling and scarification are required for germination. Acid scarification is commonly used, but duration of treatment must be established by trial for each seed

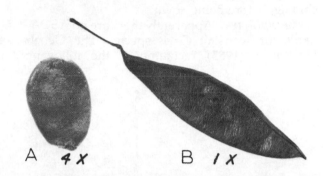

Figure 1. *Cercis canadensis*, eastern redbud: (A) seed ×4; (B) pod, ×1 (Roy 1974b).

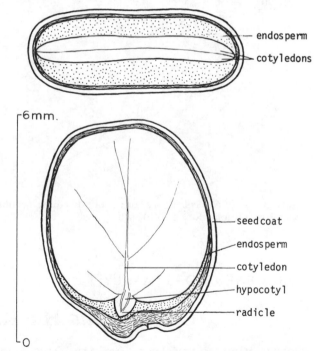

Figure 2. *Cercis canadensis*, eastern redbud: transverse section through a seed (above) and longitudinal section (below), ×10 (Roy 1974b).

Table 1. *Cercis:* phenology of flowering and fruiting, characteristics of mature trees, yield data and seed weight (Roy 1974b).

Species	Flowering dates	Fruit ripening dates	Pod Length	Pod Color	Seed-bearing Age (years)	Seed-bearing Interval	Pod wt. 100 liters (kg)	Pods/kg	Seeds/100 liters pods (kg)	Seeds/ gram
C. canadensis	Mar.–May	July–Aug.	5–10	Reddish purple	5	Biennially	–	–	–	39
C. occidentalis	Feb.–Apr.	July–Aug.	4–7	Dull red	–	Annually	5.1	3960	2.3	27

source. Hot water treatments have also been used for breaking dormancy of redbud seeds. The recommended treatment is 1 minute in boiling water. Seeds have also been placed in 82°C water and allowed to cool over night. Dry heat at 110°C for 9 minutes has been used to produce about 50% germination with western redbud seeds. Mechanical filing or scarification of individual seeds is possible for small seed lots.

The recommended prechill time is 5 to 8 weeks for eastern redbud and 12 weeks for western redbud at 3 to 5°C. Seeds should not be allowed to dry after prechilling, before being sown.

GERMINATION TEST Apparently there are no ASOA standards for seeds of *Cercis* species. The Genebanks Handbook (1985) recommends the following for seeds of eastern redbud: scarification followed by 5 to 8 weeks prechilling and then incubation at 20/30°C. The same procedure is recommended for seeds of *C. siliquastrum* except prechilling is not required.

NURSERY AND FIELD PRACTICE Pretreated (apparently scarified and prechilled) seeds can be sown in prepared beds in April and early May. Drilled seeds should be covered with 0.6 cm of soil. Broadcast seeds should be covered with 1.25 cm of coarse sand.

Fall sowing of freshly harvested seeds that have not thoroughly dried has been successful. Germination is epigeal (Fig. 3). The average number of seedlings per kilo of eastern redbud seed sown was 2,420. See Table 2 for propagation techniques.

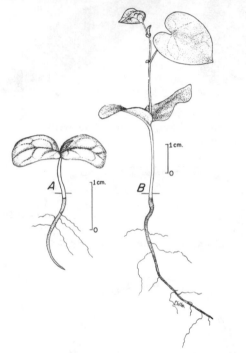

Figure 3. *Cercis occidentalis*, western redbud: young seedling about 1 month old (Roy 1974b).

Table 2. *Cercis:* propagation techniques (Dirr and Heuser 1987).

Species	Propagation techniques
C. canadensis	Seeds: many nurseries fall plant and obtain good germination. Acid scarification is usually for 30 minutes and must be followed by prechilling if not Fall planted. Cuttings: very difficult to root. Cultivars almost always not on own roots. Grafting: T-budding traditionally practiced in late July, but results vary from year to year. Tissue culture: nodal explants have best proliferation on woody plant medium containing 2 mg/L IBA.
C. chinensis	Seeds: Fall planting has been successful, otherwise acid scarify 15–30 minutes and prechill 2 months. Cuttings: timing critical, June–July softwood may work.
C. occidentalis	Seeds: many nurseries Fall plant and obtain good germination. Acid scarification is usually 30 minutes and must be followed by prechilling if not Fall planted.
C. reniformis	Grafting: T-bud in July and August on *C. canadensis* seedlings.
C. siliquastrum	Seeds: Fall sowing will work. Does not have embryo dormancy, but does require scarification for hard seedcoat.

ROSACEAE — ROSE FAMILY

Cercocarpus H B K. — Mountain mahogany

GROWTH HABIT, OCCURRENCE, AND USE The genus *Cercocarpus* consists of about 10 species of shrubs to small trees, generally having persistent leaves that are native to the dry interior and mountainous regions of western North America (Deitschman et al. 1974a). Mountain mahogany (*C. montanus*) may be a shrub 1 m high or a tree to 6 m tall. Curlleaf mountain mahogany (*C. ledifolius*) can be a tree 2 to 5 m tall and can attain the age of 1100 years in extreme cases. The *Cercocarpus* species provide shelter and browse for wildlife. Lack of reproduction is a major problem with curlleaf mountain mahogany. It characteristically grows in relatively pure stands in the higher elevations of the sagebrush zone, extending into juniper and pine woodlands. Often these stands have limbs out of reach of browsing animals and lack reproduction. Curlleaf mountain mahogany seedlings do often occur spontaneously on extremely harsh mine spoils.

FLOWERING AND FRUITING The small, greenish white to reddish brown bisexual flowers have no petals and are borne individually or in twos or threes in the axils of the leaves. Flowering generally occurs during May and June and sometimes extends into July. The fruits ripen and are dispersed by wind and occasionally by animals during August and September. The individual fruit, a soft hairy, cylindrical achene, is distinguished by a feathery persistent style, 5 to 7.5 cm long (Fig. 1). Hairs from these styles can be extremely irritating to the skin and are called "hell feathers" by cowboys who have ridden through stands of mahogany when the fruits were mature. The minimum fruit-producing age of the mountain mahoganies averages 10 years, but may be 15 years or more for curlleaf mountain mahogany. Good seed crops are produced at irregular intervals ranging from 1 to 10 years.

Figure 2. *Cercocarpus montanus,* mountain mahogany: achene with style removed, ×4 (Deitschman et al. 1974a).

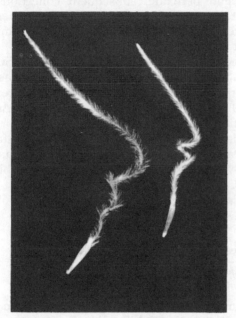

C. montanus
mountain cercocarpus

C. ledifolius
curlleaf cercocarpus

Figure 1. *Cercocarpus:* achenes with feathery style, ×1. Size of the achene varies greatly within each species (Deitschman et al. 1974a).

COLLECTION, EXTRACTION, AND STORAGE OF SEED When seeds are ripe they can be collected by shaking the branches onto a ground cloth or hoppers. Collectors need to wear protective clothing and breathing and eye protection. Time of collection is critical as the fruits can be lost in a single windy day.

For ease in handling, the fruits may be passed through as hammer mill to remove the styles (Figs. 2, 3). Seed weight is apparently highly variable among years of production from the same stands (Deitschman et al. 1974a): *C. ledifolius* yields 115 seeds per gram, *C. montanus* yields 130. Seeds with moisture content of 7 to 10% can be stored in ventilated containers for 5 years.

GERMINATION Dormancy is variable among species and among ecotypes within species. Some collections of mountain mahogany have limited dormancy, but generally seeds of this species require prechilling. Birchleaf mountain mahogany (*C. betuloides*) is a valuable browse plant found in cis-montane California. Seeds of this species have no dormancy and germinate readily. Seeds of curlleaf mountain

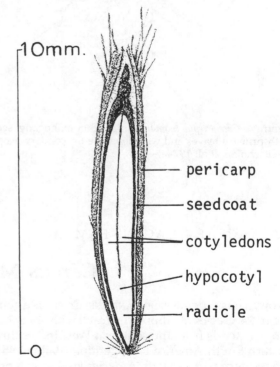

Figure 3. *Cercocarpus ledifolius,* curlleaf mountain mahogany longitudinal section through an achene, ×8 (Deitschman et al. 1974a).

mahogany are extremely dormant. Dealy (1978) was the first to consistently obtain germination of seeds of this species through 9-month prechilling treatments. We obtained germination at a wide range of constant and alternating incubation temperatures by prechilling curlleaf mountain mahogany seeds in aeriated solutions enriched with GA₃ and potassium nitrate for 3 weeks. If these seeds were dried, enhanced germination would only occur at a limited range of temperatures (Young et al. 1978).

Dealy (1978) also discovered that dormancy of

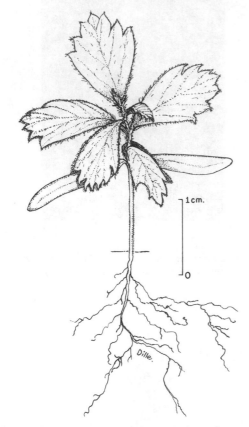

Figure 4. *Cercocarpus montanus*, mountain mahogany: seedling with primary leaves and well-developed secondary leaves, ×2 (Deitschman et al. 1974a)

curlleaf mountain mahogany seeds could be broken by soaking them in hydrogen peroxide solutions. He suggested a 30% solutions for 15 minutes. We subsequently determined that solutions of hydrogen peroxide as dilute as 3% were effective in breaking dormancy with corresponding prolonged soaking periods. Testing of numerous collections of seeds have shown that not all sources will respond to the hydrogen peroxide treatment. This seems to be an inherent difference, because once a source is found that will respond with enhanced seed germination to hydrogen peroxide, subsequent collections in different growing seasons always respond when treated and repeated collections fron nonresponding shrubs never change.

GERMINATION TEST Recently established ASOA standards for curlleaf mountain mahogany are as follows: substrata, paper; incubation temperature, 15 or 10/30°C; and duration of test, 28 days. Additional directions include prechilling at 1 to 2°C for 70 days. Standards for mountain mahogany are substrata, B, P; incubation temperature, 15 or 10/30°C; and duration, 28 days. Additional directions include prechilling at 1 to 2°C for 60 days.

NURSERY AND FIELD PRACTICE Seeds should be sown in the fall (Deitschman et al. 1974a):

Species	Density (m²)	Depth (cm)	Mulch	Plant (%)	Outplant (year)
C. ledifolius	210	0.6	Straw	40	1
C. montanus	270	0.6	Straw	50	1

We have had excellent success with container grown stock. With both prechill and hydrogen peroxide treatment of curlleaf mountain mahogany seeds, we have obtained about 20% seedlings that will not elongate from a rosette form without chilling. Seedlings are illustrated in Figure 4.

CACTACEAE — CACTUS FAMILY

Cereus Mill. — Giant cactus

GROWTH HABIT, OCCURRENCE, AND USE *Cereus* is a genus of about 24 species of arborescent cacti with spiny ribs. Its range extends from the southern West Indies through eastern South America to Argentina (Bailey 1951). In the western part of North America species extend into southeastern Arizona. Often *Cereus* species are tall with erect stems, but sometimes they are low and spreading. Usually they are much branched and strongly angled or ribbed with spiny areoles along the sides of the stem.

The saguaro (*C. giganteus*) is the largest columnar cactus growing naturally in the United States (Alcorn and Martin 1974). As an indicator-plant of the Sonoran Desert, it ranges in restricted elevations below 1230 m

in southeastern Arizona. Papago Indians use saguaro wood in the construction of fences and hogans and as firewood. The fruit pulp can be eaten or used in the preparation of jellies or liquor. Seeds have been used in the preparation of flour, and an excellent honey is derived from the nectar.

FLOWERING AND FRUITING The flowers are large, elongated, and bloom nocturnally. The perianth is deciduous soon after anthesis. The outer segements of the flowers are thick and dull green in color, the inner ones thin and white or red, rarely yellow, in color. The fruit is a fleshy red or rarely yellow berry that splits down one side when ripe. The seeds are black. In the vicinity of Tucson, Arizona, flowering occurs from the latter part of April through early June and reaches a peak in mid-May. Unless effectively cross-pollinated, most flowers drop within 3 or 4 days of opening. The fruits ripen about 37 days after flowering.

COLLECTION, EXTRACTION, AND STORAGE OF SEED The green fruits turn red as they ripen, starting at the flower end. They contain about 2500 black seeds (Figs. 1, 2) enmeshed in the pulp. This pulp must be removed before the seeds will germinate. There are about 990 cleaned seeds per gram. Apparently seeds can be stored for long periods.

GERMINATION Germination studies have shown that light is required for germination. Warm incubation temperatures are also required (Heit 1970b). Recommended test procedures for germination are 20/30°C incubation temperatures with at least 8 hours of light. Reducing incubation temperatures to 15°C reduces germination from about 50% to less than 5% (Heit 1973a). Williams and Arias (1978) determined the light requirement for germination of seeds of *C. griseus* could be partially replaced with gibberellin.

Seedlings can be transplanted at any season of the year. They always should be protected from frost. Plants up to 30 cm should be protected from intense sunlight.

Figure 1. *Cereus giganteus,* saguaro: seeds, ×8 (Alcorn and Martin 1974).

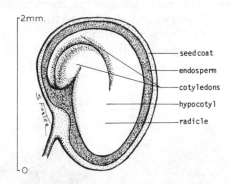

Figure 2. *Cereus giganteus,* saguaro: longitudinal section through a seed, ×20 (Alcorn and Martin 1974).

ROSACEAE — ROSE FAMILY

Chaenomeles Lindl. — Flowering quince

GROWTH HABIT, OCCURRENCE, AND USE The genus *Chaenomeles* consists of 3 species of usually thorny, hardwood shrubs that are grown for their brilliant, very early flowers. The leaves are deciduous or more or less persistent into the winter, arranged alternately and simple in outline, with serrate margins.

FLOWERING AND FRUITING The flowers are solitary or in small clusters from lateral winter buds. The flowers typically have 5 sepals, 5 petals, and multiple stamens.

The fruit is a pome. Chinese quince (*C. sinensis*) bears a fruit 10 to 15 cm long that is used for culinary purposes. Japanese flowering quince (*C. japonica*), which is sometimes called dwarf quince, bears a small pome 0.5 to 2.5 cm long. Most of the horticultural varieties of flowering quince belong to *C. speciosa*, the common flowering quince. The fruit of this species is a pome 3 to 5 cm long.

COLLECTION, EXTRACTION, AND STOREAGE OF SEED The ripe pomes should be collected in the fall, the pulp mashed and the seeds extracted, dried, and stored under cold temperatures in sealed containers (Dirr and Heuser 1987).

GERMINATION Seeds should be planted in the fall or prechilled for 2 to 3 months. Cuttings taken in June-July from softwood and treated with 1000 to 5000 ppm of IBA-solution will root (Dirr and Heuser 1987).

ROSACEAE — ROSE FAMILY

Chamaebatia foliolosa Benth. — Bearmat

GROWTH HABIT, OCCURRENCE, AND USE Two varieties of *Chamaebatia foliolosa* are recognized (Magill 1974a). The typical variety, bearmat, is an evergreen shrub 0.15 to 0.6 m tall that grows between 615 and 2150 m elevation on the western slopes of the Sierra Nevada in California. It grows in open ponderosa and red fir forests.

Southern bearmat, *C. foliolosa* var. *australis,* grows to 2 m on dry slopes in chaparral type from San Diego County to Baja California, Mexico.

The typical variety is normally regarded as a pest because it inhibits the establishment and growth of trees. From an aesthetic viewpoint, the plants provide an attractive ground cover, but the glutinous leaves are highly aromatic. This species is useful for watershed stabilization.

FLOWERING AND FRUITING Bearmat produces perfect flowers throughout its range from May through July, while southern bearmat flowers from November through May. The fruit is a brown achene 5 mm long (Figs. 1, 2).

GERMINATION Seeds require from 1 to 3 months of prechilling to break dormancy.

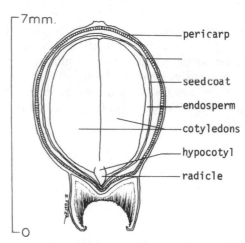

Figure 1. *Chamaebatia foliolosa,* bearmat: achene (left) and extracted seed (right), ×8 (Magill 1974a).

Figure 2. *Chamaebatia foliolosa,* bearmat: longitudinal section through an achene (Magill 1974a).

ROSACEAE — ROSE FAMILY

Chamaebatiaria (Porter) Maxim. — Fern bush

GROWTH HABIT, OCCURRENCE, AND USE Fern bush (*Chamaebatiaria millefolium*) is a monotypic representative of the rose family found growing on the eastern slopes of the Sierra Nevada and across the northern intermountain area to Wyoming, south to northern Arizona. This unusual species is worth growing as a specimen plant.

The fruit is a leathery follicle.

GERMINATION No pregermination treatment is necessary for germination of fresh seeds. Stored seeds require 3 months of prechilling (Young and Young 1985).

CUPRESSACEAE — CYPRESS FAMILY

Chamaecyparis Spach — White cedar

GROWTH HABIT, OCCURRENCE, AND USE The genus *Chamaecyparis* consists of 6 species: 3 are native to North America, 2 to Japan, and 1 to Taiwan (Harris 1974). Port Orford cedar (*C. lawsoniana*) is a long-lived evergreen that attains 3.5 m in diameter and 75 m in height. Its branching is distinctive with the many-branched twigs and small, paired scalelike leaves arranged in fernlike sprays. Port Orford is perhaps the most beautiful of the North American conifers. Because of their beauty and variety of forms, white cedars are often used for ornamental plantings, hedges, and wind breaks. These species are important timber species, and the wood is very durable.

GEOGRAPHIC RACES AND HYBRIDS Geographic varieties of Atlantic white cedar (*C. thyoides*) are var. *henryae* in Georgia, Florida, Alabama, and Mississippi, and var. *thyoides* in the area from South Carolina to Maine. There are over 200 recognized cultivars of Port Orford cedar, 20 of Alaska cedar (*C. nootkatensis*), and 19 of Atlantic white cedar. A hybrid between Alaska cedar and *Cupressus macrocarpa* has been extensively planted in Great Britain.

FLOWERING AND FRUITING The tiny, yellow or reddish male pollen-bearing flowers and greenish female flowers are borne in the spring on the tips of branchlets. Mature cones of *Chamaecyparis* are 0.6 to 1.25 cm in diameter, spherical, and erect on branchlets (Fig. 1). They ripen in September and October (Table 1). Cones mature at the end of the first growing season, except for Alaska cedar in the northern end of its range where the cones require 2 years to mature. Cones have

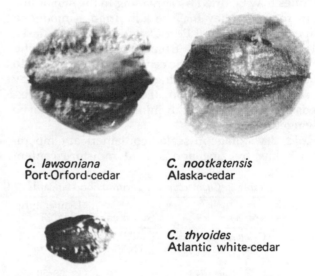

C. lawsoniana
Port-Orford-cedar

C. nootkatensis
Alaska-cedar

C. thyoides
Atlantic white-cedar

Figure 2. *Chamaecyparis*: seeds, ×8 (Harris 1974).

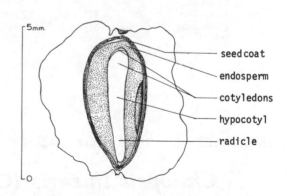

Figure 1. *Chamaecyparis nootkatensis*, Alaskan cedar: mature cones, ×2 (Harris 1974).

Figure 3. *Chamaecyparis lawsoniana*, Port Orford cedar: longitudinal section through a seed, ×8 (Harris 1974).

Table 1. *Chamaecyparis:* phenology of flowering and fruiting, and characteristics of mature trees, cleaned seed weight (Harris 1974).

Species	Flowering dates	Cone ripening dates	Seed dispersal dates	Mature tree height (m)	Year first cultivated	Seed-bearing Age (years)	Seed-bearing Interval (years)	Ripe cone color	Seeds/ gram
C. lawsoniana	Spring	Sept.–Oct.	Sept.–May	75	1854	5	3–5	Red-brown	460
C. nootkatensis	May–June	Sept.	Oct.–Spring	54	1851	–	4+	Red-brown	240
C. thyoides	March	Sept.–Oct.	Oct.–March	28	1727	3–20	1+	Red-brown	100

from 6 to 12 scales, each scale bearing from 1 to 5 seeds with thin marginal wings (Figs. 2, 3). Seed dispersal starts in the fall and may not be completed until the next spring. Characteristics of mature trees are given in Table 1.

COLLECTION, EXTRACTION, AND STORAGE OF SEEDS Cones may be collected by hand or raked from the branches of standing or felled trees. If possible, collections should be limited to years with good seed crops. Studies of Port Orford cedar seed production have indicated a range over 3 years of 20,000 to 4,600,000 seeds produced per hectare (Zobel 1979). This wide variation in seed production is apparently common and in low production years seed quality is often low (Morita 1979).

Cones may be dried by spreading in the sun or in a warm room, or they may be kiln-dried at moderate temperatures. Seeds of Port Orford cedar and Alaska cedar can be extracted from dry cones by gentle tumbling. Seeds of Atlantic cedar are more difficult to extract, and overnight soaking followed by drying may be required for extraction of seeds of this species. Seeds of all species are easily injured and should not be dewinged.

Cold, dry storage in sealed containers at temperatures below freezing and a seed moisture content below 10% are recommended. Viability of Port Orford cedar seeds dropped from an initial 93% to 43% after 7 years of storage at 8% moisture content and 0°C temperature, while seeds stored at room temperature lost all viability over the same time period. Under natural conditions seeds of Atlantic cedar remain viable for at least 2 years on the forest floor.

PREGERMINATION TREATMENT Germination of *Chamaecyparis* species seed is characteristically low, but it is not clear if this is due to embryo dormancy or generally low quality seed.

GERMINATION Germination standards are provided in Table 2. Of the North American species, Alaska cedar is the most difficult to germinate. Prolonged prechilling is often necessary. Studies of the germination of *C. formosensis* indicate that an incubation temperature of 15°C coupled with 16 hours of light produced the best germination, but this was only 38% (Hu et al. 1978).

NURSERY AND FIELD PRACTICE Seeds of Port Orford cedar are broadcast in the spring and covered with 0.3 to 0.6 cm of soil. Sowing rate is calculated to produce a seedling density of 310 to 530 per m². Most nurseries prechill seeds for uniform germination. Shade over the beds may be desirable until mid season. Transplanting stock is usually 2-0. A nursery yield of 283,800 plants per kilo of seed planted has been reported.

Table 2. *Chamaecyparis*: germination standards (ASOA 1985, Genebank Handbook 1985).

Species	Substrata	Temperature (°C)	Duration (days)	Additional intructions	Source
C. lawsoniana	TP	20/30	28		GB
C. lawsoniana	TB, P	20	28	Paired test, use potassium nitrate.	ASOA
C. nootkatensis	TP	20/30	28	Prechill 21 days 3-5°C.	GB
C. nootkatensis	TB, P	20	28	Incubate 28 days at room temperature, then prechill 120 days.	ASOA
C. obtusa	TP	20/30	21		GB
C. pisifera	TP	20/30	21		GB
C. thyoides	TP	20	28	Prechill 90 days at 3-5°C.	GB

BIGNONIACEAE — BIGNONIA FAMILY

Chilopsis linearis (Cav.) Sweet — Desert willow

GROWTH HABIT, OCCURRENCE, AND USE Desert willow grows along dry washes and streams in southwestern United States and northwestern Mexico (Magill 1974b). It is a deciduous shrub or small tree that attains heights of 3 to 8 m. The plant is found in desert environments from 450 to 1500 m in elevation.

FLOWERING AND FRUITING Desert willow produces perfect flowers between April and August throughout its range. The fruit is a 2-celled capsule about 0.6 cm in diameter and 10 to 30 cm long. It ripens from late summer to late fall and persists through winter. The numerous light brown oval seeds are about 0.8 cm long and have a fringe of soft white hairs on each end (Figs. 1, 2).

COLLECTION, EXTRACTION, AND STORAGE OF SEED Seed pods can be handpicked after late September through the winter months. Care must be taken not to pick unripe fruit because of the range in maturity due to indeterminate flowering.

Seed extraction simply requires that the pods be

spread, dried, beaten lightly, shaken, and the seeds recovered by screening. Each 100 kg of dried fruits should produce 66 to 110 g of cleaned seeds that number from 88 to 280 per gram and average 190 per gram (Magill 1974b). Commercial seed has averaged 92% purity and 87% soundness. A cold, dry place is recommended for storage.

GERMINATION Seeds of desert willow are not dormant, but prechilling for several days on wet sand will speed germination. Germination testing should be conducted at 20/30°C.

NURSERY AND FIELD PRACTICE Seed should be sown in the spring. Sowing depth is 0.6 cm. A ratio of 7 times as much seed as the number of plants needed is commonly used. Damping-off is a problem. Desert willow can also be propagated by cuttings.

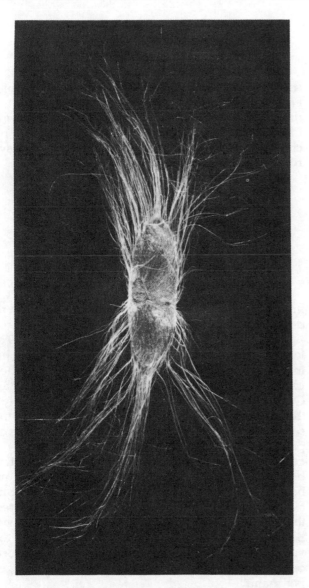

Figure 1. *Chilopsis linearis,* desert willow: seed, ×4 (Magill 1974b).

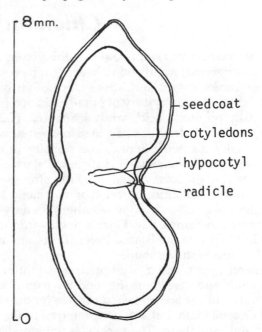

Figure 2. *Chilopsis linearis,* desert willow: longitudinal section through a seed, ×10 (Magill 1974b).

CALYCANTHACEAE — CALYCANTHUS FAMILY

Chimonanthus Lindl. — Wintersweet

GROWTH HABIT, OCCURRENCE, AND USE The genus *Chimonanthus* consists of 2 species of evergreen shrubs native to China and Japan. Fragrant wintersweet (*Chimonanthus praecox*) is grown as an ornamental. It reaches 3 m in height and has leaves 7 to 15 cm long that are bright green lustrous above and glabrous beneath.

FLOWERING AND FRUITING The flowers are 1.5 to 4 cm across, fragrant, and yellow in color. The inter sepals are striped purplish brown (Bailey 1951). Flowering occurs in January to March with fruits ripening in June. The achenes are enclosed in small, cylindrical to urn-shaped receptacles.

GERMINATION If achenes are collected in the green-to-brown color stage and sown immediately germination is rapid with no dormancy. If the achene wall is allowed to harden (July or later collection) germination may be reduced to 5%. Seeds that are collected when fully mature require 3 months of prechilling to break dormancy (Dirr and Heuser 1987).

Cuttings are difficult to root. Cuttings taken in July and treated with 3000 ppm IBA-solution had 70% rooting success (Dirr and Heuser 1987).

VERBENACEAE — VERBENA FAMILY

Clerodendrum L. — Glorybower

GROWTH HABIT, OCCURRENCE, AND USE *Clerodendrum* is a genus of shrubs and trees, often scandent, that are cultivated in greenhouses and out-of-doors as ornamentals. There are about 100 species native to the tropics, mostly in the Eastern Hemisphere.

FLOWERING AND FRUITING The fruit is a drupe often enclosed with the calyx (Bailey 1951).

GERMINATION Seeds of glorybower (*Clerodendrum trichotomum*) are dormant and require 3 months prechilling for germination. Softwood cuttings are easy to root.

OLEACEAE — OLIVE FAMILY

Chionanthus L. — Fringetree

GROWTH HABIT, OCCURRENCE, AND USE Fringetree (*Chionanthus virginicus*) occurs on rich, well-drained soils of stream banks, coves, and lower slopes, but is most abundant in the understory of pine-hardwood forests, especially on moist, acid, sandy loam soils (Gill and Pogge 1974b). It develops best in semiopen situations, but is tolerant, being found occasionally in dense understories. Though widely distributed, it usually is a minor part of the total vegetation. Fringetree is a relatively short-lived shrub or tree that may attain 10.8 m in height. Its range is from southern Pennsylvania southward to Tampa Bay, Florida, westward through the Gulf States to the Brazos River, Texas, and northward to southern Missouri.

Fringetree is planted as an ornamental throughout the South and elsewhere beyond its natural range. Twigs and foliage are preferred browse for deer in the Gulf Coastal Plain, but are less preferred in the Piedmont and mountains. The species is only moderately resistant to browsing, and plants may die when more than one-third of the annual growth is removed. The fruit is utilized by wildlife. The date of earliest cultivation is 1736.

FLOWERING AND FRUITING Although the flowers appear to be bisexual, individual plants are functionally either male or female. Flowering occurs during May-June in the middle of the range, but as early as March in the deep South.

The fruit is a dark blue to purple drupe about 1.9 cm long. It is 1-seeded (Figs. 1, 2). Fruits ripen during July in eastern Texas and as late as October in the northern part of the range. Fruit drop occurs in September and October. Seed dispersal beyond the limited vicinity of the parent tree is by birds and rodents. Plants produce some fruit at 5 to 8 years of age. Good seed crops are produced every year.

COLLECTION, EXTRACTION, AND STORAGE OF SEED In September or October, after the fruit has turned purple, it may be handpicked from the branches. For small lots, pulp may be macerated by rubbing the fruit over a screen that retains the seeds. The pulp may be washed away. Seeds retain some viability in cold storage for 1 to 2 years. Fruits average 1386 per kg. About 3300 seeds can be extracted from 1 kilo of fruit. Seeds average 4 per gram.

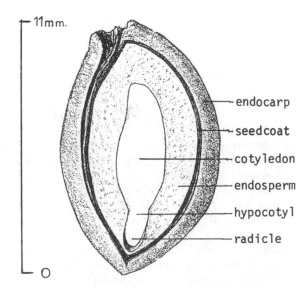

Figure 1. *Chionanthus virginicus*, fringetree: fruit (drupe) and stone (seed), ×2 (Gill and Pogge 1974b).

Figure 2. *Chionanthus virginicus*, fringetree: longitudinal section through the embryo of a stone, ×6 (Gill and Pogge 1974b).

GERMINATION Germination is hypogeal (Fig. 3). Natural germination usually occurs the second year following seedfall. This delay is caused by double dormancy that requires both warm stratification and prechilling (Dirr and Heuser 1987) (Table 1). The embryo excision method is recommended for testing germination (Heit 1955).

NURSERY AND FIELD PRACTICE Seed may be sown in either fall or spring. Fall sowing should be done soon after the seeds are cleaned. Seed rows should be 20 to 30 cm apart, and the seeds covered with 0.6 cm of firm soil. Beds should be protected from frost until after the danger of hard freezing. Spring sowing requires prechilling of the seeds.

Table 1. *Chionanthus:* propagation techniques (Dirr and Heuser 1987).

Species	Propagation techniques
C. retusus	Seeds: double dormant, requiring warm stratification and prechilling. Fall planting produces germination the second spring. Seeds picked early and planted immediately may produce some germination. Cuttings: difficult species to root. Cuttings taken in June to mid-July, as growth hardens, and treated with 1% IBA solution may root.
C. virginicus	Seeds: plant in Fall and wait for second year emergence, or 3 months warm and 3 months prechilling as preplanting treatments. Cuttings: very difficult to root. Grafting: species has been grafted on bare root *Fraxinus ornus* using side grafts.

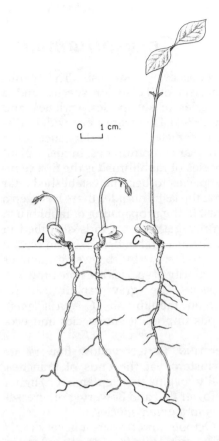

Figure 3. *Chionanthus virginicus*, fringe tree: seedling development at 1, 4, and 7 days after germination (Gill and Pogge 1974b).

SAPOTACEAE — SAPODILLA FAMILY

Chrysophyllum L. — Star apple

GROWTH HABIT, OCCURRENCE, AND USE The genus *Chrysophyllum* consists of about 60 species mostly native to the American tropics. Star apple (*Chrysophyllum cainito*) is grown for its edible fruit (Bailey 1951). The evergreen foliage of these species is covered with hairs that produce an iridescent effect. The trees reach 17 m in height. The oblong leaves are pointed and 10 cm in length.

FLOWERING AND FRUITING The flowers are small and white in color, and go almost undetected beneath the leaves. The fruit is a 5- to 10-celled, globose berry. The 3 to 8 shining seeds are surrounded by a translucent whitish pulp. A related species is the Jamaican plum (*C. oliviforme*).

GERMINATION *Chrysophyllum* species are propagated by seeds and cuttings, but little is published about either technique.

COMPOSITAE — COMPOSITE FAMILY

Chrysothamnus **Nutt.** — Rabbitbrush

GROWTH HABIT, OCCURRENCE, AND USE The genus *Chrysothamnus* consists of 12 major species and a virtually endless series of subspecies, varieties, and forms (Anderson 1986). These are very widely distributed plants that characterize secondary succession in a wide variety of plant communities. In many plant communities a species of rabbitbrush is the first semi-woody perennial species to become established after disturbance. During the last 2 decades there has been a great deal of interest in the propagating of rabbitbrush species for use in revegetation of lands disturbed in mining. There are forms of rabbitbrush within certain species that are preferred by browsing animals although the vast majority of species are note browsed.

The major species are gray rabbitbrush (*C. nauseosus*) and green rabbitbrush (*C. viscidiflorus*). There are numerous important subspecies and ecotypes of both.

FLOWERING AND FRUITING Perfect yellow flowers are borne in heads clustered at the ends of branches. Flowering generally begins in September. Higher elevations plants flower first and flowering progresses to lower elevations in a given species.

Fruits ripen in October and dispersal is very rapid for most species. The minimum fruiting age is 2 years. The fruit, an achene, is crowned by a ring of pappus that is soft and white for gray rabbitbrush and somewhat rigid for green rabbitbrush (Figs. 1, 2).

COLLECTION, EXTRACTION, AND STORAGE OF SEED Seedheads turn golden to grayish brown when ripe. Achenes that abort or are damaged by insects begin to disperse almost as soon as flowering is complete. Care is required to collect filled achenes rather than aborted fruits. The seed heads can be stripped by hand or mechanically.

Deitschman et al. (1974b) suggested running bulk seed collections through a hammer mill and sowing the resulting trashy mixture, which would have about 10% purity. They considered cleaning rabbitbrush seed too expensive. Now that rabbitbrush seed is sold in commerce, screening and air screening are used to produce a relatively pure product.

Deitschman et al. (1974b) reported that seeds of rabbitbrush stored in warehouses had severely reduced viability after 2 years. Kay et al. (1988) reported that storage at 4 or −15°C protected viability of seeds of several species of rabbitbrush for as long as 8 years. Yield and seed weight for seeds of green and gray rabbitbrush are as follows:

	C. nauseosus	*C. viscidiflorus*
Fruit wt./100 liters	5 kg	3 kg
Seeds/gram	1530	1720

GERMINATION Germination standards have been developed by Susan Meyer for seeds of gray rabbitbrush (ASOA 1989). These standards call for the use of a paper substrate, with an incubation temperature of 25 or 20/30°C, for 28 days.

McArthur et al. (1987) investigated the germination of 27 collections of gray rabbitbrush at 3°C incuba-

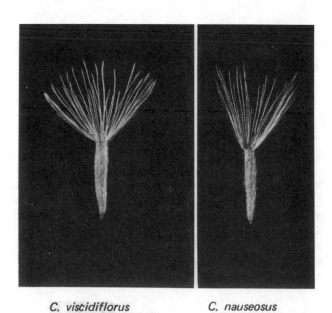

C. *viscidiflorus*
Douglas rabbitbrush

C. *nauseosus*
rubber rabbitbrush

Figure 1. *Chrysothamnus*: achenes with pappi intact, ×4 (Deitschman et al. 1974b).

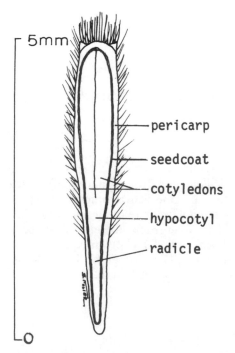

Figure 2. *Chrysothamnus viscidiflorus*: longitudinal section through an achene , ×16 (Deitschman et al. 1974b).

tion temperature. Germination after 14 days varied from 0.4 to 100% depending on the collection. Khan et al. (1987) also determined that seeds of gray rabbitbrush would germinate at a range of constant and alternating temperatures. They also found that germination decreased rapidly with decreasing osmotic potential.

NURSERY AND FIELD PRACTICE Establishment in the field

from direct seeding is not easy. A study by Stevens et al. (1986) for seed of the *albicaulis* subspecies of gray rabbitbrush suggests that orientation of the seed is very important in establishment. Removal of the achene pappus reduced the number of seeds that landed on the seedbed with the radicle end down, and emergence was subsequently reduced.

LAURACEAE — LAUREL FAMILY

Cinnamomum Blume

GROWTH HABIT, OCCURRENCE, AND USE Camphor tree (*Cinnamomum camphor*) is a handsome, medium-sized evergreen tree, 6 to 13 m in height, with an enlarged base. The spreading branches form a round-topped crown of dense, glossy light green foliage. The leaves are simple, arranged alternately, with the odor of camphor when crushed.

Camphor wood is a light, pale, and insect-resistant wood. The Chinese have always used this wood for a type of storage chest. Other valuable species of *Cinnamomum* are as follows (Macoboy 1982):

Species	Use
C. burmanii	Inferior spice
C. cassia	Source of cassia and cinnamomum
C. loureirii	Spice
C. zeylanium	Commercial cinnamomum

FLOWERING AND FRUITING The flowers are small, yellowish and borne in branched axillary clusters. The fruit is a small one-seeded berry, about 0.75 cm in diameter.

GERMINATION Seeds of *C. camphor* can be recovered by maceration in water and flotation (Willan 1985). Little is published on the germination of these seeds despite the commercial importance of the plants. The seeds of other genera of Lauraceae require light for germination (Genebank Handbook 1985).

CISTACEAE — ROCK ROSE FAMILY

Cistus L. — Cistus

GROWTH HABIT, OCCURRENCE, AND USE The genus *Cistus* consists of about 20 species of shrubs native to the Mediterranean region (Bailey 1951) that are important components of the chaparral-type vegetation. Some species are grown as ornamentals in California, and some are naturalized in the southern coastal area of that state.

FLOWERING AND FRUITING The flowers are showy, borne

in terminal, simple or compound cymes. The fruit is a capsule that dehisces into 5 to 10 valves.

GERMINATION Seeds remain dormant for years in soil seedbanks. Burning in wildfires conditions germination. The Genebank Handbook (1985) suggests germinating seeds of *Cistus incanus* at 20°C for a 21-day test period. They should be presoaked in hot water for 24 hours.

RUTACEAE — RUE FAMILY

Citrus L.

GROWTH HABIT, OCCURRENCE, AND USE The genus *Citrus* consists of a dozen or more small, glossy-leaved evergreen, often spiny trees. Most species are native to tropical and subtropical Asia and Malaysia. *Citrus* species are widely grown for their fruits and for ornamental purposes. The common orange, the sweet orange (*C. sinensis*), probably originated in China.

FLOWERING AND FRUITING *Citrus* flowers are borne singly or clustered in axils, or sometimes in lateral cymes or panicles. They are usually bisexual, white after expansion, and commonly 5-merous.

The fruit is a large, globose or ovoid berry. The compartments of the fruit contain pulp-vescicles and 1 to 8 large seeds.

PREGERMINATION TREATMENT In one test of sweet orange seed, untreated seeds had only 2% germination. Acid-scarified seeds had significantly increased germination (Brown et al. 1984). Acid scarification is not as effective in enhancing germination as is removing the pericarp (Cohen 1956). Gibberellin enrichment has been used to enhance germination of these seeds (Burns and Coggus 1969). Germination pretreatments for several species of *Citrus* are shown in Table 1.

Table 1. *Citrus*: germination techniques (Genebank Handbook 1985).

Species	Germination techniques
C. aurantifolia	Remove pericarp, prechill, or enrich with gibberellin, incubate at 25°C.
C. aurantium	Remove pericarp and incubate at 25°C.
C. jamheri	Enrich with gibberellin at 500 ppm.
C. karna	Enrich with potassium nitrate at 750 ppm.
C. limon	Remove pericarp, incubate at 25°C.
C. limonia	Enrich with gibberellin at 10–40 ppm.
C. sinensis	Soak in ascorbic acid, 100 ppm.
C. trifoliata	Prechill 5–12 weeks at 5°C, incubate at 25°C.

LEGUMINOSAE — LEGUME FAMILY

Cladrastis lutea (Michx. f.) K. Koch — Yellow wood

GROWTH HABIT, OCCURRENCE, AND USE Yellow wood is a small deciduous tree attaining a height of 12 to 18.5 m at maturity (Olson and Barnes 1974a). The restricted native range of yellow wood extends from western North Carolina to eastern and central Tennessee, northern Alabama, Kentucky, and southern Illinois and Indiana. It also occurs in the glade country of southern Missouri and central and northern Arkansas. The species was first cultivated in 1812. Locally it grows on limestone cliffs in rich soil, and its greatest abundance is in Missouri. It is hardy as far north as New England and is often planted for its ornamental value.

The wood is hard, close-grained, and bright yellow, turning to light brown with exposure. Commercially it is used as a substitute for walnut in gun stocks and as a source of yellow dye.

FLOWERING AND FRUITING The fragrant, perfect, white, showy flowers bloom in June, usually in alternate years, and the fruit ripens in August to September.

The fruit is a legume 7.5 to 10 cm long (Fig. 1) that falls and splits open soon after maturity. Each fruit contains 4 to 6 short, oblong, compressed seeds with thin, dark brown seedcoats and without endosperm (Fig. 2). Good seed crops are produced on alternate years.

COLLECTION, EXTRACTION AND STORAGE OF SEED The pod may be picked at maturity or dropped onto ground cloths. Seed separates readily from the pod when it opens.

If necessary the pods can be run through a thresher to free the seeds. The seeds can be cleaned with an air screen and stored in sealed containers at 5°C.

PREGERMINATION TREATMENT Natural germination of yellow wood seed is epigeal (Fig. 3) and takes place in the spring following natural seedfall. Dormancy is caused chiefly by an impermeable seedcoat and to a

Figure 1. *Cladrastis lutea*, yellow wood: pods, ×0.5 (Olson and Barnes 1974a).

lesser degree by embryo dormancy (Heit 1967c). Acid scarification for 30 to 60 minutes followed by prechilling for 90 days at 5°C is recommended for enhancement of germination. Hot water and hydrostatic pressure have been used in attempts to crack the hard seedcoats.

GERMINATION TEST Germination tests have been conducted in sand for 30 days at 20/30°C.

NURSERY AND FIELD PRACTICE Seeding may be done in the fall or spring. Beds should be prepared and drilled with seeds placed 0.6 cm deep. Dirr and Heuser (1987) did not consider embryo dormancy to be a problem with yellow wood.

Cuttings taken in December root readily. Cultivars can be budded on seedling root stock.

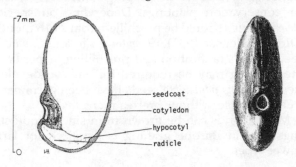

Figure 2. *Cladrastis lutea*, yellow wood: right, exterior view of seed: left, longitudinal section, ×5 (Olson and Barnes 1974a).

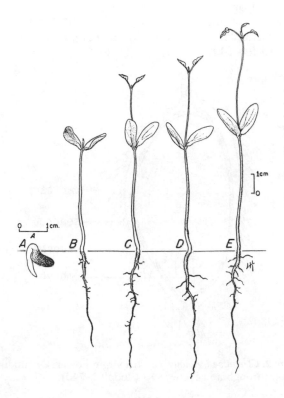

Figure 3. *Cladrastis lutea*, yellow wood: seedling development at 1, 6, 16, and 20 days after germination (Olson and Barnes 1974).

RANUNCULACEAE — BUTTERCUP FAMILY

Clematis L.

GROWTH HABIT, OCCURRENCE, AND USE The genus *Clematis* includes more than 200 species of climbing vines or somewhat woody herbs widely distributed throughout the temperate region, chiefly in the Northern Hemisphere (Rudolf 1974d). Many horticultural varieties are grown as ornamentals, but many native species are useful for erosion control and wildlife habitat plantings.

FLOWERING AND FRUITING Clematis flowers are perfect and fruits are borne in heads of one-seeded achenes with persistent feathery styles (Figs. 1, 2) (Doniushkina and Novikova 1984). Achenes are dispersed by the wind (Table 1).

COLLECTION, EXTRACTION, AND STORAGE OF SEED Fruits are brown when ripe and may be gathered from the plants by hand and then dried and shaken to remove the seeds from the heads. Seeds per gram are given in Table 1. Viability of dry seed has been maintained for 2 years without refrigeration.

PREGERMINATION TREATMENT Clematis seeds require prechilling for germination. Prechilling in moist sand or peat for 60 to 180 days at 3 to 5°C has been satisfactory.

Research in Australia with *C. microphylla* indicated that germination was much quicker with the pericarp

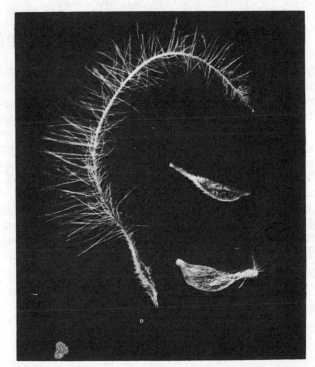

Figure 1. *Clematis virginiana*, eastern virgin-bower: achene with complete style and achenes with style removed, ×4 (Rudolf 1974d).

removed or if the seeds were previously exposed to cycles of wetting and drying (Lush et al. 1984). GERMINATION TEST Germination test can be conducted in sand for 40 to 60 days at 20/30°C.

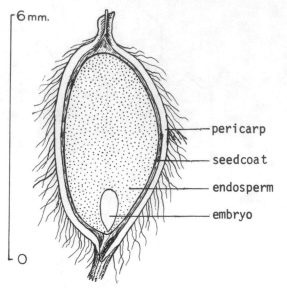

Figure 2. *Clematis virginiana*, eastern virgin-bower: longitudinal section through an achene, ×10 (Rudolf 1974d).

Table 1. *Clematis:* phenology of flowering and fruiting, and cleaned seed weight (Rudolf 1974d).

Species	Flowering dates	Achene ripening dates	Seeds/ gram
C. drummondii	Mar.–Sept.	Aug.–Oct.	–
C. flammula	Aug.–Oct.	Aug.–Oct.	55
C. ligusticifolia	Mar.–April	Aug.–Nov.	205
C. pauciflora	Mar.–April	May–July	190
C. verticillaris	May–June	July–Aug.	140
C. virginiana	July–Sept.	July–Sept.	420
C. vitalba	July–Sept.	July–Sept.	704
C. viticella	June–Aug.	June–Aug.	60

NURSERY AND FIELD PRACTICE The Soil Conservation Service, Plant Material Center, U.S.D.A., Pullman, Washington, has been developing cultivars of *Clematis* for conservation plantings. Depending on species, *Clematis* seeds should be prechilled from 2 to 6 months (Dirr and Heuser 1987). *Clematis vitalba* and *C. viticella* need warm stratification and prechilling. A period of afterripening may be required for embryo development. *Clematis paniculata* seeds have been germinated on a mist table without pretreatment.

Most hybrids can be grown from single node cuttings, and in Europe grafting is practiced for large flower types.

CORNACEAE — DOGWOOD FAMILY

Clethra L.

GROWTH HABIT, OCCURRENCE, AND USE Clethras are shrubs or small trees with handsome foliage and white fragrant flowers followed by dehiscent capsules fruits (Bailey 1951). Summer sweet (*Clethra alnifolia*) is native to the east coast of North America. It is a shrub to 3 m in height. Cinnamon clethra (*C. acuminata*) is native to the east coast from Virginia to Georgia. It reaches 6 m in height. The third species is Japanese clethra (*C. barbinervis*).

COLLECTION, EXTRACTION, AND STORAGE OF SEED Seeds should be collected in the fall. They can be shaken from the capsule and stored under refrigeration until sown. GERMINATION Propagation techniques for clethras follow (Dirr and Heuser 1987):

Species	Propagation technique
C. acuminata	Seeds: no pretreatment necessary. Seeds are very small and should be sown on the surface of flats under mist. Cuttings: treat July cuttings with 1000 ppm IBA-solution.
C. alnifolia	Seeds: no pretreatment necessary. Seeds are very small and should be sown on the surface of flats under mist. Cuttings: very easy to root.
C. barbinervis	Seeds: sow as soon as collected. Cuttings: treat June cuttings with 3000 ppm IBA-talc.

POLYGONACEAE — BUCKWHEAT FAMILY

Coccoloba R. Br. — Sea grape

GROWTH HABIT, OCCURRENCE, AND USE The genus *Coccoloba* consists of about 125 species of evergreen trees and shrubs. Most species are native to the American tropics and subtropics. Sea grape (*Coccoloba uvifera*) is used throughout the tropical world in seaside gardens because of its great salt tolerance. This rapidly growing species reaches 7 m in height and has a thick, squat base. The leaves are round and may reach 20 cm across. The hard timber takes a fine polish and is used in the West Indies for cabinet making.

FLOWERING AND FRUITING The small, greenish white flowers appear in dense racemes up to 25 cm long and are followed by strings of edible purplish red fruits.

GERMINATION *Coccoloba* species are propagated by seeds and cuttings.

ROSACEAE — ROSE FAMILY

Coleogyne Torr. — Blackbrush

GROWTH HABIT, OCCURRENCE, AND USE Blackbrush (*Coleogyne ramosissima*) characterizes a vegetation zone across the southern intermountain area between the pinyon/juniper woodlands and warm desert communities.

FLOWERING AND FRUITING The fruit is an achene with persistent villous styles.

GERMINATION The seeds require prechilling for germination (Young and Young 1985).

LEGUMINOSAE — LEGUME FAMILY

Colutea arborescens L. — Common bladder senna

GROWTH HABIT, OCCURRENCE, AND USE Common bladder senna (*Colutea arborescens*) is a deciduous shrub or small tree 2.1 to 4.6 m in height and native to southern Europe and North Africa. This species has been cultivated as an ornamental since 1570. It has been widely used in North America for conservation plantings and wildlife habitat enhancement (Rudolf 1974e).

FLOWERING AND FRUITING The perfect, bright yellow flowers are 1.9 cm long and occur in 6- to 8-flowered axillary racemes, blooming from May to July. The fruits are indehiscent, inflated pods about 6.3 to 7.5 cm long that ripen from July to October. Each pod contains several small seeds.

COLLECTION, EXTRACTION, AND STORAGE OF SEED Ripe pods are picked from the shrubs in late summer and spread in well-aerated sheds to dry. The pods are threshed to free the seeds, which are recovered with an air screen. Cleaned seeds average 75 per gram.

GERMINATION Dormancy results from a hard seedcoat. Acid scarification for 1 hour will enhance germination. An alternative is steeping seeds in 88°C water for 24 hours (Dirr and Heuser 1987). Seed germination can be tested in dishes, on paper, with incubation temperatures of 20/30°C for 30 days. There have been germination studies conducted in Spain with this species (Allue-Andrade 1983).

NURSERY AND FIELD PRACTICE Untreated seed may be sown in the fall. About 10% of the seed sown will produce usable plants. Softwood cuttings root easily (Dirr and Heuser 1987).

MYRICACEAE — SWEET GALE FAMILY

Comptonia peregrina (L.) Coult. — Sweet fern

GROWTH HABIT, OCCURRENCE, AND USE The genus *Comptonia peregrina* consists of a single species, a deciduous shrub that reaches 1.2 m in height (Bailey 1951). The branchlets are pubescent, and the leaves alternate and pinnatifid. The fruit is a 0.6 cm long ovoid nutlet that is subtended by bracts. The natural range is the east coast of North America from Canada to North Carolina. This genus was named for Henry Compton, Bishop of Oxford, 1632–1713.

GERMINATION Seeds are very difficult to germinate. Scarified seeds should be soaked for 24 hours in 5000 ppm gibberellic acid. Nonscarified seeds treated with gibberellin have about 20% germination (Dirr and Heuser 1987).

Stem cuttings from mature plants root poorly. Root cuttings will root (Dirr and Heuser 1987).

LEGUMINOSAE — LEGUME FAMILY

Copaifera L.

GROWTH HABIT, OCCURRENCE, AND USE The genus *Copaifera* consists of about 30 species confined to the tropics. It is sometimes listed in Caesalpiniaceae by those who split the legume family. *Copaifera officinalis* is a tree that reaches 22 m in height. Balsam of copaiha is obtained from this species in Trinidad (Menninger 1962). The tree is conspicuous in bloom with masses of white flowers. The Brazilian species *C. langsdorffi* also furnishes balsam and its white flowers are overlaid with pink.

GERMINATION Apparently the only germination work on seeds of this species has been done by Borges et al. (1982) in Brazil. The seeds appear to be typical of woody members of the legume family and require scarification.

BORAGINACEAE — BORAGE FAMILY

Cordia L.

GROWTH HABIT, OCCURRENCE, AND USE This genus *Cordia* consists of about 250 species of trees, shrubs, and vines native to subtropical and tropical areas, mainly in the Western Hemisphere (Bailey 1951). The flowers are bisexual or unisexual and borne in dense, headlike clusters or scorpioid cymes. The tubular or campanulate corolla is usually white or orange. Several species are planted for their showy flowers. The fruit is a 4-celled, usually 4-seeded drupe, that is more or less enclosed by the persistent calyx.

Geiger tree (*C. sebestena*) is an evergreen tree or shrub native to the Florida Keys. Its flowers are 2.5 to 5 cm long and orange or scarlet in color. The fruit is a white drupe enclosed in a hazelnutlike husk. About 20 species of *Cordia* are grown in southern Florida as ornamentals. *Cordia boisseri* occurs in southern Texas and is hardy enough to stand some frost.

Opler (1975) described the reproductive biology of 8 species of *Cordia* native to the seasonally dry Pacific slope of Costa Rica. Shrubby species of section *Varronia* (*C. curassavica*, *C. inermis*, and *C. pringlei*) have relatively thin testa and red-colored fleshy fruits that are distributed by birds.

Three species that attain the size of trees in section *Myxa* (*C. collococca*, *C. dentata*, and *C. panamensis*) have extremely astringent and mucilaginous fruits with seeds with thick testa. These fruits are dispersed by arboreal mammals. The 2 tall trees in section Gerascanthus (*C. allio* and *C. gerascanthus*) have persistent, scarious corollas that act like parachutes in dispersing seeds by wind.

All the species of *Cordia* have host-specific bruchid weevils that destroy large numbers of seeds.

GERMINATION Specific germination information is very limited, but the seeds of *C. africana* require prechilling for several weeks in damp sand (Willan 1985).

CORNACEAE — DOGWOOD FAMILY

Cornus L. — Dogwood

GROWTH HABIT, OCCURENCE, AND USE The genus *Cornus* consists of about 40 species of shrubs or small trees native to the temperate region of the Northern Hemisphere, except for 1 species found in Peru (Brinkman 1974d). Most species are deciduous and used chiefly for their ornamental qualities: flowers, fruit, foliage, or color of twigs. The wood is hard and heavy; that of the tree species is used for turnery and charcoal. Some species produce edible fruits, and the bark of others contains a substitute for quinine.

FLOWERING AND FRUITING The small, perfect flowers, white, greenish white or yellow in color, are borne in terminal clusters in the spring. In flowering dogwood (*C. flordia*) and Pacific dogwood (*C. nuttallii*), the clusters are surrounded by a conspicuous enlarged involucre of 4 to 6 white or pinkish petallike, enlarged bracts.

C. alternifolia
alternate-leaf dogwood

C. amomum
silky dogwood

C. Xcalifornica
California dogwood

C. drummondii
roughleaf dogwood

C. florida
flowering dogwood

C. nuttallii
Pacific dogwood

C. racemosa
gray dogwood

Figure 1. *Cornus*: cleaned stones (seeds), ×6 (Brinkman 1974d).

The fruits are globular or ovoid drupes 0.3 to 0.6 cm in diameter, with a thin, succulent or mealy flesh containing a single 2-celled and usually 2-seeded bony stone (Figs. 1, 2). In many stones, only 1 seed is fully developed. Fruits ripen in the late summer or fall (Table 1). Seed dispersal is largely by birds and animals.

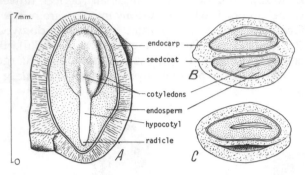

Figure 2. *Cornus stolonifera*, red osier dogwood: (A) longitudinal section through an embryo of a stone; (B) transverse section of a stone containing two embryos; (C) transverse section of a stone having a single embryo, ×6 (Brinkman 1974d).

COLLECTION, EXTRACTION, AND STORAGE OF SEED To reduce losses to birds, fruits should be collected as soon as ripe by stripping or shaking from the branches. Seeds of flowering dogwood should not be collected from isolated trees because these may be self-sterile and the fruits empty.

Dogwood stones can be sown without extracting them from the fruit; commercial seed may consist of either dried fruit or cleaned stones (Table 1). After collection, the fruit may be sown immediately or prechilled for spring planting. Seed to be stored usually is cleaned to reduce bulk. If cleaning cannot be done soon after collection, the fruit should be spread out in shallow layers to prevent excessive heating, but slight fermentation facilitates removal of the pulp. The stones can be readily extracted by macerating the fruits in water or by running them through a hammer mill, allowing the pulp to float away. Clean, air-dried stones may be stored in sealed containers at 3 to 5°C.

PREGERMINATION TREATMENT Natural germination of most species occurs in the spring following seedfall, but some seeds do not germinate until the second spring. Germination is epigeal (Fig. 3). Seeds of all species show delayed germination due to dormant embryos; in most species hard pericarps also are present. Where both types of dormancy exist, warm stratification for at least 60 days in a moist environment, followed by a longer period at a much lower temperature, is required (Table 1). Acid scarification enhances germination of seeds that require warm scarification. Enrichment with gibberellin has been used to enhance germination of seeds of some species.

The predation of dogwood fruits and their fate have been the subject of several research efforts (Borowicz 1986, Krusi and Debussche 1988).

GERMINATION ASOA (1985) standards exist for several species of dogwood (Table 2).

Table 1. *Cornus*: phenology of flowering and fruiting, characteristics of mature trees, yield data, stone weight and pregermination treatments (Brinkman 1974d).

Species	Flowering dates	Fruit ripening dates	Seed dispersal dates	Mature tree height (m)	Year first cultivated	Seed-bearing Age (years)	Seed-bearing Interval (years)	Stones/ 100 kg fruit	Stones/ gram	Substrata	Warm stratification Temperature (°C)	Warm stratification Duration (days)	Cold stratification Temperature (°C)	Cold stratification Duration (days)
C. alba	May–June	Aug.–Sept.	–	3.1	1741	–	–	–	–	–	–	–	–	–
C. alternifolia	May–July	July–Sept.	July–Sept.	7.7	1760	–	1	41	27	Sand	20/30	60	5	60
C. amomum	May–July	Aug.–Sept.	Sept.	3.1	1658	4–5	1	–	74	–	–	–	3–5	21
C. × californica	Apr.–Aug.	July–Nov.	–	4.6	–	–	–	–	148	Sand	–	–	5	120
C. canadensis	May–June	Aug.	Aug.–Oct.	0.3	–	–	–	50	35	Sand	25	60	1	150
C. drummondii	May–June	Aug.–Oct.	–	13.8	1836	–	–	72	10	Sand	22/28	1	5	30
C. florida	Mar.–Apr.	Oct.	Aug.–Winter	12.3	1731	6	1–2	33	21	Sand	–	–	5	120
C. kousa	June	Aug.	Nov.	3.1	1875	–	2	–	5	–	–	–	3–5	120
C. mas	Mar.–Apr.	Aug.–Oct.	Nov.	8.0	Old	–	–	26	10	Soil	20/30	4	4	120
C. nuttallii	Apr.–May	July–Oct.	Sept.–Oct.	24.6	1835	10	2	47	29	Peat	–	60	5	90
C. racemosa	May–July	Aug.–Sept.	–	3.7	1758	–	–	39	–	Sand	20/30	60	5	60
C. rugosa	May–June	Aug.–Oct.	–	7.1	1784	–	–	–	41	Soil	–	–	5	90
C. stonifera	June–Aug.	Aug.–Oct.	May	6.2	1656	–	–	39	41	Sand	–	–	5	90

Table 2. *Cornus:* germination standards (ASOA 1985, Genebank Handbook 1985).

Species	Substrata	Temperature (°C)	Duration (days)	Additional directions	Source
C. alba	–	20/30	–	Light 8hr/day, prechill 12–14 weeks at 1–5°C	GB
C. amomum	–	20/30	–	Light 8hr/day, warm stratification 8–12 weeks at 25°C, then prechill 8–12 weeks at 1–5°C	GB
C. controversa	–	20/30	28	Prechill 90–100 days at 3–5°C	GB
C. florida	TB	18/22	10	Excise embryo.	ASOA
C. kousa	–	20/30	–	Light 8hr/day, prechill 12–14 weeks at 1–5°C	GB
C. mas	–	20/30	–	Light 8hr/day, warm stratification 16 weeks at 25°C, then prechill 4–16 weeks at 1–4°C	GB
C. nuttallii	–	20/30	21	Light 8hr/day, prechill 12–14 weeks at 1–5°C	GB
C. nuttallii	C, TB	20/30	28	Prechill 90–120 days; excise embryo.	ASOA
C. sanguinea	–	20/30	–	Light 8hr/day, warm stratification 8 weeks at 25°C, then prechill 8–12 weeks.	GB
C. stolonifera	TB	12/22	10	Prechill 90 days at 3-5°C.	GB
C. stolonifera	TB	20/30	28	Prechill 120–160 days.	ASOA

NURSERY AND FIELD PRACTICE Best results are obtained from fall sowing of freshly harvested seeds. Fruits collected too late to sow in the fall should be stored, prechilled until the next season, and sown the following fall. Seeds sown in nursery beds should be covered with 0.6 to 1.25 cm of soil. Fall-sown beds should be mulched during the winter (Fig. 3). Propagation techniques are given in Table 3.

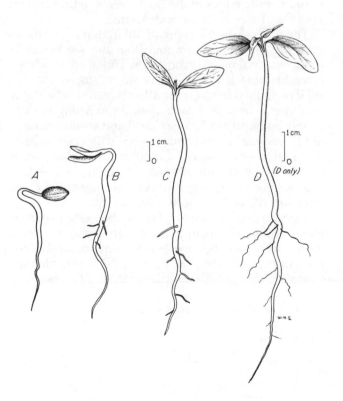

Figure 3. *Cornus florida,* flowering dogwood: seedling development at 2, 4, 8, and 31 days after germination (Brinkman 1974d).

Table 3. *Cornus:* propagation techniques (Dirr and Heuser 1987).

Species	Propagation techniques
C. alba	Seeds: fall plant or prechill 2–3 months. Cuttings: summer softwood easy to root.
C. amomum	Seeds: fall plant or prechill 3–4 months.
C. canadensis	Seeds: acid scarify 1 hour, then prechill 2–3 months. Cuttings: ground cover species best propagated by dividing sod.
C. controversa	Seeds: sow as soon as collected or warm stratification 5 months and prechill 3 months. Cuttings: treat late August cuttings with 8000 ppm IBA solution.
C. florida	Seeds: fall plant or prechill 3–4 months. Cuttings: take softwood cuttings in late June–July. Grafting: shield budding in summer on seedling understock.
C. kousa	Seeds: fall plant or prechill 3–4 months. Cuttings: take softwood cuttings in late June–July.
C. macrophylla	Seeds: warm stratification 3 months, followed by prechilling 3 months. Cuttings: easy to root from softwood.
C. mas	Seeds: warm stratification 4–5 months, then prechill 3 months; fall-planted seeds germinate the second year. Cuttings: not easy to root. Softwood cuttings treated with IBA give best results.
C. officinalis	Seeds: warm stratification 4–5 months, then prechill 3 months; fall-planted seeds germinate the second year.
C. racemosa	Seeds: plant immediately at maturity or prechill. Cuttings: treat July cuttings with IBA.
C. sanguinea	Seeds: fall plant or prechill 2–3 months. Cuttings: easy to root from midsummer cuttings.
C. walteri	Seeds: difficult to germinate; both warm stratification and prechilling required. Cuttings: difficult to root; take softwood cuttings in late June–July.

HAMAMELIDACEAE — WITCH HAZEL FAMILY

Corylopsis Sieb. et Zucc. — Winter hazel

GROWTH HABIT, OCCURRENCE, AND USE The genus *Corylopsis* consists of about 12 species of deciduous shrubs or small trees native to east Asia and the Himalayas (Bailey 1951). The leaves are alternate, prominently veined, and with deciduous stipules.

FLOWERING AND FRUITING The flowers are yellow, appearing before the leaves, borne in nodding racemes that have large bracts at the base. The fruit is a capsule with two shiny black seeds. These are beautiful, elegant shrubs, especially in flower, when the soft-to-deep-yellow flowers flutter in early spring breezes (Dirr and Heuser 1987).

COLLECTION, EXTRACTION, AND STORAGE OF SEED The capsule ripens in late summer, and fruits must be collected before it dehisces. A yellow-green to brown color change indicates the fruit should be picked. They should be brought indoors and placed in a cardboard box with a lid. Seeds will be expelled from the capsules and can be collected from the box and stored in sealed containers at low temperatures.

GERMINATION A major problem is finding good seed. The seeds are dormant and 5 months warm stratification followed by 3 months of prechilling is required for germination. Freshly collected seeds that are sown immediately will produce limited emergence (Dirr and Heuser 1987). If seed is allowed to dry before planting in the fall it will take 2 years for emergence to occur. Softwood cuttings taken in the summer are easy to root.

BETULACEAE — BIRCH FAMILY

Corylus L. — Hazel, filbert

GROWTH HABIT, OCCURRENCE, AND USE The genus *Corylus* includes about 15 species of large deciduous shrubs, rarely small trees, that occur in the temperate parts of North America, Europe, and Asia (Brinkman 1974e). European filbert (*C. avellana*) has been cultivated for the commercial production of nuts in Europe and the United States. Several native species have potential value for wildlife, shelterbelt, and environmental plantings.

FLOWERING AND FRUITING Male and female flowers are borne separately on one-year-old lateral twigs of the same plant. They are formed late in the summer and open the following spring before the leaves appear (Table 1).

By late summer or early fall, the fertilized female flowers develop into fruits. These are round or egg-shaped, hard-shelled, brown to dark tan nuts. Each nut is enclosed in an involucre or husk consisting of 2 more or less united hairy bracts (Figs. 1, 2). Natural seed dispersal is chiefly by birds and mammals. Large seed crops are produced at irregular intervals, usually every 2 or 3 years.

COLLECTION, EXTRACTION, AND STORAGE OF SEED Hazel nuts may be eaten by rodents or birds even before they are mature. To reduce such losses, fruits should be picked as soon as the edges of the husks begin to turn brown. This may be as early as mid-August.

The fruits should be spread out in thin layers to dry until the husks open enough that the seeds can be removed by flailing. Production and seed yields are given in Table 1. Filbert seeds can be stored until the fall if prechilled in moist sand. Seeds can also be stored for a year in unsealed containers at room temperature. American hazel nut (*C. americana*) and California hazel nut (*C. cornuta* var. *californica*) seeds will retain viability if stored at 5°C. Seeds should not be dried before storage and high humidity should be maintained.

Even partial drying reduces post-prechilling germination (Kowalski and Kowecki 1982).

PREGERMINATION TREATMENT Freshly harvested seeds of filbert are not dormant although dry storage of seeds induces dormancy (Jarvis 1975). Under natural conditions dormancy is broken by a period of prechilling, a treatment that potentiates the synthesis of gibberellins

Table 1. *Corylus:* phenology of flowering and fruiting, yield data, and seed weight (Brinkman 1974e).

Species	Flowering dates	Fruit ripening dates	Seeds/100 kg fruits	Seeds/gram
C. americana	Mar.–May	July–Sept.	61	1.1
C. avellana	Feb.–Mar.	Sept.–Oct.	132	0.8
C. cornuta				
var. *californica*	Jan.–Feb.	Sept.–Oct.	–	0.9
var. *cornuta*	Jan.–Feb.	Aug.–Sept.	–	1.2

Figure 1. *Corylus cornuta* var. *californica*, California hazel: mature fruit including husk, ×1.5 (Brinkman 1974e).

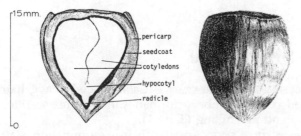

Figure 2. *Corylus cornuta* var. *californica*, California hazel: (left) longitudinal section through a nut; (right) exterior view of nut with husk removed, both views ×2 (Brinkman 1974e).

(Ross and Brandbeer 1971).

Experimentally, dormancy is readily broken by applications of gibberellin (Jarvis and Wilson 1978). Brandbeer et al. (1978) demonstrated a stimulatory effect of light on germination of freshly harvested seeds, and on naked embryos taken from seeds previously dry stored for 4 weeks. Phytochrome has been shown to play a role in the germination of dry seeds (Shannon et al. 1983). Pretreatment of hazel seeds with ethanol alone does not break dormancy even when the seeds are light treated (Jeavons and Jarvis 1984). The combination of ethanol and mercuric chloride soaking and then exposure to light does induce filbert seeds to germinate. The anaesthetic properties of mercuric

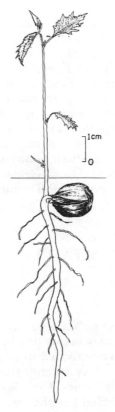

Figure 3. *Corylus cornuta* var. *californica*, California hazel: seedling development 30 days after germination (Brinkman 1974e).

Table 2. *Corylus*: propagation techniques (Dirr and Heuser 1987).

Species	Propagation techniques
C. americana	Seeds: fall plant or prechill 2–6 months. Cuttings: timing is critical; take cuttings from mid-June to mid-July, when they are in active growth stage; treat with IBA.
C. avellana	Seeds: fall plant or prechill 2–6 months. Cuttings: timing is critical; take cuttings from mid-June to mid-July, when they are in active growth stage; treat with IBA. Grafting: nut-producing cultivars are grafted on C. avellana seedlings in winter using 3–4 bud scions and a whip graft. Tissue culture: immature embryo explants through callus culture produced meristemoids.
C. clurna	Seeds: fall plant or prechill 2–6 months. Cuttings: timing is critical; take cuttings from mid-June to mid-July, when they are in active growth stage; treat with IBA.
C. cornuta	Seeds: fall plant or prechill 2–6 months.
C. maxima var. purpurea	Seeds: fall plant or prechill 2–6 months. Cuttings: timing is critical; take cuttings from mid-June to mid-July, when they are in active growth stage; treat with IBA.

chloride may be inducing germination.

During dry storage the germination inhibitors present in the testa and pericarp move through the cotyledons to the embryonic axis, the site of dormancy (Brandbeer 1968). This movement is accompanied by a 97% loss of endogenous gibberellin, and a decrease in moisture content (Ross and Brandbeer 1971). Brandbeer and Pinfield (1967) suggest that the dormancy of filbert seeds is determined by relative levels of gibberellin and abscissic acid.

Dormancy can be effectively overcome by either stratification, which potentiates gibberellin biosynthesis, or by exogenous application of gibberellin. Jarvis (1966) reported in his doctoral thesis that increasing the time of dry storage augmented the time required for germination. Gibberellin-treated seeds could be stored for up to 12 weeks without reduction in germination (Jarvis 1975). The influence of gibberel-lin on longer storage is not known (Nussbaum and Lagerstedt (1983).

GERMINATION TEST Germination standards provided by the Genebank Handbook (1985) for *C. avellana* are substrata, sand or soil; temperature 20/30°C; duration, 35 days. Additional directions are to remove pericarp and prechill for 2 months.

NURSERY AND FIELD PRACTICE Most nurseries plant *Corylus* seeds in the fall. Seeds are planted 2.5 cm deep and covered with 2.5 to 3 cm of sawdust. Seedling densities are kept low, from 42 to 63 per m². Seedlings are planted out after 1 year. Germination is hypogeal (Fig. 3). Propagation techniques are given in Table 2.

PALMAE — PALM FAMILY

Corypha L.

GROWTH HABIT, OCCURRENCE, AND USE The genus *Corypha* consists of 6 species native to tropical Asia (Brandis 1970). It is widely cultivated in southern Asia. These tall, monocarpic palms die after ripening their seed at the age of 17 to 40 years. Their leaves are very large, with segments folded lengthwise.

FLOWERING AND FRUITING The flowers are bisexual, small, and with tripartite calyx and petals. The fruit is a globose drupe 2 to 3 cm long. The seeds are borne erect at the summit of the hard, horny endosperm.

GERMINATION Talipot palm (*C. umbraculifera*) fruits had 32% germination over a 9-to-27-week test (Manokaran 1979). Seeds of gebang palm (*C. elata*) germinated readily in moist sand (Basu and Mukhenjee 1972). Seeds of *C. lecomtel* required 3 months to germinate (Ishihata 1974).

ANACARDIACEAE — CASHEW FAMILY

Cotinus Mill. — Smoketree

GROWTH HABIT, OCCURRENCE, AND USE The genus *Cotinus* includes 2 species of small, deciduous trees or shrubs widely distributed through southern Europe and the Himalayas to central China, and in the southern United States (Rudolf 1974f). They are cultivated primarily for ornamental purposes. The durable wood of American smoketree (*C. obovatus*) is used for posts.

FLOWERING AND FRUITING The greenish yellow flowers, which bloom in the spring or summer, are polygamous or unisexual with the female and male flowers borne on different trees. The fruit is a dry, compressed drupe about 0.8 cm long, light reddish brown, containing a thick, bony stone (Fig. 1). Seed crops are produced annually, but are often poor. Seed production and flowering are given in Table 1.

COLLECTION, EXTRACTION, AND STORAGE OF SEED The fruits may be picked from the bushes as soon as they are ripe. Common smoketree (*C. coggygria*) fruits may be picked slightly green and sown immediately. Otherwise the dry fruits should be run through a hammer mill and the debris separated by an air screen. Apparently seeds can be stored for several years.

PREGERMINATION TREATMENT *Cotinus* seeds have hard seedcoats and embryo dormancy that require scarification and prechilling (Tabe 1).

GERMINATION TEST Apparently no germination standards have been established. Pretreated seeds can be

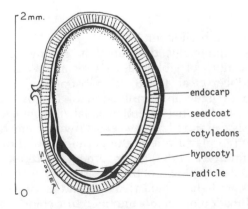

Figure 1. *Cotinus obovatus*, American smoketree: longitudinal section through a seed, ×24 (Rudolf 1974f).

Table 1. *Cotinus:* phenology of flowering and fruiting, characteristics of mature trees, cleaned seed weight, and pregermination treatments (Rudolf 1974f).

Species	Flowering dates	Fruit ripening dates	Mature tree height (m)	Year first cultivated	Seeds/ gram	Acid scarification (min)	Prechilling Substrata	Prechilling Temperature (°C)	Duration (days)
C. coggygria	June–July	Aug.–Oct.	4.9	1656	97	20/80	Peat	3	60/80
C. obovatus	Apr.–May	June–Sept.	10.8	1882	111	20/40	Naked	3	60

tested on peat or kimpak at 15°C for 30 days (Heit 1967c).

NURSERY AND FIELD PRACTICE Seeds should be fall sown immediately after collection at a rate of 420 per m². Seedbeds should be mulched with sawdust. Seedlings can be outplanted after one year. Dirr and Heuser (1987) suggest the following propagation techniques for *C. coggygria* and *C. obovatus:* Acid scarify and prechill seeds. *Cotinus coggygria* root cuttings will root, and soft wood cuttings can be rooted directly in pots. *Cotinus obovatus* cuttings are more difficult to root; timing is important.

ROSACEAE — ROSE FAMILY

Cotoneaster B. Ehrh.

GROWTH HABIT, OCCURRENCE, AND USE The genus *Cotoneaster* consists of about 50 species of deciduous and evergreen shrubs, rarely small trees, native to the temperate regions of Europe, northern Africa, and Asia except Japan (Slabaugh 1974a). They are valued as ornamentals for their glossy green foliage, attractive fruits, and neat, unique habits of growth. Fall foliage is often a showy blend of orange and red. The more hardy species are used in environmental plantings on the northern Great Plains.

FLOWERING AND FRUITING The perfect, white or pinkish flowers occur singly or in clusters at the ends of leafy lateral branchlets. The fruits are small, black or red, berrylike pomes which ripen in late summer or early fall and persist into winter (Fig. 1). Each fruit contains 1 to 5 seeds (Figs. 2, 3). The phenology of flowering for selected species is given in Table 1.

COLLECTION, EXTRACTION, AND STORAGE OF SEED The ripe fruits should be collected by hand after leaf fall. The firmness of the fruit and its color are good criteria of ripeness (Table 1). The fruit may be collected when slightly green to reduce dormancy. The minimum

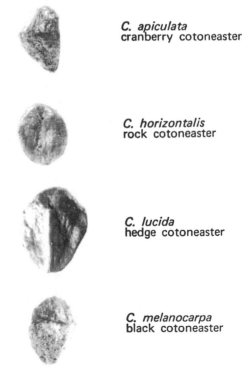

C. apiculata
cranberry cotoneaster

C. horizontalis
rock cotoneaster

C. lucida
hedge cotoneaster

C. melanocarpa
black cotoneaster

Figure 1. *Cotoneaster:* fruit, ×2 (Slabaugh 1974a).

Figure 2. *Cotoneaster:* seeds, ×4 (Slabaugh 1974a).

Table 1. *Cotoneaster:* phenology of flowering and fruiting, characteristics of mature trees, cleaned seed weight, and pregermination treatments (Slabaugh 1974a).

Species	Flowering dates	Fruit ripening dates	Seed dispersal dates	Mature tree height (m)	Year first cultivated	Ripe fruit color	Seeds/gram	Acid scarification (minutes)	Substrata	Duration (days)
C. acutifolia	May–June	Sept.–Oct.	Sept.	4	1883	Black	64	10–90	Peat	30–90
C. apiculara	May–June	Aug.–Sept.	Sept.	2	1910	Red	–	120	–	90
C. horizontalis	June	Sept.–Oct.	Sept.	1	1880	Bright red	140	90–180	–	90–120
C. lucida	May–June	Sept.	Sept.	3	1840	Black	52	5–20	Peat	30–90
C. melancarpa	–	–	–	3	1829	Black	–	10–90	Peat	30–90

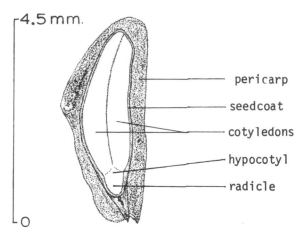

Figure 3. *Cotoneaster horizontalis*, rock cotoneaster: longitudinal section through seed, ×12 (Slabaugh 1974a).

Table 2. *Cotoneaster:* propagation techniques (Dirr and Heuser 1987).

Species	Propagation techniques
C. acutifolia	Seeds: acid scarify followed by 3 months prechilling or sow immature seeds. Cuttings: July cuttings root well.
C. adpressus	Seeds: acid scarify 2 hours and fall plant, or 3 months warm and cold pretreatment. Cuttings: June–August cuttings treated with IBA solution will root.
C. apiculatus	Seeds: acid scarify 2 hours followed by 2–3 months prechilling. Cuttings: Summer cuttings easily root.
C. conspicuus	Seeds: acid scarify 2 hours followed by 2–3 months prechilling.
C. congestus	Seeds: acid scarify 2 hours followed by 2–3 months prechilling.
C. dammeri	Seeds: fall plant or prechill 3 months. Cuttings: easiest cotoneaster to root. Tissue culture: shoot tip explants through tissue culture have led to shoot proliferation.
C. divaricatus	Seeds: fall plant or acid scarify 2 hours followed by 2–3 months prechilling. Cuttings: treat early June cuttings with 1000 ppm IBA solution.
C. glaucophyllus	Cuttings: June–August cuttings treated with IBA solution will root.
C. horizontalis	Seeds: acid scarify 1.5 hours followed by 3–4 months prechilling. Cuttings: treat late July cuttings with 3000 ppm IBA talc.
C. × hybridus	Grafting: often grafted on *C. phaenopyrum* to form small weeping tree.
C. lacteus	Grafting: easy to root.
C. lucidus	Seeds: acid scarify 5–20 minutes followed by 1–3 months prechilling. Cuttings: similar to *C. acutifolius*.
C. microphyllus	Seeds: imported seeds germinate easily.
C. multiflorus	Seeds: fall plant or prechill 3 months. Cuttings: not easy to root. Take cuttings in June.
C. racemiflorus	Seeds: acid scarify 1–2 hours followed by 3 months prechilling. Cuttings: similar to *C. multiflorus*.
C. salicifolius	Seeds: acid scarify 2 hours followed by 2–3 months prechilling. Cuttings: easy to root.

seed-bearing age is 3 years, and seed crops are produced annually.

The seeds may be extracted by running the fruits through a macerator and floating off the pulp. Unfilled seeds will also float off. Seeds may be removed from dried fruits by abrasion. Cleaned seed yields are given in Table 1. Seeds of *C. divaricata* maintained their viability for more than 2 years when stored at 5°C.

PREGERMINATION TREATMENT The seeds of many cotoneasters have hard seedcoats and embryo dormancy. Acid scarification followed by prechilling is necessary to induce germination (Slabaugh 1974a):

Species	Acid scarification (minutes)	Substrata	Duration (days)
C. acutifolia	10–90	Peat	30–90
C. apiculata	120	Sand	90
C. horizontalis	90–180	Sand	90–120
C. lucida	5–20	Peat	30–90
C. melanocarpa	10–90	Peat	30–90

GERMINATION TEST Genebank Handbook (1985) offers rather complex standards for *Cotoneaster* species: substrata, not specific; incubation temperature, 20/30°C; duration, not specified; and additional instructions, light, 8 hours/day, warm stratification at 25°C followed by prechilling at 3 to 5°C for 30 to 60 days.

NURSERY AND FIELD PRACTICE In the northern Great Plains, acid-scarified seeds of hedge cotoneaster (*C. lucida*) can be sown in midsummer or fall in sheltered seed frames. The seedbed should be mulched with hay held in place with snow fencing and kept moist until the ground is frozen. When sprouting begins in the spring, half the mulch should be removed. The remaining half should be removed when the plants begin growth. The seeds should be planted 0.5 cm deep. Prechilled seeds should not be permitted to dry before planting. For hedge cotoneaster, average seedling success is 30%. Seedlings are outplanted when 2 years old. For propagation techniques see Table 2.

ROSACEAE — ROSE FAMILY

Cowania mexicana var. *stansburiana* (Torr.) Jepson — Cliffrose

GROWTH HABIT, OCCURRENCE, AND USE The native range of cliffrose is southern Colorado and throughout most of Utah, west to southern California, and southeast to northern New Mexico, Arizona, and Sonora and Chihuahua in Mexico (Alexander et al. 1974). It grows in exposed, rocky, well-drained situations, such as the south-facing slopes of mesas and canyons. This evergreen shrub ranges in height from 1 to 8 m and is a very valuable browse species for domestic livestock and wildlife. Conspicuous fragrant flowers, plumed fruits, and aromatic foliage make cliffrose a desirable species for ornamental plantings.

HYBRIDIZATION *Cowania* hybridizes readily with *Purshia tridentata*, and *P. glandulosa*, and to a limited extent with *Fallugia paradoxa*.

FLOWERING AND FRUITING The white to sulphur-yellow perfect flowers appear from early May to late June depending on the location. Frequently there is a later flowering in July and August, especially when there are summer storms that wet the soil appreciably. Under these conditions, there may be continued blooming along with some dispersal of fruits until frost.

The fruit is an achene with a persistent feathery style about 2.5 to 5 cm long (Fig. 1), borne in clusters of 4 to 10 on a flat disk (Fig. 2). The first and usually best crop of fruits ripens from the middle of July through August. Fruits maturing from later flowers may be dispersed from August through October. These late

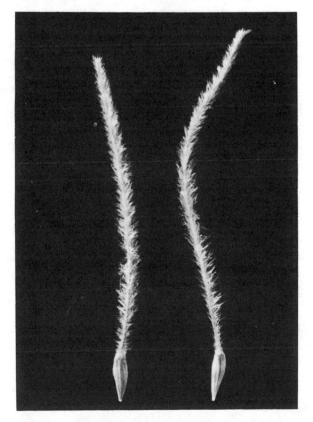

Figure 1. *Cowania mexicana* var. *stansburiana*, cliffrose: achene, ×2 (Alexander et al. 1974).

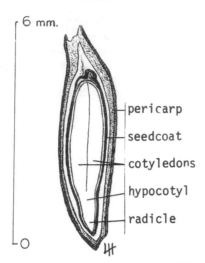

Figure 2. *Cowania mexicana* var. *stansburiana*, cliffrose: longitudinal section through achene, ×10 (Alexander et al. 1974).

fruits are usually of poor quality and are not worth harvesting.

Plants began bearing seed at 5 years. Good crops are borne annually or every 2 years. Seeds are dispersed by wind and animals. Rodent caching of seeds is common.

COLLECTION, EXTRACTION, AND STORAGE OF SEED Fruits may be collected as soon as they are mature and dry. They are rapidly dispersed by wind at maturity so timing of collection is important. They can be stripped by hand or shaken into containers.

The dried flower parts remain attached to the achene in contrast to seeds of *Purshia* that thresh clean. The styles can be readily removed by rubbing-threshing and separating with an air screen.

Fruits weigh 31 kg per 100 liters. Cleaned seed numbers 140 per gram. Seeds remain viable for as long

as 7 years when stored in metal containers at room temperatures.

PREGERMINATION TREATMENT Without pregermination treatment, seed of cliffrose have limited germination. When germination was tested at 55 constant or alternating temperatures ranging from 0 through 40°C, the mean germination of optima (defined as not significantly lower than the maximum observed germination and its confidence interval at the 0.01 level of probablity) was 13% (Young and Evans 1981). Soaking seeds of cliffrose in a 3% solution of thiourea increased mean optimal germination to 66%. We do not recommend the use of this very toxic chemical suspected of producing cancer in laboratory animals. Prechilling cliffrose seeds for 2 weeks at 5°C produced mean optimal germination of 55%. In contrast to seeds of *Purshia* species, soaking seeds of cliffrose in hydrogen peroxide did not enhance germination (Young and Evans 1981). The incubation temperatures that most frequently supported optimum germination were 10/20, 10/25, 10/30, and 15/25°C.

GERMINATION TEST ASOA suggested germination standards are as follows: substrata, B, P; incubation temperature, 15, 10/30°C; duration, 28 days; and additional directions, prechill for 30 days at 2°C or use TZ (Meyers et al. 1989).

NURSERY AND FIELD PRACTICE Cliffrose has been successfully direct seeded on range lands in Utah. Seeds were drilled or broadcast in the fall on sites that had the competing vegetation reduced and some seedbed preparation. About 5.5 kg of seed was planted per hectare.

Container-grown and bare rootstock of cliffrose have been successfully transplanted to range land sites. Alexander et al. (1974) suggested transplanting relatively young seedlings. We have had excellent results with 12- to 18-month-old, container-grown stock that was fully hardened off when transplanted.

ROSACEAE — ROSE FAMILY

Crataegus L. — Hawthorn

GROWTH HABIT, OCCURRENCE, AND USE Hawthorns in North America consist of 100 to 200 species of small trees and shrubs, mostly in the eastern half of the United States (Brinkman 1974f). Their taxonomy is difficult and confusing; some 1100 specific names have been published, but most are no longer accepted. Hybrids no doubt exist and many varieties are recognized. Hawthorns furnish food and cover for wildlife; species with fruits that persist over winter are especially valuable. Many species are useful for environmental plantings. The fruits are sometimes used to make jellies. Because they tolerate a wide variety of sites, hawthorns have also been planted to stabilize banks, for shelterbelts, and for erosion control.

FLOWERING AND FRUITING The white and pink perfect flowers appear with or after the leaves (Table 1). The fruit, a pome containing from 1 to 5 nutlets, ripens 3 to 6 months later. The fruit of many species remains on the tree over winter. Hawthorn fruits usually are red, although they are yellowish or black in some species (Table 1).

COLLECTION, EXTRACTION AND STORAGE OF SEED Fruits that persist on the tree until winter usually must be picked from the tree. In species such as dotted hawthorn (*Crataegus punctata*), however, fruits drop early and can be gathered from the ground. The number of seeds per fruit is variable among trees, so frequent cutting tests are necessary during fruit collection.

Table 1. *Crataegus*: phenology of flowering and fruiting, characteristics of mature trees, yield data, seed weight and pregermination treatments (Brinkman 1874f).

Species	Flowering dates	Fruit ripening dates	Seed dispersal dates	Mature tree height (m)	Year first cultivated	Ripe fruit color	Seeds/kg fruit	Seeds/ gram	Acid scarification (hours)	Warm stratification Temperature (°C)	Warm stratification Duration (days)	Cold stratification Temperature (°C)	Cold stratification Duration (days)
C. arnoldiana	May	Aug.–Sept.	Winter	9	1900	Bright crimson	–	–	4.5	–	–	2–8	180
C. chrysocarpa	May–June	Sept.	–	9	1906	Red to orange	–	24	–	–	–	–	–
C. crus-galli	May–June	Oct.	Winter	9	1656	Bright to dull red	–	–	2–3	22/28	21	2–8	180
C. douglasii	May–June	July–Sept.	Fall	8	1828	Black	33	50	0.5–3	–	–	5	112
C. mollis	Apr.–May	Aug.–Oct.	Fall	–	–	–	–	–	0	30	21	10	180
C. phaenopyrum	May–June	Oct.–Nov.	–	8	1738	Bright red	–	66	0	–	–	10	135
C. punctata	May–June	Sept.–Oct.	Fall	9	1746	Dull red to yellow	25	10	0	22	120	10	135
C. sanguinea	May	Aug.–Sept.	–	7	1822	Bright red	33	–	2	20/25	21	5	21
C. succulenta	May–June	Sept.–Oct.	–	8	1830	Red	–	45	0.5	–	–	5	140

Unless prompt extraction is planned, the fruit should be spread out in thin layers to avoid excessive heating. Extraction of nutlets is readily accomplished by macerating the ripe fruits in water and allowing the pulp to float away. Nutlets of most species have hard, bony endocarps (Figs. 1, 2). Because empty seeds of such species do not float off with the pulp during

C. crus-galli
cockspur hawthorn

C. douglasii
black hawthorn

C. mollis
downy hawthorn

C. phaenopyrum
Washington hawthorn

C. punctata
dotted hawthorn

C. succulenta
fleshy hawthorn

Figure 1. *Crataegus*: cleaned nutlets, ×4 (Brinkman 1974f).

Table 2. *Crataegus:* germination standards (ASOA 1985, Genebank Handbook 1985).

Species	Substrata	Temperature (°C)	Duration (days)	Additional directions	Source
C. mollis	C, TB	20/30	14	Soak 2 hours in concentrated H_2SO_4 then incubate 90 days at room temperature and prechill 120 days; TZ may be used.	ASOA
C. monogyna	S	20/30	28	Warm stratification 3 months at 25°C, then prechill 9 months at 3–5°C.	GB
Crataegus sp.	–	–	40	Abrade with sharp sand, or file or nick seedcoat, prechill 30–60 days at 1–5°C; warm stratification 2–4 weeks at 25°C, then prechill 16 weeks.	GB

extraction, the percentage of sound seeds after cleaning is often low. Nutlets of Washington hawthorn (*C. phaenopyrum*) (Fig. 1), however, have thin endocarps and most empty seeds of this species float off with the pulp. Nutlets should be air dried before storage. Number of seeds per gram is shown in Table 1.

PREGERMINATION TREATMENT Seeds of all hawthorn species have embryo dormancy and require prechilling before germination will occur (Table 1). Seeds of species such as cockspur hawthorn (*C. crus-galli*) are enclosed in a thick, bony endocarp (Fig. 2), that inhibits germination. Under natural conditions such seed would not germinate until the second spring after planting. Acid scarification, followed by warm stratification and then prechilling, often increases germination. The acid treatment should not be used until seeds have been dried for several weeks at room temperature, because acid penetrates fresh seeds and destroys the embryo. Some lots of seed fail to germinate regardless of the treatments used.

Recent research has shown that seeds of *C. monogyna* require between 30 minutes and 2 hours acid scarification while *C. crus-galli* and *C. prunifolia* require up to 4 hours (St. John 1982).

GERMINATION TEST Both ASOA (1985) and Genebank Handbook (1985) list germination standards for *Crataegus* species (Table 2).

NURSERY AND FIELD PRACTICE Species that do not require stratification-prechilling can be fall planted. Best germination often appears when seeds are planted early in the fall before the first frost. Seed should be seeded in drill rows 20 to 30 cm apart and covered with 0.6 cm of soil. Seedlings rapidly develop a tap root and should not be kept in the nursery more than a year. Propagation techniques are given in Table 3.

Table 3. *Crataegus:* propagation techniques (Dirr and Heuser 1987).

Species	Propagation techniques
C. arnoldiana	Seeds: acid scarify 4.5 hours, then prechill 6 months. Grafting: good hardy understock.
C. crus-galli	Seeds: acid scarify 2–3 hours followed by 3 months warm stratification and 3 months prechilling. Grafting: var. *inermis* and 'Crusader' are budded on seedling understock in August and September.
C. laevigata	Seeds: acid scarify, then prechill. Grafting: bench grafted in January on understock of this species.
C. mollis	Seeds: relatively fresh seeds can be planted in the fall or prechilled 3 months.
C. monogyna	Seeds: acid scarify 0.5–2 hours, then prechill.
C. phaenopyrum	Seeds: fall plant or prechill 3 months. Grafting: universal rootstock for most hawthorns; budded in August.
C. prunifolia	Seeds: acid scarify 1–3 hours, then fall plant or prechill 3–4 months.
C. punctata	Seeds: acid scarify 1–2 hours, then prechill 4 months.
C. succulenta	Seeds: acid scarify 30 minutes, then prechill 3 months.

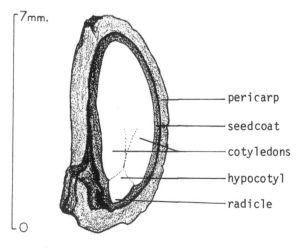

Figure 2. *Crataegus* species: longitudinal section through a nutlet, ×8 (Brinkman 1974f).

CROSSOSOMATACEAE — CROSSOSOMA FAMILY

Crossosoma Nutt.

GROWTH HABIT, OCCURRENCE, AND USE *Crossosoma* is a rare genus consisting of 3 or 4 species found in the southwestern United States. *Crossosoma californicum* is a tall shrub found on the channel islands of southern California. Some of these isolated species have become very successful ornamentals.

The fruit is a follicle. The seeds have a conspicuous fringed aril. No pregermination treatment is necessary (Young and Young 1985).

EUPHORBIACEAE — SPURGE FAMILY

Croton L. — Croton

GROWTH HABIT, OCCURRENCE, AND USE *Croton* is a genus of small trees or shrubs widely distributed in warmer areas. The genus may include 500 species.

The flowers are quite small and greenish. The fruit is a 3-shouldered capsule, which splits into 3 to 6 parts.

Seeds of *C. argyatum* tested in Malaysia had 47% germination over a 7-week period (Ng and Sahah 1979).

TAXODIACEAE — TAXODIUM FAMILY

Cryptomeria japonica D. Don. — Japanese cryptomeria

GROWTH HABIT, OCCURENCE, AND USE *Cryptomeria japonica* is a monospecific genus in the family that contains the redwoods (Walters 1974b). The single species native to China and Japan is an evergreen pyramidal tree with a straight, slender trunk. The stiff branches are whorled and spreading, arranged like long tassels, and the bark is reddish brown. The leaves are linear-subulate and spirally arranged. Trees of this species reach 50 m in height.

COLLECTION, EXTRACTION, AND STORAGE OF SEED Cones should be collected when they change from grayish brown to reddish brown, and spread to dry. The seeds can be extracted by shaking cones on screens (Figs. 1, 2).

GERMINATION The seeds should be soaked in cold water for 12 hours and then placed in plastic bags for 2 to 3 months of prechilling. Bags should be left open for adequate aeration. A germination rate of 30% is considered normal.

The Genebank Handbook (1985) suggests testing *Cryptomeria* seeds on top of blotter paper, at 20/30°C incubation temperatures for 28 days.

Cuttings taken from August to March will root. Plantlets have been successfully regenerated using cotyledons as the explant source in tissue culture (Dirr and Heuser 1987).

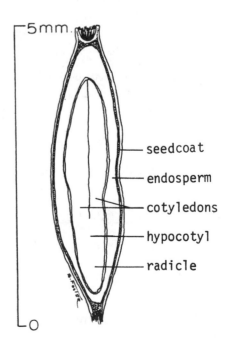

Figure 1. *Cryptomeria japonica*: seed, ×4 (Walters 1974b).

Figure 2. *Cryptomeria japonica*: longitudinal section through a seed, ×16 (Walters 1974b).

TAXODIACEAE — TAXODIUM FAMILY

Cunninghamia R. Br. — China fir

GROWTH HABIT, OCCURRENCE, AND USE The genus *Cunninghamia* consists of 2 species native to China and Formosa (Bailey 1951). These evergreen trees can reach 35 m in height. Their branches are whorled and spreading, and the rigid needles are spirally arranged and 2-rowed. The pistillate cones are globose, the scales leathery, serrate, and pointed. The seeds are narrowly winged. The genus is named for James Cunningham who discovered the trees in 1702.

Cunninghamia is related to Japanese *Cryptomeria* and the genus *Araucaria* of Australia. It has been called China's most indispensable tree. *Cunninghamia lanceolata* is cultivated in the warmer parts of the United States.

COLLECTION, EXTRACTION, AND STORAGE OF SEED Cones should be collected when turning brown and air dried inside.

GERMINATION Nursery growers commonly prechill the seeds for 1 month (Dirr and Heuser 1987). The germination of seeds of *C. lanceolata* has been researched by Ma and Liu (1986) and Chem (1985) in China. However, we have access to few details of this work. Research in Taiwan indicated that prechilling does not improve the germination of seeds of *C. lanceolata* seeds (Kung 1976).

Cuttings taken in November and treated with IBA-talc have a reasonable chance of rooting. Cuttings from lateral branches grow horizontally. Cuttings taken from vertical shoots grow upright (Dirr and Heuser 1987).

CUPRESSACEAE — CYPRESS FAMILY

Cupressus L. — Cypress

GROWTH HABIT, OCCURRENCE, AND USE The cypresses are evergreen trees or shrubs native to the warm-temperate areas of the Northern Hemisphere. The genus comprises about 20 species distributed throughout the western United States, Mexico, northern Central America, the Mediterranean region, North Africa, and from southwestern Asia to Japan (Johnson 1974a).

Most New World cypress are restricted in their occurrence. MacNab (*Cupressus macnabiana*) and Sargent (*C. sargentii*) cypress are associated with serpentine soils. Arizona and Arizona smooth cypress (*C. arizonica* var. *arizonica* and var. *glabra*, respectively) form large stands confined to north slopes. Of the 10 species and varieties found in California, none grow in large pure stands.

Cypresses are widely used in landscaping and environmental plantings. Cypress canker (*Coryneum cardinale*) has limited such plantings, except for resistant species. Arizona cypress is a popular Christmas tree. In Africa and New Zealand, Mexican (*C. lusitanica*) and Monterey (*C. macrocarpa*) cypress are planted for lumber and pulp production. Italian cypress (*C. sempervirens* var. *sempervirens*) is the most widely planted of all the cultivated cypresses.

FLOWERING AND FRUITING Cypresses are monoecious. Staminate and ovulate strobili are produced on the ends of short twigs or branchlets. The staminate strobili are 0.3 to 0.7 cm long, cylindrical or oblong, and light green or, rarely, red. They become yellow as pollen-shedding time nears. Ovulate strobili at time of pollination are less than 0.6 cm long, subglobose to cylindrical, erect, greenish, and with 6 to 12 (rarely 14) distinctly arranged scales.

Pollen is shed in late fall, winter, and spring. Seeds mature 15 to 18 months after pollination. Thus the ovulate cones and their seeds ripen the second season after pollination. Mature cones (Fig. 1) are up to 3 cm in diameter, woody or leathery, and the pellate seeds usually have a central mucro. Each cone produces 12 to 15 seeds.

Cone production starts in most species at an early age. Most cypresses have serotinous cones. Exceptions are San Pedro Martin (*C. arizona* var. *montana*) and Mexican cypress; they have cones that open and shed their seeds at maturity (Wolf and Wagener 1948). Cones on some trees within a stand will open and shed

Figure 1. *Cupressus goveniana* var. *goveniana*, Gowen cypress: cones, ×1 (Johnson 1974a).

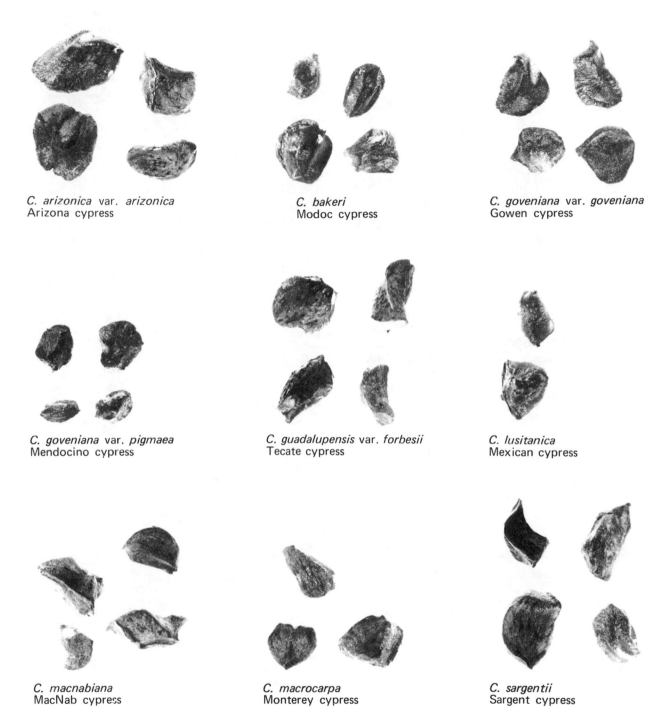

C. *arizonica* var. *arizonica*
Arizona cypress

C. *bakeri*
Modoc cypress

C. *goveniana* var. *goveniana*
Gowen cypress

C. *goveniana* var. *pigmaea*
Mendocino cypress

C. *guadalupensis* var. *forbesii*
Tecate cypress

C. *lusitanica*
Mexican cypress

C. *macnabiana*
MacNab cypress

C. *macrocarpa*
Monterey cypress

C. *sargentii*
Sargent cypress

Figure 2. *Cupressus*: seeds, ×4 (Johnson 1974a).

their seeds in July (Posey and Goggans 1967). For best seed viability it is desirable to collect mature cones not older than 4 years.

Cypress seeds vary widely in shape and size (Figs. 2, 3). Length with wings attached ranges from 0.2 to 0.8 cm; width dimensions are slightly less. Seeds are flattened and lens-shaped, and the wings are tegumentary extensions of the seedcoat. Seed color is an important criterion for determining species (Table 1).

COLLECTION, EXTRACTION, AND STORAGE OF SEED A tree can

be damaged if one or more cones are pulled by hand. Clusters of cones should be cut with pruning shears. According to Wolf and Wagener (1948), care must be taken to collect fully mature cones that have matured the season of collection and have darkened seeds. They note that 2-year-old seeds collected from January to March will have mature seeds. Cone and seed color help to determine maturity (Table 1).

Tight clusters of cones should be cut apart to allow them to open. Cypress cones can be dried at room

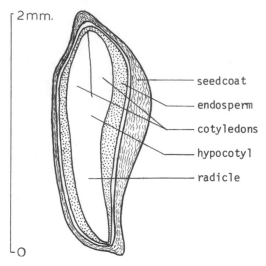

Figure 3. *Cupressus arizonica* var. *arizonica*, Arizona cypress: longitudinal section through a seed, ×32 (Johnson 1974a).

temperature, allowing 1 to 2 months for the cones to open. The process can be speeded up by boiling the cones for 30 to 60 seconds or by cutting each cone in half. Case hardening is a problem with sun drying. Storing the cones under refrigeration to reduce moisture content before drying reduces case hardening. Seeds readily fall from open cones.

The percentage of filled seeds varies among species, varieties, and collections within species. Seeds can be stored far as long as 20 years at room temperatures.

PREGERMINATION TREATMENT *Cupressus* seeds usually require prechilling for about 3 weeks before germination.

GERMINATION TEST ASOA (1985) list standards for Arizona cypress; the Genebank Handbook (1985) is slightly more specific:

Species	Substrata	Temperature (°C)	Duration (days)	Source
C. arizonica[1]	TB	20/30	28	ASOA
C. macrocarpa	TB	20/30	35	GB
C. sempervirens	TB	20	28	GB

[1]Additional instructions: Some lots need 21 days prechilling.

Germination capacity of most species that have been tested is quite low.

NURSERY AND FIELD PRACTICE Seedlings are very susceptible to damping-off. Fall sowing or spring sowing with prechilled seeds is recommended. Seeds are sprinkled on beds and given a very thin covering of soil. Seedbed densities of 310 to 630 per m^2 are commonly used. Germination beds are often mulched with decomposed granite. Cypress seedlings can be out planted as 1- or 2-year-old stock. One-year-old seedlings have only juvenile foliage.

Cuttings of Arizona cypress are difficult to root. Cuttings should be taken from juvenile plants and treated with 2000 to 5000 ppm of IBA. Cuttings taken in November-December have the best chance of rooting (Dirr and Heuser 1987). Untreated cuttings of Italian cypress taken in April root readily.

Horticultural cultivars are often grafted on seedling rootstocks (Dirr and Heuser 1987). Pollen explants have been induced to produce callus through tissue culture.

Table 1. *Cupressus:* growth form, mature tree height, ripe cone and fruit colors, cleaned seed weight, and scales/seeds per cone (Johnson 1974).

Species	Growth form	Mature tree height (m)	Ripe cone color	Ripe seed color	Seeds/ gram	Scales/ cone	Seeds/ cone
C. arizonica							
var. *arizonica*	straight	21	dull gray-brown	dark brown	180	6–8	90–120
var. *glabra*	straight	15	glaucous bloom	purplish brown	120	5–10	90–100
var. *montana*	spreading	21	rich brown	red-brown	440	8–12	60–70
var. *nevadanensis*	pyramidal	15	silver-gray	rich tan	130	6–8	90
var. *stephonsonii*	narrow	15	dull gray	dark brown	110	6–8	100–125
C. bakeri	narrow	15	dull brown	light tan	360	6–8	50–85
C. goveniana							
var. *goveniana*	shrub	18	brown	black	280	3–5	90–110
var. *pygmaea*	shrub	9	weathered gray	jet black	240	88–10	130
C. guadalupensis							
var. *forbesii*	irregular	9	dull brown	dark brown	100	6–10	–
var. *guadalupensis*	broad	19	–	glaucous brown	60	8–10	100
C. lusitanica	drooping	30	dull brown	rich light tan	260	6–10	75
C. macnabiana	multiple	12	brownish	brown	120	6–8	75–105
C. sargentii	variable	23	dull brown	dark brown	120	6–10	100
sempervirens							
var. *sempervirens*	columnar	46	brown	–	140	8–14	60–280

CYRILLACEAE — CYRILLA FAMILY

Cyrilla L. — Leatherwood

GROWTH HABIT, OCCURRENCE, AND USE *Cyrilla* is a monospecific genus consisting of *C. racemiflora*. This species is a deciduous shrub or small tree to 10 m in height. The leaves are 4 to 8 cm long and lustrous bright green above and reticulate beneath.

FLOWERING AND FRUITING The small flowers are borne in slender racemes that become pendulous in fruit.

The flowers appear in July and August and are followed by the capsule fruits in September and October. The fruit is a 2-celled, loculicidally dehiscent capsule. The fruits should be collected in the yellow-brown transition stage. The seeds will dehisce from drying fruits.

GERMINATION Seeds require no pretreatment for germination (Dirr and Heuser 1987).

Cuttings taken in August and treated with 1.0% IBA will root. Cuttings transplant readily (Dirr and Heuser 1987).

LEGUMINOSAE — LEGUME FAMILY

Cytisus scoparius (L.) Lk. — Scotch broom

GROWTH HABIT, OCCURRENCE, AND USE Scotch broom is an evergreen, bushy shrub, generally 0.9 to 2.2 m tall, but occasionally as tall as 4 m (Gill and Pogge 1974c). The plant is native to Europe, but has escaped from cultivation in North America. It favors dry, sandy soils in full sunlight. Scotch broom is considered poisonous to livestock, but is seldom browsed. It is considered a pest in California and is subject to a biological control program.

FLOWERING AND FRUITING The flowers are bisexual. Flowering occurs in May-June; fruits ripen in August and disperse in September. The fruit, a narrow, oblong pod 3 to 5 cm long, is brown to black at maturity. The dark seeds are 0.3 cm long (Fig. 1).

COLLECTION, EXTRACTION, AND STORAGE OF SEED After the fruits are ripe, from late July to September, the pods may be picked from the shrubs or gathered from the ground. Pods should be air-dried after collection. The dry pods can be threshed and the seeds recovered by air screening. Cleaned seed yield per 100 kg of pods is 35 to 48 kg, and seeds average 127 per gram.

GERMINATION *Cytisus* seeds have hard seedcoats that require scarification. Hot water or acid scarification are recommended (Heit 1967c). The Genebank Handbook (1985) standards suggest substrata, TP; temperature, 20/30°C; duration, 28 days; and special instructions, chip or file cotyledon end of testa, and presoak for 3 hours.

NURSERY AND FIELD PRACTICE Seed should be sown in the spring after acid scarification or hot water treatment. Dirr and Heuser (1987) suggest 30 minutes in acid. Cuttings should be container grown because they do not transplant well.

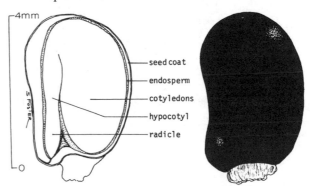

Figure 1. *Cytisus scoparius*, Scotch broom: longitudinal section through a seed (left) and exterior view (right), both at ×10 (Gill and Pogge 1974c).

LEGUMINOSAE — LEGUME FAMILY

Dalea L. — Indigo bush

GROWTH HABIT, OCCURRENCE, AND USE The genus *Dalea* is often considered in modern treatments to be *Psorothamnus*. It consist of over 100 species native to North and South America. Mozingo (1987) considered smokebush (*D. polyadenia*) to be the most fragrant of the temperate desert shrubs. Its characteristic fragrance is the result of volatile substances secreted from numerous pinhead-sized orange or yellow glands located on the stems. In addition to their glandular stems, these shrubs are easily recognized by their angular, stiff stems, which are sparsely leafy at best. Often these shrubs have leaves only during periods of available soil moisture; they can be leafless for most of the year. Most species of *Dalea* are confined to the warm deserts of southwestern United States.

The fruit of smokebush is a legume from which the seeds do not initially dehisce. It contains a germination inhibitor (Mozingo 1987). The seedcoats are hard and require scarification. Once the seeds are scarified, germination is rapid.

THYMELAEACEAE — MEZEREUM FAMILY

Daphne L.

GROWTH HABIT, OCCURRENCE, AND USE The genus *Daphne* consists of about 50 species of erect or prostrate shrubs native to temperate and subtropical Europe and Asia (Bailey 1951). It contains both deciduous and evergreen shrubs, some with showy, fragrant flowers.

FLOWERING AND FRUITING The fruit is a fleshy drupe of various colors. The fruit of *D. mezereum* is bright red and ripens in June to July. It is relished by birds.

GERMINATION Seeds are rarely used in propagation because they are thought to be very dormant. Seeds of *D. giraldii* apparently will germinate in the spring if sown as soon after collection as possible (Dirr and

Heuser 1987). Old seed of this species requires 3 months prechilling for germination.

Apparently there is a water-soluble inhibitor in the drupe that can be removed by soaking. The pulpy matrix should be removed from the seed.

Daphne odora usually does not set seed in the United States. Seeds of *D. retusa* had 100% germination after 1 month warm stratification followed by 2 months of prechilling and incubation at low temperatures (Dirr and Heuser 1987).

Daphne species are difficult to root from cuttings.

NYSSACEAE — NYSSA FAMILY

Davidia Baill. — Dove tree

GROWTH HABIT, OCCURRENCE, AND USE The genus *Davidia* consists of one species of deciduous tree native to China (Bailey 1951). Dove tree (*D. involucrata*) reaches 20 m in height and has broad, ovate leaves.

FLOWERING AND FRUITING The flowers are unisexual, apetalous, and borne in dense, nearly globose heads. The fruit is a rather large drupe 4 cm long that ripens in October. It has an outer coat that can be softened by

parial fermentation in a plastic bag. The endocarp is rigid and extremely hard.

GERMINATION Dormancy is caused by the hard wall and epicotyl dormancy. Warm stratification for 5 months followed by 3 months prechilling is necessary for germination. Seeds planted in the fall will emerge the second year (Dirr and Heuser 1987).

PAPAVERACEAE — POPPY FAMILY

Dendromecon rigida Benth. — Stiff bushpoppy

GROWTH HABIT, OCCURRENCE, AND USE Stiff bushpoppy is an openly branched evergreen shrub 0.6 to 2.5 m high, sometimes to 6 m (Neal 1974b). It has a woody base with gray or white shreddy-barked stems. It grows on dry chaparral slopes, ridges, and washes below 1850 m in the Coast Range, from Baja California to Sonoma County, California, and in the west foothills of the Sierra Nevada, from Shasta County to Tulare County.

The species is useful for watershed protection and browse production. Goats are especially fond of bushpoppy. Deer and goats browse on the sprouts after a fire.

FLOWERING AND FRUITING Flowers are bisexual, yellow,

showy, and solitary on stalks. They appear in April through June and sometimes into August. Fruits are linear, grooved capsules, 5 to 10 cm long with 2 valves separating incompletely at maturity. Ripe fruits may be collected in May, June, and July.

Seeds are black and have minute embryos (Fig. 1). In 2 samples of cleaned seeds, purity was 77% and soundness 97%. There are 92 to 115 seeds per gram.

GERMINATION Seeds have been sown in a moist medium at temperatures alternating diurnally from 5 to 22°C. Germination started after 50 days at these temperatures and reached 21% 102 days after sowing (Mirov and Kraebel 1939).

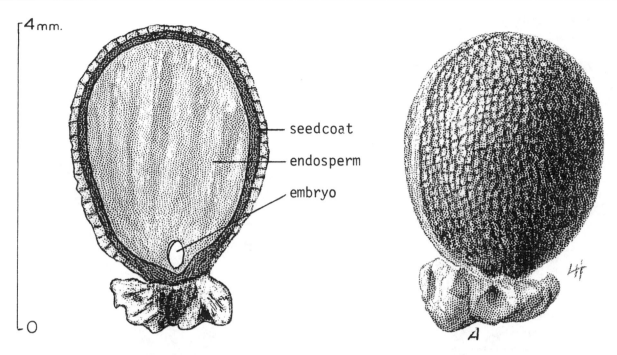

Figure 1. *Dendromecon rigida*, stiff bushpoppy: longitudinal section through a seed (left) and exterior view (right), ×20 (Neal 1974b).

SAXIFRAGACEAE — SAXIFRAGE FAMILY

Deutzia Thunb. — Deutzia

GROWTH HABIT, OCCURRENCE, AND USE *Deutzia* is a genus of about 50 shrubs native to eastern Asia and the Himalayan region, with 2 species in Mexico (Bailey 1951). Several species are widely planted as ornamentals.

FLOWERING AND FRUITING The white or rose-pink flowers are followed by capsular fruits.

GERMINATION The seeds require no pretreatment for germination (Dirr and Heuser 1987).

These species are usually propagated from hardwood cuttings.

CAPRIFOLIACEAE — HONEYSUCKLE FAMILY

Diervilla Mill. — Bush honeysuckle

GROWTH HABIT, OCCURRENCE, AND USE The genus *Diervilla* consists of 3 species of deciduous stoloniferous shrubs native to North America. Dwarf bush honeysuckle (*D. lonicera*) reaches 0.75 to 1.5 m in height and occurs from eastern Canada to North Carolina. Georgia bush honeysuckle (*D. rivularis*) reaches 2 m in height and is found growing naturally from North Carolina to Alabama. Southern bush honeysuckle (*D. sessilifolia*), 1.5 m in height, is native to southeastern United States.

The fruit is a capsule. The seeds require no pretreatment for germination (Dirr and Heuser 1987).

EBENACEAE— EBONY FAMILY

Diospyros L.— Persimmon

GROWTH HABIT, OCCURRENCE, AND USE The common persimmon (*D. virginiana*) is a small- to medium-sized deciduous tree, normally attaining a height of 10 to 20 m at maturity (Olson and Barnes 1974b). It occurs in open woods and as an invader of old fields from Connecticut, west through southern Ohio to eastern Kansas, and south to Florida and Texas. Common persimmon develops best in the rich bottom lands of the Mississippi River and its tributaries and in coastal river valleys. In these optimum habitats, the species often attains a height of 21 to 24 m and a diameter of 50 to 60 cm.

In past years, persimmon wood was used extensively for weaver's shuttles, golf club heads, and other products requiring hard, smooth-wearing wood. At present, such uses have diminished because of laminates and other materials.

The fruit, which is exceedingly astringent when green, but delicious when thoroughly ripe, is eaten by mammals and birds. The common persimmon is a handsome foliage tree and a valuable honey plant. It has been cultivated since 1629. Several varieties have been developed for fruit production.

There are nearly 200 widely distributed species of *Diospyros* (Bailey 1951). *Diospyros melanoxylon* is an important tropical forest tree. *Diospyros texana* is a shrub found on the range lands of central and south Texas and adjacent Mexico. It is a pest on range lands, but is a valuable species for wildlife (Everitt 1984).

FLOWERING AND FRUITING The small, dioecious, axillary flowers of common persimmon are borne after the leaves, from March to mid-June, depending on the latitude. The insect-pollinated flowers are most common in April and May.

The fruit is green before ripening and may vary in color when ripe from green, yellow, orange, or yellowish brown to dark reddish purple-black (Fig. 1). It is a plum-like berry 1.9 to 5 cm wide, glaucous, with a conspicuous, persistent calyx, and contains 3 to 8 seeds. The fruits ripen from September to November; the flat, brown seeds, about 1.25 cm long, are dispersed from ripening to late winter (Fig. 2). The seed is disseminated by birds and animals that feed on the fruits, and to some extent, by floods in bottom lands. Seed bearing may begin at age 10, but the optimum is 25 to 50 years. Good seed crops are borne about every 2 years, with light crops in intervening years.

COLLECTION, EXTRACTION, AND STORAGE OF SEED The fruit of common persimmon may be picked or shaken from the trees as soon as it is ripe and soft in texture. It may also be picked from the ground after natural fall. The seeds are easily removed by running the fruits with water through a macerator and allowing the pulp to float away. Dirr and Heuser (1987) suggest storing ripe fruits in a plastic bag until they turn liquid and then

Figure 1. *Diospyros virginiana*, common persimmon: mature fruit and a single seed, both ×2 (Olson and Barnes 1974b).

pouring off the pulp.

After cleaning, seeds should be spread out to dry. Dry seed can be stored in sealed, dry containers at 5°C. The seeds are large with 1.5 to 3.9 seeds per gram. Seed yield from fruits is about 4.5 to 13.6 kg of seeds per 45 kg of fruits.

PREGERMINATION TREATMENT Natural germination of common persimmon usually occurs in April or May, but 2- to 3-year delays have been noted. The cause of delayed germination is the seedcoat, which caps the radicle. Removal of this cap results in complete germination.

Seed dormancy also may be broken by prechilling for 60 to 90 days. Acid scarification has also been used. Lalman and Misra (1980a) considered phenol compounds in the seedcoat were responsible for the dormancy of seeds of *D. melanoxylon*. The compounds could be leached from the seeds with water. In a sep-

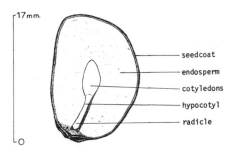

Figure 2. *Diospyros virginiana*, common persimmon: longitudinal section through a seed, ×2 (Olson and Barnes 1974b).

arate study they determined that acid scarification and enrichment with gibberellin enhanced germination of seeds (Lalman and Misra 1980a).

Fruits of *D. texana* produce extracts that are highly inhibitory to root growth and germination of other plant species (Myer et al. 1970). The inhibitor occurs at all stages of fruit development.

Seeds of *D. kaki* have highest germination when planted immediately after collection (Dirr and Heuser 1987). Seeds of *D. marmorata* collected in the Andaman Islands could be stored for about 20 days if left in the fruit. Emergence of 14 to 28% was obtained with these seeds (Sharma 1977).

GERMINATION TEST Seeds of common persimmon can be tested on sand or peat substrates at incubation temperatures of 20/30°C. Germination capacities seldom exceed 60%.

NURSERY AND FIELD PRACTICE Persimmon seed may be sown in the fall, or prechilled and sown in the spring. In Missouri, fall sowing at a depth of 5 cm is the normal practice, and seedbeds are mulched. An average yield of 25 to 33% trees is easily obtainable. Seedlings have a

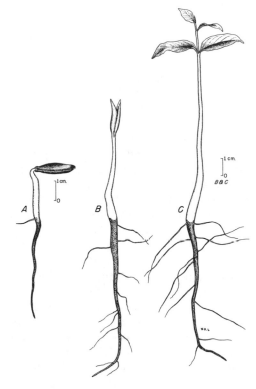

Figure 3. *Diospyros virginiana*, common persimmon: seedling development at 4, 6, and 8 days after germination (Olson and Barnes 1974b).

strong tap root (Fig. 3) and should be field planted at the end of the first season.

Common persimmon roots sucker naturally, and root cuttings have been used for vegetative propagation (Dirr and Heuser 1987). Budding with chip buds on seedlings can be conducted in late July and August.

DIPTEROCARPACEAE — DIPTEROCARP FAMILY

Dipterocarpus Gaertn.

GROWTH HABIT, OCCURRENCE, AND USE The dipterocarps are the main tree species in the tropical rain forests of southeast Asia. The family contains many genera besides *Dipterocarpus* (Table 1). Seeds that do not germinate within a few days of falling to the forest floor rapidly lose viability. These species are vulnerable to extinction by logging because there is no reserve of seeds on the forest floor (Ng 1975, 1977a).

FLOWERING AND FRUITING Studies in Malaysia have shown that at least one species of dipterocarp is flowering every month except August (Tamari 1976). Most species flower for a period of 2 to 5 months. It is estimated that 90% of the flowers are lost to physiological stress or predation before they set viable seeds.

COLLECTION, EXTRACTION, AND STORAGE OF SEED Most species are huge trees from which it is difficult to collect fruits. As dipterocarp seeds mature, the moisture content reduces to 50 to 60% of dry weight. Fully mature seeds germinate faster and with greater capacity than immature seeds. Therefore, timing of seed collection has practical silvicultural significance (Sasaki 1980).

Dipterocarp seeds lose viability at low moisture content, below 20 to 30% of their dry weight (Sasaki 1980). For storage the seeds must be maintained above the critical level for each species. Seeds can be stored in plastic bags or closed containers at 95% relative humidity and 25°C. Storage is very difficult for some species (Tang and Tamari 1973).

Table 1. *Dipterocarpus:* genera and germination (Ng 1977a).

Genus and species	Germination period	Weeks to germination
Balanocarpus hemii	Rapid	2–7
Dipterocarpus baudii	Rapid	3–4
D. oblongifolia	Rapid	2–3
D. oblongifolius	Rapid	2–4
Dryobalanops aromatica	Rapid	1–2
Hopea dyeri	Rapid	1–5
H. helferi	Rapid	2
H. nervosa	Rapid	3–6
H. nutans	Rapid	2–3
H. odorta	Rapid	1–3
H. subalata	Rapid	2–9
Parashorea densiflora	Prolonged	4–30
Shorea assamica	Rapid	2–4
S. leprousula	Rapid	1–5
S. maxima	Rapid	1–3
S. ovalis	Rapid	3–5
S. parvifolia	Rapid	1–4
S. platycladus	Rapid	1–3
S. resinosa	Rapid	–
S. singkawang	Rapid	2–4
S. sumatrana	Rapid	2–3
S. talura	Rapid	3–7
Vatica lowii	Rapid	3–10
V. stapfiana	Rapid	3–6
V. wallichii	Rapid	4–5

THYMELAEACEAE — MEZEREUM FAMILY

Dirca L. — Leatherwood

GROWTH HABIT, OCCURRENCE, AND USE The genus *Dirca* consists of 2 species of shrubs native to North America (Bailey 1951). *Dirca palustris* reaches 2 m in height. Natural distribution is in eastern North America from Canada to Florida.

FLOWERING AND FRUITING The fruit is a yellow-tinged drupe that ripens in June and July and is either consumed by birds or falls to the ground.

GERMINATION Freshly cleaned seeds, if sown immediately, have about 50% germination the following spring (Dirr and Heuser 1987). Most germination treatments to these seeds only lower emergence.

APOCYNACEAE — DOGBANE FAMILY

Dyera costulata Hook. — Jelutong

GROWTH HABIT, OCCURRENCE, AND USE Jelutong (*Dyera costulata*) has long been known for its latex which is used as a base for chewing gum (Williams 1963). It is one of the most important timber species in southeast Asia. Accelerated logging has nearly exhausted natural stands and artificial regeneration is necessary to ensure a future supply.

FLOWERING AND FRUITING In Malaysia most jelutong trees flower from July to December (Yap 1980). Anthesis occurs at night and abundant white petals are found on the ground the next morning. Flowering lasts for 2 to 3 weeks on a tree, and young fruits are detected after 2 to 3 months. Ripe fruit can be collected 8 to 9 months after anthesis.

Jelutong seeds are enclosed in paired woody pods, varying from 28 to 41 cm in length. An average pod contains 18 seeds. Each seed is extremely flattened, with the edges expanded into membranous wings, and weighs only 0.137 g.

COLLECTION, EXTRACTION, AND STORAGE OF SEED Ripening of the fruits begins with gradual flattening of the pods and reduction of latex in the pericarp. Collection can be done by climbing the trees and breaking off the ripe fruits with bamboo poles. The pods will split and dehisce the seeds after a week of air drying. Fully mature seeds are brownish with a lower moisture content than greenish immature seeds.

GERMINATION Fully mature seeds are highly viable when tested in dishes on tissue paper at room temperature. Germination capacities of 75 to 92% were obtained in 14 to 28 days.

Low-temperature storage was found to be deleterious for jelutong seed. Storage at 20 to 40°C with a relative humidity of 60% retained viability for 3 months and resulted in a loss of only 20% viability in 8 months.

COMPOSITAE — SUNFLOWER FAMILY

Dyssodia Cav.

GROWTH HABIT, OCCURRENCE, AND USE The 40 species of genus *Dyssodia* are found in the deserts of southwestern United States and Mexico. *Dyssodia cooperi* is a semishrub 0.3 to 0.5 m in height, found in the Mojave Desert and in the temperate deserts of the Great Basin.

The fruit is a bristly achene.

GERMINATION In studies conducted by Kay et al. (1988), initial germination of achenes was very low. After 4 years of storage at room temperature, germination exceeded 20%.

COMPOSITAE — SUNFLOWER FAMILY

Eastwoodia Bdg.

GROWTH HABIT, OCCURRENCE, AND USE *Eastwoodia* is a monotypic, xerophytic genus named in honor of Alice Eastwood, a California botanist. It is a rounded desert shrub of central and southern California.

The fruit is a silky-pilose achene.

GERMINATION In studies by Kay et al. (1988) seeds of *E. elegans* had about 35% germination, and under cold or laboratory storage maintained this level of germination for 7 years.

BORAGINACEAE — BORAGE FAMILY

Ehretia R. Br. — Anacua

GROWTH HABIT, OCCURRENCE, AND USE Anacua (*Ehretia anacua*) is a tree found on bottom land sites in central and south Texas. It is evergreen in extreme southern Texas, but is partially deciduous in the northern part of its range. Anacua reaches a height of 15 m and has a rounded crown. Puna (*E. acuminata*) is found in sub-Himalayan areas to 1500 m in elevation. The leaves are mixed with tea as a beverage and also serve as a fodder source.

FLOWERING AND FRUITING It flowers in March and April and sometimes again in September after rains. The flowers are white and fragrant and are followed by showy, yellow-orange fruits.

GERMINATION Seeds of anacua have hard seedcoats and require acid scarification for 2 hours (Alaniz and Everitt 1980).

Soaking seeds of *E. acuminata* as a pregermination treatment results in the formation of seed mucilage which interferes with sowing (Sagwal 1986). Acid scarification for 10 minutes resulted in 87% germination.

ELAEAGNACEAE — OLEASTER FAMILY

Elaeagnus L.

GROWTH HABIT, OCCURRENCE, AND USE The genus *Elaeagnus* includes about 40 species of shrubs or trees (Olson 1974b). These species are grown as ornamentals, produce edible fruits, and serve as valuable wildlife plants. Russian olive (*E. angustifolia*) has been widely planted in North America, has escaped cultivation, and is naturalized in several areas.

FLOWERING AND FRUITING The fragrant, small, perfect flowers are borne in late spring and are pollinated by insects (Table 1). The fruit is a dry, indehiscent achene

enveloped by a persistent fleshy perianth (Figs. 1, 2, 3). Color of the ripe fruit varies with species. Seeds are distributed by birds.

COLLECTION, EXTRACTION, AND STORAGE OF SEED Ripe fruits can be handpicked or flailed onto ground sheets. Seeds can be recoverd by maceration with water and flotation, followed by air screening of dried seeds. Commercial seed may be either cleaned seeds or dried fruits. Cleaned seeds dried to 6 to 14% moisture can be stored at cold temperatures, and viability can be maintained for 1 to 3 years. Seed weight and ripe fruit color are given in Table 1.

PREGERMINATION TREATMENT Elaeagnus seeds are dormant and require prechilling for germination. The duration of prechilling is usually 60 days. Russian olive seeds are sometimes considered to have hard seedcoats and are acid scarified. Hamilton and Carpenter (1975) found a coumarinlike germination inhibitor in

Figure 1. *Elaeagnus angustifolia*, Russian olive: fruit, ×2 (Olson 1974b).

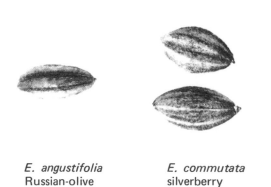

E. angustifolia
Russian-olive

E. commutata
silverberry

Figure 2. *Elaeagnus*: achenes with fleshly perianth removed, ×2 (Olson 1974b).

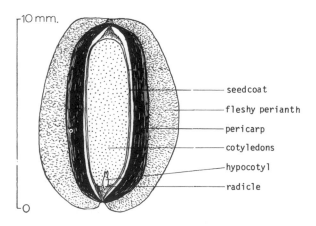

seedcoat
fleshy perianth
pericarp
cotyledons
hypocotyl
radicle

Figure 3. *Elaeagnus angustifolia*, Russian olive: longitudinal section through an achene enclosed in its fleshy perianth, ×5 (Olson 1974b).

Table 1. *Elaeagnus:* phenology of flowering and fruiting, characteristics of mature trees, ripe fruit color, and cleaned seed weight (Olson 1974b).

Species	Flowering dates	Fruit ripening dates	Seed dispersal dates	Mature tree height (m)	Year first cultivated	Seed-bearing Age (years)	Interval (years)	Ripe fruit color	Seeds/ gram
E. angustifolia	June	Aug.–Oct.	Winter	9	Long	3–5	1	Silver gray	11.4
E. commutata	June–July	Aug.–Sept.	Sept.–Nov.	5	1813	–	–	Silver	8.4
E. umbellata	May–June	Aug.–Oct.	Sept.–Nov.	4	1830	6	–	Red-pink	60.8

the endocarp, testa, and embryo of dormant seeds of Russian olive. Activity of this inhibitor did not decrease during 90 days of prechilling, but its effect was reduced by a growth-promoting substance formed during prechilling.

Germination of seeds of Silverberry (*E. commutata*) was enhanced by soaking in warm water for 2 days to remove a soluble inhibitor (Fung 1984). These seeds are photoblastic in that germination is reduced by light.

Seeds of autumn elaeagnus (*E. umbellata*) were shown to be similar to Russian olive seeds in having a coumarinlike inhibitor (Hamilton and Carpenter 1975). Fowler and Fowler (1987) found that 16 weeks prechilling followed by incubation at 10/20°C produced optimum germination for autumn elaeagnus seeds.

GERMINATION TEST The Genebank Handbook (1985) lists the following standards for germination of seeds of *Elaeagnus* species:

Species	Additional instructions
E. angustifolia	Light, acid scarification, 4 weeks of warm stratification, prechill 8–12 weeks. Incubate at 23/30°C.
E. multiflora	Presoak 24 hours, prechill 30–60 days.
E. philippensis	Presoak 24 hours.
E. pungens	Presoak 24 hours, prechill 30–60 days.
E. umbellata	Prechill 8–12 weeks.

NURSERY AND FIELD PRACTICE Seeds may be sown 1.25 to 2.5 cm deep in late summer or fall without prechilling or in the spring after prechilling. In areas with large populations of mice, the pulp should be removed from the seeds before sowing. Seedbeds should be mulched to prevent soil splash which coats the seedling leaves. A seedling density of 125 to 315 plants per m^2 is desirable.

PALMAAE — PALM FAMILY

Elaeis Jacq. — Oil palm

GROWTH HABIT, OCCURRENCE, AND USE The genus *Elaeis* consists of a single species, *E. guineensis*, with many botanical varieties and forms. It is native to tropical Africa where it yields nuts and seeds for the extraction of oil.

GERMINATION The seed coverings—pericarp, endocarp, and operculum—are apparent causes of dormancy.

There have been reports that the embryo may also be dormant (Genebank Handbook 1985). The form *pisifera* has been reported to be more dormant than the forms *dura* and *tenera*. In preparation for planting some form of pregermination heat treatment is used. Predrying is the term used for dry heat, and warm stratification is used for moist heat.

LABIATAE — MINT FAMILY

Elsholtzia Willd.

GROWTH HABIT, OCCURRENCE, AND USE The genus *Elsholtzia* consists of about 20 species native to eastern and central Asia, south to Java, and extending to Europe and the Abyssinian highlands. *Elsholtzia stauntonii* is a shrub growing up to 1.3 m in height.

FLOWERING AND FRUITING The flowers are lilac-purple in color and are borne in dense, one-sided spikes. The fruit is a nutlet.

GERMINATION Seeds germinate without pretreatment (Dirr and Heuser 1987). Cuttings taken in June and July and treated with IBA root readily.

COMPOSITAE — SUNFLOWER FAMILY

Encelia Adans.

GROWTH HABIT, OCCURRENCE, AND USE The genus *Encelia* consists of about 14 species distributed from southwestern United States to South America and the Galápagos Islands. Virgin River encelia (*E. virginensis*) is found in the Colorado and Mojave deserts. This species has potential for use in environmental plantings.

GERMINATION The fruit is an achene that germinates without pretreatment (Young and Young 1985). In studies conducted by Kay et al. (1988) germination was initially about 75%. Under storage, germination rapidly declined, but over time it increased gradually so that after 14 years germination of seed stored under cold temperatures or in the laboratory was back to the 60 to 70% range.

ERICACEAE — HEATH FAMILY

Enkianthus Lour.

GROWTH HABIT, OCCURRENCE, AND USE The genus *Enkianthus* consists of about 10 species native to Japan, China, and the Himalayas. These are elegant shrubs with refined foliage and lovely, white to pink flowers. The fall color of the foliage is flaming red. The fruit is a 5-valved dehiscent that capsule should be collected in the yellow-brown stage and stored inside.

GERMINATION Seeds are very small, like those of azaleas. Care must be taken not to sow them too densely. Sow seeds on milled sphagnum in flats and place them under mist. Germination occurs in 2 to 3 weeks, and seedlings can be transplanted in 6 to 8 weeks. *Enkianthus campanulatus*, *E. cernuus*, and *E. deflexus* germinate readily without pretreatment. Seeds of *E. perulatus* have been reported to be very dormant.

EPHEDRACEAE — EPHEDRA FAMILY

Ephedra L.

GROWTH HABIT, OCCURRENCE, AND USE The order Ephedrales of the Coniferophyta contains a single family with only one genus. *Ephedra* consists of about 40 species of shrubs or small trees with jointed, grooved, green stems and scalelike, opposite leaves. The plants are usually dioecious. The cones are borne axillary, with small scales. Ephedra species are found in the arid portions of central Asia and western North America. In western United States they are important browse species and valuable landscape plants.

FLOWERING AND FRUITING North American species flower in midspring and the seeds mature in midsummer. Seed crops tend to be irregular. Seeds can be collected by beating branches over the edge of a container. Some of the species from Asia have a beautiful red aril (Ali et al. 1979, Martynchuk 1984).

GERMINATION The seeds of green (*E. viridis*) and gray (*E. nevadensis*) ephedra germinate over a wide range of temperatures without pretreatment (Young et al. 1977). Gray ephedra seeds did not maintain high viability after a year of storage at room temperature or under cold temperatures (Kay et al. 1988). Seeds of green ephedra stored at −15°C retained viability for 14 years. Seeds of *E. trifurca* and *E. californica* remained viable for 7 years under all storage conditions tested. Seeds of *E. funera* had low initial germination that did not improve or decrease with storage.

ERICACEAE

Epigaea L. — Trailing arbutus

GROWTH HABIT, OCCURRENCE, AND USE Trailing arbutus (*Epigaea repens*) is an evergreen, prostrate, creeping shrub that grows in patches up to 0.6 m in diameter (Blum and Krochmal 1974). It is found growing in woodlands on acid, sandy soils from New England, southeast to New York, Pennsylvania, West Virginia, and Ohio. The variety *glabrifolia* ranges north from the higher parts of the Appalachian Mountains to Newfoundland, Nova Scotia, Labrador, and west to Saskatchewan. The species is noted for its fragrant blossoms. It has been used occasionally as an ornamental since 1736. The fruits of trailing arbutus are eaten by small mammals.

FLOWERING AND FRUITING The flowers are spicy smelling, pink to white in color, and bloom from March to May, although specimens have been known to bloom as early as January at low elevations in the southern part of the range. Flowering usually begins when plants are 3 years old. The flowers are usually unisexual, with male and female flowers on different plants, but occasional flowers are perfect. Double-flowered forms and fall bloomers have been reported.

The fruit is is a 5-lobed, hairy, dehiscent capsule about 0.3 cm in diameter. The seeds are imbedded in a fleshy pulp within the capsule. A sample of 155 wild fruits contained an average of 241 tiny, shiny, brown, hard seeds per capsule (Fig. 1). The capsule splits when ripe and ejects most of the seeds with some force, but separation of the capsule before the seeds are ripe is difficult. In June to July the capsules become ripe, split open, and eject most of their seeds, but a few seeds remain in the capsule during the fall. The seeds of this species are distributed by ants (Clay 1983).

COLLECTION, EXTRACTION, AND STORAGE OF SEED Capsules can be collected after they are mature and before they eject their seeds. They can be air-dried in a container until the seeds are ejected. The seeds are very small,

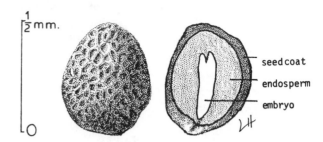

Figure 1. *Epigaea repens*, trailing arbutus: exterior view of seed (left) and longitudinal section (right), both at ×60 (Blum and Krochmal 1974).

running about 22,000 per gram. They do not store well over 1 year at room temperature.

GERMINATION Germination is epigeal and occurs readily without pretreatment. Recent research by Lincoln (1980) suggests that *E. repens* seeds should be prechilled at 5 to 8°C for 30 days and then incubated at 15/25°C for 68 days with light during this warm period for 8 hours daily.

NURSERY AND FIELD PRACTICE The extremely small seeds can be germinated in flats on an acid peat medium. The flats can be covered with plastic film to maintain a high relative humidity. When seedlings develop 3 to 5 leaves they can be transplanted to individual pots. In 1 year the plants develop into rosettes about 10 cm in diameter.

When transplanted to permanent locations, each plant should be set in a pocket of prepared soil consisting of 4 parts leaf mold, 2 parts crumbled peat moss, and 2 parts sand. These plants do best in acid soil. Care needs to be taken to ensure proper mycorrhizal inoculation by bringing in soil from around a mature healthy plant.

With care the plant can be propagated by stem cuttings (Dirr and Heuser 1987).

ERICACEAE — HEATH FAMILY

Erica L.

GROWTH HABIT, OCCURRENCE, AND USE *Erica* is a genus of nearly 600 species of shrubs and subshrubs native to South Africa and the Mediterranean region. The fruit is a capsule containing many minute seeds.

GERMINATION Dirr and Heuser (1987) suggest seeds of the common species in cultivation germinate without pretreatment. Seeds should be placed on milled peat and kept under mist until seedlings emerge.

ROSACEAE — ROSE FAMILY

Eriobotrya Lindl.

GROWTH HABIT, OCCURRENCE, AND USE The genus *Eriobotrya* consists of about 10 species of evergreen shrubs native to eastern Asia. *Eriobotrya japonica* is grown for its edible fruit, a yellow pome with a persistent calyx and a few large seeds. The seeds germinate without pretreatment.

POLYGONACEAE — BUCKWHEAT FAMILY

Eriogonum Michx.

GROWTH HABIT, OCCURRENCE, AND USE The genus *Eriogonum* consists of 150 species native to North America. Many species are herbaceous, and shrub species are semiwoody at best. A number of species have great potential as horticultural species for use in gardens and environmental plantings. Of the semiwoody species, the following are among the most important (Ratliff 1974a):

Species	Common name	Occurrence
E. fasciculatum	California buckwheat	Cis-montaine California
E. heermannii	Zigzag bush	Arid Great Basin
E. inflatum	Desert trumpet	Mojave Desert and Great Basin
E. umbellatum	Sulphur flower	Mountains of the west

California buckwheat and sulphur flower are the best-known and most widely used species. Both are extremely variable, with numerous named and ecologically important subspecies. California buckwheat is widely used in environmental plantings, and sulphur flower is increasingly used. Sulphur flower is also an important browse species for mule deer.

FLOWERING AND FRUITING The small, perfect flowers are borne in small clusters. There are so many species that the time of flowering and fruit ripening must be obtained from regional floras and experience.

The fruits of California buckwheat ripen in August and may be collected by stripping. The fruit is a 3-angled achene, enclosed by a persistent calyx (Figs. 1, 2). The achenes can be recovered by threshing through a fine screen and air screening. The seeds are small, running about 800 per gram.

GERMINATION Germination of California buckwheat seed is epigeal (Fig. 3). We have experienced considerable dormancy and low viability in collections of California buckwheat seeds. Seeds of sulphur flower are generally highly germinable and germinate at a wide range of incubation temperatures.

Seedlings can be grown in containers for transplanting into the field.

Figure 1. *Eriogonum fasciculatum*, California buckwheat: achene in calyx (left), and achene with calyx removed (right), ×12 (Ratliff 1974a).

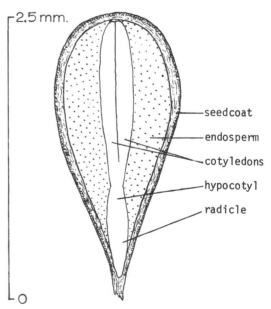

Figure 2. *Eriogonum fasciculatum*, California buckwheat: longitudinal section through a seed that was excised from an achene, ×30 (Ratliff 1974a).

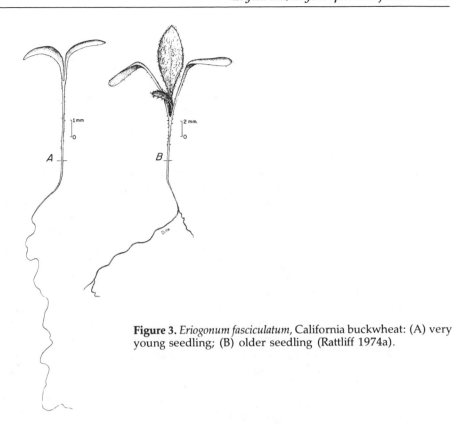

Figure 3. *Eriogonum fasciculatum*, California buckwheat: (A) very young seedling; (B) older seedling (Rattliff 1974a).

LEGUMINOSAE — LEGUME FAMILY

Erythrina L. — Coral tree

GROWTH HABIT, OCCURRENCE, AND USE The genus *Erythrina* consists of about 50 species native to warm and temperate areas around the world. The flowers are usually brilliant red. The fruit is a long, stipitate, 2-valved pod constricted between the seeds. The seeds are often brightly colored.

Suina (*E. speciosa*) is native to Brazil where it is grown as an ornamental for its beautiful red flowers. Coralbean (*E. flabelliformis*) has its most northerly distribution in southern Arizona.

GERMINATION Seeds of *E. speciosa* have hard seedcoats that require scarification. Pricking the seedcoat on the hypocotyl-radicle axis end resulted in the highest germination (Carvalho et al. 1980).

E. caffra is a South African species that also has hard seedcoats. Dormancy of this species can be overcome by mechanical abrasion, piercing the hilum, or by soaking in sulphuric acid for 120 minutes (Small et al. 1977). Hot water treatments were not effective in enhancing germination.

Seedling establishment occurs in specific habitats that apparently reflect germination-temperature relationships (Conn and Synder-Conn 1981).

LEGUMINOSAE — LEGUME FAMILY

Erythrophleum fordii Afzel.

GROWTH HABIT, OCCURRENCE, AND USE *Erythrophleum fordii* is a rare and protected species in China. Members of the genus contain the poisonous alkaloid erythropheine.

GERMINATION This species produces a high percentage of hard seeds that are difficult to germinate. Studies of the structure of the testa showed that it consists of a layer of closely arranged palisade cells with thick walls and narrow cavities (Chen and Fu 1984). Treatment with sulphuric acid destroyed this layer and enhanced germination. Extracts from the testa and embryo contained germination inhibitors that could be overcome by soaking the seeds in growth regulators.

MYRTACEAE — MYRTLE FAMILY

Eucalyptus L'Hérit

GROWTH HABIT, OCCURRENCE, AND USE The genus *Eucalyptus* comprises more than 523 species, and new species and revisions of old taxons continue to be described (Krugman 1974a). Some species are exceptionally tall trees while others are shrubs. Eucalypts are mainly native to Australia, but a few species are native to the Philippines, New Guinea, and Timor. Eucalypts are widely cultivated in warmer parts of the world, including southern Europe, Africa, Asia, and South America.

The genus was first introduced to California and Hawaii about 1853. Eucalypts have been planted to a limited extent on the Gulf Coast and as far north as coastal Oregon. About 200 species have been introduced into the United States; most of them are grown in California and Hawaii as ornamentals. *Eucalyptus globulus* has been the most extensively planted species in the United States. It was initially planted for timber production. Currently it is planted for windbreaks and for fuel wood. It has potential as a source of wood fiber.

In Hawaii several species including *E. robusta*, *E. sideroxylon*, *E. grandis*, *E. saligna*, and *E. globulus*, have been planted as windbreaks, for watershed protection, and for timber production.

FLOWERING AND FRUITING The flower clusters develop enclosed within an envelope formed by two bracteoles. These bracteoles split and are shed during development, revealing the flower buds. The perfect flowers are white, yellow, or red, often in axillary umbels or corymbose, or paniculate clusters. In a few cases the flowers develop singly, but most often they are in 5- to 10-flowered axillary umbels. Sepals and petals are united to form a cap in the bud, which drops off at anthesis. The stigma is receptive for a few days after the cap drops. Pollination is mainly by insects. The ovary has 3 to 6 locules with many ovules. There is a wide range of flowering times for such a broad array of species. In California some species such as *E. viminalis* may flower all year, while other species flower only in the spring, summer, or fall (Table 1).

The fruit is a hemispherical, conical, oblong, or ovoid hard woody capsule 0.6 to 2.5 cm in diameter, that is loculicidally dehiscent at the apex by 3 to 6 valves. Usually only a few seeds are fertile in a given capsule, and seed size can be extremely variable. When more than one seed ripens in a given locule, the seeds are variously shaped and angular (Fig. 1). When solitary, the seed will be ovate or somewhat compressed.

The seedcoat is thin and smooth, but it can be ribbed, pitted, or sculptured in various ways. Usually it is black or dark brown in color, but it can be pale brown (Table 1). The seeds are numerous and extremely small in most species (Table 1). The embryo consists of 2-lobed cotyledons which are folded or twisted over the straight radicle. There is no endosperm (Fig. 2).

Fruits ripen at various times during the year, depending on the species. Dispersal is largely by wind within a month or 2 after ripening, for most species. For other species, such as *E. viminalis*, dispersal may not take place until 10 months after ripening. Good seed is produced by most species by 10 years of age. For mature trees the interval between large seed crops varies from 2 to 5 years.

Table 1. *Eucalyptus:* phenology of flowering and fruiting (when grown under California conditions), characteristics of mature trees, and seed and chaff data (Krugman 1974a).

Species	Flowering dates	Fruit ripening dates	Seed dispersal dates	Mature tree height[1] (m)	Seed color	Chaff color	Viable seeds/g plus chaff
E. camaldulensis	Feb.–Apr.	July–Oct.	Dec.	37	Yellow-brown	Yellow-brown	780
E. citriodora	Nov.–Jan.	May–Aug.	–	40	Black	Brownish red	150
E. dalrympleana	June–Aug.	Aug.–Oct.	Oct.–Nov.	37	–	–	200
E. delegatensis	Apr.–June	Apr.–July	May–July	85	Pale brown	Pale brown	80
E. fastigata	Apr.–May	July–Aug.	–	62	Pale brown	Pale brown	150
E. glaucescens	July–Aug.	May–Sept.	Nov.–Feb.	12	Black or brown	Pale red-brown	70
E. globulus	Nov.–Apr.	Oct.–Mar.	Oct.–Mar.	55	Dark brown	Brownish red	90
E. grandis	Sept.–Nov.	–	–	55	–	–	710
E. microcorys	Dec.–Feb.	–	–	46	–	–	240
E. nitens	Apr.–July	May–June	May–June	92	Black or brown	Pale red-brown	390
E. obliqua	Apr.–July	May–Aug.	–	77	Dark brown	Orange-brown	90
E. paniculata	Feb.–May	–	–	43	–	–	180
E. pilularis	Dec.–Mar.	Jan.–Apr.	All year	62	–	–	40
E. regnans	Apr.–July	June–Sept.	–	107	Pale brown	Pale brown	320
E. robusta	Jan.–Mar.	–	–	28	Dark brown	Brownish-red	390
E. rudis	Jan.–Mar.	–	–	15	–	–	610
E. saligna	Apr.–June	Oct.–Dec.	–	46	Black	Brownish-red	460
E. sideroxylon	June–Sept.	Oct.–Dec.	–	31	Dark brown	Orange brown	240
E. viminalis	All year	All year	–	46	Black or brown	Pale red-brown	360

[1]May refer to native habitat.

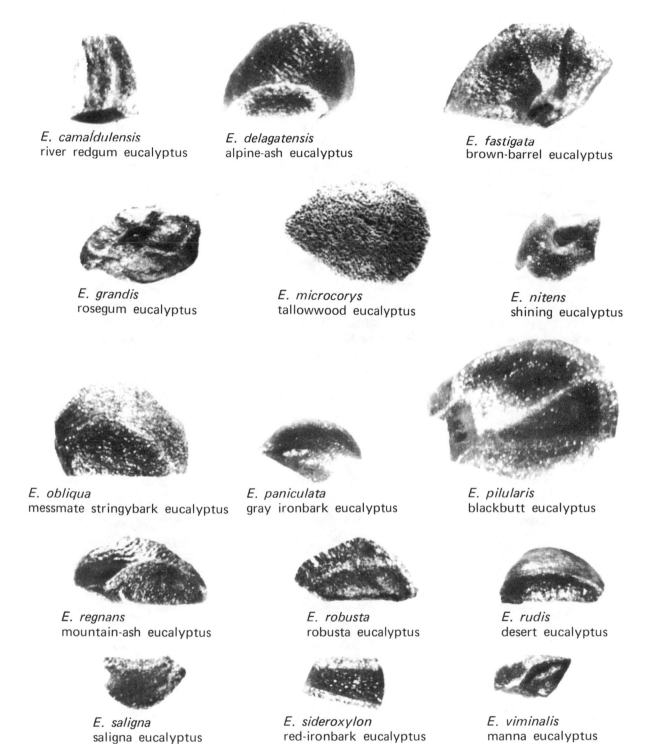

E. camaldulensis
river redgum eucalyptus

E. delagatensis
alpine-ash eucalyptus

E. fastigata
brown-barrel eucalyptus

E. grandis
rosegum eucalyptus

E. microcorys
tallowwood eucalyptus

E. nitens
shining eucalyptus

E. obliqua
messmate stringybark eucalyptus

E. paniculata
gray ironbark eucalyptus

E. pilularis
blackbutt eucalyptus

E. regnans
mountain-ash eucalyptus

E. robusta
robusta eucalyptus

E. rudis
desert eucalyptus

E. saligna
saligna eucalyptus

E. sideroxylon
red-ironbark eucalyptus

E. viminalis
manna eucalyptus

Figure 1. *Eucalyptus*: seeds, ×20 (Krugman 1974a).

COLLECTION, EXTRACTION, AND STORAGE OF SEED   Collecting mature fruits of eucalypts is made easy by the long interval between seed ripening and opening of the capsule. Care should be taken to collect only well-developed seeds. The capsules should be spread in thin layers to permit rapid drying. Artificial drying temperatures should not exceed 38°C because high temperatures for prolonged periods strengthen the dormancy of seeds of some species. The capsules of most species are air-dried within a week and kiln drying requires 3 to 6 hours.

Seeds are recovered by vigorously shaking open capsules. If shaking is not done, only unfertile seeds will be recovered. In immature capsules the fertile seeds are attached at the base of the capsule and may not come loose unless shaken. Viable seeds can be extracted from infertile seeds with a specific gravity separator such as an air column. The proportion of chaff to viable seed ranges from 5:1 to 30:1. Seed size is highly variable among species (Table 1).

Eucalypt seeds have germinated after 30 years of storage at room temperature. The seeds of most species can be stored for as long as 10 years in sealed containers at a moisture content of 4 to 6% at temperatures from 0 to 5°C.

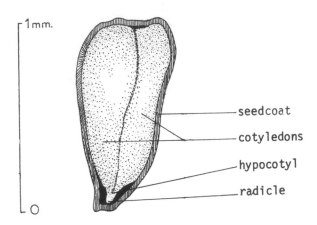

Figure 2. *Eucalyptus rudis*, desert eucalyptus: longitudinal section through a seed, ×50 (Krugman 1974a).

PREGERMINATION TREATMENT   Viable mature seeds of most eucalypt species germinate under favorable conditions without pretreatment. Most species require light for germination. A few species, particularly alpine species, require prechilling to break dormancy. In some species prechilling can be substituted for light requirements. Both *Eucalyptus Seed* (Bolland et al. 1976)

Table 2. *Eucalyptus:* partial listing of recent literature concerning seed germination and related topics.

Author—Date	Subject	Species
Agiar and Bisarro 1978	Arboretum	Several
Agiar and Nakane 1983	Seed size	E. citriodora
Bachelard 1985	Soil moisture	–
Bell et al. 1987	Understory species	–
Beltrati 1978	Morphology	E. maidenii
Boado 1976	Substrata	E. deglupa
Bolte et al. 1985	Inhibitors	E. grandis
Borges et al. 1980	Drying	–
Bowen 1980a	Collection, storage	E. deglupa
Bowen and Eusebio 1983	Collection, storage	E. deglupa
Brice-Bruce 1977	Nursery stock	E. grandis
Christensen and Schuster 1979	Germination	E. diversicolor
Cremer and Mucha 1985	Temperature	E. microtheca
Doran et al. 1987	Storage	–
Drake 1975	Hybrids, seed set	–
Edgar 1977	Moisture stress	E. camaldulenses
Ferreira et al. 1977	Flowering	E. grandis
Geary and Miller 1982	Precision sowing	E. robusta
Gibson and Bachelard 1986	Variation	–
Gibson and Bachelard 1987	Substrate temperature	E. gieberi
L. M. Hodgson 1977a	Grafting	–
T. J. Hodgson 1977	Seed orchard	E. grandis
King and Krugman 1980	Species test, California	–
Lisbao and Filho n.d.	Storage humidity	E. saligna
Loneragan 1979	Phenology	E. diversicolor
Merwin 1987	California	–
Moura 1982	Temperature	–
Pereira and Garrido 1975	Seed size	E. grandis
Scott 1983	Temperature	–
Silvia 1979	Light	E. grandis
Singh and Bawa 1982	Allelopathic	E. globulus
Suiter and Lishao 1973	Relative humidity	E. saligna
Surech and Rai 1987	Allelopathic	–
Turnpull and Doran 1987a	Comprehensive	–
Valahos and Bell 1986	Seed banks	–
Yap and Wong 1983	Seed biology	–
Zohar et al. 1975	Light, temperature	E. occidentalis

and *Germination of Australian Native Plant Seed* (Langkamp 1987) contain lists of germination requirements for hundreds of species of eucalypts.

GERMINATION TEST Germination standards for *E. deglupta* include substrata, creped cellulose paper or top of paper; temperature 20/30°C; duration, 14 days (ASOA 1985). For *E. grandis* germination standards are substrata, C or TB; temperature, 25°C; duration, 14 days (ASOA 1985). The Genebank Handbook (1985) adds that according to International Seed Testing Association rules state seeds should be weighed out for testing rather than counted.

The literature on the germination of *Eucalyptus* species is extensive. We provide a partial listing of recent literature in Table 2.

NURSERY AND FIELD PRACTICE Except in Hawaii, eucalypt seeds are not directly sown in the United States. Seedlings are usually started in flats and transplanted. We observed a nursery in Spain where several million seedlings were produced annually by direct seeding in liter-sized plastic bags filled with a mixture of soil and sheep manure. The bags were thinned to one seedling each after emergence. Germination is epigeal (Fig. 3). Emergence begins 7 to 10 days after planting and is completed in a couple weeks. Seedlings are generally transplanted to containers and grown out before being transplanted to the field.

Several species of eucalypt can be rooted from shoots that have juvenile leaves.

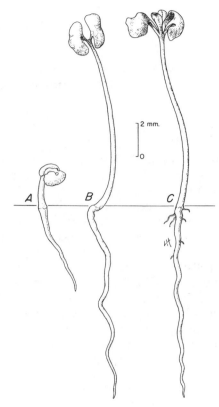

Figure 3. *Eucalyptus* seedling development: (A) *E. viminalis* at 1 day; (B) at 8 days; (C) *E. rudis* at 42 days (Krugman 1974a).

EUCOMMIACEAE — EUCOMMIA FAMILY

Eucommia Oliv.

GROWTH HABIT, OCCURRENCE, AND USE A single monotypic genus constitutes Eucommia family. *Eucommia ulmoides*, the hard rubber tree, is grown for its foliage and as a curiosity because of the latex in its sap.

The fruit is a one-seeded, winged nutlet. Fruits should be collected in the fall and sown at once, or dry fruits should be prechilled for 3 months (Dirr and Heuser 1987).

CELASTRACEAE — STAFF TREE FAMILY

Euonymus L.

GROWTH HABIT, OCCURRENCE, AND USE The genus *Euonymus* consists of about 170 species of deciduous or evergreen shrubs or small trees native to North and Central America, Europe, Asia, Madagascar, and Australia (Rudolf 1974g). Because of their attractive fruits, euonymus species are planted widely for ornamental purposes, and they also have value as wildlife species. Species with potential value for environmental plantings are listed in Table 1.

FLOWERING AND FRUITING The usually perfect flowers, borne in clusters, bloom in the spring. The fruit, which ripens in late summer or fall, is a 4- to 5-celled capsule that is usually lobed and sometimes winged (Fig. 1). Each fruit cell contains 1 to 2 seeds enclosed in a fleshy, usually orange aril (Fig. 2). Good fruit crops are borne almost annually. Natural seed dispersal usually takes place soon after the fruits are ripe. Flowering and fruiting phenology are given in Table 1.

COLLECTION, EXTRACTION, AND STORAGE OF SEED Seeds may be collected in late summer and early fall by picking ripe fruits from the bushes or trees by hand or by shaking them onto ground sheets. Fruits should be air-dried for several days.

The seeds can be extracted by flailing the fruits in bags or by rubbing them through a coarse screen. The seeds can be recovered by air screening. The arils should be removed if the seed is going to be stored (Fig. 3). Commercial seed usually has at least part of the aril still attached (Fig. 4).

Seed yields of 4 to 9 kg of seeds per 45 kg of fruits have been reported. Seeds per gram are given in Table 1.

Moist, cold storage seems to be the most practical way to store these seeds. Drying reduces subsequent germination.

PREGERMINATION TREATMENT Most species have dormant embryos that require both warm stratification and prechilling (Table 1).

GERMINATION The Genebank Handbook (1985) offers the following standards for *E. europaeaus* seeds: substrate, TP; temperature, 20/30°C; duration, 10–14 days; and additional instructions, prechill for 45 days or warm stratification for 8–12 weeks and the prechill. Germination capacities for *Euonymus* species have been given (Table 1). The results in Table 1 were for prechilled and stratified seeds, indicating that for these seed lots germination capacity was low for some species.

NURSERY AND FIELD PRACTICE For best results cleaned seeds should be sown in fall soon after collection and before the seeds dry out. They should be sown 0.6 cm deep with a seedling density of 420 per m^2 (Fig. 5). The seedbeds should be mulched with pine needles.

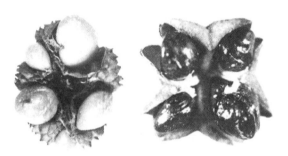

E. americanus
brook euonymus

E. atropurpureus
eastern wahoo

Figure 1. *Euonymus*: top views of open capsule, ×2 (Rudolf 1974g).

Figure 2. *Euonymus americanus*, brook euonymus: seeds enclosed in their fleshy arils, ×4 (Rudolf 1974g).

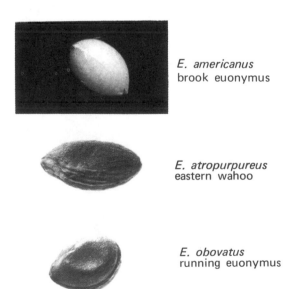

E. americanus
brook euonymus

E. atropurpureus
eastern wahoo

E. obovatus
running euonymus

Figure 3. *Euonymus*: seeds with arils removed, ×4 (Rudolf 1974g).

Table 1. *Euonymus*: phenology of flowering and fruiting, characteristics of mature trees, cleaned seed yields, pregermination treatments and germination test conditions and results (Rudolph 1974g).

Species	Flowering dates	Fruit ripening dates	Aril color	Mature tree height (m)	Year first cultivated	Seeds/ gram	Warm stratification (days)	Prechilling (days)	Substrata	Temperature (°C)	Germination (%)
E. alatus	May–June	Sept.–Oct.	Orange-red	3	1860	55	0	90–100	Paper	20	45
E. americanus	May–June	Sept.–Oct.	Scarlet	2	1697	77	0	139	Sand	20/30	15
E. atropurpureus	May–June	Sept.–Oct.	Scarlet	6	1756	37	60	60	–	20/30	40
E. bungeanus	June	Sept.–Oct.	Orange	6	1883	30	0	180	Sand	0/10	20
E. europaeus	May–June	Aug.–Oct.	Orange	7	Long	29	90	120	–	12/15	71
E. maackii	June	Oct.	Orange	5	1880	–	–	–	–	–	–
E. obovatus	April–June	Aug.–Oct.	Orange to scarlet	0.6	1820	56	0	120	–	–	–
E. verrucosus	May–June	Aug.–Oct.	Orange to red	2	1763	45	0	90	Paper	15	70

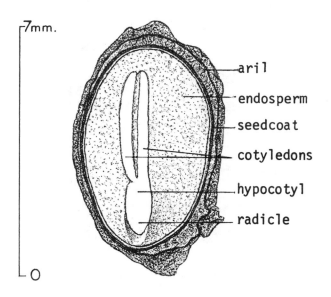

7mm.

aril
endosperm
seedcoat
cotyledons
hypocotyl
radicle

0

Figure 4. *Euonymus europaeus*, European euonymus: longitudinal section through a seed, ×10 (Rudolf 1974g).

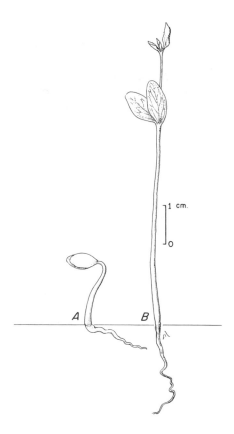

1 cm.

0

A   B

Figure 5. *Euonymus europaeus*, European euonymus: seedling development at 1 and 12 days after germination (Rudolf 1974).

THEACEAE — TEA FAMILY

Eurya Thunb.

GROWTH HABIT, OCCURRENCE, AND USE The genus *Eurya* consists of more than 50 species native to the Eastern Hemisphere. One species, *E. japonica*, is adapted to the southeastern United States. It is a shrub or small tree growing to 5 m in height. The fruit is a berry.

This species is propagated by cuttings. Apparently, little is known about seed germination (Hunt 1971; March 1976).

ROSACEAE — ROSE FAMILY

Exochorda Lindl. — Pearl bush

GROWTH HABIT, OCCURRENCE, AND USE The genus *Exochorda* consists of 3 or 4 shrubs native to Asia. The shrubs are spiraea-like, but have individual flowers as large as apple blossoms. The fruit is a 5-angled or winged capsule that separates into 5 bony 1- or 2-seeded carpels.

The capsular fruits should be collected in the yellow-brown stage and dried either for storage or prechilling (Dirr and Heuser 1987). The seeds of most species require prechilling:

Species	Prechilling (months)	Alternative treatment
E. giraldii	1–2	Sow immediately
E. korelkowii	—	Sow immediately
E. racemosa	Brief	

Cuttings taken in June-July and treated with IBA will root.

LEGUMINOSAE — LEGUME FAMILY

Eysenhardtia Kunth — Kidneywood

GROWTH HABIT, OCCURRENCE, AND USE Kidneywood (*Eysenhardtia texana*) is a shrub that grows from 1 to 4 m tall on calcareous soils in south and west Texas and adjacent Mexico. It is considered an excellent browse species for desert mule deer.

GERMINATION The seeds germinate without pretreatment (Wisenant and Ueckert 1982).

FAGACEAE — BEECH FAMILY

Fagus L. — Beech

GROWTH HABIT, OCCURRENCE, AND USE The beeches consist of about 10 species of medium-sized, deciduous trees native to the temperate regions of the Northern Hemisphere (Rudolf and Leak 1974). They are valuable for their ornamental qualities and nuts; some are important timber species. American beech (*Fagus grandifolia*) and European beech (*F. sylvatica*) are used in reforestation and ornamental plantings.

FLOWERING AND FRUITING The male and female flowers of beech appear with the leaves and are borne separately on the same tree; they bloom in the spring after the leaves unfold (Table 1). The flowers are quite vulnerable to spring frost. The male flowers are borne as long-stemmed heads, while the female flowers occur in clusters of 2 to 4.

The fruits consist of 2 or 3 one-seeded nuts surrounded by a prickly bur or husk developed from the floral involucre. The husk turns brown and opens soon after maturity in the autumn (Fig. 1), allowing the nuts to fall to the ground. Commercial seed consists of the ovoid, unequally 3-angled, chestnut-brown, shining, thin-shelled nuts without endosperm (Figs. 2, 3). Natural seed dispersal is chiefly by animals.

COLLECTION, EXTRACTION, AND STORAGE OF SEED Beech nuts may be shaken from the trees after frost has opened the burs, or they can be raked from the ground. Closed burs can be picked from fallen trees. The cleaned nuts should be sown or prechilled as soon after collection as

Table 1. *Fagus:* phenology of flowering and fruiting, and characteristics of mature trees (Rudolf and Leak 1974).

Species	Flowering dates	Fruit ripening dates	Seed dispersal dates	Mature tree height (m)	Year first cultivated	Seed-bearing Age (years)	Seed-bearing Interval (years)
F. grandifolia	Mar.–May	Sept.–Nov.	After frost	37	1800	40	2–3
F. sylvatica	Apr.–May	Sept.–Oct.	Heavy frost	31	Long	40–80	2–20

F. grandifolia
American beech

Figure 1. *Fagus:* nuts enclosed in a partially open husk, ×1 (Rudolf and Leak 1974).

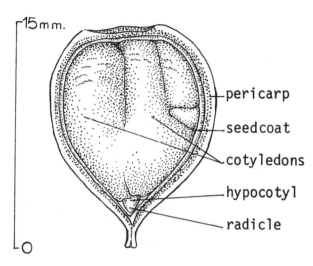

—pericarp
—seedcoat
—cotyledons
—hypocotyl
—radicle

Figure 3. *Fagus grandifolia,* American beech: longitudinal section through a seed, ×4 (Rudolf and Leak 1974).

F. sylvatica
European beech

Figure 2. *Fagus:* nut, ×4 (Rudolf and Leak 1974).

possible. The average weight per seed for American and European beech is 3.5 and 4.6 g, respectively.

Nuts stored for long periods should be dried to 8 to 10% moisture and stored at −15°C. Nuts to be spring planted can be dried to 20 to 30% moisture and stored in plastic bags at 2 to 5°C.

PREGERMINATION TREATMENT Beech seeds require prechilling for prompt germination.

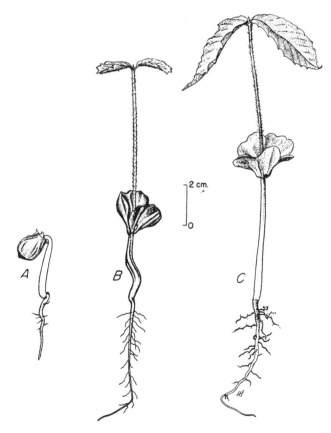

Figure 4. *Fagus grandifolia,* American beech: seedling development at 2, 5, and 7 days after germination (Rudolf and Leak 1974).

GERMINATION The Genebank Handbook (1985) suggests the following standards for European beech seeds: substrate, TB; temperature 20/30°C; duration, up to 24 weeks; additional directions, prechill at 3 to 5°C for 28 days.

There is very little recent literature on the germination ecology of beech seeds from the United States, but a great deal from the rest of the world (Table 2).

NURSERY AND FIELD PRACTICE Beech seeds can be sown in the fall as soon as collected, or prechilled seeds can be sown in the spring. Sufficient seed should be sown to provide 450 first year or 270 second year seedlings per m². They should be covered with 1.25 cm of soil. Fall-sown beds should be mulched. Seedling beds should have half-day shade the first year (Fig. 4). A tree percent of 15 can be expected.

Table 2. *Fagus:* recent literature on seed germination and related topics.

Author—Date	Subject	Location
Asanuma et al. 1984	*F. crenata*, herbicides, and germination	Japan
Bonnet-Masimbert and Muller 1973	*F. sylvatica*, prospects for research on storage of seed	France
Bonnet-Masimbert and Muller 1975	*F. sylvatica*, feasibility of storage	France
Bonnet-Masimbert and Muller 1976	*F. sylvatica*, rapid germination test	France
Brinar 1976	*F. sylvatica*, viability and seed production by provenance	–
Engler et al. n.d.	*Fagus* spp., predation	France
Falusi 1982	*F. sylvatica*, dormancy in relation to provenience of seed production	Italy
Fuhrer and Pall 1984	*F. sylvatica*, fertilizer, seed production	Hungary
Gregorius et al. 1986	*F. sylvatica*, genetics	–
Hashizume 1979	*F. japonica*, seed development, biochemistry	Japan
Hashizume and Fukutomi 1978	*F. crenata*, seed development	Japan
Khutortsov and Anikeeko 1981	*Fagus* spp., predation, chemical composition	USSR
Majer 1982	*F. sylvatica*, seed production, and periodicity	Hungary
Mishnev 1984	*F. orientalis*, seed production	USSR
Muller and Bonnet-Masimbert 1983	*F. sylvatica*, seed dormancy	France
Muller and Bonnet-Masimbert 1985	*F. sylvatica*, pretreatment to break seed dormancy	France
Perrin et al. n.d.	*Fagus* spp., seed storage	France
Pontailler n.d.	*F. sylvatica*, seed production	France
Sabeeve and Olisaev 1983	*F. orientalis*, seed sources	USSR
Staafilt et al. 1987	*F. sylvatica*, seedbanks	–
Sulli 1975	*F. sylvatica*, seed testing	Italy
Suszka 1975	*F. sylvatica*, prechilling	Poland
Suszka n.d.	*F. sylvatica*, prechilling	Poland
Suszka and Kluczzynska 1980	*F. sylvatica*, prechilling	Poland
Suszka and Zieta 1976	*F. sylvatica*, prechilling	Poland
Suszka and Zieta 1977	*F. sylvatica*, prechilling	Poland
Tokarz 1974	*F. sylvatica*, seed storage	–
Weissen 1980	*F. sylvatica*, storage	Belgium
Weissen n.d.	*F. sylvatica*, predation	Belgium

ROSACEAE — ROSE FAMILY

Fallugia paradoxa (D. Don) Endl. — Apache plume

GROWTH HABIT, OCCURRENCE, AND USE   Apache plume is an attractive, often evergreen, many-branched shrub, found in a wide variety of sites from western Texas and southwestern Colorado through southern Nevada to southeastern California, and southward into Mexico (Deitschman et al. 1974c). This species is related to *Purshia* and *Cowania*, with which it may form hybrids.

Apache plume is an important browse species. Its attractive flowers and plumed seeds make it desirable for ornamental plantings.

FLOWERING AND FRUITING   Large, white, rose-like flowers are borne singly or in clusters on long stalks. Flowering can occur as early as April or as late as August.

The fruits, each a small hairy achene tipped with a persistent, feathery style 2.5 to 5 cm in length (Fig. 1), usually form dense clusters of 20 to 30 or more. As the fruit ripens, it changes from a greenish to a reddish color. Ripening of the seeds and wind dispersal occurs about 1 to 2 months after flowering. Ripe achenes (Figs. 2, 3) comprise commercial seed. The species is a consistent seed producer at 1- to 3-year intervals.

COLLECTION, EXTRACTION, AND STORAGE OF SEED   Seeds may be collected when the reddish color of the hairy style whitens and the plump seeds fall readily. They can be stripped or shaken onto ground sheets. The style can be removed by rubbing and the seeds recoverd by air screening. Seed with a moisture content of 7 to 12% can be stored in cloth bags in warehouses for 2 to 3 years. According to Deitschman et al. (1974c), 35 liters of fruit weigh 900 g and yield 110 to 150 g of cleaned seed; 45 kg of fruits yield 5 to 7 kg of cleaned seed; and there are 1190 seeds per gram.

GERMINATION   Seed of Apache plume germinate without pretreatment. Germination is epigeal (Fig. 4).

Figure 1. *Fallugia paradoxa*, Apache plume: achene with style (tip broken), ×4 (Deitschman et al. 1974c).

Figure 2. *Fallugia paradoxa*, Apache plume: achene with style removed, ×8 (Deitschman et al. 1974c).

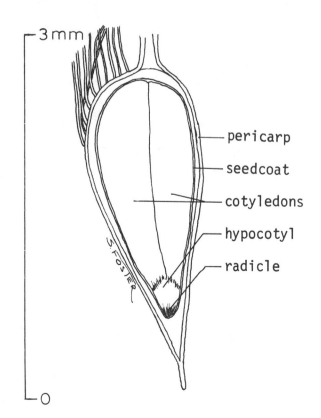

Figure 3. *Fallugia paradoxa*, Apache plume: longitudinal section through an achene, ×32 (Deitschman et al. 1974c).

NURSERY AND FIELD PRACTICE Seeds can be sown in the Southwest either in the spring or during the summer rainy period. They should be covered with a thin layer of soil on a firm seedbed. Germination occurs within 4 to 10 days after sowing.

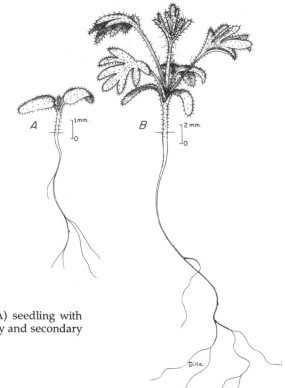

Figure 4. *Fallugia paradoxa,* Apache plume: (A) seedling with primary leaves only; (B) seedling with primary and secondary leaves (Deitschman et al. 1974c).

ARALIACEAE — GINSENG FAMILY

Fatsia Decne. et Planch.

GROWTH HABIT, OCCURRENCE, AND USE The monospecific *Fatsia* genus is represented by *F. japonica,* an evergreen, unarmed, large bush or small tree. It is often planted in mild regions for its subtropical effect.

GERMINATION Seeds germinate without pretreatment if removed from the fruit (Dirr and Heuser 1987).

MYRTACEAE — MYRTLE FAMILY

Feijoa sellowiana O. Berg. — Pineapple guava

GROWTH HABIT, OCCURRENCE, AND USE The genus consists of 2 species of shrubs or small trees. Pineapple guava is widely planted for its edible fruit, an oblong berry.

GERMINATION Seeds removed from the fruit pulp will germinate without pretreatment. Cuttings will root, but timing and care is important (Dirr and Heuser 1987).

MORACEAE — MULBERRY FAMILY

Ficus L. — Fig

GROWTH HABIT, OCCURRENCE, AND USE *Ficus* is a huge genus of over 2000 species. One species is grown for its fruit and many others are grown as ornamentals. Two species are indigenous to southern Florida (Bailey 1951). Besides the common fig (*F. carica*) several species are grown in Florida, California, and Hawaii.
FLOWERING AND FRUITING The fig fruit is a unique structure formed by the receptacle, with the flowers that form achenes located within the receptacle. The flowers are often pollinated by special insects.

GERMINATION Seeds of *F. aurea* and *F. religiosa* are reported to require light for germination, whereas seeds of *F. populnea* will germinate in the light or dark. Seeds of *F. religiosa* are sensitive to the drying of germination test substratum (Galil and Meri 1981). The Genebank Handbook (1985) suggests that seeds of *Ficus* species be tested for germination on top of filter paper at 25 or 30°C in the light and that the filter paper should not be allowed to dry out.

RUTACEAE — RUE FAMILY

Flindersia brayleyana F. Muell. — Queensland maple

GROWTH HABIT, OCCURRENCE, AND USE Queensland maple is a native of Queensland, Australia, and was introduced to Hawaii in 1935 (Wick 1974). It is a broadleaf, tropical hardwood tree that attains a height of 25 to 30 m at maturity.

Queensland maple ranks with mahogany, walnut, cedar, and blackwood as being among the world's best cabinet woods and is one of the most valuable on the Australian market. It is also used for plywood, laminated panels, and doors. It is a wood of medium density.
FLOWERING AND FRUITING Queensland maple has small, white, fragrant, 5-petaled flowers that generally form large panicles occurring from August to September. The fruit is a hard-shelled, warty, 5-valved capsule. In Hawaii, it generally ripens from June to July, with the discharge of 2-winged seeds in July through September (Figs. 1, 2). It usually starts bearing seeds at 8 years of age and produces a crop annually.
COLLECTION, EXTRACTION, AND STORAGE OF SEED When capsules turn from green to brown they are ripe and should be picked. In Hawaii they are picked by hand from felled or standing trees, and spread on trays to dry. When the capsules open, they release fairly large-winged, light brown seeds, about 5 cm long. There are on the average 11 seeds per gram. Seeds are stored in air-tight containers at 3 to 5°C. They do not store well and lose their viability within a year. Seeds are easily damaged.
GERMINATION Good germination is obtained without pregermination treatments.
NURSERY AND FIELD PRACTICE In Hawaii seeds are sown as soon as they are collected, at a rate of 160 to 210 per m². Young seedlings are provided overhead shade for the first 2 months.

Figure 1. *Flindersia brayleyana*, Queensland maple: seed, ×1 (Wick 1974).

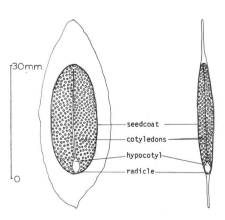

Figure 2. *Flindersia brayleyana*, Queensland maple: longitudinal section through two planes of a seed, ×1 (Wick 1974).

OLEACEAE — OLIVE FAMILY

Fontanesia fortunei Carr. — Fortune fantanesia

GROWTH HABIT, OCCURRENCE, AND USE The genus *Fontanesia* consists of 2 species of shrubs native to Asia. Fortune fantanesia (*Fontanesia fortunei*) is planted as an ornamental species.

The fruit is a flat, winged nutlet.

Seeds should be collected and sown immediately at maturity. Stored seeds require prechilling (Dirr and Heuser 1987).

OLEACEAE — OLIVE FAMILY

Forsythia Vahl.

GROWTH HABIT, OCCURRENCE, AND USE Forsythias are popular, yellow-flowered, early spring-blooming deciduous shrubs. There are about 8 species native to China, Japan, and southeast Asia.

The fruit is a septicidally dehiscent capsule with many winged seeds.

GERMINATION Germination varies with the species (Dirr and Heuser 1987):

Species	Propagation techniques
F. europaea	Seeds: germinate without pretreatment. Cuttings: difficult to root.
F. mandschurica	Cuttings: treat June cuttings with IBA.
F. ovata	Seeds: germinate without pretreatment.
F. suspensa	Seeds: require 1–2 months of prechilling. Cuttings: take cuttings in June.
F. viridissima	Seeds: fresh seeds germinate about 50%, but prechilling improves germination. Cuttings: take cuttings from May through September

HAMAMELIDACEAE — WITCH HAZEL FAMILY

Fothergilla L.

GROWTH HABIT, OCCURRENCE, AND USE The genus *Fothergilla* consists of 4 species of shrub native to the southeastern United States where they are often planted as ornamentals.

The capsule ripens in September and should be collected before the seeds dehisce, because the seeds are ejected from the capsule and will be lost if collection is delayed.

GERMINATION Seeds are dormant and require both warm stratification and prechilling. *Fothergilla major* seeds require a fluctuating temperature regime during warm stratification which may be as long as 12 months. These species are relatively easy to root from cuttings (Dirr and Heuser 1987).

OLEACEAE — ASH FAMILY

Fraxinus L. — Ash

GROWTH HABIT, OCCURRENCE, AND USE The ashes comprise a large genus of deciduous trees that are valued for many reasons (Bonner 1974d). Practically all ashes have been planted for use in landscaping. They make excellent shade trees. A dozen varieties of European ash (*Fraxinus excelsior*) are cultivated. Native favorites for landscaping are white (*F. americana*) and green (*F. pennsylvanica*) ash in eastern and central United States and velvet ash (*F. velutina*) in the Southwest.

FLOWERING AND FRUITING The small, usually inconspicuous flowers appear in the early spring (Table 1) with, or just before, the leaves in terminal or axillary clusters. Flowering habit varies with species. Ash fruits are elongated, winged, single-seeded samaras that are borne in clusters (Figs. 1, 2, 3). The fruits are mature by late summer or fall, and are dispersed by most species shortly afterwards. Samaras of black (*F. nigra*) and blue (*F. quadrangulata*) ash have a characteristic spicy order. Fruiting data are given in Table 1.

COLLECTION, EXTRACTION, AND STORAGE OF SEED Ash fruits are usually collected in the fall when their color has faded from green to yellow or brown. Collection of

Table 1. *Fraxinus*: phenology of flowering and fruiting, characteristics of mature trees, cleaned seed weight, and pregermination treatments (Bonner 1974d).

Species	Flowering dates	Fruit ripening dates	Seed dispersal dates	Year first cultivated	Seed-bearing Age (years)	Seed-bearing Interval (years)	Mature tree height (m)	Seeds/ gram	Warm stratification Temperature (°C)	Warm stratification Duration (days)	Prechilling Temperature (°C)	Prechilling Duration (days)
F. americana	Apr.–May	Oct.–Nov.	Sept.–Dec.	1724	20	3–5	25	29	20/30	30	5	60
F. caroliniana	Feb.–Mar.	Aug.–Oct.	–	–	–	–	12	13	–	–	3	60
F. dipetala	Apr.–May	July–Sept.	–	–	–	–	6	15	–	–	3–5	90
F. exelsior	Apr.–May	Aug.–Sept.	Winter–Spring	Long	15	1–2	39	13	20	60–90	5	90
F. nigra	May–June	June–Sept.	July–Oct.	1800	–	–	25	18	20/30	60	5	90
F. ornus	May–June	–	–	Long	20	1	20	–	Warm	60	Cool	90
F. pennsylvanica	Mar.–May	Sept.–Oct.	Oct.–Spring	1824	–	–	22	42	20	60	0–5	60–150
F. profunda	Apr.–May	Sept.–Oct.	Oct.–Dec.	–	10	–	37	7	–	–	5	60
F. quadrangulata	Mar.–Apr.	June–Oct.	–	1823	25	3–4	9	14	20/30	60	5	90
F. uhdei	Mar.–May	July–Sept.	July–Sept.	1900	15	1	37	36	–	–	–	–
F. velutina	Mar.–Apr.	Sept.	–	1900	–	–	15	45	–	–	3–5	90

Figure 1. *Fraxinus americana*, white ash: cluster of samaras, ×1 (Bonner 1974d).

European and flowering ash (*F. ornus*) in Europe is recommended when the samaras are slightly green and sowing can be done immediately. Another index of maturity is a firm, crisp, white, fully elongated seed within the samara. Clusters should be picked by hand or with pruners and seed hooks. Fully dried samaras may be shaken or whipped from limbs of standing trees onto ground sheets. Samaras can also be swept up from paved streets or other hard surfaces after they fall.

Samaras should be spread in shallow layers for complete drying, especially when collected early. Dried clusters may be broken apart by hand, by flailing sacks of clusters, or by running them through a dry macerator. Stems and trash can then be removed by an air screen. Dewinging the samaras is not necessary. Seed yield data are given in Table 1.

Complete data on the long-term storage of seeds of all *Fraxinus* species are not available, but methods used for white and green ash apparently work for the entire genus. No loss of viability for 7 years was found when seeds of these 2 species were stored in sealed containers at 5°C with a moisture content from 7 to 10%. PREGERMINATION TREATMENT Most species of ash exhibit dormancy that is apparently due to both internal factors and to seedcoat effects. White and blue ash have

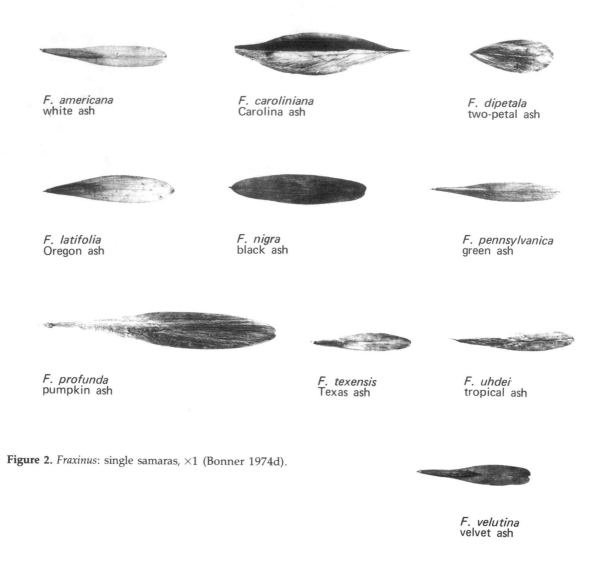

F. americana
white ash

F. caroliniana
Carolina ash

F. dipetala
two-petal ash

F. latifolia
Oregon ash

F. nigra
black ash

F. pennsylvanica
green ash

F. profunda
pumpkin ash

F. texensis
Texas ash

F. uhdei
tropical ash

Figure 2. *Fraxinus*: single samaras, ×1 (Bonner 1974d).

F. velutina
velvet ash

immature embryos that must grow during after-ripening for good germination. The degree of dormancy is also related to seed age; older seeds appear more dormant than freshly collected ones. The epigeal germination (Fig. 4) may occur the spring following seedfall, or seeds may lie dormant in the litter for several years before germinating. The most successful pretreatments are combinations of warm stratification and prechilling (Table 1). These treatments are necessary for spring sowing.

GERMINATION Germination standards for *Fraxinus* spp. given by ASOA (1985) are substrate, paper; temperature, 18 to 22°C; duration, 10 to 14 days; and additional instructions, use embryo excision method or TZ method. Selected recent literature on the germination for seeds of several species are listed in Table 2.

NURSERY AND FIELD PRACTICE Ash seeds may be sown in the fall without prechilling, especially in the northern part of the United States. They should be planted as soon as collected, hopefully before October 15 and never after November 1. Fall sowing of European and flowering ash is said to be essential in Europe if good germination is to be obtained the following spring.

Fall-sown beds should be mulched with burlap or straw, and the mulch removed as germination starts in the spring. Pretreated seeds must be sown in the spring. Seeds of most species should be drilled in rows 15 to 30 cm apart at a rate of 80 to 90 seeds per meter of row. Broadcast seeding should be at the rate of 100 to 150 per m². The seeds should be covered with 0.6 to 0.8 cm of soil. Shading of the beds for a short time after germination may be desirable. The normal out-planting age for North American ashes is 1-0, or, in some cases, 2-0 stock.

Table 2. *Fraxinus:* recent literature on seed germination.

Author—Date	Subject	Location
Bonner 1975	*F. americana*, prechilling and incubation temperatures	USA
Bonner 1977	*Fraxinus* spp.	USA
Borland 1986	*F. anomala*, propagation	USA
Chappelka and Chevone 1986	*F. americana*, acid rain	Canada
Cram 1984	*F. pennsylvanica*, presowing treatments, and storage	Canada
Daniels n.d.	*F. americana*, phenotypic variation in seed	USA
Flowerdew and Gardner 1978	*Fraxinus* spp., rodent seed predation	UK
Gardner 1987	*F. excelsior*, reproductive capacity	UK
Gendel et al. 1977	*F. americana*, embryo dissection	USA
Houston 1976	*F. americana*, X-ray determination of seed quality	USA
Huluta and Tomescu 1973	*F. excelsior*, pretreatment to enhance germination	–
Kamenick and Rypak 1985	*F. americana*, influence of growth regulators on germination	Romania
Kazadaev 1985	*F. pennsylvanica*, winter cuttings	USSR
Krass and Kohler 1985	*Fraxinus* spp., prechilling	–
Ling and Dong 1983	*F. mandschurica*, seed physiology	China
Marinov 1977	*F. oxycarpa*, prechilling	Bulgaria
Marinov 1977b	*F. oxycarpa* and *F. excelsior* embryo development during ripening and prechilling	Bulgaria
Marinov 1977a	*F. excelsior* and *oxycarpa*, influence of molybdenum on germination	Bulgaria
Marinov 1978	*F. oxycarpa*, prolonged storage and subsequent germination	Bulgaria
Marshall 1981	*F. pennsylvanica*, germination enhancement	Bulgaria
Marshall and Kozlowski 1976	*F. pennsylvanica*, importance of endosperm to seedlings	USA
McBride and Dickson 1972	*F. americana*, influence of growth regulators on germination	USA
McCutchen 1977	*Fraxinus* spp., spinning of samaras	USA
Nikolaeva and Vorabeva 1978	*F. excelsior*, variation in seed biology	USSR
Rypak 1979	*F. excelsior*, morphological and biochemical changes during prechilling	Czechoslovakia
Simeonov 1974	*F. excelsior*, molybdenum influence on germination	Bulgaria
Simeonov 1975	*F. excelsior*, molybdenum influence on germination	Bulgaria
Stinemetz and Roberts 1984	*F. americana*, gibberellic and abscisic acid content of seeds	USA
Tinus 1987	*F. pennsylvanica*, influence of dewinging, soaking, prechilling, and growth regulators on germination	USA
Vanstone and LaCroix 1975	*F. nigra*, embryo immaturity and dormancy	Canada
Vorchleva 1981	*F. excelsior*, embryo growth	Canada
Wcislinska 1977	*F. excelsior*, temperature influence on germination	Canada
Zhoa 1983	*F. mandshurica*, growth regulator influence on germination	China

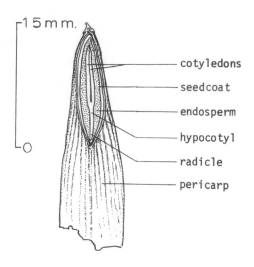

Figure 3. *Fraxinus pennsylvanica*, green ash: longitudinal section through the embryo of a samara, ×2 (Bonner 1974d).

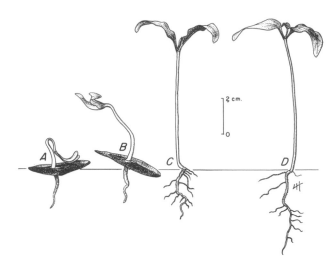

Figure 4. *Fraxinus nigra*, black ash: seedling development at 1, 2, 8, and 14 days after germination (Bonner 1974d).

STERCULIACEAE — STERCULIA FAMILY

Fremontodendron Cov. — Fremontia

GROWTH HABIT, OCCURRENCE, AND USE The fremontia species are handsome arborescent shrubs or small trees with brilliant flowers that make them desirable for environmental plantings (Nord 1974). They are drought resistant and have survived well when planted in brush fields. A number of sprouts develop from the root crowns following fire, and for several years subsequent to burning these sprouts provide valuable browse.

FLOWERING AND FRUITING Flowering occurs the second growing season following germination. The perfect flowers bloom from April to July, and fruit ripening generally occurs between July and September in California (Nord 1974):

Species	Flowering dates	Fruit ripening dates	Seed dispersal dates
F. californicum	May–July	Aug.–Sept.	Sept.–Oct.
ssp. decumbens	Aug.–Sept.	Sept.–Oct.	
F. mexicanum	Apr.–June	July–Aug.	Aug.–Sept.

The fruit is a dense, woolly or quite bristly, 4- to 5-celled, egg-shaped capsule containing numerous reddish brown seeds (Figs. 1, 2). When fully ripened, the capsule splits open at the tip and the seeds are cast from the plant when shaken by wind or other disturbance.

Figure 1. *Fremontodendron mexicanum*, Mexican fremontia: seed, × 8 (Nord 1974).

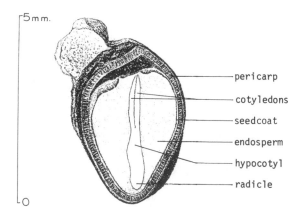

Figure 2. *Fremontodendron californicum*, California fremontia: longitudinal section through a seed, ×10 (Nord 1974).

COLLECTION, EXTRACTION, AND STORAGE OF SEED Gloves should be worn to protect hands from the bristles on the fruits. The capsules can be picked from the plants. The seeds are retained in the capsules for about a month after ripening so timing is not critical. Capsules can be shredded with a hammer mill and the seeds recovered with an air screen.

Seeds have been stored in sealed container at 5°C for 2 years without loss of viability. *Fremontodendron californicum* and *F. mexicanum* have 40 and 60 seeds per gram, respectively.

PREGERMINATION TREATMENT Fremontia seeds are initially dormant from an inhibitor in the seedcoat and from embryo dormancy. Soaking in hot water, followed by prechilling, has been effective in breaking dormancy. Germination tests can be conducted in petri dishes, with incubation at 15/25°C for 40 to 60 days.

NURSERY AND FIELD PRACTICE Fremontias have been established in the field in California by spot seeding of hot-water-treated seeds in the fall or hot-water-treated and prechilled seeds in the spring. All fremontias can be propagated from stem cuttings.

ONAGRACEAE — EVENING PRIMROSE FAMILY

Fuchsia L.

GROWTH HABIT, OCCURRENCE, AND USE The genus *Fuchsia* consists of about 100 species of shrubs and small trees. Most are native to tropical America, but a few are native to New Zealand and Tahiti. They are widely grown as ornamentals. The fruit is a soft, 4-celled berry.

GERMINATION ASOA (1985) germination standards require substrate, paper; incubation temperature, 15°C; duration, 28 days; and additional directions, light for at least 8 hours per day.

GARRYACEAE — GARRYA FAMILY

Garrya Dougl. — Silktassel

GROWTH HABIT, OCCURRENCE, AND USE The genus *Garrya* consists of 4 species native to North America (Reynolds and Alexander 1974a). These medium to large (1.8 to 3.6 m), many-branched evergreen shrubs grow in the lower elevation mountains of the far western United States from Washington to west Texas. The leaves of some species contain a bitter alkaloid— garryin. The plants are browsed by domestic livestock and wildlife. Used as an ornamental since 1842, the silktassel species are also occasionally planted for erosion control.

FLOWERING AND FRUITING The flowers are dioecious. Both appear in axillary or terminal catkin-like racemes from January through May. The fruit is a globose to ovoid, rather dry, 1- or 2-seeded berry (Figs. 1, 2, 3) that ripens to a dark purple from June through December (Reynolds and Alexander 1974a):

Species	Flowering dates	Fruit ripening dates	Seeds/gram
G. flavescens	Apr.–May	July–Aug.	55
G. fremontii	Jan.–May	Aug.–Dec.	65
G. wrightii	Mar.–Aug.	Aug.–Sept.	50

Ripe fruits can be stripped from the branches or picked up from the ground. They are often infested with insect larvae.

Figure 1. *Garrya fremontii*, Fremont silktassel: (A) berry; (B) seed, ×4 (Reynolds and Alexander 1974a).

Figure 2. *Garrya wrightii*, Wright silktassel: berry and seed, ×4 (Reynolds and Alexander 1974a).

COLLECTION, EXTRACTION, AND STORAGE OF SEED The fruits can be run through a macerator and waste removed by flotation. About 50% yield of seeds by weight is obtained from fruits. No information is available on seed storage.

PREGERMINATION TREATMENT Seeds of *G. flavescens* and *G. fremontii* have embryo dormancy. Seeds of *G. wrightii* are sometimes dormant. Seeds should be prechilled for 30 to 120 days and then soaked in 100 ppm of gibberellin before planting.

GERMINATION No standard tests have been established. A strange mixture of procedures has been tried in the past.

NURSERY AND FIELD PRACTICE Apparently information exists for *G. wrightii* only. Seeds should be sown in late winter after 90 days prechilling. This raises the ques-

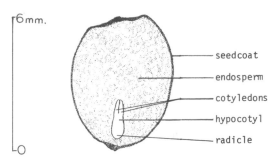

Figure 3. *Garrya fremontii*, Fremont silktassel: longitudinal section through a seed, ×6 (Reynolds and Alexander 1974a).

tion, why not fall seed? *Garrya fremontii* has been successfully propagated from cuttings.

ERICACEAE — HEATH FAMILY

Gaultheria L. — Wintergreen

GROWTH HABIT, OCCURRENCE, AND USE The genus *Gaultheria* consists of about 100 species mostly native to Asia, Australia, and South America, with 6 species found in North America (Dimock et al. 1974). Salal (*Gaultheria shallon*) has a distinctly woody stem, while creeping snowberry (*G. hispidula*) and checkerberry (*G. procumbens*) are semishrubs. All 3 attain their best development in moist, acid soils. They provide cover and food for wildlife. The leaves of checkerberry contain oil of wintergreen, which has been extracted for pharmaceutical use. Salal is a dominant shrub in coastal watersheds in the Pacific Northwest. Its fruits are an important food source for wildlife.

FLOWERING AND FRUITING The bisexual white to pinkish white flowers are borne either solitary and axillary, or in axillary or terminal racemes (Fig. 1). Flowering dates range from early spring to late summer (Dimock et al. 1974):

Species	Flowering dates	Fruit ripening dates	Ripe fruit color
G. hispidula	Apr.–Aug.	Aug.–Sept.	Bright white
G. procumbens	May–Sept.	Aug.–June	Scarlet
G. shallon	Mar.–July	Aug.–Sept.	Dark purple-black

The fruit is a many-seeded capsule surrounded by a persistent, thickened, and pulpy calyx that forms a fleshy pseudoberry (Fig. 2). Birds and mammals are the chief means of seed dispersal. Good seed crops are frequent.

COLLECTION, EXTRACTION, AND STORAGE OF SEED Fruits can be collected over a prolonged period. They can be stripped or picked individually. There are about 6 fruits of checkerberry per gram. Salal fruits average 8.5 per cluster and 125 seeds per fruit.

Either dry or wet seed extraction is possible. Dried checkerberry fruits will crumble and the seeds can be recovered by screening. Salal seeds can be recovered by wet macerating the fruits followed by flotion (Figs. 3, 4).

Cool dry storage maintains seed viability for at least moderate periods. Creeping snowberry has 6800 seeds per gram; checkerberry, 8500; and salal, 7100.

PREGERMINATION TREATMENT Apparently seeds are initially dormant and require prechilling for germination. Prechilling from 30 to 120 days has been used

Figure 1. *Gaultheria shallon*, salal: racemes of pinkish white flowers, ×1 (Dimock et al. 1974).

with a variety of substrata. The seeds of salal seem to require light for germination.

GERMINATION The Genebank Handbook (1985) recommends for salal a 30-day test with no prescribed substrate or temperature, after prechilling at 1 to 5°C for 30 to 60 days. Dimock et al. (1974) presented results that indicated prechilling lowered germination for this species when compared to incubation with light and no pretreatment. It appears that those interested in the germination of *Gaultheria* species be prepared to test and develop their own germination standards.

NURSERY AND FIELD PRACTICE It would appear that creeping snowberry and checkerberry seeds should be sown in the fall. Salal seeds should be surface sown, either under clear plastic film or some protective covering, and then transplanted to containers or beds. Salal seedlings are apparently frost susceptible (Fig. 5). All 3 species can be propagated vegetatively.

Figure 3. *Gaultheria shallon,* salal: seeds, ×20 (Dimock et al. 1974).

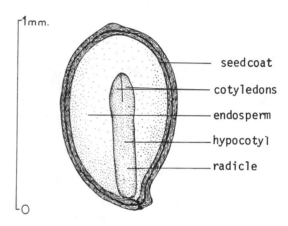

Figure 4. *Gaultheria procumbens,* checkerberry: longitudinal section through seed, ×50 (Dimock et al. 1974).

Figure 2. *Gaultheria shallon,* salal: a ripe, purplish black, somewhat fuzzy fruit, ×3 (Dimock et al. 1974).

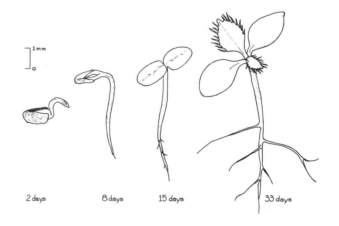

Figure 5. *Gaultheria shallon,* salal: seedling development 2, 8, 15, and 22 days after germination (Dimock et al. 1974).

ERICACEAE — HEATH FAMILY

Gaylussacia baccata (Wangh.) K. Koch — Black huckleberry

GROWTH HABIT, OCCURRENCE, AND USE Black huckleberry is a small deciduous shrub found from Louisiana to Florida and north to Maine, Iowa, and Manitoba (Bonner and Halls 1974b). It is upright, highly branched, and reaches 0.3 to 1.3 m in height at maturity. The berries are an important wildlife food. The shrub has been cultivated since 1772.

FLOWERING AND FRUITING The small, perfect, pinkish flowers bloom in May and June. The black, berrylike, drupaceous fruit, maturing in July to September, is 0.6 to 1 cm long. Each fruit contains 10 one-seeded, bone-colored nutlets (Figs. 1, 2).

COLLECTION, EXTRACTION, AND STORAGE OF SEED Fruits may be stripped from the branches by hand or with a blueberry rake any time after they thoroughly ripen. They often persist for several weeks. Seeds may be extracted by macerating the berries and allowing the pulp to float away. There are 6.6 kg of cleaned seeds per 220 kg of fruit and an average of 780 seeds per gram. Seeds have been stored in sealed bottles at 5°C for 2 years without loss of viability.

GERMINATION Untreated seeds are slow to germinate, but both germinative energy and capacity can be increased with stratification. Previously successful pregermination treatments have included 30 days warm stratification at 20/30°C followed by incubation at 10°C. This is a very strange treatment. It has the appearance of a standard germination test conducted at 20/30°C that failed to produce any germination so the incubation temperature was dropped and the seeds germinated. What is the influence of prechilling, with or without warm stratification? Germination is epigeal (Fig. 3).

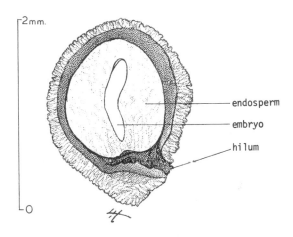

Figure 2. *Gaylussacia baccata*, black huckleberry: longitudinal section through a seed, ×25 (Bonner and Halls 1974b).

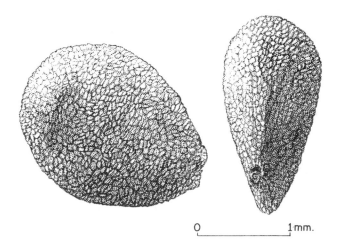

Figure 1. *Gaylussacia baccata*, black huckleberry: exterior view of seed in two planes, ×24 (Bonner and Halls 1974b).

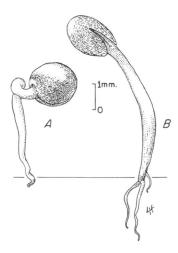

Figure 3. *Gaylussacia baccata*, black huckleberry: seedling development at 3 and 9 days after germination (Bonner and Halls 1974b).

LEGUMINOSAE — LEGUME FAMILY

Genista L. — Broom

GROWTH HABIT, OCCURRENCE, AND USE The genus *Genista* has about 140 species and is native to Europe, the Canary Islands, North Africa, and western Asia. The plants are generally low growing and broomlike, with yellow leguminous flowers. The fruit is a small pod that should be collected at color change. Many species are planted as ornamentals.

The seeds of most species require 30 minutes acid scarification (Dirr and Heuser 1987). With so many species it is probable that variable durations of acid scarification are appropriate and that hot water treatments would be effective for some.

GINKGOACEAE — GINKGO FAMILY

Ginkgo biloba L. — Ginkgo

GROWTH HABIT, OCCURRENCE, AND USE Ginkgo is a monotypic genus native to China (Alexander 1974). The single species is the sole survivor of its ancient family, Ginkgoaceae. This tall, deciduous, sparsely branched, long-lived tree has been cultivated extensively in the Far East and Europe. Introduced to North America in 1784, it has generally been successful on good sites in the moist temperate zone of the midwestern and eastern United States and along the St. Lawrence River in Canada. It is chiefly valuable as an ornamental and shade tree, particularly for parks. It is highly resistant to air pollution and can be grown in areas where air pollution damages other species.

FLOWERING AND FRUITING The species is dioecious. The catkinlike male flowers appear in late March or early April, and the pistillate flowers appear later in April before the wedge-shaped leaves. A single naked ovule ripens into a drupelike seed with an acrid, ill-smelling, fleshy outer layer and a thin, smooth, cream-colored, horny inner layer (Figs. 1, 2). The flesh-coated seeds are frquently called fruits. They are cast in the fall after the first frost, but at this time a large percentage of the seeds have immature embryos and cannot be germinated under normal test conditions. Embryo development continues on the ground as the seeds are exposed to the temperature regimes normally encountered in the fall and early winter. Embryo development is usually complete about 6 to 8 weeks after the seeds drop.

COLLECTION, EXTRACTION, AND STORAGE OF SEED Ginkgo begins bearing seeds when 30 to 40 years old. The flesh-coated seeds may be collected on the ground as they ripen or picked by hand from standing trees from late fall through early winter. Seeds may be prepared for cleaning by storing them in a warm place until the fleshy layers become soft enough to be washed off. Seed yields have been reported as 55 kg of seeds per 220 kg of fleshy seeds. The number of seeds per gram varies from 0.4 to 1.1. Cleaned seeds have been kept in ordinary dry storage in both open and closed containers, at 5 to 22+°C, without apparent loss of viability.

Figure 1. *Ginkgo biloba*, ginkgo: seeds enclosed in their fleshy outer layers, and cleaned seeds (fleshy layers removed), ×1 (Alexander 1974).

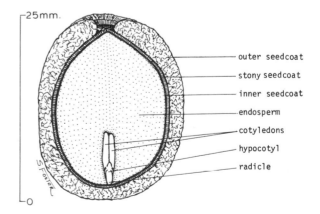

Figure 2. *Ginkgo biloba*, ginkgo: longitudinal section through a seed, ×2 (Alexander 1974).

GERMINATION TEST Germination tests have been conducted in moist sand at 20/30°C for 60 days. Under these conditions germination capacity was 40% for October-collected seeds and 90% for December-collected seeds. Prechilling probably would increase the germination of early collected seeds.

NURSERY AND FIELD PRACTICE Seeds should be sown in the late fall, preferably in furrows, and covered with 5 to 7.5 cm of soil and a sawdust mulch. About 50% of the viable seed, sown produces usable 2-0 seedlings. Ginkgo is also propagated by cuttings.

LEGUMINOSAE — LEGUME FAMILY

Gleditsia L. — Honey locust

GROWTH HABIT, OCCURRENCE, AND USE There are 12 species of honey locust (Bonner et al. 1974). All are deciduous species useful for timber and wildlife food and habitat.

FLOWERING AND FRUITING Honey locust flowers are borne in single or densely clustered axillary racemes. Those of water locust (*Gleditsia aquatica*) and honey locust (*G. triacanthos*) are greenish in color, while flowers of Texas honey locust (*G. × texana*) are orange-yellow.

The fruits are pods (Fig. 1) that ripen in the fall but often persist until winter. The small, flat, brownish seeds contain a thin, flat embryo surrounded by a layer of horny endosperm (Figs. 2, 3). Phenology of flowering and fruiting is as follows (Bonner et al. 1974):

Species	Flowering dates	Fruit ripening dates	Seed dispersal dates
G. aquatica	May–June	Aug.–Oct.	Sept.–Dec.
G. × texana	Apr.–May	Aug.–Sept.	Sept.–Dec.
G. triacanthos	May–June	Sept.–Oct.	Sept.–winter

Seed bearing starts at about age 10, and good crops are borne almost every year. The fruits turn from green to an orange-brown or deep reddish brown at maturity.

COLLECTION, EXTRACTION, AND STORAGE OF SEED Pods may be picked from the trees after they dry or from the ground. Moist pods should be spread out to dry.

Dried pods may be run through a macerator or mechanical thresher to extract the seeds. The pods can also be threshed by hand. The seeds can be recovered from the chaff with an air screen.

Seed yield per 220 kg of honey locust pods is 44 to 77 kg. The average number of seeds per gram is 6. Seeds of Texas honey locust are larger with 4 seeds per gram.

Viability of honey locust seeds stored in sealed containers at 0 to 8°C was retained for several years.

PREGERMINATION TREATMENT The hard seedcoats of *Gleditsia* species must be broken before seeds will germinate. Acid scarification and hot water treatments have been used, but the acid method has been more effective. The optimum soaking time varies from 1 to 2 hours. For the hot water treatment, seeds should be dropped into 3 to 4 times their volume of boiling water

G. *triacanthos*
honeylocust

G. *aquatica*
waterlocust

G. *Xtexana*
Texas honeylocust

Figure 1. *Gleditsia*: pods, ×0.5 (Bonner et al. 1974).

and allowed to cool overnight. The imbibed seeds must be planted; they cannot be stored.

GERMINATION TEST The ASOA (1985) standards for honey locust seeds are substrate, between blotters; incubation temperature, 20°C; duration, 21 days; and

Figure 2. *Gleditsia*: seeds, ×2(Bonner et al. 1974).

additional instructions, soak in concentrated sulphuric acid for 1 hour.

NURSERY AND FIELD PRACTICE Pretreated seeds have been

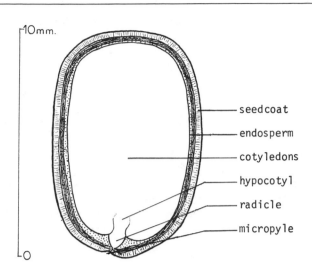

Figure 3. *Gleditsia triacanthos*, honey locust: longitudinal section through a seed, ×6 (Bonner et al. 1974).

drilled in rows 15 to 25 cm apart and covered with soil to a depth of 1.25 to 1.9 cm. A seeding rate of 32 to 48 per meter of row is standard. Seedlings reach a suitable size for transplanting in 1 year. Vegetative propagation is also possible.

VERBENACEA — VERBENA FAMILY

Gmelina L.

GROWTH HABIT, OCCURRENCE, AND USE There are many references to *Gmelina arborea* being grown in plantation agriculture in tropical Africa and Asia.

COLLECTION, EXTRACTION, AND STORAGE OF SEED Apparently the fruit is a drupe that contains seeds in a pulp matrix. The fruit pulp inhibits germination (Iwegbulam 1983) and should be removed by hand for best germination. For mass production of seeds the pulp is

removed by allowing it to ferment.

GERMINATION Green and yellow fruits of this species collected from the ground in Malaysia produce as good germination as ripe fruits picked from the tree (Aminunddin and Zakaria 1980). Seeds from brown fruits had much lower germination capacity. Bowen (1980b) reported similar results.

THEACEAE — TEA FAMILY

Gordonia Ellis — Loblolly bay

GROWTH HABIT, OCCURRENCE, AND USE *Gordonia* is a genus of about 16 species native to the southern United States and Asia. Loblolly bay (*G. lasianthus*) is a tree growing to 18 m in height that is native to the Atlantic seaboard

from Virginia to Florida. The fruit is a woody capsule. The capsules should be collected when brown and dry, and the seeds extracted and sown immediately; there will be no dormancy (Dirr and Heuser 1987).

CHENOPODIACEAE — GOOSEFOOT FAMILY

Grayia H. & A. — Hopsage

GROWTH HABIT, OCCURRENCE, AND USE Hopsage is an American genus with 2 species (Smith 1974a). They are freely branched, low shrubs, 0.3 to 1.6 m tall with alternate, rather fleshy, entire leaves. Spiny hopsage (*G. spinosa*) is the most widely distributed. It derives its name from the branch tips which are often spine tipped. Its natural range extends from eastern Washington to southern California and Arizona and east to Montana, western Wyoming, and Colorado. Smith (1974a) suggested that it is common in alkaline soil situations. While common on the margins of salt deserts, it also often occurs in nonsalt affected soils. Often spiny hopsage is a component of the plant communities of the lower portion of the *Artemisia tridentata* zone. Spiny hopsage is an important browse plant on many ranges. Often large preference differences are expressed by browsing animals among plants in a given community.

The range of spineless hopsage (*G. brandegei*) extends from Colorado and Utah to Arizona (Smith 1974a). It, too, is reported to be a very valuable browse species on winter range. Compared to spiny hopsage, the leaves of spineless hopsage are longer and narrower, and its fruits are smaller.

FLOWERING AND FRUITING The male and female flowers are usually borne on separate plants in terminal and axillary spikelike clusters from April to June. The fruit (Fig. 1) is a utricle consisting of a nutlet (Fig. 2) closely invested by 2 flattened papery bracts forming a round sac 8 to 15 mm broad, which turns whitish, with a tinge of red, as it matures in late June or July. In the nutlet, the pericarp adheres closely to the seedcoat. The curved embryo almost completely encircles a disk of endosperm (Fig. 3).

COLLECTION, EXTRACTION, AND STORAGE OF SEED The papery fruits are easily stripped from branches into containers. Smith (1974a) recommended running the fruits through a hammer mill to thresh the seeds from the bracts. As is the case with many species of chenopods, the bracts play a role in germination. They have been determined to be hydroscopic and to aid in establishment on the surfaces of seedbeds (Wood et al. 1976). There are about 370 fruits per gram and 870 to 930 cleaned seeds per gram. Seed yield per 220 kg of

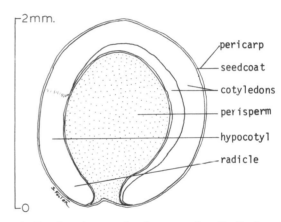

Figure 3. *Grayia spinosa*, spiny hopsage: longitudinal section through a nutlet, ×25 (Smith 1974a).

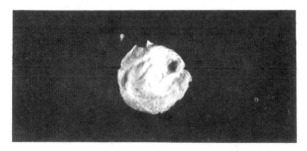

Figure 1. *Grayia spinosa*, spiny hopsage: fruit with bract, ×2 (Smith 1974a).

Figure 2. *Grayia spinosa*, spiny hopsage: nutlet with bract removed, ×12 (Smith 1974a).

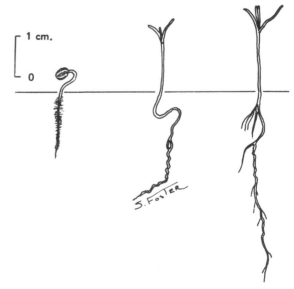

Figure 4. *Grayia spinosa*, spiny hopsage: seedling development at 1, 9, and 14 days after germination (Smith 1974a).

fruits has been estimated at about 6 kg of cleaned seeds. The winged fruits of spineless hopsage average 120 per gram (Smith 1974a). Seeds of spiny hopsage were stored at room temperature, at −15°C, or at 4°C for 14 years without loss of viability (Kay et al. 1988). Seeds from the same seed lots stored in a warehouse rapidly lost viability.

GERMINATION Wood et al. (1976) conducted detailed studies of the germination of collections of spiny hopsage seeds from the Great Basin. They determined that optimum germination after 2 weeks of incubation occurred with a temperature of 5°C during the cool period (16 hr) and temperatures of 10 to 30°C during the warm period (8 hr). Germination was rapid, with 70% germination at optimum temperatures after 1 week of incubation. One seed source had high germination at low osmotic potentials. The rapid germination permitted emergence on soils that were dried from field capacity to low matric potentials.

NURSERY AND FIELD PRACTICE Seedlings of spiny hopsage can be container grown without problems except that they require vernalization for continued growth and the seedlings are very brittle. Wood et al. (1976) obtained nominal seedling establishment by broadcasting bracted seeds on a loose seedbed. Germination is epigeal (Fig. 4). The autecology of spiny hopsage is currently under investigation by Nancy Shaw of the Forest Service, USDA, Boise, Idaho. The results of her investigations should be very valuable.

PROTEACEAE — PROTEA FAMILY

Grevillea robusta A. Cunn. — Silk-oak

GROWTH HABIT, OCCURRENCE, AND USE Silk-oak is an evergreen tree introduced from Australia in the late 1800s and planted on most of the major Hawaiian Islands (Wong 1974). If Australia were known internationally for only one tree it would be silk-oak. It is planted around the world in tropical and subtropical environments.

Grevillea robusta has adapted well to Hawaii and grows from sea level to 1230 m. It has been planted extensively in reforestation programs, and has become naturalized in certain areas. Wide dissemination of the seeds by wind and tolerance to many site conditions have enhanced its ability to proliferate. This tree attains heights of 25 to 37 m, and diameters up to 1 m. Individual trees are found in many yards and around ranches because of their showy orange blossoms. The wood has a beautiful, well-marked, silver grain, making it desirable for furniture and cabinet work. Care must be taken when machining and finishing this wood because the sawdust contains a skin irritant that produces an uncomfortable rash lasting a week or more. Hydrocyanic acid has been detected in the fruits and flowers.

Another species, Kahili flower (*G. banksii*), is less common because reforestation attempts with it have failed. Only on Kauai and Maui are remnant stands of early plantings found. It grows into a small tree 6 to 9 m high. The flower and fruits also contain cyanogenic substances that produce a rash similar to the one produced by poison ivy. A white-flowered form is also found in Hawaii. Hawaii State regulations classify this species as a noxious weed.

FLOWERING AND FRUITING Silk-oak is monoecious and flowers from early March through October, reaching its peak during May, June, and early July. Trees in Hawaii usually produce flowers and seeds when 10 to 15 years of age. In Jamaica, trees seed profusely from 10 years of age. The bright orange blossoms are borne on horizontal, racemes 8 to 12 cm long, which are on short, leafless branches of the old wood.

The fruit is a green, leathery follicle 15 to 25 mm long, tipped with a slender recurved stiff style (Fig. 1).

Figure 1. *Grevillea robusta*, silk-oak: follicle and seed, ×2 (Wong 1974).

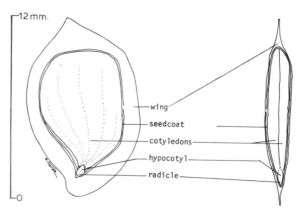

Figure 2. *Grevillea robusta*, silk-oak: longitudinal section through two planes of a seed, ×4 (Wong 1974).

The seed case remains on the tree for a year or so after the seeds have dispersed. Two brown seeds, each about 15 mm long with light, winged margins, are found in each follicle (Figs. 1, 2). Seed crops occur annually.

The flowers, fruit, and seeds of *G. banksii* resemble those of *G. robusta*. The blossoms are red instead of orange, while creamy flowers characterize the form *albifora*.

COLLECTION, EXTRACTION, AND STORAGE OF SEED The fruits of silk-oak are gathered from the trees before opening, when the first hint of brown color appears, indicating that the seeds are mature. The seeds are extracted by air drying the fruit in trays in the shade for 4 or 5 days, or until the seed cases open. The seeds which fall from the cases are then separated by means of an air screen. Purity has averaged 87%. Moisture content of fresh seed collected in Hawaii has averaged 28.5%. In Hawaii there are 65 seeds per gram, in East Africa 66 to 154, and in Australia, 80 to 100.

Seeds have been stored for 2 years at cold temperatures with germination ranging from 60 to 70% when seed moisture was maintained below 10% in air-tight containers.

PREGERMINATION TREATMENT Either a 48-hour soak in water or 30 days of prechilling at 5°C enhances germination (Wong 1974).

GERMINATION The Genebank Handbook (1985) lists germination standards for silk-oak seeds as substrate, top of paper; incubation temperature, 20/30°C; dura-tion 28 days; and additional instructions, add KNO_3 and prechill. ASOA (1985) standards are similar except light is specified. Germination suggestions for other species of *Grevillea* are given in Table 1.

NURSERY AND FIELD PRACTICE In Hawaii, silk-oak seeds are sown in the spring at a depth of 1.25 cm without mulch. Seedling densities range fron 210 to 315 per m². Outplanting is done when the seedlings are 9 months old. Seedlings grown in Ceylon are outplanted when about 38 cm high; those in Jamaica are outplanted when about 60 cm.

Table 1. *Grevillea:* germination procedures (Langkamp 1980).

Species	Pretreatment	Temperature (°C)	Germination (%)
G. banksii	KNO_3	25/33	70
G. berryana	Scarify	–	50
G. bipinnatifida	–	–	10
G. bracteosa	Scarify	–	25
G. dielsiana	Hot water	–	70
G. dryandri	Scarify	–	80
G. eriostavhya	Scarify	–	50
G. fasciculata	Testa removed	–	87
G. heliosperma	–	25	75
G. johnsonii	Scarify	–	43
G. leucopteris	Testa removed	–	60
G. longistyla	Scarify	–	10
G. manglesioides	Scarify	–	66
G. polybotrya	Scarify	–	20
G. pteridifolia	Scarify	–	67
G. stenobotrya	Scarify	–	84
G. trifida	Scarify	–	25
G. wickhamii	Fresh seed	–	Good

LECYTHIDACEAE — LEAYTHIS FAMILY

Gustavia superba (Kunth) O. Berg.

GROWTH HABIT, OCCURRENCE, AND USE *Gustavia superba* is a neotropical forest tree native to Central America.

GERMINATION Sork (1985) investigated the germination of its seeds with field planting experiments. Seeds were planted in 20-, 100-, and 300-year-old forests, either buried or unburied, and in 3 microsites—light gap, gap edge, and understory. Seeds germinated readily under most conditions (84% germination); burial had a minor influence. Germination was lowest and required the longest time in large light gaps. The author concluded that light is not required for germination and that the seeds can germinate under a range of microsite conditions.

COMPOSITAE — SUNFLOWER FAMILY

Gutierrezia Lag. — Snakeweed

GROWTH HABIT, OCCURRENCE, AND USE The genus *Gutierrezia* consist of about 25 species of perennial herbs or subshrubs native to North and South America. The flower heads are very small, with yellow ray flowers, scattered or crowded in cymes or panicles. Broom snakeweed (*Gutierrezia sarothrae*) and thread-leaf snakeweed (*G. microcephala*) are heavily branched subshrubs distributed throughout the western United States. Both are poisonous to livestock and compete with desirable forage species.

FLOWERING AND FRUITING Flowering occurs through late summer into fall. Most seeds are dispersed beneath the plant in late fall. The fruit is an oblong achene with a pappus.

GERMINATION Seeds of both species of snakeweed were dormant at maturity (Mayeux and Leotta 1981). An afterripening period of 4 to 6 months with storage at room temperature was necessary before seeds would germinate. An alternating temperature regime of 10/20°C (16 hours cold and 8 hours warm) favored germination during the afterripening period, but 20/30°C was optimum after 24 months of storage. Light for 8 hours daily roughly doubled germination. The afterripening dormancy could be partially overcome by imbibing seeds at 50°C. Kruse (1970) determined that germination of seeds of broom snakeweed rapidly declined as osmotic potentials decreased.

LEGUMINOSAE — LEGUME FAMILY

Gymnocladus dioicus (L.) K. Koch — Kentucky coffee tree

GROWTH HABIT, OCCURRENCE, AND USE Kentucky coffee tree is a medium to large deciduous tree whose natural range extends from New York and Pennsylvania west to Minnesota, southward to Oklahoma, and east to Kentucky and Tennessee (Sander 1974b). It is used chiefly for timber and fence posts, and has been used to some extent as an ornamental in the eastern United States. The species was first introduced into cultivation in 1784.

FLOWERING AND FRUITING The greenish white, dioecious flowers appear in May and June, after the leaves, and are borne in terminal, racemose clusters. The fruit is a tardily dehiscent, flat, thick, woody legume that ripens in September or October and usually persists unopened on the tree until late winter or early spring. The dark brown or red-brown pod is 15 to 25 cm long, 2.5 to 5 cm wide, and usually contains 4 to 8 dark brown or almost black seeds separated by a mass of brown pulp (Figs. 1, 2).

The seeds are oval, about 1.9 cm long, with a very thick, very hard, and bony coat. They generally remain in the pod until it falls and is broken up by decay, a process that may take 2 years or longer.

COLLECTION, EXTRACTION, AND STORAGE OF SEED The fruit can be collected at any time during the late fall, winter, or spring by picking from the tree or from the ground if they have fallen. Sometimes the pods can be dislodged by vigorously shaking or flailing the branches.

Seeds may be extracted by hand, either dry or after maceration. The number of clean seeds per gram varies from 0.4 to 0.7 and averages 0.5. Kentucky coffee tree seeds apparently remain viable for several years in

Figure 1. *Gymnocladus dioicus*, Kentucky coffee tree: pod and seed, ×0.5 (Sander 1974a).

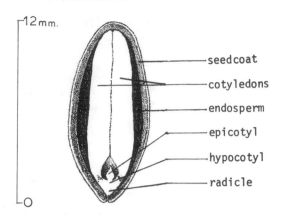

Figure 2. *Gymnocladus dioicus*, Kentucky coffee tree: longitudinal section through a seed, ×4 (Sander 1974a).

nature. Cold dry storage is recommended for over winter storage.

PREGERMINATION TREATMENT Kentucky coffee tree seeds have a hard, impermeable seedcoat that prevents or delays germination. The best germination can be obtained by acid scarification. These seeds are large enough that small amounts of seed can be individually filed.

Yeiser (1983) compared acid scarification of seeds of Kentucky coffee tree that had been presoaked in water for 24 hours before the acid treatment with unsoaked seeds. Seeds soaked in water and acid scarified for 150 minutes germinated slightly faster than seeds similarly treated, but scarified for 120 minutes. Ball and Kisor (1985) also investigated the acid scarification requirements for seeds of this species.

GERMINATION Apparently standards do not exist for germination of seeds of this species. A suggested methodology is substrate, sand; incubation temperature, 20/30°C; and special instructions, acid scarify seeds first.

NURSERY AND FIELD PRACTICE Pretreated seeds should be sown in spring in rows 45 to 75 cm apart. The sowing rate should be 40 to 60 seeds per meter of row. Seeds should be covered with 2.5 cm of firm soil. About 60 to 75% of the seeds planted will produce seedlings. The seedlings can be planted in the field after 1 year. Germination is hypogeal (Fig. 3).

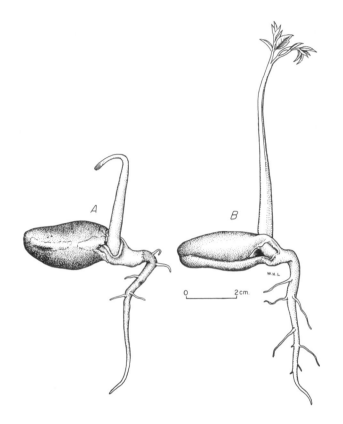

Figure 3. *Gymnocladus dioicus*, Kentucky coffee tree: seedling development at 2 and 5 days after germination (Sander 1974a).

PROTEACEAE — PROTES FAMILY

Hakea Schrad. — Pincushion tree

GROWTH HABIT, OCCURRENCE, AND USE Popular ornamental species in Australia, southern France, and California (Macoboy 1982), pincushion trees are more cold tolerant than many Australian trees. In nature they are loose, gangling trees that may reach 7 m in height. *Hakea saligna* is a drought-resistant Australian shrub to 2 m tall that is planted in Hawaii.

Hakea species have escaped from cultivation and become naturalized in South Africa. Richardson et al. (1987) considered the high rate of seed production and resistance of the woody fruit to burning in wildfires to be important characteristics in the colonizing success of these species.

FLOWERING AND FRUITING The small red flowers are packed in dense 5 cm globular heads. The inflorescence resembles brightly colored pincushions or sea urchins. The fruit is a hard, woody, rough, oval, follicle, 1.25 cm in diameter, with a curved beak. It contains 2 seeds, each with a long wing on one end (Neal 1965).

GERMINATION Langkamp (1980) lists germination characteristics for about 15 species under Australian conditions. Seeds should be sown in April in Australia, in pots of moist, sandy, loam-textured soil. Incubation temperatures should be about 30°C. Emergence occurs in 5 to 60 days and for most species ranges from 65 to 100%.

STYRACACEAE — SNOWBELL FAMILY

Halesia carolina L. — Carolina silverbell

GROWTH HABIT, OCCURRENCE, AND USE Carolina silverbell is a small deciduous tree that may reach 12 m at maturity (Bonner and Mignery 1974). It is found naturally from Virginia and southern Illinois south to northwestern Florida, western Tennessee and Alabama. It has been planted in the eastern United States as far north as Massachusetts and, to some extent, in northern and central Europe. The species was first cultivated in 1756. It is valued as an ornamental and a wildlife species.

FLOWERING AND FRUITING The white, perfect, axillary flowers of Carolina silverbell are borne in fascicles of 1 to 5 in March to May. The fruit, which matures in autumn, is an oblong or ovate, dry, 4-winged, reddish brown, corky drupe, 2.5 to 5 cm long. The ovary is a 4-celled ellipsoid stone, 1.25 to 1.7 cm long (Figs. 1, 2). The fruits are persistent on the branches, and dispersal occurs well into the winter.

COLLECTION, EXTRACTION, AND STORAGE OF SEED Carolina silverbell fruits may be collected from the trees in late fall and early winter. Dewinging can be done mechani-cally and is recommended to reduce bulk and facilitate handling. Complete extraction of stones from the fruit is not necessary. There are 2.6 to 5.5 dewinged fruits per gram. Dry cold storage of seeds is recommended.

PREGERMINATION TREATMENT Untreated seeds will not germinate satisfactorily the same year they are planted. Successive treatments of warm stratification and pre-chilling have induced adequate germination. Dirr and Heuser (1987) suggest 3 months of prechilling followed by 6 months of warm stratification, followed by an additional 4.5 months of prechilling. Seed lots vary widely in the degree of dormancy and prechilling alone has been satisfactory for some seed sources.

GERMINATION TEST Pretreated seeds have been germinated in flats of sand or sand-peat mixtures for 60 to 90 days at 20/30°C. Germination is epigeal (Fig. 3).

NURSERY AND FIELD PRACTICE Pretreated seeds should be sown in the spring. Some growers have planted seeds in flats and kept them in the greenhouse during the early winter months. In January the flats are moved to outdoor cold frames for the cold part of the treatment. They are protected by mulch in the cold frames. Plants can also be propagated by layering, root cuttings, and cuttings.

Figure 1. *Halesia carolina* var. *carolina*, Carolina silverbell: fruit and stone (seed), ×1 (Bonner and Mignery 1974).

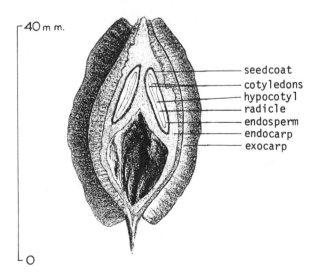

Figure 2. *Halesia carolina* var. *carolina*, Carolina silverbell: longitudinal section through 2 embryos of a stone, ×1½ (Bonner and Mignery 1974).

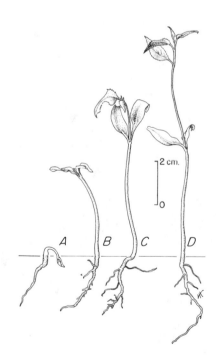

Figure 3. *Halesia carolina* var. *carolina*, Carolina silverbell: seedling development at 1, 4, 16, and 49 days after germination (Bonner and Mignery 1974).

CHENOPODIACEAE — GOOSEFOOT FAMILY

Haloxylon Bunge

GROWTH HABIT, OCCURRENCE, AND USE Native to central and western Asia, *Haloxylon* species are the most woody or treelike of the chenopods. Despite a wide range and number of species, this genus remains little known in North America.

A sampling of recent Soviet literature illustrates something of the habitat and regeneration of *Haloxylon*. Georgievskii (1974) reported on the natural reproduction of *H. ammodendron* in the Kara-Kum desert of Turkmenistan. An area cleared in 1932 had completely regenerated in 20 years.

Georgievskii and Khodzamkuliev (1977) artificially planted seeds of this species in natural stands.

Emergence and survival were best in the peripheral part of the subcanopy and on north aspects. Field emergence was 15 to 20% which was 3 or 4 times laboratory germination.

Smurova (1975) reported on the long-term storage of *H. aphyllum* seeds: Those stored in cloth bags maintained viability for 6 to 7 months, while those stored under refrigeration maintained viability for 2 years. For storage, seeds should be dried to 3% moisture content.

Suleimanov (1980) studied the seeding of this species in Kazakbstan.

HAMAMELIDACEAE — WITCH-HAZEL FAMILY

Hamamelis virginiana L. — Witch-hazel

GROWTH HABIT, OCCURRENCE, AND USE Witch-hazel is a deciduous shrub or small tree attaining a height of 5 to 6 m (Brinkman 1974g). It is native from Nova Scotia to southeastern Minnesota, south to Missouri, southeastern Oklahoma and Texas, and east to Florida. First cultivated in 1736, witch-hazel is used in environmental plantings largely because it flowers in late autumn. It also provides food and cover for wildlife. Bark, leaves, and twigs have been used medically in the form of extracts.

FLOWERING AND FRUITING The flowers of witch-hazel open in September or October, but the fruits do not ripen until early the next autumn. Capsules (Fig. 1) burst open when dry, each discharging 2 shiny black seeds (Fig. 2). There is limited seed dispersal by birds.

COLLECTION, EXTRACTION, AND STORAGE OF SEED Witch-hazel fruits should be picked before they split and discharge the seeds. Ripe fruits are dull orange-brown, with blackened, adhering fragments of floral bracts. Seeds apparently mature as early as August, before the fruit appears mature. In a study of seed production of witch-hazel, Steven (1982) found 4 factors governing seed production: (1) physiological abortion, (2) damage by caterpillars, (3) infestation by a host specific weevil, and (4) consumption by squirrels.

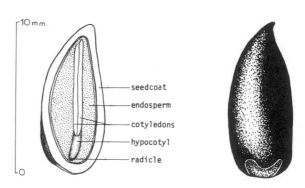

Figure 1. *Hamamelis virginiana*, witch-hazel: fruits (capsules) before and after seeds are discharged, ×2 (Brinkman 1974g).

Figure 2. *Hamamelis virginiana*, witch-hazel: longitudinal section through a seed exterior view, both at ×4 (Brinkman 1974g).

Fruits should be spread to allow drying before screening to recover seeds. Seeds per gram range from 19 to 24. Fresh seeds can be stored dry in sealed containers at 5°C. For overwinter storage prior to spring planting, seeds can be stored by prechilling.

GERMINATION Some prechilled seeds germinate the first spring, but many remain dormant until the following year. Dormancy is due to the seedcoat and the embryo. Pregermination treatments consist of warm stratification for 60 days at 20/30°C, followed by 90 days prechilling. Germination tests are conducted on pretreated seeds at 20/30°C for 60 days.

More recent studies conducted by Gaut and Roberts (1984) serve to reinforce how complex dormancy is for seeds of *H. virginiana*. Optimum conditions for breaking dormancy in fresh seeds are at least 8 weeks at about 20°C, followed by at least 20 weeks below 4°C, both in moist peat. Neither warm nor cold temperatures alone will break dormancy, but warm temperatures in some way sensitize the embryo to react to subsequent cold. While dormant, the embryo produces a lipase inhibitor that presumably diffuses into the endosperm and prevents premature mobilization of fatty storage reserves. For warm treatment, a soak in gibberellic acid can be substituted, but a substitute for cold cannot be found.

NURSERY AND FIELD PRACTICE Witch-hazel seed may be fall sown in the nursery as soon as collected, or pretreated and sown in the spring. Early collected seeds may avoid dormancy problems associated with later collections. Fall-sown seedbeds should be mulched over the winter and uncovered at germination time in the spring. Sowing in drill rows spaced 20 to 25 cm apart is recommended. Secondary leaves develop on seedlings within 21 days after emergence (Fig. 3). Propagation by layering and cuttings is possible (Dirr and Heuser 1987).

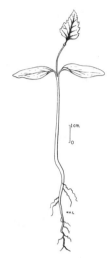

Figure 3. *Hamamelis virginiana*, witch-hazel: seedling 21 days after germination (Brinkman 1974g).

COMPOSITAE — SUNFLOWER FAMILY

Haplopappus parishii (Greene) Blake — Parish goldenweed

GROWTH HABIT, OCCURRENCE, AND USE Parish goldenweed is an erect shrub with a mature height of 1 to 2.5 m (Ratliff 1974b). This species occurs in the lower parts of the chaparral belt in California. The genus contains about 150 species native to the western United States, Mexico, and Chile. Many are herbaceous perennials. Parish goldenweed is primarily of value for erosion control on dry slopes.

FLOWERING AND FRUITING The plants will flower and produce seeds at 2 years of age. Flowering takes place from July to October, and ripe seeds may be collected in October and November.

The fruit of Parish goldenweed is an achene (Figs. 1, 2). There are about 3590 seeds per gram. Seeds can be

Figure 1. *Haplopappus parishii*, Parish goldenweed: achene with pappus removed, ×12 (Ratliff 1974b).

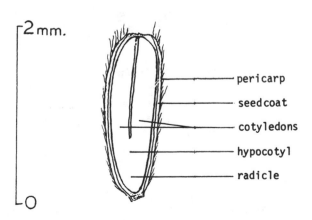

Figure 2. *Haplopappus parishii*, Parish goldenweed: longitudinal section through an achene, ×24 (Ratliff 1974b).

collected by hand and recovered by rubbing and air screening.

GERMINATION The achenes germinate without pretreatment, but germination capacity is usually only near 20%. Parish goldenweed can also be propagated by cuttings.

Galli et al. (1975) and Galli and Miracca (1979) conducted detailed biochemical studies of the influence of abscissic acid, gibberellin, and fusicoccin on the germination of seeds of *Haplopappus gracilis*.

LEGUMINOSAE — LEGUME FAMILY

Hardwickia Roxb.

GROWTH HABIT, OCCURRENCE AND USE *Hardwickia* is a small genus of trees native to India that belongs to the legume family. These species are used in fodder tree and reforestation projects in arid areas.

GERMINATION Pathak and Gupta (1984, 1985) investigated germination of *Hardwickia* seeds after storage as open seeds and in pods at ambient tempera- tures or 10°C for 3 to 24 months. There was little loss of viability up to 14 months, after which slight losses occurred. Seeds of *H. binata* had 60% germination at greatly reduced osmotic potentials. Pretreating seeds with ascorbic acid increased germination of large seeds. Small seeds did not germinate well.

TILIACEAE — LINDEN FAMILY

Heliocarpus L.

GROWTH HABIT, OCCURRENCE, AND USE *Heliocarpus donnell- smithii* is a fast-growing tropical pioneer tree from Mexico.

GERMINATION The germination of seeds of this species is favored by high temperatures, a thermoperiod, and the presence of light (Vazques-Yanes 1981). Germination of dormant seeds can be enhanced by heating.

ROSACEAE — ROSE FAMILY

Heteromeles arbutifolia (Ait.) M. Roem. — Christmas berry

GROWTH HABIT, OCCURRENCE, AND USE Christmas berry grows below 1230 m on the foothills and canyon bottoms of the Sierra Nevada, Coast, and Transverse ranges in California (Magill 1974c, as *Photinia*). It is also found on San Clemente and Santa Catalina islands, in the Rocky Mountains, and in Baja California and other parts of Mexico. Christmas berry is an evergreen shrub or small tree 2 to 10 m tall. It is useful in erosion con- trol, and is a source of honey. Its foliage and berries are consumed by wildlife. The attractive foliage and berries are also used for ornamental purposes.

FLOWERING AND FRUITING Christmas berry produces per- fect flowers from June through July, and the fruit ripens from October through January. Good seed crops are produced annually. The fruit is a pome about 6 mm long, containing 1 or 2 seeds. The variety *cerina* has larger fruits ranging from 9 to 12 mm long. The seeds have no endosperm (Figs. 1, 2). The green pomes turn bright red when ripe, except the variety *cerina*, which

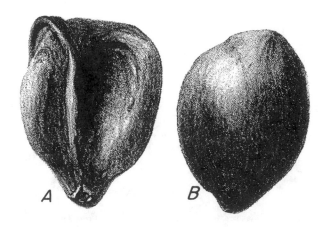

Figure 1. *Heteromeles arbutifolia*, Christmas berry: exterior views of seed from two planes, ×16 (Magill 1974c).

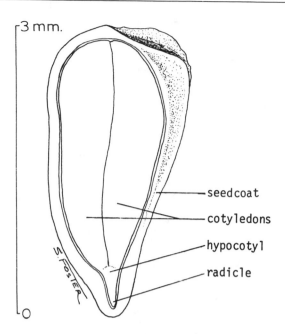

Figure 2. *Heteromeles arbutifolia*, Christmas berry: longitudinal section through a seed, ×25 (Magill 1974c).

turns yellow. Plants have been difficult to grow from seed of the yellow fruit, the color was only rarely retained, and it is thought to indicate an physiological condition.

COLLECTION, EXTRACTION, AND STORAGE OF SEED Pruning shears are needed to collect pomes in fall. The pomes should be soaked in water and allowed to ferment in a warm place until the seed can be separated from the mash. Fruits can be run through a macerator and the seeds recovered by flotation. Cleaned seeds average 52 per gram. They can be dried and stored at low temperatures in sealed containers.

GERMINATION Christmas berry will germinate under natural conditions in about 36 to 40 days in the greenhouse. Germination is epigeal (Fig. 3). Fresh seeds do not require prechilling, but stored seeds need to be prechilled for 3 months. In nursery beds, untreated seeds can be sown in the fall, or prechilled seeds in the spring. Christmas berry can be propagated by cuttings.

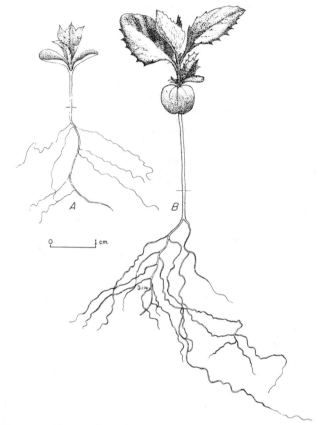

Figure 3. *Heteromeles arbutifolia*, Christmas berry: A, young seedling; B, older seedling (Magill 1974c).

EUPHORBIACEAE — SPURGE FAMILY

Hevea brasiliensis Muell. — Rubber

GROWTH HABIT, OCCURRENCE, AND USE The first seeds of rubber were sent to Kew Gardens in 1873, and out of several hundred, less than a dozen germinated. In 1876 Henry Wick collected 70,000 seeds in Brazil and upon arrival at Kew Gardens only 2700 germinated.

GERMINATION *Hevea* seeds lose their viability rapidly and if stored without protection, germination drops in a month to 45% or less. Apparently they will germinate initially without pretreatment. *Hevea* seeds are classified as recalcitrant. They are quite large, 2.5 × 2 cm, so germination test are usually conducted in sand.

Chin et al. (1981) investigated harvest and storage

procedures for rubber seeds. Freshly harvested *Hevea* seeds of 36% moisture content were dried by different methods, and some were kept moist for a period of 96 hours. Irrespective of the method of drying, dehydration resulted in loss of viability. The critical moisture content level was around 15 to 20%; below this level seeds were killed. Over a period of 96 hours seeds that retained their original moisture content showed little or no loss of viability at mean temperatures of 22 and 28°C, but those stored at −5°C or 45°C lost viability completely. Thus, *Hevea* seeds are killed by dehydration, high temperatures, and freezing.

MALVACEAE — MALLOW FAMILY

Hibiscus L. — Rose mallow

GROWTH HABIT, OCCURRENCE, AND USE The genus *Hibiscus* consists of about 200 species native to both tropical and temperate regions. One of the more popular ornamental species is rose of Sharon (*Hibiscus syriacus*).

The fruit is a 5-valved capsule that can be collected in the fall, dried, and the seeds recovered by screening.

GERMINATION Seeds germinate without pretreatment, to the point that some species can become weeds. Cuttings taken in June and July, and treated with IBA, can be rooted (Dirr and Heuser 1987).

ELAEAGNACEAE — ELAEAGNUS FAMILY

Hippophae rhamnoides L. — Common sea buckthorn

GROWTH HABIT, OCCURRENCE, AND USE The genus *Hippophae* consists of 2 species native to Europe and Central Asia (Slabaugh 1974b). Common sea buckthorn is a very hardy, deciduous shrub or small tree used primarily for ornamental purposes. In Europe and Asia it is used in hedges and screens and in environmental plantings. A tendency to form thickets from root suckers has limited its use in shelterbelts. The berries, a rich source of vitamins, have been used for food in the USSR. The plant stem bears many sharp, stiff thorns, and provides cover for wildlife.

FLOWERING AND FRUITING The species is dioecious; its very small, yellowish, pistillate flowers appear in March and April before the leaves. Acid, orange-yellow, drupelike fruits about the size of a pea (Fig. 1) ripen in September and October, and frequently persist on the shrubs until the following March. Each fruit contains 1 bony, ovoid stone (Figs. 1, 2). Seed crops are borne annually.

COLLECTION, EXTRACTION, AND STORAGE OF SEED Common sea buckthorn fruits are soft and cling tenaciously to the brittle twigs. Fruit may be picked from the bushes at any time between late fall and early spring. Seeds may be extracted and recovered by maceration in water with flotation. Prompt cleaning is important because viability drops if seeds are stored in the fruit. From 220 kg of fruit, 22 to 66 kg of cleaned seed may be expected. The average number of seeds per gram ranges from 55 to 139. Dry seeds have been kept satisfactorily for 1 or 2 years at room temperature.

GERMINATION Slabaugh (1974b) considered seeds to require prechilling before they would germinate. Heit (1976b) reported that in his 10 years of experience with seeds of the species pretreatment was not necessary. He recommended incubating seeds at 20/30°C in light with a good moisture-supplying substrate for a duration of 28 days. Il'ina (1982) investigated the influence of incubation temperature on the germination of *H. rhamnoides* seeds, but details of the results are not available.

NURSERY AND FIELD PRACTICE Untreated seed may be sown in the fall. Either broadcast or drilling is satisfactory. Seeds need to be covered with 0.6 cm of soil. Shading during the early stages of germination is beneficial.

Seedling development is shown in Figure 3. This species can also be propagated by cuttings.

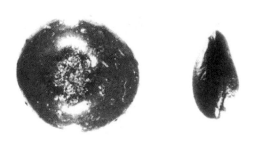

Figure 1. *Hippophae rhamnoides*, common sea buckthorn: fruit and stone, ×4 (Slabaugh 1974b).

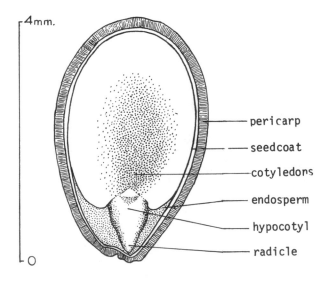

Figure 2. *Hippophae rhamnoides*, common sea buckthorn: longitudinal section through a stone, ×16 (Slabaugh 1974b).

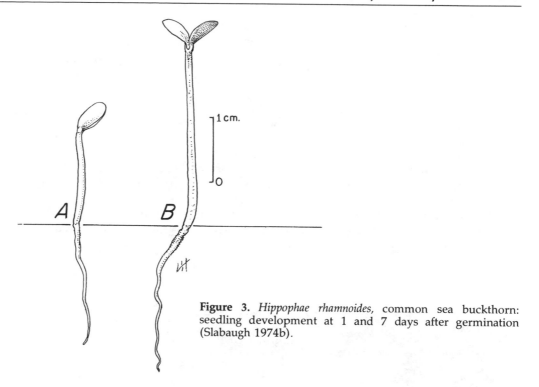

Figure 3. *Hippophae rhamnoides,* common sea buckthorn: seedling development at 1 and 7 days after germination (Slabaugh 1974b).

COMPOSITAE — SUNFLOWER FAMILY

Hofmeisteria pluriseta Gray — Arrowleaf

GROWTH HABIT, OCCURRENCE, AND USE Native to the deserts of the Southwest, this bushy, suffrutescent shrub reaches 0.3 to 0.8 m in height, with a rounded outline.

GERMINATION Older stems have white, shreddy bark. The fruit is an achene. Seeds collected in the deserts of southern California initially had 2% germination. After 1 year of storage, germination increased to about 10% (Kay et al. 1988).

ROSACEAE — ROSE FAMILY

Holodiscus discolor (Pursh) Maxim. — Ocean-spray

GROWTH HABIT, OCCURRENCE, AND USE Ocean-spray is a deciduous shrub, 1 to 3 m tall, native to the Pacific crest mountains from British Columbia to southern California (Stickney 1974a). This shrub is extremely wide in ecological amplitude, ranging from coastal rain forests, through the fir woodlands of the interior Pacific Northwest, to isolated remnant stands on the higher peaks of the Great Basin mountains. Normally symmetrical, it is a very attractive shrub in full bloom when it bears large panicles of creamy-white flowers on slender, arching branches. It was first cultivated in

1827. This species is browsed by deer and elk in the fall and winter.

FLOWERING AND FRUITING Flower bud appear in the spring; full flowering occurs in June and July and may continue into August. Fruits ripen in late August, and seed dispersal continues through November. The fruits are light yellow achenes, 2 mm long (Figs. 1, 2). The number of seeds per gram is 11,700.

GERMINATION Seeds are initially dormant and require prolonged periods of prechilling for germination.

Figure 1. *Holodiscus discolor*, ocean-spray: achene, ×16 (Stickney 1974c).

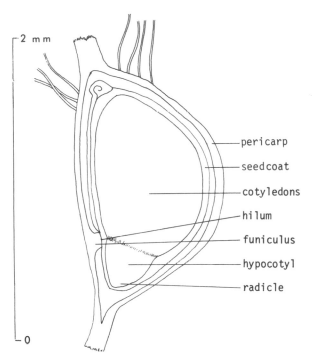

Figure 2. *Holodiscus discolor*, ocean-spray: longitudinal section through an achene, ×40 (Stickney 1974c).

DIPTEROCARPACEAE — DIPTEROCARP FAMILY

Hopea Roxb.

GROWTH HABIT, OCCURRENCE, AND USE *Hopea* is a genus of tropical trees native to southeastern Asia.

GERMINATION Initial germination is not a problem with *Hopea* seeds, but maintaining seed viability with storage is a problem. Song et al. (1983, 1984) have conducted detailed studies of the storage of seeds of *H. hainanensis*. Seeds were dried at 20 and 25°C to different moisture contents. Drying from 34.9 to 25.6% moisture content reduced germination and vigor of the

seeds and caused darkening and wilting of the embryos and cotyledons. Changes also occurred in the ultrastructure of the radicle tips. When moisture content was reduced to 20.3%, all seeds died and the cell structure was destroyed. High-quality seeds stored at 18°C in a slowly ventilated atmosphere, with moisture content controlled to 35 to 38%, retained a germination potential of above 80% after storage for a year or more.

RHAMNACEAE — BUCKTHORN FAMILY

Hovenia dulcis Thunb. — Japanese raisin tree

GROWTH HABIT, OCCURRENCE, AND USE Japanese raisin tree is grown as an ornamental, for medicinal and secondary products, and for fruit production. It is analogous to cashews in that the edible portion of the plant is the peduncle.

Frett (1989) determined that seeds of *Hovenia dulcis* should be collected in the fall, treated with sulfuric acid for 45 minutes, and prechilled at 5°C for 90 days.

SAXIFRAGACEAE — SAXIFRAGE FAMILY

Hydrangea L.

GROWTH HABIT, OCCURRENCE, AND USE The genus *Hydrangea* consists of about 80 species of shrubs native to North and South America. Some species are valuable ornamentals. The fruit is a capsule.

GERMINATION Propagation techniques are listed here (Dirr and Heuser 1987):

Species	Propagation techniques
H. anomala	Seeds: prechill for uniform germination. Cuttings: difficult to root.
H. arborescens	Seeds: no pretreatment necessary. Seeds are very small and should be seeded in flats. It is difficult to water the tiny, weak seedlings. Cuttings: soft wood easy to root.
H. macrophylla	Seeds: prechill for uniform germination. Cuttings: take cuttings from May to June from terminal growth and treat with IBA.
H. paniculata	Seed: no pretreatment necessary. Cuttings: take cuttings from May to June from terminal growth and treat with IBA.
H. quercifolia	Seeds: very germinable, with complete germination in 2 weeks. Cuttings: difficult to root.

FLACOURTIACEAE — FLACOURTIA FAMILY

Idesia polycarpa Maxim.

GROWTH HABIT, OCCURRENCE, AND USE The genus *Idesia* consists of single species (*Idesia polycarpa*) that is native to China and Japan (Bailey 1951). It is grown in the southeastern United States as an ornamental. It is a tree that reaches 15 m in height.

The fruit is a many-seeded berry. It may be planted entire or the seeds recovered by flotation after macerating the fruit.

Seeds germinate without pretreatment. Cuttings taken from young plants in early August and wounded will root if then treated with IBA.

AQUIFOLIACEAE — HOLLY FAMILY

Ilex L. — Holly

GROWTH HABIT, OCCURRENCE, AND USE The hollies include about 500 species of evergreen (or rarely deciduous) shrubs and trees that occur in temperate and tropical regions of both hemispheres (Bonner 1974e). About 20 species are native to eastern North America. Most are highly valued for ornamental plantings, and all are good food sources for wildlife. The wood of American holly (*Ilex opaca*) is used in cabinetry and for construction of novelties and specialized wood products.

FLOWERING AND FRUITING The small, axillary, greenish white flowers appear in spring (Table 1). Holly fruits are rounded, berrylike drupes, 0.6 to 1.3 cm in diameter (Fig. 1); they contain 2 to 9 bony, one-seeded, flattened nutlets (pyrenes) (Fig. 2). They mature in the fall, turning from green to various shades of red, yellow, or black (Table 1). The nutlets contain a very small embryo in a fleshy endosperm (Fig. 3) and are used as seeds. Fruiting phenology is given in Table 1.

COLLECTION, EXTRACTION, AND STORAGE OF SEED Ripe fruits may be picked by hand or flailed from the branches onto ground sheets. Seeds should be extracted by running the fruits through a macerator and separating the seeds by flotation. American holly seeds can be run through a hammer mill and allowed to ferment in warm water. Hand rubbing on a screen can be used to remove the pulp from small quantities of seed. If the seeds are prechilled immediately, drying is not necessary. If the seeds are to be stored they should be dried

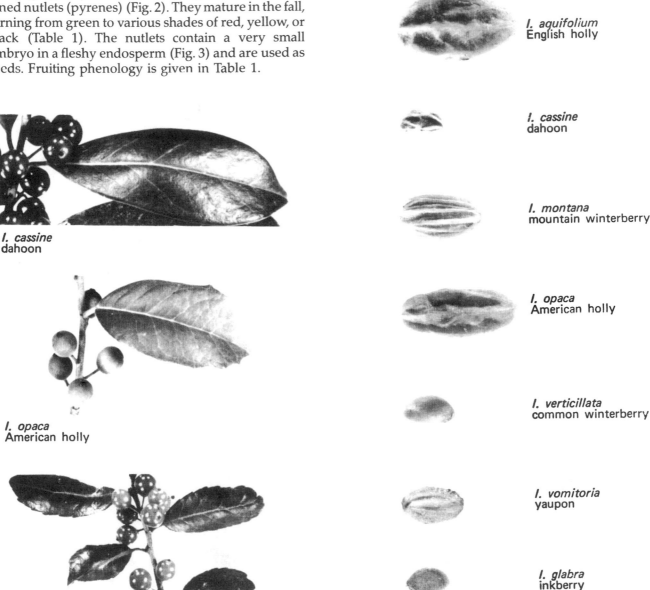

I. cassine
dahoon

I. opaca
American holly

I. vomitoria
yaupon

I. aquifolium
English holly

I. cassine
dahoon

I. montana
mountain winterberry

I. opaca
American holly

I. verticillata
common winterberry

I. vomitoria
yaupon

I. glabra
inkberry

Figure 1. *Ilex*: fruits and leaves, ×1 (Bonner 1974e).

Figure 2. *Ilex*: nutlets (pyrenes) containing seeds, ×4 (Bonner 1974e).

Table 1. *Ilex*: phenology of flowering and fruiting, characteristics of mature trees, and cleaned seed weight (Bonner 1974).

Species	Flowering dates	Fruit ripening dates	Seed dispersal dates	Mature tree height (m)	Year first cultivated	Seed-bearing age (years)	Ripe fruit color	Seeds/ 100 kg fruit	Seeds/ gram
I. aquifolium	May–June	Sept.	Winter–Spring	25	Ancient	5–12	Light red	–	130
I. glabra	Mar.–June	Fall	Winter–Spring	4	1759	–	Black	5	60
I. montana	May–June	Sept.	–	12	1870	–	Orange	–	80
I. opaca	Apr.–June	Sept.–Oct.	March	31	1744	5	Red	20/27	60
I. verticillata	June–July	Sept.–Oct.	Fall–Winter	8	1736	–	Red	11–20	200
I. vomitoria	Apr.–May	Sept.–Oct.	Winter	8	–	–	Scarlet	–	80

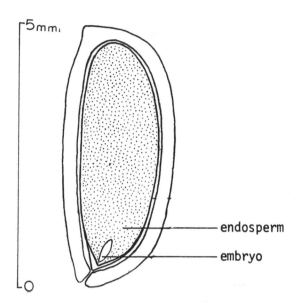

Figure 3. *Ilex montana*, mountain winterberry: longitudinal section of a nutlet (pyrene), ×14 (Bonner 1974e).

and placed in sealed containers. Seed yield data is given in Table 1.

PREGERMINATION TREATMENT Holly seeds exhibit a deep dormancy caused partly by the hard endocarp surrounding the seedcoat (Fig. 3) and partly by conditions in the embryo. American holly seeds contain an immature embryo that must complete its development after apparent fruit maturity. In nature, germination is commonly delayed for 16 months and may require 3 years for completion. Some benefit may be obtained by stratifying seeds at the warm temperatures of 20/30°C for 60 days followed by 60 days of prechilling at 5°C. This treatment is probably most beneficial with common winterberry (*I. verticillata*) seeds, which have an endocarp softer than those of the other *Ilex* species. Prechilling can replace overwinter storage for species planted in the spring.

GERMINATION TEST The Genebank Handbook (1985) suggests the Kew Garden algorithm, which is to imbibe two samples of seed at 26°C for 4 weeks with light applied for 12 hours daily. One sample is moved to a constant temperature regime of 2°C and the other sample to a constant temperature regime of 6°C. In both cases, light is applied for 12 hours daily. If these treatments are not successful in promoting full germination then the second step is to alter these regimes. For example, the warm pretreatment can be lengthened. Another option is to try alternating temperature regimes of 9/23°C or 19/33°C on a 12/12 hour basis with light during the warm temperature period.

NURSERY AND FIELD PRACTICE *Ilex* seeds may be broadcast or sown in drills in fall or spring. Sowing immediately after collection has been recommended for inkberry (*I. glabra*) and American holly. Complete germination will not occur until the second or third spring. Seeds should be covered with 0.5 to 1.25 cm of soil, and fall-sown beds should be mulched. Half shade is recommended for English holly (*I. aquifolium*) beds during the first 2 summers. *Ilex* species are often propagated by cuttings.

ILLICIACEAE — ILLICIUM FAMILY

Illicium L. — Anise tree

GROWTH HABIT, OCCURRENCE, AND USE *Illicium* is a genus consisting of about 40 species of shrubs and small trees. Some 5 species are native to the southeastern United States. *Illicium floridanum* is the most widely cultivated of the native species. It is native to the Gulf Coast. *Illicium anisatum* is native to Japan and Korea and it is also cultivated in the southeastern United States.

FLOWERING AND FRUITING The fruit is a star shaped aggregate of follicles that contain small, brown seeds. It should be collected in the fall. The seeds can be recovered by shaking the fruits.

GERMINATION Apparently the seeds germinate without pretreatment, but results can be variable. Cuttings taken in July and treated with IBA will root (Dirr and Heuser 1987).

LEGUMINOSAE — LEGUME FAMILY

Indigofera L.

GROWTH HABIT, OCCURRENCE, AND USE *Indigofera* is a genus of about 350 species native to tropical and warm temperate regions of the world, and extending to the Cape region of South Africa. Several species are native to the United States. Most are low-growing shrubs that are relatively unknown outside botanical gardens.

GERMINATION The seeds of most species have hard seedcoats that require hot water treatment for germination. Seeds should be dropped into 4 times their volume of water heated to 88°C and allowed to cool overnight (Dirr and Heuser 1987).

SAXIFRAGACEAE — SAXIFRAGE FAMILY

Itea L.

GROWTH HABIT, OCCURRENCE, AND USE The genus *Itea* consists of about 11 species of shrubs and small trees. Several are cultivated as ornamentals. Virginia sweetspire (*I. illicifolia*) is native to the United States; the other species are found in temperate and tropical Asia.

GERMINATION No pretreatment is necessary, but the seeds are very small and seedlings must be watered with care (Dirr and Heuser 1987).

OLEACEAE — OLIVE FAMILY

Jasminum L. — Jasmine

GROWTH HABIT, OCCURRENCE, AND USE *Jasminum* is a genus of some 200 species native to tropical and subtropical regions, and planted as ornamentals for their flowers and fragrance.

The fruit is a 2-lobed berry; each carpel contains 1 or 2 seeds.

Members of this genus rarely set seeds in the United States, but supposedly the seeds germinate without pretreatment (Dirr and Heuser 1987). Plants can also be propagated by cuttings.

JUGLANDACEAE — WALNUT FAMILY

Juglans L. — Walnut

GROWTH HABIT, OCCURRENCE, AND USE The walnuts include about 15 species of deciduous trees or large shrubs that occur in the temperate regions of North America, northwestern South America, and from northeastern Europe to eastern Asia (Brinkman 1974h). The wood of most species is used to some extent, and that of many species is highly valued for cabinet work, gunstocks, and interior trim. The nuts provide food for humans as well as for wildlife. Persian walnut (*Juglans regia*) is extensively planted for nut production. Black walnut (*J. nigra*), butternut *J. cinerea*), Hinds walnut (*J. hindsii*), and little walnut (*J. microcarpa*) are also planted for nut production or timber.

GEOGRAPHICAL RACES There are three distinct races of Persian walnut: Turkestanian, Himalayan, and Central Asian. They differ in frost hardiness. Many horticultural varieties have been developed of both Persian and Siebold walnut (*J. ailantifolia*).

FLOWERING AND FRUITING The male and female flowers are borne separately on the same tree, but mature at different times. Male flowers are slender catkins that develop from axillary buds on the previous year's outer nodes. Female flowers occur in few- to many-flowered short terminal spikes borne on the current year's shoots. These flowers appear with, or shortly after, the leaves (Table 1).

The ovoid, globose, or pear-shaped fruits ripen in the first year. The fruit is a nut enclosed in an indehiscent, thick husk that develops from a floral involucre

Table 1. *Juglans*: phenology of flowering and fruiting, characteristics of mature trees, cleaned seed weight, and pregermination treatments (Brinkman 1974h).

Species	Flowering dates	Fruit ripening dates	Seed dispersal dates	Year first cultivated	Mature tree height (m)	Seed-bearing Age (years)	Seed-bearing Interval (years)	Fruit color Preripe	Fruit color Ripe	Seeds/kg	Prechilling (days)	Germination (%)
J. ailantifolia	May–June	Aug.–Oct.	Oct.	1860	20	10	1–3	–	–	150	0	75
J. californica	Mar.–Apr.	Fall	Fall	1889	12	5–8	–	Light green	Dark brown	110	156	58
J. cinerea	Apr.–June	Sept.–Oct.	Sept.–Oct.	1633	31	20	2–3	Greenish bronze	Greenish brown	70	90–120	65
J. hindsii	Apr.–May	Aug.–Sept.	Sept.–Oct.	1878	25	9	–	Light yellowish	Dark brown	100	156	41
J. major	Spring	–	–	1894	15	–	–	–	–	200	120–190	64
J. microcarpa	Mar.–Apr.	Aug.–Sept.	Fall	1868	6	20	2–3	–	–	200	190	46
J. nigra	Apr.–June	Sept.–Oct.	Oct.–Nov.	1686	46	12	–	Light green	Dark brown	90	90–120	50
J. regia	Mar.–May	Sept.–Nov.	Fall	Long	28	8	–	Light yellowish	Black	90	30–156	82

(Fig. 1). The nut (Fig. 2) is incompletely 2- or 4-celled, and has a bony, rugose shell (pericarp) (Fig. 3). The 2- to 4-lobed seeds remain within the shell during germination. Available data on seeding habits are given in Table 1.

COLLECTION, EXTRACTION AND STORAGE OF SEED Walnut fruits are collected from the ground in fall or early winter, either after they have fallen from the trees naturally or after they have been knocked from the trees by flailing or shaking. Mechanical tree shakers are used with Persian walnuts. Collection should begin promptly after the nuts mature in the fall to prevent excessive losses to squirrels or other animals. Husks turn brownish to black when ripe. Thrifty butternut trees will yield from 9 to 35 liters of clean nuts, those of Persian walnut will yield many liters. Yield, size, and number of fruits per gram vary widely among species (Table 1). One hundred liters of black walnut fruits yields about 35 liters of sound seed. Similar yields have been reported for Hinds walnut.

Nuts are easy to extract when the husks are in the early stages of softening (i.e., firm but slightly soft). In later stages, the husks become mushy and cannot be removed completely. If allowed to dry thoroughly, they become very hard and removal is not feasible. In the slightly soft stage, husk can be removed by hand or by running the fruits through a corn sheller, but mechanical hullers are more efficient for cleaning large quantities of fruits. After cleaning, unfilled seeds can be separated by floating them in water. Seeds enclosed in a

J. cinerea
butternut

J. nigra
black walnut

Figure 1. *Juglans*: nuts enclosed in their husk, ×1 (Brinkman 1974h).

J. californica
California walnut

J. hindsii
Hinds walnut

J. microcarpa
little walnut

J. cinerea
butternut

J. nigra
black walnut

Figure 2. *Juglans:* nuts with husk removed, ×1 (Brinkman 1974h).

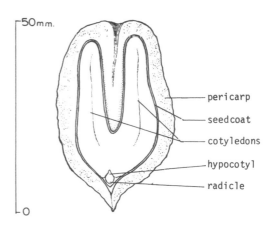

Figure 3. *Juglans cinerea,* butternut: longitudinal section through a nut, ×1 (Brinkman 1974h).

husk will germinate, but most nurseries find it easier to control seedling density with cleaned seeds. Husking is necessary if the seeds are to be treated with a fungicide.

Seeds of most species can be stored with or without husks. Siebold walnut, butternut, and little walnut seeds can be stored for long periods at relative humidities form 80 to 90% and temperatures from 3 to 4°C. Cleaned Persian walnut seeds with a moisture content of 20 to 40% have been stored successfully at 4°C for a year in plastic bags, and seeds with 50% moisture content were placed in screen containers in an outdoor pit for 4 years without significant loss in germination capacity.

PREGERMINATION TREATMENT Natural germination usually occurs in the spring following seedfall. Seeds of most *Juglans* species have a dormant embryo and species native to North America also have hard seedcoats. Dormancy can be broken by prechilling at 3 to 5°C (Table 1). In the case of Seibold walnut soaking seed in water is adequate. In practice, walnut seeds are either sown in the fall soon after collection, or prechilled outdoors over winter in moist sand, covered by 60 cm of soil mulch. Screening is necessary to

Table 2. *Juglans:* selected recent literature on germination.

Author—Date	Subject
Campbell and Martin 1976	Moisture determinations of *J. regia*
DeHayes and Waite 1982	Spring sowing of *J. nigra*
Dorn and Mudge 1985	Gibberellin and germination of *J. nigra*
Frutostomas 1981	Seed moisture and gibberellin in germination of *J. regia*
Kawecki 1973	Prechilling physiology of *J. regia*
Lebeck and Skofitsch 1984	Serotonin in germination of *J. regia*
Rietveld 1983	Juglone and allelopathy
Sholokhov and Bogoroditskii 1978	Accelerated preparation of *J. regia* for sowing
Sinha et al. 1977	Germination of *J. regia*
Velkov and Popov 1976	Moisture absorbtion and germination of *J. regia*
Velkov and Popov 1978	Pretreatment of *J. regia*
Williams 1980	Prechilling of *J. nigra*

exclude rodents, and fungicides are often applied to prevent diseases during prechilling. Small lots of seed may be prechilled in plastic bags with sand or peat moss for 90 to 120 days.

GERMINATION The Genebank Handbook (1985) offers the following suggestions for testing germination of seeds of *Juglans* species: After prechilling, a germination test should be conducted at alternating temperatures of 20/30°C, either in moist sand or between paper towels; the removal of a small part of the shell may reduce the time necessary for the test. Examples of germination techniques for various species are shown in Table 1. Selected recent literature concerning the germination of *Juglans* species is given in Table 2.

NURSERY AND FIELD PRACTICE Nuts can be sown in the fall soon after collection with husks removed. A hot water treatment preceding fall sowing of Hinds walnut seeds has been used. Seeds of this species were soaked in water at 88°C for 1.5 to 2 minutes.

To prevent alternating freezing and thawing, seedbeds are mulched with sawdust, grass hay, or straw. The heavier mulches must be removed when germination begins in the spring.

Prechilled seeds can be sown in the spring. Some nurseries broadcast seeds on tilled seedbeds and press the nuts into the ground with a roller, but the usual practice is to hand seed the seed in drill marks at the rate of 160 per m². Heavier seeding rates produce seedlings too small for field planting. For black walnut a seedling density of 84 per m² seems best. Seeds should be covered with 2.5 to 5 cm of nursery soil and screened against rodents, especially for fall-sown seeds.

Hinds walnut seeds often are sown directly into growing beds with the seedlings thinned to leave 20 cm between plants in a row. A special technique is used in some nurseries: (a) unhulled nuts are air-dried to reduce moisture to about 50% and kept indoors until January; (b) the partially dried nuts are put into sprout beds containing as many as 3 layers of nuts, with 2.5 cm of sand below and 2.5 cm of vermiculite above each layer; (c) about March 15th the beds are opened up and the sprouted nuts are hand transferred to growing beds in rows spaced 1.5 m apart with the nuts 20 cm apart in the row.

Root pruning to a depth of 20 to 25 cm sometimes is used to produce a more compact seedling. Some nurseries successfully regulate the length of tap roots by controlled irrigation. Seedlings are usually outplanted as 1-0 stock. A summary of nursery practices is given in Table 3.

Etiolated shoots from juvenile plants of black walnut can be induced to root, but it is difficult (Dirr and Heuser 1987). There has been considerable research on the tissue culture of Persian walnut (Dirr and Heuser 1987).

Table 3. *Juglans:* nursery practices (Brinkman 1974h).

Species	Prechilling medium	Season	Depth (cm)	Mulch
J. californica	Peat	Spring	5	–
J. cinerea	Sand	Spring	2.5–5	Sawdust
J. hindsii	–	Fall	2.5	–
J. microcarpa	Sand or peat	Spring	5	Vermiculite
J. nigra	Sand	Spring	2.5–5	–
J. regia	Sand	Spring	5	–

CUPRESSACEAE — CYPRESS FAMILY

Juniperus L. — Juniper

GROWTH HABIT, OCCURRENCE, AND USE The junipers include from as few as 50 to as many as 140 species, with 70 species probably being a reasonable figure (Johnsen and Alexander 1974). In the recent geologic past junipers may have been the most rapidly evolving conifers. They are the conifers' answer to the development of midlatitude deserts. Two characteristics of species from arid areas—scales instead of leaves, with the stomata protected beneath the scales, and a pseudo-berry that is dispersed by birds—are rare among other conifers (Salomonson 1978). Juniper species are widely distributed throughout the temperate and subtropical regions of the Northern Hemisphere and south of the equator in East Africa. Generally, 13 species are considered native to the United States. Eastern red cedar (*Juniperus virginiana*) is the most widespread and common juniper in the eastern United States, and Rocky Mountain (*J.*

scopulorum), Utah juniper (*J. osteosperma*), and western juniper (*J. occidentalis*) are the most common species in the western part of the country.

All the native junipers are valuable ornamental species, and many horticultural varieties have been developed. Several species such as Chinese juniper (*J. chinensis*) have been introduced into the United States as ornamentals. Many junipers are so similar in appearance that most people do not readily distinguish between species or varieties. Both eastern red cedar and Rocky Mountain juniper are widely used in shelterbelts and wildlife plantings. The close-grained, aromatic, and durable wood of tree-sized junipers is used for furniture, interior paneling, novelties, and fence posts. The fruits and young branches contain an aromatic oil that is used in medicines and to flavor gin.

GENETIC VARIATION AND HYBRIDIZATION There is considerable variation in growth habit and appearance within species, particularly the horticultural varieties developed as ornamentals. So many cultivars have been developed, named, and renamed by nursery trade that it is difficult to determine relations among the ones currently used as ornamentals. Marked differences have been noted in growth rate, color, crown form, disease resistance, and browsing animal preference in eastern red cedar, Rocky Mountain juniper, western juniper, and common juniper (*J. communis*). Numerous hybrids have been reported among the native species, especially when ranges overlap.

FLOWERING AND FRUITING The small, inconspicuous flowers are borne in the spring on the ends of short branchlets or along the branchlets. The flowers are dioecious or occasionally monoecious in one seed juniper (*J. monosperma*) and western juniper. The male flowers are yellow and form a short catkin; the greenish female flowers are composed of 3 to 8 pointed scales, some or all of which bear 1 to 2 ovules.

Scales of the female flowers become fleshy and fuse to form small, indehiscent strobili commonly called berries (Fig. 1), that ripen the first, second, or occasionally third year, depending upon the species (Table 1). Soviet literature uses the term *aril* instead of berry for these fruits. Immature berries are usually greenish; ripe berries are blue-black to red-brown, and are usually covered with a conspicuous white, waxy bloom. The fruit coat may be thin and resinous as in eastern red cedar, Rocky Mountain, and one seed juniper, or nearly leathery or mealy as in alligator (*J. deppeana*) and Utah juniper.

There are usually 1 to 4, rarely as many as 12, brownish seeds per fruit. The seeds are rounded or angled, often with longitudinal pits (Fig. 2). The seed-

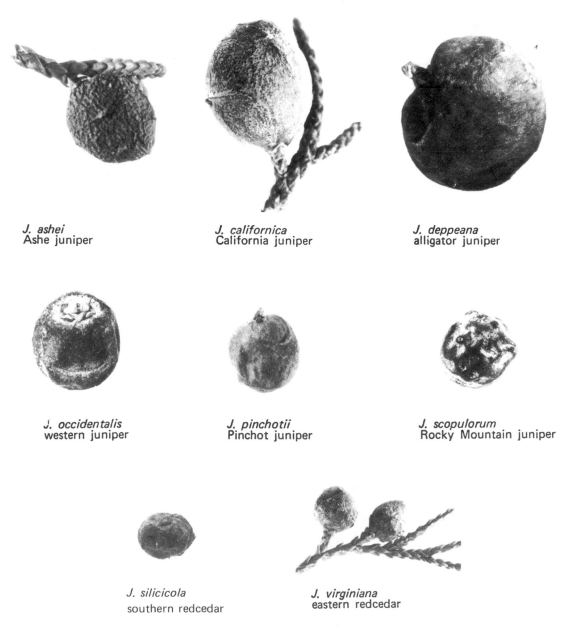

J. ashei
Ashe juniper

J. californica
California juniper

J. deppeana
alligator juniper

J. occidentalis
western juniper

J. pinchotii
Pinchot juniper

J. scopulorum
Rocky Mountain juniper

J. silicicola
southern redcedar

J. virginiana
eastern redcedar

Figure 1. *Juniperus*: strobili (berries), ×3 (Johnsen and Alexander 1974).

coat has two layers—the outer layer thick and hard, the inner thin and membranous (Fig. 3). Embedded within the fleshy, white to creamy-colored endosperm is a straight embryo with 2 to 6 cotyledons. Many seeds from a given tree contain no endosperm or embryo.

Junipers begin bearing seed when about 10 to 20 years old (Table 1). We have noted western junipers more than 30 years old still with juvenile foliage and not yet bearing fruit. Seeds are dispersed largely by birds, but ripe fruits will some times persist on the plant. Pollination is apparently highly uncertain in many juniper species. Occasionally crops will be heavy throughout a woodland, and occasionally there are years when few berries are produced throughout a large geographic region. Almost every year a tree can be found in a stand that is so loaded with berries it appears covered with wax; such trees are popularly known as candle trees.

COLLECTION, EXTRACTION AND STORAGE OF SEED Juniper berries are usually collected in the fall by stripping or picking by hand from the trees, or by flailing the fruits to ground cloths. The larger fruits, such as those from alligator or Utah juniper, may be picked from the ground. Care should be taken not to pick immature berries. Since the number of filled seeds varies widely from tree to tree, it is important to test the seeds by cutting to determine percent fill. For some species it is possible to collect over winter, but delayed collection risks losses to birds. Fresh berries must be spread to reach moisture equilibrium. Seeds may be stored as berries or cleaned seed.

The seeds can be recovered by macerating the seeds and flotation. Johnsen and Alexander (1974) suggested presoaking the fruits of resinous species in a solution of weak lye before macerating. We have found that the addition of a detergent to the maceration water

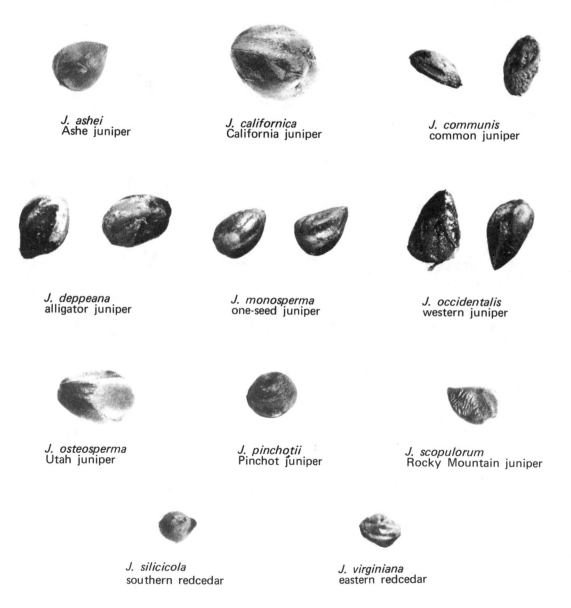

J. ashei
Ashe juniper

J. californica
California juniper

J. communis
common juniper

J. deppeana
alligator juniper

J. monosperma
one-seed juniper

J. occidentalis
western juniper

J. osteosperma
Utah juniper

J. pinchotii
Pinchot juniper

J. scopulorum
Rocky Mountain juniper

J. silicicola
southern redcedar

J. virginiana
eastern redcedar

Figure 2. *Juniperus*: seeds, ×3 (Johnsen and Alexander 1974).

Table 1. *Juniperus:* phenology of flowering and fruiting, characteristics of mature trees, and cleaned seed weight (Johnsen and Alexander 1974).

Species	Flowering dates	Fruit ripening dates	Seed dispersal dates	Mature tree height (m)	First year cultivated	Seed-bearing Age (years)	Interval (years)	Seeds/ 110 kg fruit	Seeds/ gram	Ripe fruit color
J. ashei	Jan.–Apr.	Sept.–Nov.	Fall–Winter	6	1925	20	–	13	22	Deep blue
J. californica	–	–	–	5	–	–	–	–	–	–
J. communis	Apr.–May	Aug.–Oct.[1]	Persist 2 years	15	1560	–	Irregular	16	80	Bluish black
J. deppeana	Feb.–Mar.	Aug.–Oct.[2]	Persist two seasons	20	1873	–	–	36	28	Bluish or reddish brown
J. monosperma	Mar.–Apr.	Aug.–Sept.	Oct.–Nov.	8	1900	10–20	2–5	15	40	Copper to dark blue
J. occidentalis	Apr.–May	Sept.	Persist 2 years	15	1840	–	–	20	27	Bluish black
J. osteosperma	Mar.–Apr.	Sept.[2]	Persist 2 years	12	1900	–	2	25	11	Reddish brown
J. pinchotii	Oct.–Nov.	Oct.–Nov.[2]	Year round	6	–	10–20	–	19	24	Copper to red
J. scopulorum	Apr.–June	Sept.–Dec.	Persist 2–3 years	15	1936	10–20	2–5	11–28	60	Blue with wax
J. silicicola	Jan.–Feb.	Oct.–Nov.	–	6	–	–	–	–	–	–
J. virginiana	Mar.–May	Sept.–Nov.	Feb.–Mar.	30	1664	10	2–3	14–18	96	Blue

[1]Second to third year
[2]Second year

accomplishes the same purpose with less hazard. Small amounts of seed can be cleaned with a blender and flotation. The Soviets have devoted a considerable literature to processes for the cleaning of seeds from juniper berries. Abseitov (1983) conducted detailed studies of the size variability of berries of *J. semiglobosa*, *J. seravschanica*, and *J. turkestanica*. Later (Abseitov and Osipov 1985) he reported on a prototype machine for extracting juniper seeds. This machine was perfected as the MIS-0,2 extractor, capable of processing 97 kg per hour (Osipov et al. 1985). Seed yields and seeds per gram are listed in Table 1.

Juniper seeds store quite well. They should be dried to 10 to 12% moisture and stored in sealed containers at cold temperatures.

PREGERMINATION TREATMENT Germination is often delayed in juniper seeds. Virtually every treatment known to seed technology has been tried with seeds of various species. We have just completed a 10-year study of the germination of seeds of western and Utah juniper. Seeds of both species are consistently highly dormant. Prechilling for 14 weeks was required for substantial germination (Young et al. 1988). We also found that prechilling in a solution of gibberellin at 5°C, with oxygen maintained at saturation, was an effective treatment. As a part of this study we sowed seeds in flats outdoors which we recovered at progressive dates during the winter for emergence testing in the greenhouse and then stored dry until the next fall when we again placed them outdoors and recovered them in the same order. These seeds had a carryover in emergence enhancement from season to season. Critics have suggested that prechilling was not effectively carried over, but rather the influence of warm stratification in the greenhouse during the emergence testing period was carried from year to year. The fact

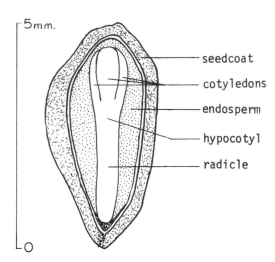

Figure 3. *Juniperus scopulorum*, Rocky Mountain juniper: longitudinal section through seed, ×12 (Johnsen and Alexander 1974)

that either influence would be apparent after several months of dry storage is remarkable.

Djavanshir and Fechner (1976) subjected seeds of eastern red cedar and Rocky Mountain juniper to various pretreatments before incubating them at 8/18°C on damp paper. Treatments included removing the tip of the seed at the hilum end, extracting the embryo followed by chilling at 5°C, soaking all or part of the seed in sulphuric acid, and cold storage at −20°C. Chilling seeds was required for hypocotyl development, but not for epicotyl development. Softening of the seedcoat by sulphuric acid for 35 to 120 minutes for eastern red cedar and Rocky Mountain juniper, respectively, increased the rate of germination provided the carbonized surface was removed. It was concluded that slow germination was

Table 2. *Juniperus:* laboratory germination methods (Genebank Handbook 1985).

Species	Substrata	Temperature (°C)	Duration (days)	Additional instructions
J. communis	TB, S	20	28	Prechill 90 days at 3–5°C.
J. occidentalis	TB, S	15	42	Prechill 30–60 days at 1-5°C.
J. scopulorum	TB, S	15	28	Warm stratification 60 days at 20°C, then prechill 40 days at 3–5°C.
J. virginiana	TB, S	15	28	Warm stratification 60 days at 20°C, then prechill 45 days at 3–5°C.

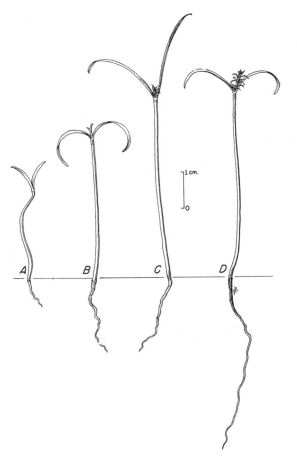

Figure 4. *Juniperus deppeana*, alligator juniper: seedling development at 2, 17, 43, and 96 days after germination (Johnsen and Alexander 1974).

Table 3. *Juniperus:* propagation techniques (Dirr and Heuser 1987).

Species	Propagation techniques
J. chinensis	Seeds: acid scarify 60 minutes, then prechill 3 months.
	Cuttings: dependent on cultivar.
J. communis	Seeds: fall plant; seedling appears in second year.
	Cuttings: easy to root, but slow.
J. conferts	Cuttings: easiest juniper to root.
J. davurica	Cuttings: successfully rooted from summer, fall, and winter cuttings.
J. excelsa	Cuttings: easy to root.
J. horizontalis	Seeds: acid scarify 60 minutes, then prechill 3 months.
	Cuttings: late summer, fall, and winter cuttings easy to root.
J. procumbens	Cuttings: varying success has been obtained.
J. sabina	Cuttings: most forms root easily.

caused by a combination of dormancy and seedcoat impermeability. Van Haverbeke and Comer (1985) reported on the results of extensive studies on the germination of seeds of eastern red cedar. They concluded that highest germination was obtained by soaking seeds for 96 hours in 10,000 ppm of citric acid followed by 6 weeks of warm stratification and 10 weeks of prechilling. Geographically separated seed sources responded differently to treatment. Use of fresh seed reduced the warm stratification time and the interaction between seed source and warm stratification.

Working with seeds of *J. procera* collected from plantations in Tanzania, Laurent and Chamshama (1987) determined that hot water treatments were the most effective way to enhance germination. Emergence rates of 60% were obtained 2 weeks after sowing seeds treated with hot water, while control seeds failed to emerge. Research in Soviet Armenia suggested the following procedure for seeds of *J. polycarpos* and *J. foetidissima*: Dip freshly harvested berries in water, then store them in bags in a cool place for 5 months. At the end of storage, dip the bags of berries alternately in hot and cold water 5 or 6 times and wash the berries in 60°C water before removing the fleshy material. Following cleaning the seeds should be warm stratified and prechilled (no duration given) before planting.

Sheikh (1983), working in Pakistan, reported that ripe juniper berries were collected in November and December, cleaned through a screen, and recovered by flotation. Initial germination was only 6% at most, but germination started again with the return of the monsoon and continued for 2 years (species not given). Pretreatment did not enhance germination. Sheikh (1980) also worked on flotation systems to improve the quality of juniper seeds through seed cleaning.

Ayaz (1980) conducted an excellent study of the anatomy of seeds of *J. excelsa*, a species native to Pakistan. (This may be the species Sheikh discusses.) Pardos and Lazaro (1983) studied moisture imbibition by seeds of *J. oxycedrus* in relation to prechilling and gibberellin enrichment. They concluded that the lack of radicle elongation was the cause of dormancy.

GERMINATION   The Genebank Handbook (1985) provides germination standards (Table 2).

NURSERY AND FIELD PRACTICE   Juniper seeds are usually sown in the nursery in late summer or fall, but may be sown in spring or summer. The seeds of most species should be sown in fall to take advantage of natural prechilling.

Juniper seeds are usually drilled in well-prepared seedbeds in rows 15 to 20 cm apart and covered with 0.6 cm of soil. In nurseries with severe climates, such as those in the Great Plains, considerable care must be taken to protect the beds with mulch and snow fences. Seedling development of alligator juniper is shown in Figure 4. Propagation techniques for juniper species are given in Table 3.

ERICACEAE — HEATH FAMILY

Kalmia L. — Mountain laurel

GROWTH HABIT, OCCURRENCE, AND USE   The genus *Kalmia* consists of about 8 evergreen or rarely deciduous shrubs native to North America and Cuba (Bailey 1951). The most noted species is mountain laurel (*K. latifolia*). Mountain laurel is a broad-leaved, evergreen shrub, occasionally reaching tree size (Olson and Barnes 1974c). It attains a height of 3 to 10 m at maturity. This species has a wide range from New Brunswick to Florida, primarily along the Appalachian mountain range, westward to Louisiana, and northward into southern Ohio and Indiana. It is a common associate of *Rhododendron maximum*. Together these two species cover 1.2 million hectares of the southern Appalachians. These stands are often nearly impenetrable thickets, known as laurel slicks or rhododendron hells. *Kalmia latifolia* is extensively cultivated as an ornamental in the eastern United States and Europe, and was introduced into cultivation in 1734.

The foliage of mountain laurel is used by deer as browse in winter, but is poisonous to them if they are forced to subsist on it exclusively. It is also toxic to cattle and sheep. The laurels have value for their beauty when in flower, and as watershed species. The hard wood is used to a minor degree.

FLOWERING AND FRUITING   The perfect flowers appear from March to July, depending on latitude and altitude, and range in color from white to crimson, or almost chocolate-purple. The flowers are pollinated by insects.

The fruits ripen in September and October of the same year and are long persistent. They are 5-celled, globular capsules about 0.6 cm in diameter, borne in clusters (Fig. 1). The minute, oblong, terminally winged seeds (Figs. 2, 3) are dispersed upon splitting of these dry, dehiscent capsules. *Kalmia* seeds are extremely small, averaging 59,000 per gram.

COLLECTION, EXTRACTION, AND STORAGE OF SEED   To collect seed, the capsules are picked from the plants at maturity, dried if necessary, and rubbed or beaten to open them. The seeds can then be shaken out and the remnants of the capsule removed by screening. Seeds can be stored dry at room temperature and sown within a year.

Jaynes (1971) tested the germination of seeds of 10 hybrids of *K. latifolia* that were stored for 2 to 4 years. Only the seeds of 1 hybrid showed a decrease in germination, while 7 had increased germination, and 2 stayed the same. In a later study Jaynes (1982) reported on the germination of *Kalmia* stored for 20 years. Only

0.5 X

4 X

Figure 1. *Kalmia latifolia*, mountain laurel: a cluster of capsules, ×0.5, and a single capsule, ×4 (Olson and Barnes 1974c).

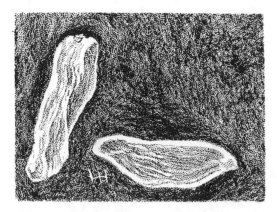

Figure 2. *Kalmia latifolia*, mountain laurel: seeds, ×40 (Olson and Barnes 1974c).

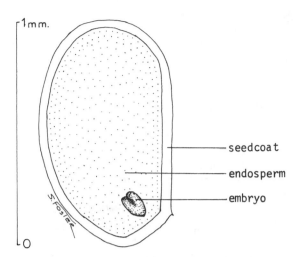

Figure 3. *Kalmia latifolia*, mountain laurel: longitudinal section through a seed, ×60 (Olson and Barnes 1974c).

seeds of *K. hirsuta* failed to germinate after prolonged storage.

GERMINATION Jaynes (1971) conducted extensive experiments on the germination of seeds of various species of *Kalmia*, and his proposed treatments are much more detailed than those given by Olson and Barnes (1974c).

Kalmia hirsuta is a species of laurel native to the southeastern United States. Jaynes used a substrate of sand, peat, and ground sphagnum with a pH of 4.2 rather than petri dishes and paper. (He had problems with mold in petri dishes). For *K. hirsuta* Jaynes used the unusual pretreatment of high temperature and high relative humidity. The seeds were incubated 2 to 30 minutes at 80 or 90°C. Older seeds required shorter or lower temperature pretreatments for maximum germination enhancement. This may be a form of accelerated afterripening, similar to treatments that are applied to grass caryopses. Dry heat was not effective in breaking the dormancy of this species. *Kalmia latifolia* seeds usually do not require pretreatment for germination, but some seed lots may be dormant.

Prechilling will often lead to more uniform germination (Dirr and Heuser 1987).

Kalmia cuneata has a limited natural range in North and South Carolina. Freshly harvested seeds have limited germination that can be enhanced by prechilling or enrichment with gibberellin (Jaynes 1971).

Seeds of *K. angustifolia*, *K. polifolia*, and *K. microphylla* germinated after 2 weeks of incubation at 22°C without pretreatment (Jaynes 1971). Jaynes believed that the seeds of all species of *Kalmia* required light for germination. It was not clear if he tried to germinate seeds of the various species in the dark.

NURSERY AND FIELD PRACTICE The small seeds are commonly planted in flats. As soon as the seedlings are large enough to lift they are transplanted into pots. Usually they are grown in containers for a year before being transplanted.

Vegetative propagation of these species is very complex and requires the assistance of a propagation expert.

ARALIACEAE — GINSENG FAMILY

Kalopanax pictus (Thunb.) Nakai — Kalopanax

GROWTH HABIT, OCCURRENCE, AND USE Kalopanax, the only species in the genus, is a deciduous tree from 8 to 31 m tall. It is native to China, eastern Siberia, Manchuria, Korea, and Japan, and has been cultivated since about 1865, chiefly in environmental forestry (Rudolf 1974h).

FLOWERING AND FRUITING The white perfect flowers occur in terminal racemes and bloom in July and August. The fruits—2-seeded, subglobose drupes about 0.5 cm across—are bluish black when they ripen in September and October.

Figure 1. *Kalopanax pictus*, kalopanax: seed, ×8 (Rudolf 1974h).

COLLECTION, EXTRACTION, AND STORAGE OF SEED The fruits should be collected by hand or shaken onto ground cloths as they ripen. They can be run through a macerator and the seeds recovered by flotation. From 100 kg of fruit about 8 to 10 kg of seeds will be recovered. The number of cleaned seeds per gram is 116. The seeds (Figs. 1, 2) have small embryos and contain endosperm tissue. They can be kept for 1 year under ordinary storage. For longer periods cold storage in sealed containers is recommended.

GERMINATION Seeds planted in the fall without pretreatment will germinate the second year (Dirr and Heuser 1987). An alternative is to warm stratify seeds for 5 months, followed by prechilling for 3 months. Some growers acid scarify instead of using the warm stratification treatment.

Plants can be propagated by root cuttings taken in early spring (Dirr and Heuser 1987).

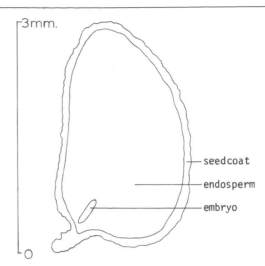

Figure 2. *Kalopanax pictus,* kalopanax: longitudinal section through a seed (Rudolf 1974h).

CHENOPODIACEAE — GOOSEFOOT FAMILY

Kochia Roth — Kochia

GROWTH HABIT, OCCURRENCE, AND USE *Kochia* is genus of about 35 species native to North America, Europe, and Asia. Most are annual or perennial herbaceous species, but several low or half-shrubs exist. Red molly (*K. americana*) is an important browse species in the Great Basin of western North America. *Kochia prostrata* is native to Central Asia and has been widely planted in western North America as a forage species.

COLLECTION, EXTRACTION, AND STORAGE OF SEED *Kochia prostrata* seed consists of a tightly coiled embryonic plant that begins to unwind as soon as it imbibes moisture (Young et al. 1981). Small quantities of seeds can be stored in paper bags without difficulty, but commercial quantities of seeds of *K. prostrata* have been known to lose viability rapidly.

NURSERY AND FIELD PRACTICE Seedlings can be transplanted to extremely arid and saline/alkaline environments where they have a good chance of establishment. These established plants will produce seeds that will establish as naturalized seedlings on the site. Despite this often-demonstrated naturalization, it is very difficult to establish seedlings from direct seeding in the field. One factor that contributes to problems with direct seeding is the difficulty of storing large quantities of seeds.

SAPINDACEAE — SOAPBERRY FAMILY

Koelreuteria paniculata Laxm. — Panicled golden raintree

GROWTH HABIT, OCCURRENCE, AND USE Native to China, Korea, and Japan, the panicled golden raintree is a small, deciduous tree ranging from 5 to 11 m tall that has been cultivated since 1763 as an ornamental (Rudolf 1974i).

FLOWERING AND FRUITING The irregular yellow flowers occur in broad, loose, terminal panicles and bloom from July to September.

The fruits are bladdery, triangular, 3-celled capsules about 3 to 5 cm long (Fig. 1); when they ripen in September and October they change from a reddish color to brown. Within the papery walls of the ripe fruit are 3 round, black seeds (Fig. 2). The seeds are naturally dispersed from fall to the next spring. Good seed crops are borne almost annually.

COLLECTION, EXTRACTION, AND STORAGE OF SEED The ripe capsules can be collected in the fall, then threshed and air screened to recover the seeds. The yield from 100 kg of fruits is about 72 kg of seeds. The average number of seeds per gram is 6. Seeds can be stored for prolonged periods.

PREGERMINATION TREATMENT Seeds require scarification to break an impermeable seedcoat. Soaking in acid for 1 hour is the recommended treatment.

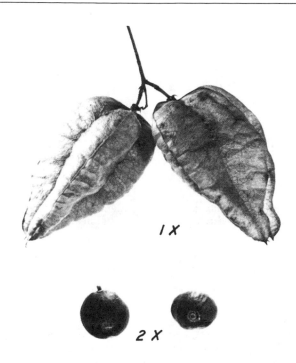

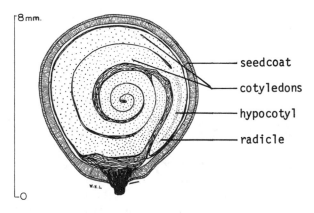

GERMINATION Germination is epigeal (Fig. 3). It should be tested in sand at 20/30°C.

NURSERY AND FIELD PRACTICE Fall plant seeds for best results. Acid-scarified seeds planted in the spring may also require prechilling. Seeds sown immediately after collection in the fall usually give the best results. Sow to produce 300 to 315 seedling per m². This species should only be planted in sunny locations. Normally 2-0 stock is outplanted. Panicled golden raintree can also be propagated from root cuttings.

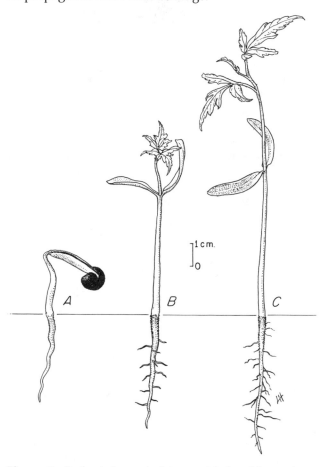

Figure 1. *Koelreuteria paniculata*, panicled golden raintree: capsule, ×1, and seeds, ×2 (Rudolf 1974i).

Figure 2. *Koelreuteria paniculata*, panicled golden raintree: longitudinal section through a seed, ×6 (Rudolf 1974i).

Figure 3. *Koelrueteria paniculata*, panicled golden raintree: seedling development at 1, 3, and 5 days after germination (Rudolf 1974i).

LEGUMINOSAE — LEGUME FAMILY

Laburnum anagyroides Med. — Goldenchain laburnum

GROWTH HABIT, OCCURRENCE, AND USE Native to southern Europe, goldenchain laburnum is a deciduous shrub or tree 5 to 9 m tall that has been cultivated since 1560 for ornamental purposes and fuel wood production (Rudolf 1974j). All parts of the tree, especially the young fruits, are poisonous. This species grows on many kinds of soil, including limestone, but does poorly on wet sites. It grows well in both shaded and sunny sites and is largely free from pests.

FLOWERING AND FRUITING The perfect, showy, golden-yellow flowers are about 1.9 cm long and occur in pendulous racemes up to 30 cm long; they bloom from May to June.

The fruits are compressed, linear pods about 5 cm long, with a thick keel. The pods are tardily dehiscent, ripening from late August to October (Fig. 1). Each

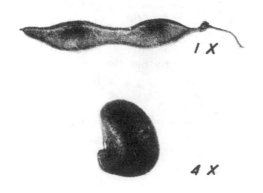

Figure 1. *Laburnum anagyroides*, goldenchain laburnum: fruit, ×1, exterior view of seed, ×4 (Rudolf 1974j).

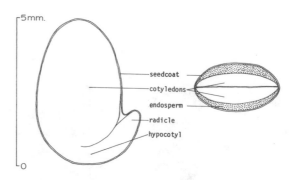

Figure 2. *Laburnum anagyroides*, goldenchain laburnum: longitudinal section through a seed, ×8 (Rudolf 1974j).

pod contains several black seeds (Figs. 1, 2). Good seed crops are borne almost annually.

COLLECTION, EXTRACTION, AND STORAGE OF SEED Pods should be picked from the trees in September or October and spread in a well-aerated shed to dry. The dried pods may be stored for the winter or threshed and the seed recovered with an air screen. The seed yield from 100 kg of pods is about 25 kg. There are about 17 seeds per gram. Seeds retain viability for 2 years in dry warehouse storage.

PREGERMINATION TREATMENT Seeds require scarification for rapid germination. Information is limited on the best scarification treatment. Acid and mechanical scarification have been used, with mechanical giving the best results. There are no reports of the use of hot water.

GERMINATION The Genebank Handbook (1985) recommended substrate, top of paper; incubation temperature, 20/30°C; duration 21 days; and additional directions, clip or file seedcoat or acid scarify, and then presoak for 3 hours.

NURSERY AND FIELD PRACTICE Scarified seeds can be drilled or broadcast in the late spring at a rate to produce 210 to 315 seedlings per m². Field planting is done with 2-0 stock. This species can be propagated by layering.

LYTHRACEAE — LOOSESTRIFE FAMILY

Lagerstroemia L. — Crape myrtle

GROWTH HABIT, OCCURRENCE, AND USE *Lagerstroemia* is a genus of about 30 species which are native to southern and eastern Asia and widely grown in the southern United States for their flowers. They are shrubs or small trees, with gorgeous summer flowers colored white through red.

FLOWERING AND FRUITING The fruit is a 6-valved dehiscent capsule that ripens in October. The capsules are borne upright so all the seeds do not fall at maturity.

GERMINATION The seeds require no pretreatment and germinate within 2 to 3 weeks after sowing. A brief period of prechilling may unify germination (Dirr and Heuser 1987). Seeds should be stored in sealed containers. Common crape myrtle (*L. indica*) is easily rooted from soft and hardwood cuttings.

PINACEAE — PINE FAMILY

Larix Mill. — Larch

GROWTH HABIT, OCCURRENCE, AND USE The larches include 10 species of cone-bearing, deciduous trees widely distributed over the cooler regions of the Northern Hemisphere. (Rudolf 1974k). Tamarack (*Larix laricina*) and western larch (*L. occidentalis*) have been used in reforestation. Because of their rot-resistant wood, the larches are useful for posts, poles, and railroad ties. The beautiful fall coloring of most larch species makes them important forest species in many environments.

HYBRIDS The larches hybridize readily. *Larix leptolepis* × *L. decidua*, known as *L. eurolepis*, originated about 1900 and is commonly called Dunkeld larch. It has been planted extensively in northwestern Europe, and to a limited extent in the eastern United States and Canada, because it combines desirable characteristics of both parent species and grows faster than either parent.

Larix leptolepis × *L. sibirica*, known as *L. marschlinsii*, originated in 1901, and *L. laricina* × *L. decidua*, known as *L. pendula*, originated prior to 1800. Many other larch hybrids are known.

GEOGRAPHIC RACES Geographic races have developed in many widely distributed larch species, and these often exhibit marked differences in growth rates and other characteristics.

European larch (*L. decidua*) includes at least 5 geographic races, often considered to be subspecies or varieties, that roughly coincide with major distributional groups of the species. These races are as follows: Alpine in south central Europe, Sudeten principally in Czechoslovakia, Tatra in Czechoslovakia and Poland, Polish principally in Selesia, and Rumanian. The races differ in seed size and viability, survival after planting, growth rate, phenology, form, and resistance to insects and diseases. The races respond differently in different localities, but in the northeastern United States and Canada the Polish and Sudeten races grow most rapidly and are recommended for planting.

Some varieties of Dahurian larch (*L. gmelini*) that are confined to definite areas appear to be geographic races. In Minnesota significant variation was exhibited among seedlings of tamarack from different seed sources.

Japanese larch (*L. leptolepis*) is native to a limited area in the mountains of Honshu. It grows in scattered stands from 1230 to 2460 m. Despite the limited native range, test plantings in other parts of the world have demonstrated considerable variability depending on the seed source.

FLOWERING AND FRUITING Male and female flowers of the larches are borne separately on the same tree. They occur randomly with the leaves on the sides of twigs or branches and usually open a few days before needle elongation. The male flowers are solitary, yellow, globose-to-oblong bodies that bear wingless pollen. The female flowers are small, usually short-stalked, erect, red or greenish cones that ripen the first year. A ripe cone is made up of brownish, woody scales, each of which bears 2 seeds at the base. The seeds are chiefly wind dispersed and the empty cones remain on the trees for an indefinite period. Larch seed is winged and nearly triangular in shape. It has a crustaceous, light brown to reddish brown outer coat, a membraneous, pale chestnut-brown, lustrous inner coat, a light-colored endosperm, and a well-developed embryo (Figs. 1, 2).

During poor seed years much of the seed of some larch species is destroyed by weevils. The spruce budworm (*Choristoneura fumiferana*) may hinder seed production of western larch. Flowering and fruiting phenology is given in Table 1.

COLLECTION, EXTRACTION, AND STORAGE OF SEED Larch cones should be collected in the fall as soon as they are ripe. They may either be picked from the trees, gathered from felled slash, or obtained from squirrel

Table 1. *Larix*: phenology of flowering and fruiting, characteristics of mature trees, cone and seed weight, and nursery practices (Rudolf 1974k).

Species	Flowering dates	Fruit ripening dates	Seed dispersal dates	Mature tree height (m)	Year first cultivated	Seed-bearing Age (years)	Seed-bearing Interval (years)	Ripe cone color	Cone wt./ 35 liters cones (kg)	Seeds/ 35 liters cones (kg)	Seeds/ gram	Season of sowing	Density (m²)	Tree (%)
L. decidua	Mar.–May	Sept.–Dec.	Sept.–Spring	40	1629	10	3–10	Light brown	53	2	160	Fall	470	10
L. gmelini	—	Sept.–Nov.	Feb.–Mar.	31	1827	—	—	—	—	—	260	—	—	—
L. laricina	Apr.–May	Aug.–Sept.	Sept.–Spring	20	1737	40	3–6	Brown	55	1.7	700	Fall	260	35
L. leptolepis	Apr.–May	Sept.	Winter	40	1861	15	3	Brown	64	1.2	250	Spring	780	15
L. lyallii	May–June	Aug.–Sept.	Sept.	12	1904	30	1–10	Green–purple	—	—	310	—	—	—
L. occidentalis	Apr.–June	Aug.–Sept.	Sept.–Oct.	55	1881	25	1–10	Green–purple	55	1.1	300	Spring	340	40
L. sibirica	Apr.–May	Sept.–Nov.	Sept.–Mar.	40	1806	12	3–5	Brownish	—	—	100	Spring	370	30

Figure 1. *Larix occidentalis*, western larch: seed with wing, ×4 (Rudolf 1974k)

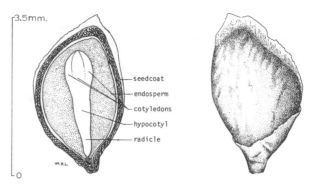

Figure 2. *Larix laricina*, tamarack: longitudinal section through a seed (left), and exterior view (right) with wing broken off, ×12 (Rudolf 1974k).

caches. In Europe larch seeds may be picked from the surface of snow. In most species ripe cones are brown, and collecting may begin as soon as tests show the seedcoats have become hard and the embryo firm. Seed-bearing information is given in Table 1.

Freshly collected cones should be spread out in thin layers to dry in the sun or in a ventilated shed. The cones may be opened by solar heat, by heating in a cone kiln, or by placing them in a heated room. Recommended kiln treatments are 8 hours at 48°C for tamarack and 7 to 9 hours at 45°C for western larch.

After opening, the cones should be run through a shaker to remove the seed. They can be dewinged in a dewinging machine or by hand. The seed should be cleaned with an air screen. Seed yield and weight is given in Table 1.

Seeds maintain viability if stored at low temperatures in sealed containers. Studies of western larch seeds conducted in Montana indicated that seeds stored in cones in a well-ventilated room produced higher viability than seeds extracted immediately (Shearer 1977). Western larch cones can be refrigerated so that the resin forms globules and the seeds are less sticky to handle.

PREGERMINATION TREATMENT The seeds of most species of larch germinate fairly well without pretreatment except for subalpine larch (*L. lyallii*) where seeds require prechilling for 21 to 60 days or soaking in a 3% solution of hydrogen peroxide.

In Kamchatka, seeds of Siberian and Japanese larch

Figure 3. *Larix laricina*, tamarack: seedling development at 1 and 8 days after germination (Rudolf 1974k).

Table 2. *Larix:* germination standards (Genebank Handbook 1985).

Species	Substrata	Temperature (°C)	Duration (days)	Additional instructions
L. decidua	TB	23/30	21	ASOA suggests light.
L. eurolepis	TB	23/30	21	ASOA suggests light.
L. gmelini	TB	23/30	21	No light.
L. kaempferi	TB	23/30	21	2 tests, with and without prechilling, at 3–5°C for 21 days.
L. laricina	TB	23/30	21	None
L. occidentalis	TB	23/30	21	2 tests, with and without prechilling, at 3–5°C for 21 days. ASOA suggests light, prechilling, or potassium nitrate.
L. sibirica	TB	23/30	21	ASOA suggests light.
L. sukaczewii	TB	23/30	21	ASOA suggests light.

are soaked in a 0.5% solution of KMnO$_4$ for 60 minutes before planting (Ostroshenko 1977). Studies of the germination of tamarack seeds in northwestern Ontario revealed that unprechilled seeds germinated in light, but prechilling was required for germination under most incubation temperatures in the dark (Farmer and Reinholt 1986).

GERMINATION Table 2 summarizes germination stan-

dards for the larches and Table 3 lists selected recent literature concerning larch seed production and germination.

NURSERY AND FIELD PRACTICE Larch seed should be sown in the fall, or in the spring using prechilled seeds, and covered with 0.6 cm of soil. Fall-sown beds should be mulched. Germination is epigeal (Fig. 3). Details of nursery practices are given in Table 1.

Table 3. *Larix:* recent literature on seed production and germination.

Author—Date	Subject	Location
Dobrin et al. 1983	*L. sukaczewii*, presoaked seeds subjected to laser radiation had increased germination.	USSR
Golyadkin et al. 1972	*L. sibirica*, UHF electromagnetic field exposure increased germination.	USSR
Hall 1981	*L. laricina*, seed yield.	Canada
Hall 1985	*L. laricina*, seed quality and yield.	Canada
Hall and Brown 1977	*Larix* hybrids, embryo development and seed yield.	UK
Hill 1976	*L. decidua*, long-term seed storage.	–
Katsuta et al. 1981	*L. leptolepis*, influence of gibberellins on strobilus production.	Japan
Kuznetsova 1978	*L. gmellini*, storage of seeds in snow.	USSR
Lobanov 1985	*L. sibirica*, phenology of cone maturity.	USSR
Logan and Polland 1981	*L. leptolepis*, relation of seed weight and germination rate to initial growth.	Canada
Loffler 1976	*L. decidua*, seed orchards.	Germany
Mohn et al. 1988	*L. laricina*, variation influence on establishment.	USA
Paves 1979	*Larix* spp., seed yield and quality.	USSR
Pintel and Cheliak 1986	*L. laricina*, biochemistry during seed germination.	–
Raevskikh 1979	*L. gmelini*, seed quality.	USSR
Saralidze and Saralidze 1976	*Larix* spp., seed extraction.	USSR
Shearer 1984	*L. occidentalis*, influence of insects on seed production.	USA
Suo 1982	*L. gmelini*, natural seed set.	China
Timofeev 1984	*L. sukaczewii*, seed production.	USSR
Trenin and Chernobrovkina 1984	*L. sibirica*, embryogenesis and seed quality.	USSR

ZYGOPHYLLACEAE — CALTROP FAMILY

Larrea tridentata Vail — Creosote bush

GROWTH HABIT, OCCURRENCE, AND USE Creosote bush is an evergreen shrub native to the arid, subtropical regions of southwestern United States, Mexico, Argentina, and Chile. There is some question whether North American *Larrea tridentata* is specifically distinct from *L. divaricata* of South America (Martin 1974). Creosote bush grows on a variety of soils. Stands vary in density and stature, depending on the relative aridity of the site. Under very low rainfall, shrubs are smaller and more widely spaced than under more favorable conditions. Creosote bush is not browsed by livestock.

FLOWERING AND FRUITING Creosote bush has perfect flowers and blooms profusely in the spring, but may also flower from time to time throughout the year. The fruit is a densely white-villous, 5-celled capsule. When the fruits are cast, they separate into individual carpels, each normally containing 1 seed (Figs. 1, 2). Carpel fill under natural conditions averages 35%. Plants may fruit sparingly at 4 to 6 years of age and reach full fruiting at 8 to 13 years. Annual production ranges from 39 to 278 fruits per 100 g of branches or 120 to 1710 fruits per plant.

Figure 1. *Larrea tridentata*, creosote bush: single carpel, ×8 (Martin 1974).

COLLECTION, EXTRACTION, AND STORAGE OF SEED Ripe fruits may be collected from the shrubs in the spring or early summer. Clean seeds, extracted from the capsule, are small, running 370 per gram. Kay et al. (1988) found that germination of *Larrea* seeds declined slowly over 14 years of storage.

PREGERMINATION TREATMENT Creosote bush seeds have a dormancy apparently induced by hard seedcoats. Mechanical scarification enhances germination.

GERMINATION Lajtha et al. (1987) determined that germination of *Larrea* seeds was inhibited by high pH and interpreted this as evidence for why creosote bush is largely absent from saline/alkaline soil areas. Lowered osmotic potential or specific ion toxicity may be a better explanation for its distribution. Tipton (1984) determined that Gompertz growth curves were most suitable for modeling the germination of seeds of creosote bush. The most comprehensive study of

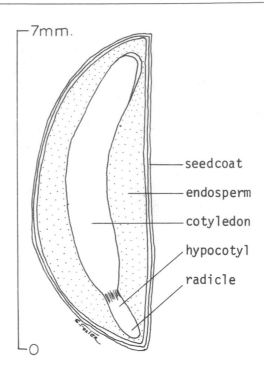

Figure 2. *Larrea tridentata*, creosote bush: longitudinal section through a carpel, ×12 (Martin 1974).

creosote bush germination was done by Barbour (1968). He determined that seeds of creosote bush readily germinated in the laboratory and that populations did vary in their germination response to temperature across the entire range of the species.

The Genebank Handbook (1985) suggests, for germination standards for creosote bush, that the test be conducted for 9 days and that the seeds be dehulled.

LABIATAE — MINT FAMILY

Lavandula L. — Lavender

GROWTH HABIT, OCCURRENCE, AND USE *Lavandula* is a genus of more than 25 species native to the area from the Canary Islands to India. These are mainly perennial herbaceous species, with some shrubs. Common lavender (*L. angustifolia*) is widely grown as an ornamental species.

The fruit is a 1-seeded nutlet.

GERMINATION No pretreatment is necessary for germination, but one nursery grower prechills seed for more uniform emergence (Dirr and Heuser 1987). An alternative method is to soak seed in 200 ppm of gibberellin and then incubate at 20/30°C. Late summer, semihardwood cuttings can be rooted in outdoor frames.

ERICACEAE — HEATH FAMILY

Ledum L. — Labrador tea

GROWTH HABIT, OCCURRENCE, AND USE The genus *Ledum* has 3 species of evergreen shrubs found in sphagnum bogs and damp places in colder parts of the Northern Hemisphere.

Labrador tea (*Ledum groenlandicum*) has very small seeds that should be sprinkled on the surface of a moisture-supplying substrate and covered with clear plastic film. Cuttings taken in mid-December root well (Dirr and Heuser 1987).

ERICACEAE — HEATH FAMILY

Leiophyllum (Pers.) R. Hedwig — Sand myrtle

GROWTH HABIT, OCCURRENCE, AND USE *Leiophyllum* is a genus of 3 species of low evergreen shrubs native to eastern North America. The fruit is a capsule.

Seeds should be sown on the surface of substrate and require light. Cuttings taken in mid-October should be treated with IBA (Dirr and Heuser 1987).

MYRTACEAE — MYRTLE FAMILY

Leptospermum Frost. — Tea tree

GROWTH HABIT, OCCURRENCE, AND USE Coastal tea tree (*Leptospermum laevigatum*) is native to New Zealand and Australia. This tree grows to 10 m in height and is often planted in coastal areas as an ornamental. The fruit is a capsule that contains many small seeds.

Manuka (*L. scoparium*) is native to the Waitakere Range near Auckland, New Zealand.

GERMINATION Seedfall was recorded from 30- to 35-year-old trees and germination was studied in the laboratory (Mohan et al. 1984a, 1984b). Viable seed was shed throughout the year. Peak seedfall occurred in October and January. Germination rate, but not final germination capacity, was influenced by temperature and periods of available moisture. Very low levels of incident radiation enhanced germination compared to darkness.

LEGUMINOSAE — LEGUME FAMILY

Lespedeza Michx.

GROWTH HABIT, OCCURRENCE, AND USE The genus *Lespedeza* consists of about 140 species of shrubs, half-shrubs, and herbaceous species (Vogel 1974). Most are native to the temperate regions of eastern Asia and only about 11 herbaceous species are considered native to North America. Several species of shrub lespedeza have been introduced into the United States, but most conservation plantings have involved only 3 species: *L. bicolor*, *L. thunbergii*, and *L. japonica*. Many of the Japanese lespedeza (*L. japonica*) material are probably variants of Thunberg lespedeza (*L. thunbergii*). Shrub lespedeza (*L. bicolor*) is the most commonly planted of the shrub lespedeza species in the United States.

The plants are adapted primarily to the southeastern two-thirds of the United States, with the exception of southern Florida. The shrub lespedezas are planted mainly for erosion control and wildlife food and cover. Some plants are grown for ornamental purposes. Mature plants are 1.1 to 2 m in height. The stems of Japanese lespedeza die back to the ground each year while stems of the other 2 species persist. When managed for seed production, the persistent stemmed species must be cut back to the ground.

Several cultivars have been selected because of the adaptive qualities for specific sites.

FLOWERING AND FRUITING The flowers are loosely arranged on elongated racemes, and are mostly rose-purple with graduation to white in some forms. The chasmogamic flowers may be self- or cross- pollinated.

Time of flowering and fruiting varies among species

and strains, and is also controlled by the latitude where the plants are grown. Flowering occurs mostly in July and August, but will begin in June in Mississippi and as late as September in Maryland. The brown-colored fruits are one-seeded indehiscent pods that mature primarily in late September and October (Fig. 1). The pods fall to the ground when ripe and most of them are down by early winter.

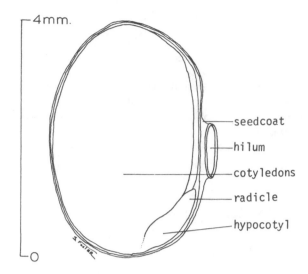

Figure 2. *Lespedeza japonica*, Japanese lespedeza: longitudinal section through a seed, ×16 (Vogel 1974).

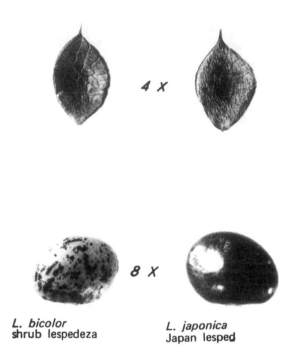

Figure 1. *Lespedeza*: pods, ×4 (above), and seeds, ×8 (below) (Vogel 1974).

A light seed crop may occur the first year from one-year-old transplants and good seed crops can be expected each successive year. Seeds of shrub lespedeza are pale brown to olive colored and copiously flecked with purple. Those of the other 2 species are solid dark purple. Seeds of lespedeza have little or no endosperm (Fig. 2).

COLLECTION, EXTRACTION, AND STORAGE OF SEED Shrub lespedeza seed is most commonly harvested with a combine as soon as the fruits are ripe and moderately dry. The combined material, which includes stems, intact pods, and hulled seeds, is air dried and then cleaned by air screening. Seed yields have reached 2700 kg/ha. A bushel (35 liters) of seeds weighs 132 kg. The number of cleaned seeds per gram is 190. Seeds can be stored at cool temperatures at low relative humidity. Unhulled seeds retain their viability longer than threshed seeds. Seeds have remained viable for 20 years in storage.

PREGERMINATION TREATMENT A high percentage of shrub lespedeza seeds have hard seedcoats and require scarification before planting. Mechanical scarification is often used. Threshing in a hammer mill will scarify about 50% of the seeds. This type of scarification will produce a good stand of plants the first year and allow for additional germination in later years. Acid scarification for 30 minutes is also possible. It causes less damage to older, brittle seed than does mechanical scarification.

GERMINATION TEST Germination tests can be made by placing seeds between blotters in a petri dish, in a rolled towel (either horizontal or vertical), or in sand. Incubation temperature recommended is 20/35°C with a test duration of 21 days. Buta and Lushy (1986) have isolated germination and growth-inhibiting chemicals from *Lespedeza* seeds.

NURSERY AND FIELD PRACTICE For producing one-year-old transplants, rows are spaced 0.9 to 1.1 m apart and 40 to 65 seeds are planted per meter of row. Seeds should be inoculated before seeding, and seed should be sown in spring after the danger of frost is passed. About 95% of one-year-old seedlings are usable for transplanting to the field. Stands can also be established by direct seeding in the field.

In the Southeast prescribed burning has been proposed to enhance the germination of *Lespedeza* species for wildlife habitat improvement (Seqelquist 1971, Martin et al. 1975).

LEGUMINOSAE — LEGUME FAMILY

Leucaena Benth. — Ipil-ipil

GROWTH HABIT, OCCURRENCE, AND USE The genus *Leucaena* includes about 10 species of trees and shrubs that grow in Central America and southeastern Asia (Whitesell 1974b). Leaves, pods, and seeds of at least 4 species have long been used for human food.

Leucaena leucocephala, the most widespread species, originated in Central America, but is now pantropical in distribution. It has become naturalized in Florida and southern Texas. This species was introduced to the Pacific Islands by the Spanish and to Hawaii about 1864. It invades cleared areas and forms dense thickets from 3 to 9 m high. Its trunk reach is 10 cm in diameter. It is evergreen where moisture is not limited seasonally.

In the tropics *L. leucocephala* is used for a shade plant in plantation agriculture and for site preparation in reforestation projects. It is widely proposed as a fodder and fuel wood species to halt desertification. However, it contains mimosine which can be toxic to nonruminants.

Strains of *L. leucocephala* can be categorized as Hawaiian or Salvadorian. The Hawaiian strain is a drought-tolerant, branching, abundant-flowering, aggressive shrub growing to 9 m. The Salvadorian strain is erect, may attain 18 m, and flowers seasonally. Salvadorian seeds average 18 per gram and those of the Hawaiian strain average 30.

FLOWERING AND FRUITING The plants have numerous, white, bisexual flowers clustered in globular heads. Flowering may begin within 2 months of planting and continue throughout the year. By the time plants are 2 years old they produce seeds year-round, but most abundantly towards the end of summer. The seeds are borne in thin, flat, acuminate pods, 12 to 18 cm long and 1.2 to 2 cm wide. They usually occur in clusters of 15 to 20 pods, but sometimes as many as 60 occur together. The pods turn brown when ripe. Each pod has 15 to 25 seeds. Seeds are elliptical, compressed, and shiny brown (Figs. 1, 2).

COLLECTION, EXTRACTION, AND STORAGE OF SEED The dry pods are gathered and the seeds threshed. They can be recovered by screening and will store for long periods in cool areas. In Hawaii, larvae of the beetle *Araecerus levipennis* can destroy the seeds.

PREGERMINATION TREATMENT Germination is delayed and reduced by the presence of a very thick and tough waxy layer that forms the seedcoat. Pouring the seeds into boiling water and allowing them to cool is a satisfactory scarification treatment. They should not be stored after hot water treatments. Ranchers in Hawaii have fed the seeds with molasses to cattle; the seeds pass through the animals' digestive tracts and establish in pastures.

GERMINATION Seeds can be soaked in water for 48 hours and then manually clipped and placed on kimpack for germination tests. Incubation temperatures are 20/30°C, with light during the warm period.

A measure of the importance of *Leucaena* around the world is the large number of recent literature entries concerning the genus:

Author and date	Subject	Location
Babeley & Kandya 1985	Pregermination treatments, mechanical scarification give best results.	India
Bhaskar 1983	Germination, soils.	India
Dalmacio 1976	Fungicides.	Philippines
Gupta et al. 1983	Seed size and acid scarification.	India
Jones et al. 1983	Nitrogen fertilization.	—
Kumar & Sharma 1982	Viability determinations.	India
Kushalapa 1981	Certified seed stocks.	India
Lulandala 1981	Viability and pregermination treatments.	—
Mamicpia et al. 1983	Storage of seeds.	—
Olvera et al. 1982	Germination.	USA
Olvera & West 1986a	Gibberellin and germination.	USA
Olvera & West 1986b	Scarification treatments.	USA
Pathak et al. 1981	Seed weight and seedling growth.	India
Rimando & Dalmacio 1978	Direct seeding.	Philippines
Seiffert 1982	Scarification.	—
Sharma et al. 1985	Incubation temperature.	India
Singh & Palical 1986	Ethylene and germination.	India
Vora 1989	Naturalized populations.	USA

Figure 1. *Leucaena leucocephala*: seed, ×4 (Whitesell 1974b).

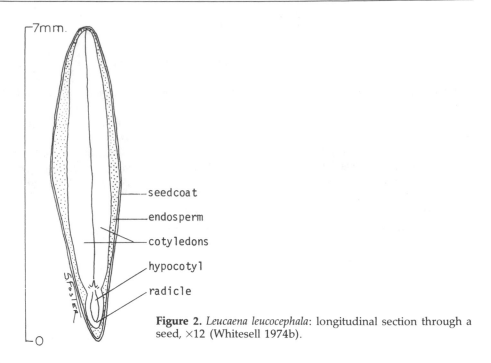

Figure 2. *Leucaena leucocephala*: longitudinal section through a seed, ×12 (Whitesell 1974b).

ERICACEAE — HEATH FAMILY

Leucothoe D. Don

GROWTH HABIT, OCCURRENCE, AND USE The genus *Leucothoe* consists of about 35 species of evergreen or deciduous shrubs native to North and South America, Madagascar, the Himalayas, and Japan. These species have urn-shaped white flowers followed by dehiscent capsules containing small dustlike seeds.

Florida leucothoe (*Leucothoe populifolia*) is used as an ornamental in southeastern gardens in the United States. Capsules of this species should be collected before they are dry and seeds recovered after the capsules dry and open. Sprinkle the small seeds on the surface of a moist substrate and cover with clear plastic film. Seeds require light for germination (Dirr and Heuser 1987).

OLEACEAE — OLIVE FAMILY

Ligustrum L. — Privet

GROWTH HABIT, OCCURRENCE, AND USE The privets include about 50 species of deciduous or evergreen shrubs or trees, native chiefly to Asia and Australia, with 1 species native to Europe and northern Africa (Rudolf 1974l). They are grown primarily for their handsome foliage and profuse white flowers. Privets are also valuable for wildlife habitat. They have been used for shelterbelts in South Africa and Europe. Glossy privet (*Ligustrum lucidum*) has been planted as a street tree in warmer parts of the United States, and is cultivated in China for wax which is an exudation of the branches caused by the insect *Coccus pe-lah*. Several varieties of European privet (*L. vulgare*) are recognized.

FLOWERING AND FRUITING The small, perfect, white flowers occur in rather dense, usually terminal panicles, and bloom in the summer (Rudolf 1974l):

Species	Flowering dates	Fruit ripening dates	Ripe fruit color
L. amurense	June–July	Sept.–Nov.	Black
L. japonicum	July–Aug.	Oct.–Nov.	Purple-black
L. lucidum	July–Aug.	Oct.–Nov.	Blue-black
L. vulgare	June–July	Sept.–Oct.	Lustrous black

The fruits are 1- to 4-seeded (Fig. 1), berrylike drupes about 0.8 to 1.25 cm long. They ripen in the fall and, in

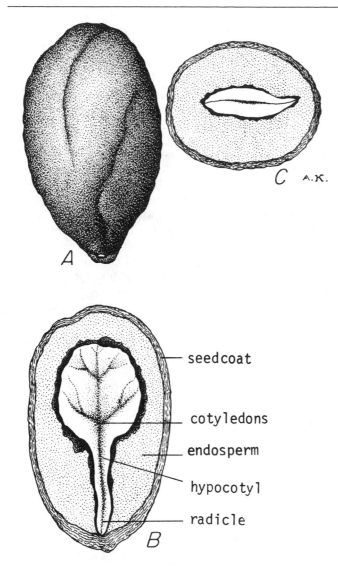

seedcoat

cotyledons

endosperm

hypocotyl

radicle

Figure 1. *Ligustrum sinense,* Chinese privet: (A) exterior view of seed; (B) longitudinal section; (C) cross section (Rudolf 1974l).

Species	Mature tree height (m)	Year first cultivated	Seeds/gram
L. amurense	5	1860	50
L. japonicum	3	1845	—
L. lucidum	10	1794	28
L. vulgare	5	Old	40

Seeds can be stored in sealed containers at low temperatures.

PREGERMINATION TREATMENT Freshly extracted seeds that are stored moist and sown promptly will germinate without pretreatment. Seeds that are stored may require 3 months of prechilling.

GERMINATION TEST Germination may be tested with prechilled seeds in sand at incubation temperatures of 20/30°C. Germination is epigeal (Fig. 2).

NURSERY AND FIELD PRACTICE Privet seeds may be sown in the fall, or prechilled and sown in the spring. Sowing immediately after collection in the early fall may be the most successful procedure. Seeding densities are usually about 410 per m². One or two-year-old seedlings are outplanted.

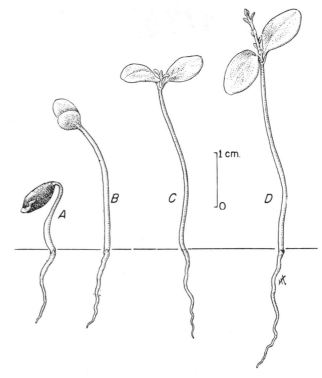

Figure 2. *Ligustrum vulgare,* European privet: seedling development at 1, 5, 50, and 132 days after germination (Rudolf 1974l).

some species, persist over winter. The privets usually bear abundant seed crops.

COLLECTION, EXTRACTION, AND STORAGE OF SEED Ripe privet fruits may be gathered from the bushes by hand in late fall or early winter. The fruits should be run through a macerator with water to separate the seeds from the pulp. Fruits yield 40 to 50% cleaned seeds by weight (Rudolf 1974l):

LAURACEAE — LAUREL FAMILY

Lindera benzoin (L.) Blume — Spicebush

GROWTH HABIT, OCCURRENCE, AND USE Common spicebush is a deciduous shrub to 5 m tall, native from Maine to Ontario and Kansas, and south to Florida and Texas (Brinkman and Phipps 1974a). Cultivated since 1683, it is valuable for wildlife food and environmental plantings.

FLOWERING AND FRUITING The flowers are dioecious or polygamous and appear from March to May, before the leaves. The fruit is a red, drupaceous berry ripening in August and September. Each fruit contains a single seed that is light violet-brown with flecks of darker brown (Figs. 1, 2).

COLLECTION, EXTRACTION, AND STORAGE OF SEED Spicebush fruit may be collected in September and October. They can be macerated and the seeds recovered by flotation. Seed yield from fruits is about 15 to 20% by weight. There are about 10 seeds per gram. The seeds apparently do not store well, but a low temperature is helpful in maintaining viability.

PREGERMINATION TREATMENT Spicebush seeds have dormant embryos that respond to warm stratification for 1 month, followed by 3 months of prechilling.

GERMINATION TEST Tests can be conducted in moist sand or peat at a constant temperature of 25°C or at 20/30°C.

NURSERY AND FIELD PRACTICE Spicebush seeds should be sown in the fall and mulched over winter. The mulch should be removed in the spring. Prechilled seeds can be sown in the spring. From 70 to 80% of the seeds sown can be expected to produce seedlings (Fig. 3).

Figure 1. *Lindera benzoin*, spicebush: seed, ×3 (Brinkman and Phipps 1974a).

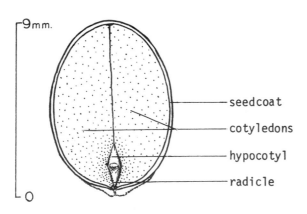

Figure 2. *Lindera benzoin*, spicebush: longitudinal section through a seed, ×5 (Brinkman and Phipps 1974a).

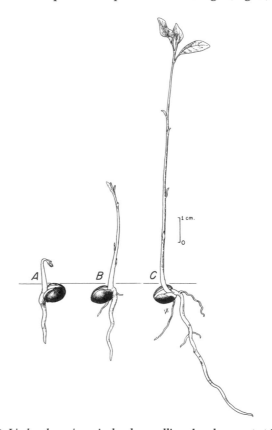

Figure 3. *Lindera benzoin*, spicebush: seedling development at 2, 3, and 10 days after germination (Brinkman and Phipps 1974a).

HAMAMELIDACEAE — WITCH-HAZEL FAMILY

Liquidambar styraciflua L. — Sweetgum

GROWTH HABIT, OCCURRENCE, AND USE Sweetgum is native to a variety of sites from Connecticut and southern Illinois, south to central Florida and northeastern Mexico (Bonner 1974f). It also occurs in Central America as far south as Nicaragua (McCarter and Hughes 1984). This large deciduous tree reaches 45 m at maturity. Sweetgum is valued for pulp, lumber, and veneer. The seeds are eaten by many species of birds, and the tree has been planted as an ornamental. It was first cultivated in 1681.

FLOWERING AND FRUITING The small, greenish, monoecious flowers bloom in March to May. The pistillate flowers are borne in axillary, globose heads that form 2.5- to 3-cm-diameter multiple heads of small 2-celled capsules (Fig. 1). The lustrous green color of the fruiting head fades to yellowish green or yellow as maturity is reached in September to November. The beaklike capsules open at this time, and the small, winged seeds (Figs. 2, 3), 1 or 2 per capsule, are dispersed. Empty fruiting heads often remain on the trees over winter. Fair seed crops occur every year and bumper crops about every 3. Late spring freezes can completely wipe out a seed crop. Trees bear good crops as they reach 20 to 30 years of age.

COLLECTION, EXTRACTION, AND STORAGE OF SEED Mature fruit heads must be picked from standing trees or logging slash before seed dispersal. The best indicator

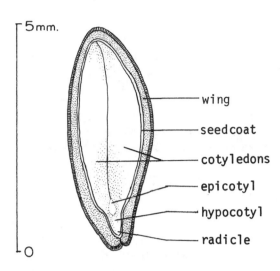

Figure 3. *Liquidambar styraciflua*, sweetgum: longitudinal section through a seed, ×12 (Bonner 1974f).

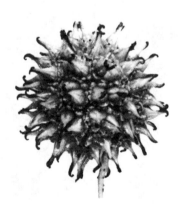

Figure 1. *Liquidambar styraciflua*, sweetgum: fruiting head, ×1 (Bonner 1974f).

Figure 2. *Liquidambar styraciflua*, sweetgum: seed, ×8 (Bonner 1974f).

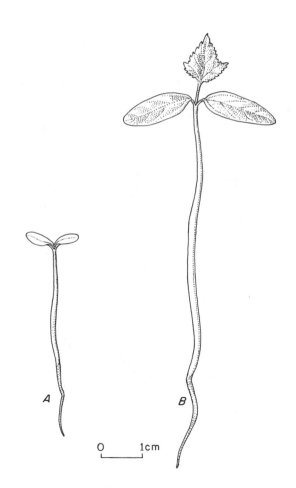

Figure 4. *Liquidambar styraciflua*, sweetgum: seedling development at 2 and 30 days after germination (Bonner 1974f).

of maturity is the fading of the green color. Fruit heads picked prematurely may ripen if stored moist at 5°C for about a month. The fruit heads should be spread to dry until they open and release the seeds. This operation may take 5 to 10 days, after which vigorous shaking will complete the extraction. The seeds can be cleaned with an air screen. Bonner (1974f) offered the following yield data from the mostly southern collections: 35 liters of fruit yielded 19 kg of seeds; by weight cleaned seed yield is 9%; there are 56 seeds per head, and 140 to 220 per gram with an average of 180 per gram. Sweetgum seeds should be stored at a moisture content of 10 to 15% in sealed bags at 3°C. Initial viability is maintained for at least 4 years.

PREGERMINATION TREATMENT Sweetgum seeds exhibit only a shallow dormancy, but germination rate is considerably increased by prechilling. Naked prechilling in plastic bags at 2°C, or mixed in wet sand at 5°C, have been successful. The duration of pretreatment ranges from 15 to 90 days, with 30 days most common. However, detailed studies of the duration of prechilling indicate that treatment as long as 63 days may be most beneficial in enhancing germination (Rink et al. 1979). Seeds from the southern portion of the range of the species require less prechilling. Barnett and Farmer (1978) compared seed collections from different elevations in terms of the influence of light, temperature, and prechilling. High altitude collections showed lower germination than collections from lower elevations.

GERMINATION Germination standards from ASOA (1985) for sweetgum seeds are substrate, top of blotters; temperature, 20/30°C; duration, 28 days; and additional instructions, light, and care with moisture because the seeds are sensitive to drying (Belcher and Miller 1975).

NURSERY AND FIELD PRACTICE Prechilled seeds should be broadcast or drilled in the spring to achieve a seedling density of 210 to 260 per m² (Fig. 4). Aluminum powder can be mixed with the wet prechilled seeds to achieve easy flow. The seeds should be sown on the surface and lightly pressed into the soil with a roller. A 0.6 to 1.25 cm mulch of sawdust, sand, or chopped pine needles should be applied. Mycorrhizal colonization of seedlings of sweetgum is important (Kormanik 1983).

MAGNOLIACEAE — MAGNOLIA FAMILY

Liriodendron tulipifera L. — Yellow poplar

GROWTH HABIT, OCCURRENCE, AND USE Yellow poplar is native to the eastern United States from Vermont and Michigan south to Louisiana and Florida. It grows from sea level to 1380 m in the Appalachians and to 300 m in the northern part of its range (Bonner and Russell 1974). This large deciduous tree attains heights of 25 to 37 m at maturity and is very valuable for lumber and veneer. It is a good honey tree and is planted extensively as an ornamental. Yellow poplar has been cultivated since 1663.

FLOWERING AND FRUITING The large, perfect, colorful flowers of yellow poplar open from April to June. The fruit is an elongated cone composed of closely overlapped carpels that are dry, woody, and winged (Fig. 1). Each carpel (samara) contains 1 or 2 seeds (Fig. 2). The cones turn from green to tan or light brown as they ripen; they mature from early August in the northern part of the range to late October in the South. Good seed crops occur almost every year; failures, as well as bumper crops occur infrequently. Although trees as young as 9 years old have been reported to bear fruit, the normal commercial seed-bearing age is 15 to 20 years. As the mature cones dry on the trees, they break apart and the samaras scatter. Peak dissemination occurs in October and November, but a few samaras fall as late as the following March. The samaras are used as seeds.

COLLECTION, EXTRACTION, AND STORAGE OF SEED Cones may be picked when ripe from standing trees or from

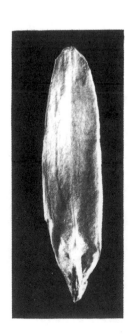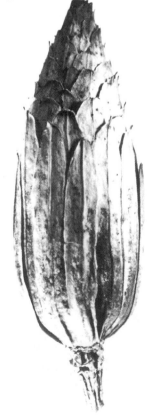

Figure 1. *Liriodendron tulipifera*, yellow poplar: cone and single samara, ×2 (Bonner and Russell 1974).

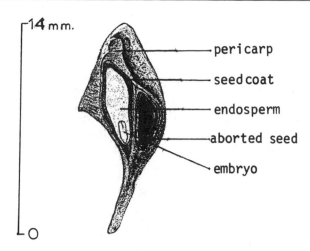

Figure 2. *Liriodendron tulipifera*, yellow poplar: longitudinal section through an embryo of a samara, ×4 (Bonner and Russell 1974)

logging slash, or collected from squirrel caches. October is the best time to collect cones (Bonner 1976a). Seeds may also be shaken onto ground cloths from standing trees. Dry weather is best for shaking, as cones are closed and less fragile in wet weather. Cones from the upper two-thirds of the crown yield more full seeds than cones from the lower one-third. The percentage of full seeds is quite low because of inefficient pollination. There is great variation between trees in this regard, but individuals regularly produce seeds of a given viability. Seeds from most trees average about 10% full, but individual trees with seeds as high as 35% full have been found.

Cones should be spread out to dry immediately after collection. Sufficient drying to separate the samaras usually requires 7 to 20 days, depending on temperature, humidity, and cone moisture content.

Thoroughly dried cones can be broken apart by hand shucking, flailing, treading, or with a hammer mill. There are about 80 to 100 samaras per cone. It appears that southern-grown samaras are larger than northern: 13 versus 31 samaras per gram. Dried samaras (seeds) may be stored in sealed cans or plastic bags at 3 to 5°C for several years without loss of viability.

PREGERMINATION TREATMENT Seeds to be sown in the spring and seeds taken from dry storage require pregermination treatment before they will germinate. Several treatments have been used: (a) storage in moist, well-drained pits or mounds of soil, sand, or peat, or mixture of these media from overwinter to as

long as 3 years; (b) prechilling in bags of peat moss or sand and peat for 60 to 90 days; and (c) cold, moist, naked prechilling in plastic bags for 140 to 168 days. Vogt (1974b) found that soaking seeds in a solution of gibberellin before prechilling effectively shortened the prechilling period.

GERMINATION TEST The Genebank Handbook (1985) provides the ASOA and ISTA standards for yellow poplar seeds as follows: substrata, top of paper (also between blotters, ASOA); incubation temperature, 20/30°C; duration, 28 days; additional directions, prechill at 3 to 5°C for 60 days or use excised embryo technique (ASOA).

NURSERY AND FIELD PRACTICE Untreated seeds may be sown in the fall, but prechilled seeds must be used for spring sowing. They may be broadcast at rates of 8.8 to 24 kg per 37 m². Seeds may be sown in rows 20 to 30 cm apart at the rate of 160 to 240 per meter of row. Seedbed densities of 260 to 315 per m² are recommended. The seeds should be covered with 0.6 cm of sawdust. Shading for 1 to 2 months from the start of germination has been recommended. Seedling development is shown in Figure 3.

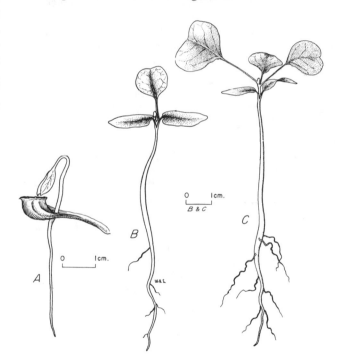

Figure 3. *Liriodendron tulipifera*, yellow poplar: seedling development at 1, 18, and 48 days after germination (Bonna and Russell 1974).

SAPINDACEAE — SOAPBERRY FAMILY

Litchi chinensis **Sonner**

GROWTH HABIT, OCCURRENCE, AND USE Litchi is a thick-foliaged, broad-crowned, medium-sized fruit tree from southern China. It was introduced to Hawaii in the 19th century. There a good tree produces 450 kg of fruit annually from May through June, after reaching 5 years of age.

GERMINATION Growing litchi from seed needs care and promptness because seeds soon lose their viability if permitted to dry after removal from the fruit (Ray and Sharma 1985). Litchi seeds germinate without pretreatment when sown fresh. They will store sealed in vials for about 5 days.

FAGACEAE — BEECH FAMILY

Lithocarpus densiflorus **(Hook. & Arn.) Rehd.** — Tanoak

GROWTH HABIT, OCCURRENCE, AND USE Tanoak is an evergreen hardwood that grows in various forms depending on the site and associated vegetation (Roy 1974c). In close stands, particularly the dense coniferous forests, tanoaks develop central axes, narrow crowns, upright branches, and long trunks which are clear for 9 to 25 m. In open stands, especially associated with other hardwoods, tanoaks are free branching, with broad crowns, large horizontal limbs, and short, thick trunks. In height, tanoak is generally a medium-sized tree from 15 to 46 m tall. The tallest tree reported was 64 m high and 1.3 m in diameter.

Tanoak ranges from southwestern Oregon southward through the Coast Range to north of Santa Barbara, California. The range extends eastward from the Humboldt Bay region to the lower slopes of Mount Shasta, thence intermittently southward along the west slope of the Sierra Nevada as far as Mariposa County. It grows from 580 to 1540 m in elevation.

Tanoak occasionally is cultivated in parks and can be used for erosion control. Although the wood now is used mainly for fuel, it has been chipped for particle board manufacturing and pulped. Other uses are flooring, furniture veneer, tool handles, and bats. Bark of tanoak has furnished the best tannage known for the production of heavy leathers. It gives excellent plumping when used to tan sole or saddle leather. Indians in the north Coast Range obtained one of their principal foods from the acorns of tanoak, and wildlife utilize the acorns as well.

GEOGRAPHIC RACES A shrubby variety of tanoak, *Lithocarpus densiflorus* var. *echinoides*, grows near Mount Shasta and in the Salmon and Klamath mountains.

FLOWERING AND FRUITING Blossoms may appear in spring, summer, or autumn. However, most tanoaks bloom in June, July, or August. Trees at lower elevations and near the coast bloom earlier than trees at higher elevations and further inland.

Almost all flowers, both male and female, are borne on new shoots of the year, where they grow from the axils of the new leaves. Flowers occasionally develop

Figure 1. *Lithocarpus densiflorus*, tanoak: acorns, ×1 (Roy 1974c).

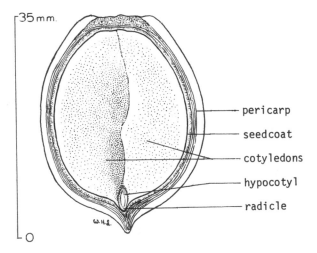

pericarp

seedcoat

cotyledons

hypocotyl

radicle

Figure 2. *Lithocarpus densiflorus*, tanoak: longitudinal section through an acorn, ×1.7 (Roy 1974c).

from buds found at the base of leaves of the previous year's growth.

Female flowers are borne at the base of erect male catkins. The profusion of yellowish blossoms that sometimes conceal the foliage suggested the tree's specific name.

The fruit is an acorn (Fig. 1) and ripens in the second autumn. Acorns usually are borne singly, in twos, or threes; occasionally more than 3 are clustered together.

COLLECTION, EXTRACTION, AND STORAGE OF SEED Tanoak is a heavy seeder. In fact, no oak on the Pacific coast produces heavier crops of acorns. Trees are heavily laden almost every alternate year, and complete crop failures are rare. Acorns may be collected from the standing trees. Cleaning involves removal of the cups.

Fruits ripen between September 20 and November 1 when their color changes from green to shades between light yellow and brown. If possible collection should be delayed until some acorns fall because the first to fall are usually infested with insects.

Tanoak acorns are large, 2.5 to 5 cm long, and 1.5 to 1.8 cm in diameter (Fig. 2). There are about 130 to 240 acorns per kg. The weight of 35 liters of cleaned seeds is 58 kg. Seed yield from collected material is about 78% by weight.

GERMINATION No pregermination treatments are necessary if acorns are planted immediately after collection. If seeds are stored, they should be prechilled. Germination is hypogeal (Fig. 3).

Tanoak does not enter the scientific literature very often. When it does, it usually is in reference to forestry practices (Tappeiner et al. 1986, McDonald et al. 1988). A more basic evolutionary reference is provided by Kaul (1986).

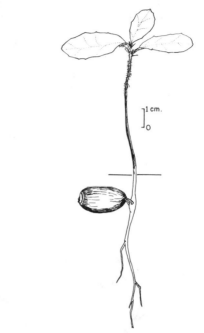

Figure 3. *Lithocarpus densiflorus*, tanoak: seedling development 2 months after germination (Roy 1974c).

CAPRIFOLIACEAE — HONEYSUCKLE FAMILY

Lonicera L. — Honeysuckle

GROWTH HABIT, OCCURRENCE, AND USE The honeysuckles include about 180 species of usually deciduous, sometimes evergreen, upright shrubs or climbing vines found throughout the Northern Hemisphere south to Mexico, north Africa, Java, and the Philippines (Brinkman 1974i). Many species are widely planted for their attractive, often fragrant flowers and ornamental fruits. Some species furnish cover and food for wildlife, while others are valuable for erosion control and shelterbelts.

Tatarian honeysuckle (*Lonicera tatarica*) has been used more than any other honeysuckle in conservation plantings, especially in shelterbelts. This species, Amur honeysuckle (*L. maackii*) and Morrow honeysuckle (*L. morrowii*) have been widely planted over much of the northern United States and have become naturalized in many areas. The larger species are especially valuable for wildlife and environmental plantings.

FLOWERING AND FRUITING The small perfect flowers of

Table 1. *Lonicera*: phenology of flowering and fruiting, characteristics of mature trees, and cleaned seed weight (Brinkman 1974i).

Species	Flowering dates	Fruit ripening dates	Seed dispersal dates	Mature tree height (m)	Year first cultivated	Ripe fruit color	Seeds/gram
L. canadensis	Apr.–July	July–Aug.	July–Sept.	1.5	1641	Red	–
L. chrysantha	May–June	July–Sept.	–	4	1854	Coral red	–
L. dioica	May–July	July–Aug.	–	2.5	1636	Red	–
L. glaucescens	May–June	July–Sept.	June–Oct.	1.5	1890	Coral red	20
L. hirsuta	May–Aug.	Sept.–Oct.	July–Oct.	2.5	1825	Purple-black	–
L. involucrata	June	Aug.	–	3.1	1828	Dark red	720
L. maackii	June	Sept.–Nov.	–	4.9	1855	Dark red	325
L. morrowii	May–June	June–Aug.	–	2.2	1875	Dark red	335
L. oblongifolia	May–June	July–Aug.	May–Aug.	1.5	1823	Orange-red	520
L. tatarica	May–June	July–Aug.	–	–	–	–	–

honeysuckle vary from white or yellow to pink, purple, or scarlet. They are borne in axillary pairs or stemless whorls, generally in the spring, but in the late summer for some species (Table 1). The attractive red, orange, or black fruits are often borne in coalescent pairs that ripen in the summer or early fall (Fig. 1). Depending on the species, each berry contains few to many small seeds (Figs. 2, 3). Commercial seed may be berries or cleaned seeds. Good seed crops of Amur and Morrow honeysuckle are borne every year. Seeds are dispersed primarily by birds. Fruits of Amur, Morrow, and Tatarian honeysuckle will remain on the plant well into the winter.

COLLECTION, EXTRACTION, AND STORAGE OF SEED   Honeysuckle fruits should be handpicked or stripped from the branches as soon after ripening as possible to reduce losses to birds. Because most species hybridize rather freely, it is better to collect fruits from isolated bushes or groups. Colors of ripe fruits and yield data are given in Table 1.

Unless seeds are to be extracted immediately, fresh fruits should be spread out in thin layers to prevent heating. Extraction is accomplished by maceration and flotation to recover seeds. After drying, the seeds can be stored or sown. Heit (1967d) found little loss of viability of *Lonicera* stored for 15 years in sealed containers at low temperatures.

PREGERMINATION TREATMENT   With the possible exception of fly (*L. canadensis*) and limber (*L. dioica*) honeysuckle, seeds of all species show some

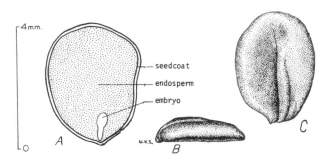

Figure 3. *Lonicera tatarica*, Tatarian honeysuckle: (A) longitudinal section through a seed; (B) and (C) exterior views of seeds in two planes, ×8 (Brinkman 1974i).

Figure 1. *Lonicera involucrata*, black twinberry: fruit, ×2 (Brinkman 1974i).

L. glaucescens
Donald honeysuckle

L. maackii
Amur honeysuckle

L. involucrata
bearberry honeysuckle

L. tatarica
Tatarian honeysuckle

Figure 2. *Lonicera*: seeds, ×8 (Brinkman 1974i).

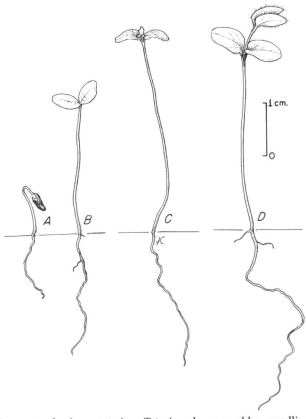

Figure 4. *Lonicera tatarica*, Tatarian honeysuckle: seedling development at 1, 3, 13, and 31 days after germination (Brinkman 1974i).

dormancy. In most species this is caused by a dormant embryo. In some, it appears the seedcoat may retard germination. Prechilling is necessary for all species, but for some, such as hairy (*L. hirsuta*) and swamp fly (*L. oblongifolia*) honeysuckle, it must be preceded by warm stratification at 20/30°C for 60 days.

GERMINATION TEST Germination can be tested with pre-treated seeds in sand or on paper. Incubation temperature is generally given as 20/30°C. Germination is epigeal (Fig. 4).

NURSERY AND FIELD PRACTICE Seeds of the honeysuckle species in which embryo dormancy is expected can be sown either broadcast or in drill rows in the fall, or pretreated seeds can be sown in the spring. Species that appear to have seedcoat-connected dormancies should be sown as soon as collected. Seeds should be sown 0.6 cm deep and the beds mulched. Germination is usually complete in 40 to 60 days. About 15% of the seeds sown produce seedlings. Seedlings are transplanted to the field after 1 or 2 years. Most species can be propagated from cuttings.

HAMAMELIDACEAE — WITCH-HAZEL FAMILY

Loropetalum chinense (R. Br.) D. Oliver

GROWTH HABIT, OCCURRENCE, AND USE *Loropetalum chinense*, a monotypic genus, comes from China and India and contains a species that is much-branched shrub to 4 m in height. The branchlets are densely reddish pubescent.

FLOWERING AND FRUITING The flowers, which occur in early spring, are white to yellowish green and feathery in outline. The fruit is a woody, dehiscent, 2-seeded capsule.

GERMINATION There are no reports of anyone ever having successfully germinated seeds of this species (Dirr and Heuser 1987). Suggested pretreatments are 3 months of warm stratification followed, by 3 months of prechilling. Cuttings taken in late July can be rooted.

LEGUMINOSAE — LEGUME FAMILY

Lupinus L. — Lupine

GROWTH HABIT, OCCURRENCE, AND USE The genus *Lupinus* includes more than 100 species, a few of which are considered shrubs (Ratliff 1974c). Shrub species of lupine that are commonly planted in California are Pauma (*L. longifolius*) and whiteface lupine (*L. albifrons*). The former occurs in southern California. It reaches a height of 1.5 m, has been cultivated since 1928, and has proven to be a valuable ornamental species. Some plants may live for 10 years, but it is generally a short-lived species. Whiteface lupine is found in northern California in the Coast Range and Sierra Nevada. It often reaches a height of 2 m and has been planted for erosion control and wildlife habitat.

FLOWERING AND FRUITING The blue, purple, or yellow flowers are bisexual, irregular, and borne in racemes. Pauma lupine will bear viable seeds at 1 year of age. It flowers from April to May and the seeds ripen from May to August. Whiteface lupine flowers from March to June and the seeds mature from early June to late July.

COLLECTION, EXTRACTION, AND STORAGE OF SEED The pods of both species of shrub lupine open at maturity and disperse 2 to 12 seeds (Figs. 1, 2). It is necessary to collect seeds when the pods are somewhat green.

L. albifrons
whiteface lupine

L. longifolius
Pauma lupine

Figure 1. *Lupinus*: seeds, ×4 (Ratliff 1974c).

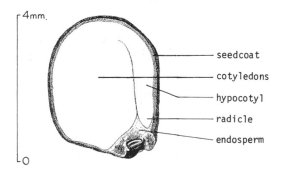

4mm.

seedcoat

cotyledons

hypocotyl

radicle

endosperm

0

Figure 2. *Lupinus longifolius*, Pauma lupine: longitudinal section through a seed, ×10 (Ratliff 1974c).

Immature pods should be air dried until the pods open. Seeds can be recovered by air screening. There are 46 seeds of Pauma lupine per gram. They can be stored for prolonged periods without loss of viability. Seed predation by insects in the pod is a problem with most species of lupine.

GERMINATION Both species of shrub lupine are reported to have hard seedcoats that require scarification. However, Ratliff (1974c) reported the seeds of the shrub lupines had high germinative capacity without pretreatment. Either mechanical scarification or hot water treatments have been effective in enhancing germination of various species of *Lupinus*. ASOA (1985) standards for *Lupinus* species are substrata, toweling or paper; incubation temperature, 20/30°C; and duration, 18 days.

Lupines are important ornamental and forage species world-wide. A brief sampling of the recent literature on the germination of *Lupinus* species follows:

Author—date	Subject	Location
Davidson & Barbour 1977	Germination, coastal bush lupine.	California
Elkinauy 1982	Physiology of germination.	—
Horn & Hill 1974	Chemical seed scarification.	Australia
Knypl et al. 1985	Cytokinins and germination.	Czechoslovakia
Quinlivan 1970	Hard seedcoats.	Australia
Wassermann & Agenbas 1978	Harvesting and seed quality.	South Africa
Wink 1983	Allelopathy.	—

SOLANACEAE — NIGHTSHADE FAMILY

Lycium L. — Wolfberry

GROWTH HABIT, OCCURRENCE, AND USE The wolfberries include about 100 species of deciduous or evergreen, thorny or unarmed shrubs native to temperate and subtropical regions in both hemispheres (Rudolf 1974m). They are ornamental shrubs valued chiefly for their showy berries, but they also provide wildlife habitat, and watershed protection; at least 2 species are grown for shelterbelt hedges. Several species are important spiny shrubs in the deserts of western North America. The introduced species, Chinese wolfberry (*Lycium chinense*) and matrimony vine (*L. halimifolium*) have been grown for the longest time and most extensively.

FLOWERING AND FRUITING The perfect purplish flowers usually bloom in the summer and are followed by bright red (rarely black or yellow) berries, each with a few to many seeds (Figs. 1, 2) (Rudolf 1974m):

Species	Flowering dates	Fruit ripening dates	Ripe fruit color
L. andersoni	Apr.–June	Aug.Sept.	Red
L. chinense	June–Sept.	June–Sept.	Scarlet-orange
L. exsertum	Aug.–Nov.	—	Orange-red
L. halimifolium	June–Sept.	Aug.–Oct.	Scarlet-red
L. richii	May–Sept.	May–Sept.	Bright red

Good crops of seed are borne by matrimony vine almost every year. *Lycium exsertum* produces seeds abundantly.

COLLECTION, EXTRACTION, AND STORAGE OF SEED Ripe berries may be picked from bushes in the fall. The berries are soft and can be pulped by forcing them through a screen and the seeds recovered by flotation. Large amounts of seed may be partially fermented and

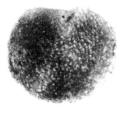

Figure 1. *Lycium halimifolium*, matrimony vine: cleaned seed, ×12 (Rudolf 1974m).

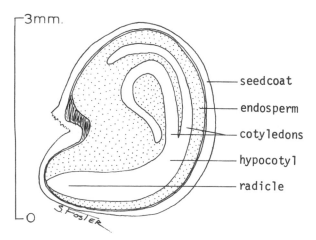

Figure 2. *Lycium halimifolium*, matrimony vine: longitudinal section through a seed, ×18 (Rudolf 1974m).

run through a macerator. After extraction the seeds should be dried and stored in sealed containers at low temperatures. Kay et al. (1988) found that seeds of Anderson wolfberry (*L. andersonii*) increased in germination after 9 years' storage at room temperature. Seeds per gram of the *Lycium* species are as follows (Rudolf 1974m):

Species	Mature tree height (m)	Year first cultivated	Seeds/gram
L. andersoni	3	1935	—
L. chinense	2	1709	380
L. exsertum	4	1935	—
L. halimifolium	6	long	570
L. richii	4	1935	3000

GERMINATION Dormancy in wolfberry seeds is variable. Seeds of Anderson wolfberry and *L. exsertum* germinate without pretreatment. Matrimony vine seeds require prechilling for 120 days. After prechilling, germination tests of seeds of this species can be conducted in sand at 20/30°C for 18 days. Seeds of Rich wolfberry (*L. richii*) have similar requirements to matrimony vine.

NURSERY AND FIELD PRACTICE Seeds may be sown in the fall as soon as collected or prechilled seeds can be sown in the spring. For matrimony vine and Chinese wolfberry the tree percentage has been given as 10 to 15%. Two-year-old seedling are usually outplanted.

ERICACEAE — HEATH FAMILY

Lyonia Nutt.

GROWTH HABIT, OCCURRENCE, AND USE The genus *Lyonia* consists of about 30 species of evergreen or deciduous shrubs in North America and Asia. The flowers are white to pinkish and are borne in axillary clusters, racemes, or terminal panicles.

He-huckleberry (*Lyonia ligustrina*) is native to eastern North America from Maine to Texas. Seeds of this species germinate without pretreatment (Dirr and Heuser 1987).

LEGUMINOSAE — LEGUME FAMILY

Maackia Rupr.

GROWTH HABIT, OCCURRENCE, AND USE The genus *Maackia* consists of 6 species of deciduous trees native to eastern Asia. *Maackia amurensis* is planted as an ornamental in the United States. The white flowers, borne in panicle racemes 15 to 20 cm long, bloom in July and August. The fruit is a pod that should be collected in the fall and dried. The small brown seeds have hard seedcoats that require scarification.

Dry seeds store well at low temperatures. Dropping seeds in boiling water and allowing them to cool overnight effectively breaks dormancy (Dirr and Heuser 1987). Cuttings taken in the fall will root.

MORACEAE — MULBERRY FAMILY

Maclura pomifera (Raf.) Schneid. — Osage orange

GROWTH HABIT, OCCURRENCE, AND USE Osage orange is native to southern Arkansas, southeastern Oklahoma, and eastern Texas, and has been widely planted in the United States (Bonner and Ferguson 1974). It is a small deciduous tree chiefly valued as a source of post and as a windbreak. Normal height at maturity is 9 m, but some trees reach 21 m.

FLOWERING AND FRUITING The small, green, dioecious flowers open from April to June. The large, globose, aggregate fruit (Fig. 1) ripens in September and October and soon falls to the ground. The yellowish green fruits, 10 to 12.5 cm in diameter, are composed of one-seeded drupelets (Fig. 2). Trees bear good crops at age 10, and good seed crops occur annually.

COLLECTION, EXTRACTION, AND STORAGE OF SEED Fruits should be picked as soon as they fall from the trees, but collections can be made throughout fall and winter. Seeds may be extracted by macerating the fruit in water and recovered by flotation. Cleaning is easy if the fruits are allowed to ferment for several months before extraction. One way to accomplish this is to store the fruits in a pile outdoors. By early spring they will be soft

Figure 1. *Maclura pomifera*, osage orange: aggregate fruit composed of one-seeded drupelets, ×0.5 (Bonner and Ferguson 1974).

and mushy, and only a brief maceration will be required. Seeds collected in this manner have a pronounced purple streak and pleasant fragrance, and germinate promptly. Seed yield is as follows (Bonner and Ferguson 1974): 35 liters yields 80 fruits and 24,500 seeds weighing 5.5 kg. There are 20 seeds per gram. Seeds can be stored in sealed containers at low temperatures.

PREGERMINATION TREATMENT AND GERMINATION Osage orange seeds should be prechilled for 30 days, and then can be tested in flats of sand at 20/30°C. Germination is epigeal (Fig. 3).

NURSERY AND FIELD PRACTICE Untreated seeds can be sown in the fall, but prechilled seeds are necessary for spring sowing. Seeds extracted from fruits fermented over winter do not require prechilling. Seeds may be drilled in rows 20 to 30 cm apart, or in bands 7.5 to 10 cm wide. They should be covered with 0.6 to 1.25 cm of soil, and fall-seeded beds should be mulched.

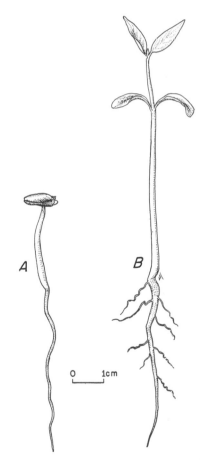

Figure 3. *Maclura pomifera*, osage orange: seedling development at 1 and 8 days after germination (Bonner and Ferguson 1974).

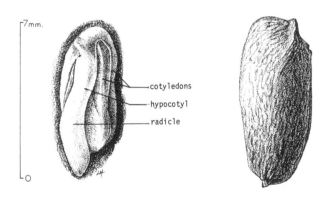

Figure 2. *Maclura pomifera*, osage orange: excised embryo and nutlet (seed), ×6 (Bonner and Ferguson 1974).

MAGNOLIACEAE — MAGNOLIA FAMILY

Magnolia L.

GROWTH HABIT, OCCURRENCE, AND USE The genus *Magnolia* comprises about 35 species (as high as 85 according to some authors) of deciduous or evergreen trees that occur in North and South America, eastern Asia, and the Himalayas (Olson et al. 1974). Some species produce valuable lumber, and medicinal products are extracted from others. Many of the magnolias are valuable ornamental species. Nine species are native to the United States. Cucumber tree (*M. acuminata*), Fraser magnolia (*M. fraseri*), southern magnolia (*M. grandiflora*), and sweetbay (*M. virginiana*) are the most commonly propagated species. Cucumber tree and southern magnolia are large trees, 18 to 28 m tall, and sweetbay and Fraser magnolia are 6 to 12 m tall.

FLOWERING AND FRUITING The large, solitary, perfect flowers are borne singly at the ends of branches in the spring and summer, and range in color from greenish yellow to white. Pollination is carried out by insects. The red or rusty brown conelike fruits, 5 to 12 cm long and 1.25 to 3.75 cm in diameter, consist of several coalescent 1-or 2-seeded follicles (Fig. 1) that ripen in late summer to early fall. At maturity the drupelike seeds are 0.6 to 1.8 cm long (Fig. 2) and red or scarlet. The removable outer portion of the seedcoat is fleshy, oily, and soft; the inner portion is stony. The inner seedcoat is thin and membranous, and encloses a large, fleshy endosperm in which is embedded a minute embryo (Fig. 3). The seed usually is suspended from the open follicle for some time by a slender, elastic thread. Seed dispersal is by wind and birds, and occurs soon after ripening. Southern magnolia is a prolific seed producer, and is one of the most fruitful of forest trees. Trees often produce seeds 10 years after planting, and bear good crops annually.

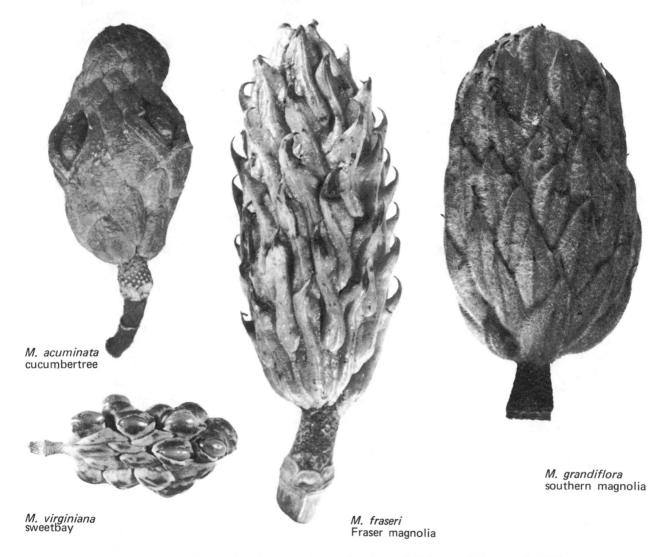

M. acuminata
cucumbertree

M. virginiana
sweetbay

M. fraseri
Fraser magnolia

M. grandiflora
southern magnolia

Figure 1. *Magnolia*: conelike fruits consisting of coalesent follicles, ×1 (Olson et al. 1974).

COLLECTION, EXTRACTION, AND STORAGE OF SEED The fruits may be picked from standing trees, or from trees recently felled in logging, as soon as they turn red to rusty brown, but picking can be delayed until the follicles have begun to open. The fruits should be spread out to dry in shallow layers (one fruit deep). After a few days, the seeds can be shaken out of the open fruits. Seeds to be used in the near future should have the fleshy part of the outer seedcoat removed by maceration in water or by rubbing on screens. The average number of seeds per gram does not vary widely among species of *Magnolia* (Olson et al. 1974):

Species	Seeds/gram
M. acuminata	12
M. fraseri	10
M. grandiflora	14
M. virginiana	17

Magnolia seed can be stored cleaned or in the dried pulp for several years with little loss of viability if in sealed containers at low temperatures. Seeds stored at higher temperatures should not be cleaned. Southern magnolia seed loses its viability if stored over winter at room temperature. Seeds of big leaf magnolia (*M. macrophylla*) can be stored under water at room temperature for 180 days (Seitner 1981).

PREGERMINATION TREATMENT Optimum germination occurs with the fleshy pulp removed. Several investigators have shown that it contains germination inhibitors. Magnolia seeds exhibit embryo dormancy, which can be overcome by 3 to 6 months of pre-chilling. Fall sowing provides natural prechilling. Seed leaf emergence will occur with as little as 42 days of prechilling and radicle emergence will usually occur without prechilling (Deltredion 1981, Browse 1987). Dirr and Brinson (1985) obtained 75% emergence of southern magnolia seedlings after 4 months in soil in a warm greenhouse without prechilling. Excessive drying after maceration can lead to loss of viability. Fresh seed shows extreme variability in seedling production potential (Dirr and Heuser 1987). It is thought that improper or incomplete fertilization is the cause.

GERMINATION In horticulture the main reason for growing magnolias from seed is to provide under-stock (Dirr and Heuser 1987). Although most species can be propagated from seed, they are slow to flower and will often be different from their parents. In the United States southern magnolia and Kobus magnolia (*M. kobus*) are often used as understock. Germination is epigeal and occurs rapidly with prechilled seeds. A fairly close estimate of viability in cucumber tree seeds can be made by removing the seedcoats, lightly scratching the surface of the endosperm, and then

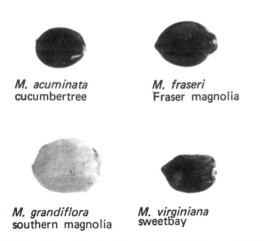

M. acuminata
cucumbertree

M. fraseri
Fraser magnolia

M. grandiflora
southern magnolia

M. virginiana
sweetbay

Figure 2. *Magnolia*: seeds showing outer fleshy layer, ×1 (Olson et al. 1974).

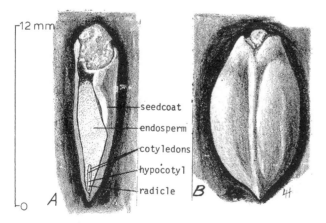

seedcoat
endosperm
cotyledons
hypocotyl
radicle

Figure 3. *Magnolia grandiflora*, southern magnolia: seeds after removal of fleshy layer, ×3; (A) longitudinal section through the embryo; (B) surface of stony layer of seedcoat (Olson et al. 1974).

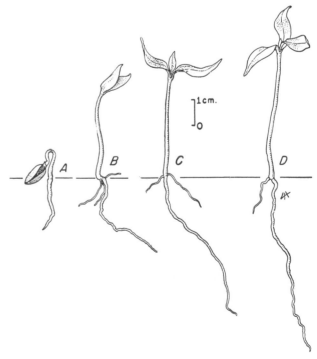

Figure 4. *Magnolia grandiflora*, southern magnolia: seedling development at 1, 5, 13, and 31 days after germination (Olson et al. 1974).

placing the seeds on moist cotton or blotting paper in covered dishes at germination temperatures. Viable seeds produce green pigment in the vicinity of the scratches in 2 to 3 days. It is not clear if light is required to get the green color. Seeds of *Magnolia* have also been tested using the excised embryo method (Heit 1955). The ASOA (1985) standards for southern magnolia are substrata, creped cellulose paper or the top of blotters; incubation temperature, 20/30°C; duration 42 days; and additional directions, prechill for 45 days or use TZ method.

NURSERY AND FIELD PRACTICE Magnolia seeds may either be sown in the fall with mulching, or prechilled seeds may be sown in the spring. Seeds should be sown in rows 20 to 30 cm apart and covered with 0.6 cm of soil. Young seedlings (Fig. 4) need half-shade during the first summer. Spring sowing is recommended if rodents are a problem.

Cutting is the preferred method of asexual propagation (Dirr and Heuser 1987). Cutting wood selected from young, previously rooted, stock plants provides the best source of propagation material. The importance of juvenile stock plants can not be over emphasized.

ROSACEAE — ROSE FAMILY

Malus Mill. — Apple

GROWTH HABIT, OCCURRENCE, AND USE The apples include about 25 species of deciduous trees and shrubs native to the temperate regions of North America, Europe, and Asia (Crossley 1974a). In this genus are some of the most important fruit bearers and ornamental trees, as well as valuable species for wildlife and shelterbelts. Many cultivated varieties of apple (*Malus pumila*) and crab apple (*M. baccata*) have been developed, but these varieties are usually propagated vegetatively.

FLOWERING AND FRUITING The pink to white perfect flowers appear in the spring with or before the leaves. Flowering times vary among species from March to June (Table 1).

The fruit is a fleshy pome (Fig. 1) in which 3 to 5 carpels, usually 5, are embedded. Each carpel contains 2 seeds or only 1 by abortion (Fig. 2). The seeds have a thin lining of endosperm (Fig. 3), except in apple which has none. Fruits of some species ripen as early as August and drop to the ground. Large amounts of common apple seed can be obtained from cores food discarded by processing plants. Seeds from cider mills are often damaged. An accepted, though cumbersome, method of extraction involves partial fermentation of the fruit in a large container. The fermented fruit is then covered with water and mashed. The seeds can be recovered by flotation. Fresh fruits can be put

M. floribunda
Japanese flowering crab apple

Figure 1. *Malus floribunda*, Japanese flowering crab apple: fruits and leaves ×1 (Crossley 1974a).

through a macerator and the seeds recovered by flotation. The number of seeds per gram varies with species (Table 1). Seeds dried to a moisture content of less than 11% have been stored in sealed containers at low

Table 1. *Malus*: phenology of flowering and fruiting, characteristics of mature trees, cleaned seed weight, and pregermination treatments (Crossley 1974a).

Species	Flowering dates	Fruit ripening dates	Ripe fruit color	Mature tree height (m)	Year first cultivated	Seeds/ gram	Prechilling Duration (days)	Prechilling Temperature (°C)
M. baccata	May	Aug.–Oct.	Red or yellow	9	1784	145	30	20/30
M. coronaria	Mar.–May	Sept.–Nov.	Yellow-green	9	1724	30	120	10
M. diversifolia	–	Late Fall	Variable	9	1836	120	90	–
M. floribunda	May	–	Red	9	1862	130	60–120	–
M. ioensis	May–June	Sept.–Oct.	Greenish waxy	–	1885	70	60	20/30
M. pumila	May	Aug.–Oct.	Yellow to red	15	Old	45	60	20/30
M. × robusta	Apr.–May	–	Red or yellow	–	1815	40	60–120	–

temperatures for 2 years without loss of either viability or subsequent seedling vigor.

GERMINATION Apple seeds display dormancy which has been overcome by prechilling (Table 1): ASOA (1985) standards for *Malus* species are substrate, paper; incubation temperatures, 18/22°C; duration, 7 to 10 days; and additional directions, use embryo excision method or TZ.

There is a huge amount of literature on the germination of apple seeds. Selected recent literature on *Malus* seed germination is given in Table 2.

NURSERY AND FIELD PRACTICE Apple rootstocks are often grown from seed in nurseries. Untreated seeds can be sown in the late fall and prechilled seeds in the spring. Seeds are sown 0.6 to 1.25 cm deep in rows and mulched with a thin layer of sawdust. Germination is epigeal (Fig. 4).

Table 2. *Malus:* selected recent literature on germination.

Author—Date	Subject	Location
Bouvier-Durant et al. 1984	Embryo biochemistry.	Canada
Come et al. 1985	Embryo oxygen supply.	France
Dickie and Bower 1985	Viability estimates.	USA
Sinska and Gladen 1984	Ethylene and dormancy.	USA
Thevenont et al. 1983	Temperature and dormancy.	Israel

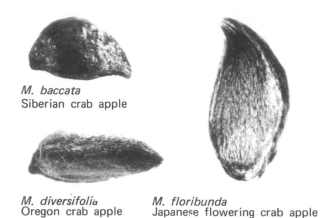

M. baccata
Siberian crab apple

M. diversifolia
Oregon crab apple

M. floribunda
Japanese flowering crab apple

Figure 2. *Malus*: seed, ×8 (Crossley 1974a).

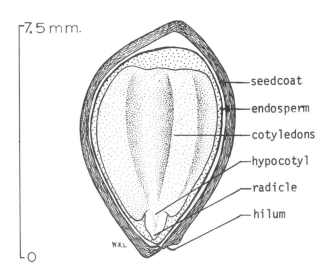

Figure 3. *Malus coronaria*, sweet crab apple: longitudinal section through a seed, ×8 (Crossley 1974a).

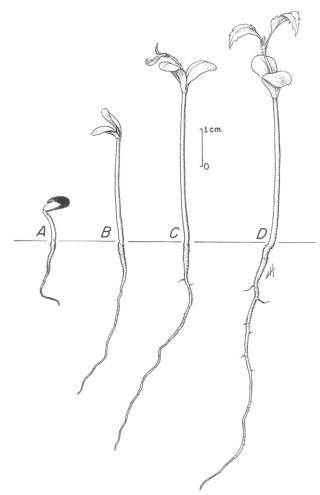

Figure 4. *Malus coronaria*, sweet crab apple: seedling development at 1, 3, 9, and 16 days after germination (Crossley 1974a).

ANACARDIACEAE

Mangifera L. — Mango

GROWTH HABIT, OCCURRENCE, AND USE Mangos are tropical trees of great economic importance. They probably rank fifth in world fruit production.

Mature mango seeds have a high moisture content and can not stand desiccation. Fresh seeds germinate at temperatures between 5 and 40°C with germination most rapid between 25 and 40°C (Corbineau et al. 1986).

MELIACEAE — MAHOGANY FAMILY

Melia azedarach L. — Chinaberry

GROWTH HABIT, OCCURRENCE, AND USE Chinaberry is a short-lived deciduous tree, native to southern Asia and Australia, that reaches a maximum height of 15 m (Bonner and Grano 1974). It has been cultivated since the 16th century, chiefly for ornamental purposes, and has become naturalized in many tropical and subtropical countries, including the United States. In India the wood is used for furniture, agricultural implements, and the manufacturing of paper. Extracts of the leaves and the fruits have insecticidal properties, and the fruits are valuable food for livestock and wildlife.

FLOWERING AND FRUITING The flowering habit is either perfect or polygamodioecious. The pretty, lilac-colored, perfect flowers are borne in axillary panicles 10 to 15 cm long, in March to May. The fruit is a subglobose, fleshy, round drupe that ripens in September and October and persists on the trees well into winter. It turns yellow and wrinkled as it ripens (Fig. 1). Inside the fleshy mesocarp is a single, fluted, light brown stone that contains 4 to 5 pointed, smooth, black seeds (Figs. 2, 3). Abundant seed crops are borne almost annually.

COLLECTION, EXTRACTION, AND STORAGE OF SEED Fruits can be collected by hand after the leaves have fallen in late autumn or early winter. They may be either run wet through a macerator and the seeds recovered by flotation, or the entire fruit may be planted immediately. There are about 1.4 seeds per gram. The seed may be stored dry for a year without loss of viability.

GERMINATION Pregermination treatments are not necessary. In nature epigeous germination usually occurs during the spring following dispersal. Incubation temperature of 20/30°C is recommended for germination testing, with sand as a substrate.

NURSERY AND FIELD PRACTICE Stones are usually sown intact immediately after collection in the fall or they can be sown the following spring. They should be sown 5 to 7.5 cm apart and covered with 2.5 cm of soil. Germination takes place about 3 weeks after spring sowing. Chinaberry can also be propagated by root or stem cuttings, and can be direct seeded in the field.

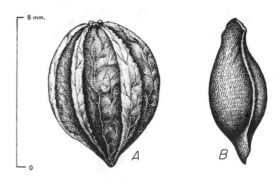

Figure 2. *Melia azedarach*, Chinaberry: (A) stone; (B) seed, ×5 (Bonner and Grano 1974).

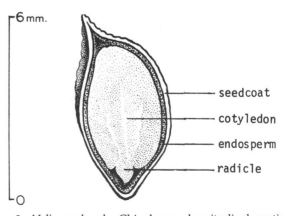

Figure 3. *Melia azedarach*, Chinaberry: longitudinal section through a seed (Bonner and Grano 1974).

Figure 1. *Melia azedarach*, Chinaberry: fruit and stone, ×1 (Bonner and Grano 1974).

MENISPERMACEAE — MOONSEED FAMILY

Menispermum canadense L. — Common moonseed

GROWTH HABIT, OCCURRENCE, AND USE Common moonseed is a climbing, woody vine that grows to a height of 3.7 m and is capable of spreading from underground stems (Brinkman and Phipps 1974b). It is native from Quebec and western New England to southeastern Manitoba, and south to Georgia, Alabama, and Oklahoma. The plants are seldom eaten by livestock, but the fruits are valuable for wildlife, although reportedly poisonous to humans. This species has been cultivated since 1646 for its attractive foliage and fruits.

FLOWERING AND FRUITING The dioecious flowers appear from May to July and the bluish black drupes ripen from September to November. The seeds are flattened stones in the form of a half-moon or ring (Figs. 1, 2).

COLLECTION, EXTRACTION, AND STORAGE OF SEED Fruits may be collected from September to November. Seeds may be extracted by maceration and then recovered by flotation. There are about 16.7 seeds per gram.

GERMINATION Seeds require from 2 to 4 weeks of prechilling. There is a suggestion that they require light for germination (Brinkman and Phipps 1974b).

NURSERY AND FIELD PRACTICE Common moonseed is readily propagated by sowing prechilled seeds in the spring or planting seeds in the fall as soon as they are ripe. Cuttings can be rooted.

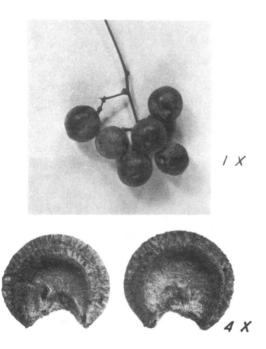

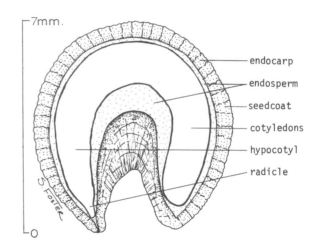

Figure 1. *Menispermum canadense*, common moonseed: fruit, ×1, and seed, ×4 (Brinkman and Phipps 1974b).

Figure 2. *Menispermum canadense*, common moonseed: longitudinal section through a seed, ×8 (Brinkman and Phipps 1974b).

OLEACEAE — OLIVE FAMILY

Menodora scabra A. Gray — Rough menodora

GROWTH HABIT, OCCURRENCE, AND USE Rough menodora is a low, semiwoody shrub 0.2 to 1.5 m in height (Krugman 1974b). It is native to dry, rocky areas and desert grasslands from 460 to 2150 m in southern California, western Texas, New Mexico, and Colorado. It provides browse for livestock and wildlife.

FLOWERING AND FRUITING Its often showy, yellow flowers appear from May through August. The fruit, a bispherical thin-walled capsule with 2 seeds in each cell, ripens in September through October.

COLLECTION, EXTRACTION, AND STORAGE OF SEED Seed collections should be made in the fall after the capsules ripen. The mature seeds are approximately 4 to 5 mm in length and 3 mm wide, flat, and greenish to brownish, with a yellowish narrow wing (Fig. 1).

Good seed crops usually occur each year. The number of cleaned seeds per gram is 235. Seeds can be stored in dry places. No pretreatment is required for germination.

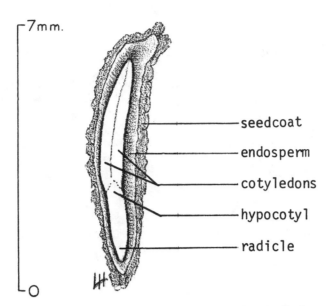

Figure 1. *Menodora scabra*, rough menodora: longitudinal section through a seed, ×10 (Krugman 1974b).

TAXODIACEAE — TAXODIUM FAMILY

Metasequoia glyptostroboides Hu & Cheng — Dawn redwood

GROWTH HABIT, OCCURRENCE, AND USE Dawn redwood is often called a living fossil because until 1941 the genus was known only from fossil records (Johnson 1974b). The present natural range of the only known living species of *Metasequoia* is quite restricted. The original type tree is found near the village of Modaogi in southeastern Hubei Province, China; the bulk of the native groves are found in Shuishaba village in what is known to Westerners as the *Metasequoia* valley.

Since its introduction into the United States in 1948, dawn redwood has mostly been planted as an ornamental. It is not a good timber tree because the wood is weak, brittle, and soft. Several cultivars of dawn redwood have been described.

FLOWERING AND FRUITING *Metasequoia* is monoecious. Planted specimens have produced male and female cones. Male cones are formed in the leaf axis or terminally on the branch. The small, male bud cones are visible just prior to leaf drop in the fall. Female cones are borne singly, and often opposite, along branches. Male and female cones begin enlargement in late January and are readily seen by early or mid-February. They are susceptible to late frost.

COLLECTION, EXTRACTION, AND STORAGE OF SEED Cones are ready to collect in late October and into November, about the time leaves are ready to fall. Seeds can be stored in sealed containers at low temperatures (Dirr and Heuser 1987). The seedcoats of dawn redwood are thin and very fragile (Fig. 1).

GERMINATION Fresh seeds will germinate without pretreatment, but 1 month of prechilling unifies and hastens germination (Dirr and Heuser 1987). Emergence begins 5 days after planting. Young seedlings thrive in high humidity, but are very susceptible to damping-off fungi.

Dawn redwood is one of the easiest conifers to root and can be propagated from softwood or hardwood cuttings.

Figure 1. *Metasequoia glyptostroboides*, dawn redwood: hollow seed, ×8 (Johnson 1974b).

MYRTACEAE — MYRTLE FAMILY

Metrosideros Banks — Ironwood

GROWTH HABIT, OCCURRENCE, AND USE *Metrosideros* is a genus of about 25 species largely native to Australia, but widely grown around the Pacific because of their dazzling display of masses of scarlet-stamened, pincushion flowers. The fruit is a capsule enclosed in the calyx tube. The common name, ironwood, comes from the red heartwood. Most species are very salt tolerant.

GERMINATION Burton (1982) investigated the germination of seeds of *M. polymorpha*. Germination was generally poor due to the low percentage of seeds with viable embryos. Light was not necessary for germination, but resulted in a 4-fold increase in germination. The optimum light quality was determined. No germination occurred below 12°C.

RUBIACEAE — MADDER FAMILY

Mitchella repens L. — Partridgeberry

GROWTH HABIT, OCCURRENCE, AND USE Partridgeberry, also called two-eyed berry or running fox, is an evergreen vine or herb with fruits that are valuable for wildlife (Brinkman and Erdmann 1974). The natural range is from Texas to Florida, and north to southwestern Minnesota. This attractive plant was introduced into cultivation in 1761 and is often used in rock gardens.

FLOWERING AND FRUITING The dimorphous flowers appear from June to August. The fruits are scarlet, drupaceous berries that ripen in July, but usually persist over winter.

COLLECTION, EXTRACTION, AND STORAGE OF SEED Partridgeberries may be picked in late fall. Fruits should be macerated in water and recovered by flotation (Figs. 1, 2). Seed yield from fruit is about 12% by weight. There

are about 430 seeds per gram. Seeds can be stored in sealed containers at low temperatures.

GERMINATION Partridgeberry seeds have internal dormancy that requires 150 to 180 days of prechilling for germination.

NURSERY AND FIELD PRACTICE Seeds should be sown in the fall. Cuttings taken in November and treated with IBA will root (Dirr and Heuser 1987).

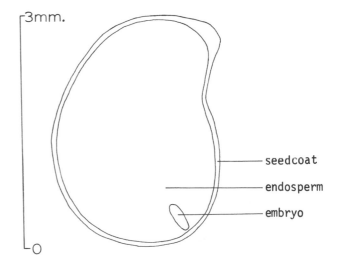

Figure 1. *Mitchella repens*, partridgeberry: seed, ×7 (Brinkman and Erdmann 1974).

Figure 2. *Mitchella repens*, partridgeberry: longitudinal section through a seed, ×20 (Brinkman and Erdmann 1974).

MORACEAE — MULBERRY FAMILY

Morus L. — Mulberry

GROWTH HABIT, OCCURRENCE, AND USE The mulberries consist of about 12 species of deciduous trees and shrubs native to temperate and subtropical regions of Asia, Europe, and North America (Read and Barnes 1974). Russian mulberry (*Morus alba* forma *tatarica*) was introduced into the United States by Russian Mennonites in 1875, and is probably the most widely planted mulberry. The Prairie States Forestry Project planted more than 1 million plants per year from 1937–1942 for windbreaks on the Great Plains from Nebraska to Texas. All mulberries are valuable food sources for birds and other wildlife. The 7 or more varieties or forms of Russian mulberry differ in their relative drought resistance and chromosome number. Its high drought resistance makes Russian mulberry well suited for shelterbelt plantings.

FLOWERING AND FRUITING Flowers are normally dioecious, but can also be monoecious on different branches of the same plant. Both types appear in stalked, axillary, pendulous catkins in April and May. The multiple fruit is composed of many small, closely appressed drupes. Fruits ripen and drop from the trees during the months of June to August, though they are often dispersed by birds and animals (Read and Barnes 1974):

M. alba
white mulberry

M. rubra
red mulberry

Figure 1. *Morus*: fruits and leaves, ×1 (Read and Barnes 1974).

Species	Flowering dates	Fruit ripening dates	Ripe fruit color
M. alba	May	July–Aug.	White, pink purplish
forma *tatarica*	May	June–Aug.	Dark red to purple, white
M. rubra	Apr.–May	June–Aug.	Dark red to purple-black

The varieties differ in the size and color of their fruit (Fig. 1). They vary in taste from insipid to sweet. Each fruit contains a dozen or more small nutlets (Figs. 2, 3), that have thin, membranous coats and endosperm. Seed bearing begins at about 5 years of age for white mulberry (*M. alba*), between 5 and 10 years for Russian mulberry, and at 10 years for red mulberry (*M. rubra*). Large crops of seeds occur nearly every year on Russian mulberry in the Great Plains.

COLLECTION, EXTRACTION, AND STORAGE OF SEED Ripe mulberry fruits may be collected by stripping, shaking, or flailing them from the trees onto ground cloths. Fruits should be collected as soon as most are ripe to avoid loss to birds. Seed yield from fruit on a weight basis is about 2%. Seed size varies (Read and Barnes 1974):

Species	Mature tree height (m)	Year first cultivated	Seeds/gram
M. alba	14	old	520
forma *tatarica*	8	1875	660
M. rubra	12	1629	690

Fruits should be run through a macerator and the seeds recovered by flotation. Fermentation at moderate indoor temperatures aids the extraction process and improves viability of Russian mulberry seeds. After drying, subfreezing temperatures are recommended for storage of mulberry seeds (Read and Barnes 1974).

PREGERMINATION TREATMENT Seeds extracted by partial fermentation germinate without pretreatment. Spring-sown seeds should be prechilled for 30 to 90 days.

GERMINATION Germination tests have been conducted on paper or sand at 20/30°C with light. Seed embryos are curved with the radicle nearly touching the cotyledons (Fig. 3). Germination is epigeal.

NURSERY AND FIELD PRACTICE Mulberry seeds may be drilled in rows 20 to 30 cm apart at the rate of 160 seeds per meter of row. Beds should be mulched and kept in half-shade for the first weeks of emergence. Seeds yield seedlings at a rate of 12 to 50%. One-year-old stock is usually used for field planting.

M. rubra	*M. alba* f. *tatarica*
red mulberry	Russian mulberry

Figure 2. *Morus*: nutlets (seeds), ×10 (Read and Barnes 1974).

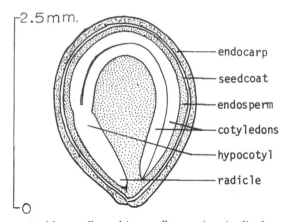

Figure 3. *Morus alba*, white mulberry: longitudinal section through a nutlet, ×20 (Read and Barnes 1974).

MYRICACEAE — WAX MYRTLE FAMILY

Myrica L. — Bayberry

GROWTH HABIT, OCCURRENCE, AND USE The genus *Myrica* consists of evergreen shrubs and small trees up to 12 m in height (Krochmal 1974). The fruits of southern (*M. cerifera*) and northern (*M. pensylvanica*) bayberry are covered with a fragrant wax that has been used to make bayberry candles. Pharmaceuticals have been extracted from the fruit, bark, and leaves of southern bayberry.

FLOWERING AND FRUITING The flowers are unisexual and mostly dioecious. Flowers of Pacific bayberry (*M. californica*), however, are unisexual and monoecious. The fruits are small, globose, dry drupes heavily coated with wax (Fig. 1). Seeds lack endosperm (Fig. 2). The color of ripe fruit varies with species (Krochmal 1974):

Species	Flowering dates	Fruit ripening dates	Ripe fruit color
M. californica	Mar.–Apr.	Sept.	Brownish purple
M. cerifera	Apr.–June	Aug.–Oct.	Light green
M. gale	Apr.–June	July	Lustrous
M. pensylvanica	Apr.–July	Sept.–Oct.	Grayish white

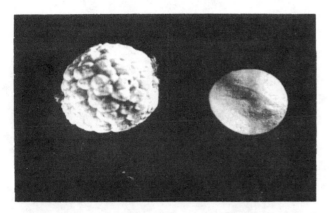

Figure 1. *Myrica cerifera*, southern bayberry: wax-coated drupe and cleaned drupe (seed), ×8 (Krochmal 1974).

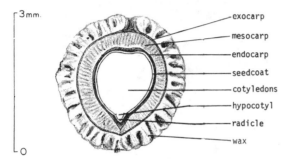

Figure 2. *Myrica cerifera*, southern bayberry: longitudinal section through the embryo of a drupe, ×12 (Krochmal 1974).

Fruits of most species of bayberry persist on the shrubs over winter, but those of Pacific bayberry are dropped early in the winter.

COLLECTION, EXTRACTION, AND STORAGE OF SEED Ripe fruits can be stripped by hand into a container or shaken onto ground cloths. The drupes are handled as seed. The only processing consists of removing the wax coating before sowing. During storage the wax should be left on the fruits. It can be removed from the dry fruits by rubbing them against a screen. The number of seeds per gram for Pacific bayberry is 48; southern bayberry, 185; and northern bayberry, 120.

GERMINATION The germination of bayberry seeds is accelerated by 30 to 90 days of prechilling. Light is apparently necessary for germination of seeds of sweetgale (*M. gale*). Dewaxed fruits show some germination enhancement from gibberellin enrichment (Dirr and Heuser 1987).

NURSERY AND FIELD PRACTICE Seeds of *Myrica* species may be sown in fall or spring. Drill rows should be spaced 20 to 30 cm apart and the seeds covered with 0.6 cm of soil. Fall-sown beds should be mulched. Spring-sown seeds need to be prechilled. Germination is epigeal (Fig. 3).

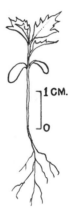

Figure 3. *Myrica californica*, Pacific bayberry: 1-month-old seedling (Krochmal 1974).

HYDROPHYLLACEAE — WATERLEAF FAMILY

Nama lobbii Gray — Woolly nama

GROWTH HABIT, OCCURRENCE AND USE One of the few perennial species in the genus, woolly nama is native to coniferous woodlands of the northern Sierra Nevada in California and western Nevada (Nord and Leiser 1974). The plants are 0.1 to 0.6 m tall and generally sparse, but when competing vegetation is removed by disturbance they rapidly form dense crowns up to 1.5 m in diameter. Fast-growing roots that extend up to 4.6 m in a single year contain a profusion of adventitious buds that sprout to form new plants.

FLOWERING AND FRUITING Numerous small purple flowers, borne in reduced, terminal cymes or in axillary angles along slightly erect stems appear from May to September. The fruit is a capsule containing 10 to 12 oval, angular, very dark brown seeds up to 1.5 mm long (Figs. 1, 2). The capsules mature in late August to October. There are about 2000 seeds per gram.

Figure 1. *Nama lobbii*, woolly nama: seed, ×20 (Nord and Leiser 1974).

COLLECTION, EXTRACTION, AND STORAGE OF SEED Mature seeds may be hand stripped or flailed directly into containers, or seedheads and some foliage may be mechanically cut and gathered with a rotary lawn mower. The seed may be recovered from dry herbage by air screening. Seed yield from this type of material by weight is about 3%. Seeds should be stored in a dry, cool place in sealed containers.

GERMINATION Woolly nama seeds are initially dormant. Leaching of seeds under mist followed by gibberellin enrichment produces about 30% germination. Excised embryos have about 60% germination.

NURSERY AND FIELD PRACTICE Leached seeds can be soaked in gibberellin and then direct seeded in the field when conditions are conducive for germination after the initial autumn rainfall.

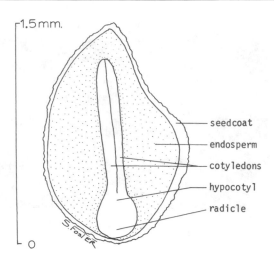

Figure 2. *Nama lobbii*, woolly nama: longitudinal section through a seed, ×40 (Nord and Leiser 1974).

AQUIFOLIACEAE — HOLLY FAMILY

Nemopanthus mucronatus (L.) Trel. — Mountain holly

GROWTH HABIT, OCCURRENCE, AND USE Mountain holly is a deciduous, branching shrub that occurs from Newfoundland to Minnesota and south to Virginia and Indiana (Schopmeyer 1974b). Introduced into cultivation in 1802, this shrub provides food and cover for wildlife.

FLOWERING AND FRUITING The fruit is a dull red berrylike drupe, 6 to 8 mm in diameter, containing 4 to 5 bony nutlets. The nutlets are crescent-shaped and bone colored with a rib on the back (Fig. 1).

COLLECTION, EXTRACTION, AND STORAGE OF SEED The fruits may be collected as late as mid-October. Small lots of seed can be extracted by rubbing the fruits through a screen and then recovering the seeds by flotation. Cleaned seed yield from fruits is about 10% by weight. There are about 100 seeds per gram.

GERMINATION The seeds require a period of afterripening before they will germinate. Seeds sown

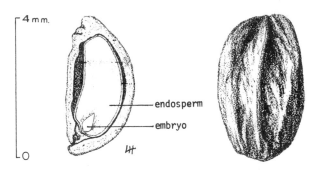

Figure 1. *Nemopanthus mucronatus*, mountain holly: (left) longitudinal section through a nutlet showing small, immature seed; (right) exterior view of nutlet; both ×9 (Schopmeyer 1974b).

in the fall will produce some emergence over a 2 year period. Softwood cuttings can be rooted.

FAGACEAE — BEECH FAMILY

Nothofagus Benth. — Southern beech

GROWTH HABIT, OCCURRENCE, AND USE The genus *Nothofagus* (meaning false beech) is related to *Fagus* of the Northern Hemisphere. Southern beeches are native to extreme southern South America, New Zealand, and Australia. There are at least 25 species growing in the southern half of the globe in climatic areas varying from cold temperate to almost tropical. The South American species are the most cold resistant. The southern beeches are very variable in habit, ranging from dwarf, almost shrubby plants in windy Tierra del Fuego to giant timber trees in the forests of Australia and New Zealand. Myrtle beech (*N. cunninghamii*) reaches heights of 70 m in Tasmania.

Studies of natural stands of *N. cunninghamii* in Tasmania showed that most seeds were dispersed relatively close to the parent trees (Hickey et al. 1983). Germination capacity was highest during seedfall and during good seed years.

GERMINATION Ledgard and Cath (1983) experimented with 82 lots of seed from several species of *Nothofagus* grown in New Zealand. Seed weight was positively correlated with viability. No correlation was found between altitude or latitude and viability.

Germination of *N. fusca*, *N. menziesii*, and *N. solandri* var. *cliffortioides* was greatly improved by 70 days of prechilling. The germination of other species was enhanced by as little as 24 hours of prechilling. Trials with seeds of *Nothofagus* grown in England showed that seeds of *N. obliqua* were very dormant while seeds of *N. procera* were not. The germination of most species can be enhanced by soaking the seeds overnight in 50 ppm of gibberellin. Work by Shafiq (1981) in Iraq also showed the value of this procedure.

NYSSACEAE — TUPELO FAMILY

Nyssa L. — Tupelo

GROWTH HABIT, OCCURRENCE, AND USE The deciduous tree species of *Nyssa* native to North America (Table 1) furnish timber, wildlife food, and honey (Bonner 1974h). Water, black, and swamp tupelo were cultivated before 1750.

FLOWERING AND FRUITING The minute, greenish white flowers that appear in spring may be perfect, or staminate and pistillate flowers may be borne separately on different trees. The fruits of *Nyssa* are thin-fleshed, oblong drupes about 1.25 to 3.75 cm long (Fig. 1). They ripen in the fall. Each fruit contains a bony, ribbed, usually one-seeded stone (Figs. 2, 3). Stones of water tupelo range in color from white to dark brown or gray; some are pinkish white. Stone color does not influence germination. Water tupelo trees bear fruit at 5 to 10 years of age and produce abundantly every year.

COLLECTION, EXTRACTION, AND STORAGE OF SEED Ripe fruit may be picked from the ground, from standing trees, or from recently felled trees. Stump sprouts of water tupelo will produce viable seeds (Prieston 1979). To extract seeds the fruits should be run through a macerator and the seeds recovered by flotation. For water tupelo there are about 350 to 600 fruits per kg. For black tupelo the cleaned seed yield from fruits by weight is about 25%. Seeds per gram are given in Table 1. *Nyssa* seeds can be kept quite satisfactorily over winter in cold storage. For short-term storage removal of the pulp is not necessary.

PREGERMINATION TREATMENT *Nyssa* seeds exhibit moderate embryo dormancy and germination is enhanced by prechilling. Both sand and naked prechilling have been used satisfactorily. Good germination has been reported with prechilling for 30 days, but some seed lots may require 120. A single source suggests that germination of seeds of black tupelo is enhanced by scarification (McGill and Whitcomb 1977).

Table 1. *Nyssa*: phenology of flowering and fruiting, characteristics of mature trees, and cleaned seed weight (Bonner 1974h).

Species	Common name	Flowering dates	Fruit ripening dates	Fruit dropping dates	Mature tree height (m)	Ripe fruit color	Seeds/gram
N. aquatica	Water tupelo	Mar.–Apr.	Sept.–Oct.	Oct.–Nov.	30	Dark purple	1
N. ogeche	Ogeechee tupelo	Mar.–May	July–Aug.	–	25	Red	3
N. sylvatica							
var. *biflora*	Swamp tupelo	Apr.–June	Aug.–Oct.	Sept.–Dec.	40	Blue-black	5
var. *sylvatica*	Black tupelo	Apr.–June	Sept.–Oct.	Sept.–Nov.	18	Blue-black	6

GERMINATION ASOA (1985) standards suggest the following for germination of *Nyssa* species:

Species	Substrata	Temperature (°C)	Duration (days)	Additional instructions
N. aquatica	TC	20/30	21	Prechill 30 days.
N. sylvatica	C	20/30	28	Prechill 28 days.

Submersion of *Nyssa* seeds should be avoided.

NURSERY AND FIELD PRACTICE Spring sowing of prechilled seeds is generally recommended. The seeds are drilled at the rate of 50 per meter of row and covered with 1.25 to 2.5 cm of soil. A mulch of sawdust is often used for water tupelo and pine needles for swamp tupelo. Beds must not be allowed to dry. Germination is epigeal (Fig. 4).

N. aquatica
water tupelo

N. sylvatica var. *sylvatica*
black tupelo

N. ogeche
Ogeechee tupelo

Figure 1. *Nyssa*: fruits, ×1 (Bonner 1974h).

N. aquatica
water tupelo

N. ogeche
Ogeechee tupelo

N. sylvatica var. *sylvatica*
black tupelo

N. sylvatica var. *biflora*
swamp tupelo

Figure 2. *Nyssa*: stones (seeds), ×1 (Bonner 1974h).

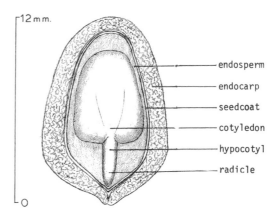

— endosperm
— endocarp
— seedcoat
— cotyledon
— hypocotyl
— radicle

Figure 3. *Nyssa sylvatica* var. *sylvatica*, black tupelo: longitudinal section through a stone, ×4 (Bonner 1974h).

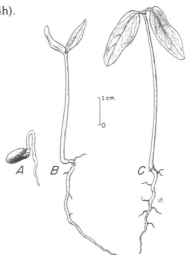

Figure 4. *Nyssa sylvatica* var. *sylvatica*: seedling development at 1, 4, and 39 days after germination (Bonner 1974h).

ROSACEAE — ROSE FAMILY

Oemleria Rchb. — Osoberry

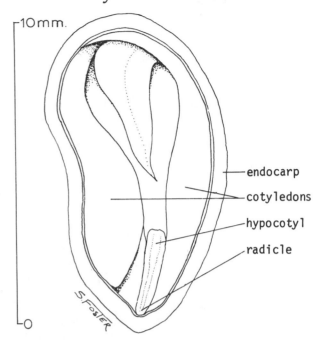

Figure 3. *Oemleria cerasiformis*, osoberry: longitudinal section through a seed showing folded cotyledons, ×8 (Dimock and Stein 1974).

GROWTH HABIT, OCCURRENCE, AND USE The genus *Oemleria* contains a single species, osoberry, *O. cerasiformis* (Dimock and Stein 1974 as *Osmaronis*). Osoberry is a deciduous shrub that grows in moist to fairly dry open woods from British Columbia southward through western Washington and Oregon to Tulare County in the Sierra Nevada and to northern Santa Barbara County in the Coast Range of California. Usually a shrub from 1.5 to 3 m tall, it may attain tree size. Roadsides, stream banks, and shaded areas are favored habitats, but its presence is conspicuous only in early spring when the light green foliage and delicate inflorescences appear before those of most associated species.

Ripening fruits are attractive to birds such as cedar waxwings. By early summer, scattered leaves throughout the crown often turn yellow, giving the shrub a handsome, variegated appearance.

FLOWERING AND FRUITING Osoberry is essentially dioecious because stamens in the bisexual flowers are rarely functional. Within its wide range, osoberry flowers from January to May, concurrent with leaf

Figure 1. *Oemleria cerasiformis*, osoberry: drooping raceme of white flowers, ×1 (Dimock and Stein 1974).

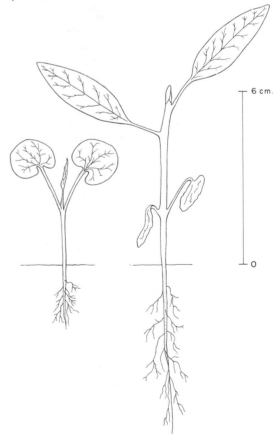

Figure 4. *Oemleria cerasiformis*, osoberry: seedling development at approximately 40 and 120 days after germination (Dimock and Stein 1974).

Figure 2. *Oemleria cerasiformis*, osoberry: seeds, ×2 (Dimock and Stein 1974).

development. The 5-petaled flowers are white, fragrant, and borne in drooping racemes (Fig. 1). Fruits are single-seeded drupes about 1 cm long. Developing fruits become peach-colored, then reddish, and finally a deep blue-black under a whitish bloom when ripe. In Oregon and Washington fruits ripen and disperse in May.

COLLECTION, EXTRACTION, AND STORAGE OF SEED Clusters of ripe fruits can be stripped from shrubs by hand. Seeds can be recovered by maceration and flotation. Osoberry seed has a bony endocarp and lacks endosperm (Figs. 2, 3). Seed yield from fruit is about 25% by weight. There are about 10 seeds per gram.

PREGERMINATION Lengthy prechilling is needed to overcome seed dormancy. Prechilling for 60 days initiates germination, but 120 days of prechilling are required for full germination.

GERMINATION Germination tests can be conducted at an incubation temperature of 20/30°C, using prechilled seeds.

NURSERY AND FIELD PRACTICE Germination is epigeal (Fig. 4). Dimock and Stein (1974) suggested that osoberry should be handled in the nursery like *Prunus* species.

OLEACEAE — OLIVE FAMILY

Olea L. — Olive

GROWTH HABIT, OCCURRENCE, AND USE The common olive (*Olea europaea*) is a broad-leaved evergreen tree reaching a height of 3 to 18 m, with a crown width often approaching that of its height (Krugman 1974c). This species was first introduced about 1800 into what is now California. Many cultivars and botanical varieties of *Olea* have subsequently been introduced to California and the Southwest. It is cultivated as a fruit tree and as a shade and ornamental species.

FLOWERING AND FRUITING In May, in the axils of the leaves on the wood of the preceding year's growth, yellowish white flowers are borne in great numbers. Both perfect (bisexual) and staminate (unisexual) flowers are found in the same inflorescence. The staminate or male flowers drop soon after blooming.

At 5 years of age olive trees start bearing annually. Some varieties bear on alternate years. The fruit is a subglobose or oblong drupe, 1.25 to 3.75 cm long, containing a single stone (Figs. 1, 2). The maturing fruit passes from a deep green to a straw color, then to a black and shining color when ripe. The fruit and seed mature from October through December. If not picked immediately, the fruit will remain on the trees until early spring.

COLLECTION, EXTRACTION, AND STORAGE OF SEED The fleshy fruits are collected by hand and the stones extracted by maceration. The number of seeds per gram varies from 1 to 4, depending on variety and fruit size. The seeds

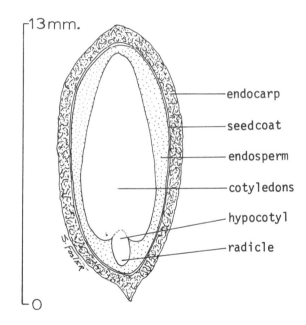

Figure 2. *Olea europaea*, olive: longitudinal section through a stone, ×6 (Krugman 1974c)

can be stored dry at room temperature for a few years.

GERMINATION The Genebank Handbook (1985) considers the endocarp and testa of *Olea* to restrict imbibition of moisture by the seed. Consequently, their removal promotes germination. Intact seeds will not germinate at 25°C or higher temperatures, but embryos excised from slightly dormant seeds will germinate at 25°C. Germination of excised and intact seeds is optimum at 13°C in light and dark; at slightly supra-optimal temperatures, light can promote germination. The quality of light is not important. Enrichment with plant growth regulators has not enhanced germination. Selected recent literature on the germination of *Olea* seeds include the following:

Figure 1. *Olea europaea*, olive: stone, ×2 (Krugman 1974c).

Author—date	Subject
Crisosta & Sutter 1985	Role of endocarp in germination.
Istanbouli et al. 1974	Rapid reproduction of olives from seed.
Istanbouli & Neville 1974	Influence of embryo coverings.
Istanbouli & Neville 1977	Light and temperature interactions in olive germination.
Lagarda, Martin, and Kester 1983	Seed development.
Lagarda, Martin, and Polito 1983	Environmental and maturity influences.
Voyiatzis & Porlingis 1987	Temperature requirements for germination.

NURSERY AND FIELD PRACTICE Seeds are sown 1.25 cm deep in well-prepared soil in the fall. Normally, the seeds of most varieties will germinate by spring. Certain varieties require cracking of the stones or acid scarification. Seedlings are usually transplanted after a year to pots for grafting.

LEGUMINOSAE — LEGUME FAMILY

Olneya tesota Gray — Tesota

GROWTH HABIT, OCCURRENCE, AND USE Tesota is an evergreen, broad-crowned tree, 5 to 8 m high, commonly found in desert washes and valleys of southeastern California, southern Arizona, and northwestern Mexico. It provides browse for cattle and can be used in native plant landscaping (Krugman 1974d).

FLOWERING AND FRUITING Flowering occurs from April to June. The white to rose-purple flowers produce a seeded, light brown, rounded hairy pod, 3.5 to 5 cm in length.

COLLECTION, EXTRACTION, AND STORAGE OF SEED Pods may be picked in August and should be dried for several days to facilitate seed extraction. The seeds are chestnut brown to black, shiny, ovoid, and 0.9 cm long with no endosperm (Fig. 1). There are about 5 seeds per gram. Many are infested with insects when collected.

GERMINATION Fresh seeds require no pretreatment prior to germination, although a 12- to 24-hour soak

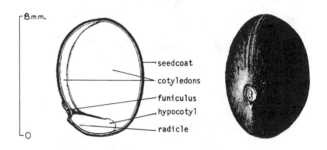

Figure 1. *Olneya tesota*, tesota: longitudinal section through a seed (left) and exterior view (right), ×4 (Krugman 1974d).

may be helpful. Stored seeds should be soaked at least 24 hours. Mild scarification before soaking may also be required.

Seedlings can be started in flats and transplanted to containers. Initial germination is prompt.

BETULACEAE — BIRCH FAMILY

Ostrya virginiana (Mill.) K. Koch — Eastern hophornbeam

GROWTH HABIT, OCCURRENCE, AND USE Eastern hophornbeam is a small deciduous tree attaining a maximum height of about 18.5 m, and occurring throughout the eastern half of the United States and into Canada (Schopmeyer and Leak 1974). Its best development occurs in Arkansas and eastern Texas.

The heavy, durable wood has been used for fence posts, tool handles, and other specialty items. It was first cultivated in 1690 and has been planted as an ornamental.

FLOWERING AND FRUITING The flowers are monoecious, the staminate in clusters of long catkins formed the

previous year, and the pistillate in small open clusters. Flowers open in April and May. The fruit is a strobile (Fig. 1) consisting of involucres, each enclosing a single nut (Fig. 2) about 0.8 cm long and 0.5 cm in diameter. The first fruits ripen as early as August and continue through the fall. Nuts are dispersed after ripening, when the strobiles fall apart. Trees do not produce abundant seeds until they are 25 years old.

COLLECTION, EXTRACTION, AND STORAGE OF SEED The strobiles may be handpicked from the trees when they are pale greenish brown in color. At this stage, they are not yet dry and hard enough to fall apart. When ripe, they are light gray to greenish brown. After drying and threshing seeds can be recovered by air screening. Seed yield from fruits is about 20% by weight. There are about 66 seeds per gram.

PREGERMINATION TREATMENT The seeds have an internal dormancy that is difficult to overcome. Warm stratification for 60 days at 20/30°C followed by 140 days of prechilling have been used as pregermination treatments.

GERMINATION Germination tests using pretreated seeds have been conducted on a sand substrate at alternating temperatures for 30 to 40 days. Germination is epigeal (Fig. 3).

NURSERY AND FIELD PRACTICE Either fall or spring sowing has been used with seed coverage of 0.6 cm. Fall sowing may be done with freshly collected, slightly green seed. Fall-sown seedbeds should be mulched. Spring-sown seedbeds with prechilled seeds need to be kept continuously wet until emergence is complete.

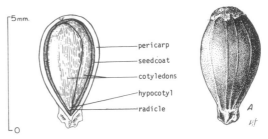

Figure 2. *Ostyra virginiana*, eastern hophornbeam: (left) longitudinal section through a seed and (right) intact seed, both ×6 (Schopmeyer and Leak 1974).

Figure 1. *Ostrya virginiana*, eastern hophornbeam: strobile, ×1 (Schopmeyer and Leak 1974).

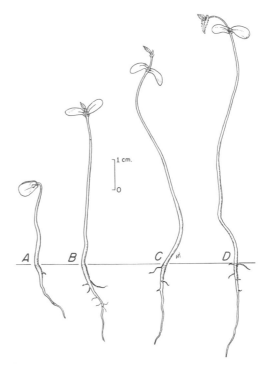

Figure 3. *Ostrya virginiana*, eastern hophornbeam: seedling development at 2, 4, 23, and 27 days after germination (Schopmeyer and Leak 1974).

ERICACEAE — HEATH FAMILY

Oxydendrum arboreum (L.) DC. — Sourwood

GROWTH HABIT, OCCURRENCE, AND USE Sourwood is a small tree, 12 to 18.5 m tall at maturity, that occurs throughout eastern United States from Pennsylvania to Indiana, south to Louisiana and east to northern Florida (Olson and Barnes 1974d). The wood is little used commercially except as fuel wood and as pulpwood in mixture with other hardwood species. Deer use twigs for winter browse. The flowers can be an important source of honey. Sourwood is used as an ornamental because of its flowers and attractive foliage. The species has been cultivated since 1747.

FLOWERING AND FRUITING Sourwood flowers appear in copious masses from late June to August. This tree is one of the latest flowering trees and shrubs to bloom.

Sourwood flowers are white and are borne in long, 1-sided racemes clustered in an open panicle that terminates the current year's branches.

The fruit is an ovoid-pyramidal, dry, dehiscent capsule 0.6 to 1.25 cm long. Borne in profuse, panicled clusters, the fruits ripen in September and October. The seeds are minute and gray to brown when ripe (Fig. 1) and are dispersed gradually through the winter by dehiscence of the capsule.

COLLECTION, EXTRACTION, AND STORAGE OF SEED Collection of fruits is made from trees during the fall and early winter; some capsules can be collected as late as early spring. When capsules are dry the seeds can be recovered by threshing and air screening. There are 4000 to 12,000 seeds per gram.

GERMINATION Sourwood seeds germinate without pretreatment. It is not clear from the literature, but light may be required for germination of these minute seeds. Germination is epigeal (Fig. 2)

NURSERY AND FIELD PRACTICE The minute seeds can be sprinkled on the surface of milled peat, covered with clear plastic film and kept moist in the presence of light. As soon as the seedlings are big enough to handle they can be transplanted to containers.

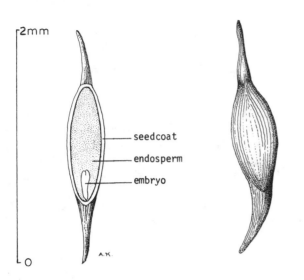

Figure 1. *Oxydendrum arboreum*, sourwood: exterior view of seed (right), and longitudinal section (left), ×30 (Olson and Barnes 1974d).

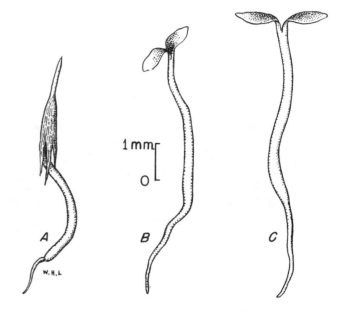

Figure 2. *Oxydendrum arboreum*, sourwood: seedlings at 2, 6, and 8 days after germination (Olson and Barnes 1974d).

BUXACEAE — BOX FAMILY

Pachysandra terminalis Siebold & Zucc. — Japanese pachysandra

GROWTH HABIT, OCCURRENCE, AND USE The genus *Pachysandra* consists of about 5 species native to eastern North America and Asia. They are grown as ground covers.

GERMINATION Fresh seeds of Japanese pachysandra (*P. terminalis*) had 20 to 30% germination capacity in 30 days of incubation. Acid-scarified seeds germinated 90% in the same time period (Dirr and Heuser 1987).

Plants are very easily rooted from cuttings taken after the flush of new growth has hardened.

PAEONIACEAE — PEONY FAMILY

Paeonia L.

GROWTH HABIT, OCCURRENCE, AND USE The genus *Paeonia* consists of 30 or more perennial, herbaceous or half-shrub species, native to the Northern Hemisphere in Asia and North America.

PREGERMINATION TREATMENT Seeds of Japanese tree peony (*P. suffruticosa*) have a double dormancy requiring warm stratification followed by prechilling.

The seeds are potted and given warm stratification for 3 months to allow the hypocotyl/root axis to grow, and then prechilled for 3 months to break epicotyl dormancy (Dirr and Heuser 1987). Emergence requires a return to warm incubation temperatures following this pretreatment.

LEGUMINOSAE — LEGUME FAMILY

Parkia clapertonia R. Br. — African locust bean

GROWTH HABIT, OCCURRENCE, AND USE African locust bean is widely grown in the savanna region of Nigeria (Etejere et al. 1982). The tree fruits from December to March and produces numerous pods. These pods contain a yellow, powdery pulp that is rich in carbohydrates and has the seeds embedded in it. They are used to prepare the spice known as iru. Livestock feed on the pods. Excessive livestock grazing and dry season burning reduce the density of this species.

GERMINATION Red-brown and dark brown seeds are produced in the same pod. Red-brown seeds germinate without pretreatment, but they are only 10% of the seeds produced. Dark brown seeds have hard seedcoats and are dormant. When heated to 80°C for 1 to 2 minutes, they show some germination. Acid-scarified dark brown seeds will germinate. The optimum temperature for germination of red-brown or acid-scarified dark brown seeds is 30°C. The yellow pulp also contains a substance that inhibits germination of both types of seeds.

LEGUMINOSAE — LEGUME FAMILY

Parkinsonia L.

GROWTH HABIT, OCCURRENCE, AND USE The genus *Parkinsonia* consists of about 5 species: 1 in South Africa, 3 in tropical North America, and 1 species in South America. *Parkinsonia aculeata* is a drought-tolerant tree that has been used for revegetation in the arid zone from the Sudan to Indo-Pakistan.

GERMINATION *Parkinsonia aculeata* produces 2 types of seeds, light and dark brown. About 25% of the light brown seeds will germinate (Mahmoud and El-Sheikh 1981). The remainder of the light brown and all the dark brown seeds have hard seedcoats. Acid scarification for 60 minutes enhances germination of both. Optimum germination of scarified seeds occurred from 15 to 40°C.

COMPOSITAE — SUNFLOWER FAMILY

Parthenium argentatum Gray — Guayule

GROWTH HABIT, OCCURRENCE, AND USE Guayule is a rubber-producing, silver-gray perennial shrub native to north-central Mexico and southern Texas. It has been the subject of a great deal of research for possible commercial exploitation as a source of rubber.

The fruit consists of an achene enclosed by 2 staminate florets and a subtending bract.

GERMINATION Freshly harvested seeds are inherently dormant, apparently because of an immature embryo. The afterripening requirement lasts 2 months. Achene coats remain impermeable to gas exchange for 6 to 12 months. Removal of the chaff, or prolonged washing, enhances germination (Naqui and Hanson 1982).

VITACEAE — GRAPE FAMILY

Parthenocissus Planch. — Creeper

GROWTH HABIT, OCCURRENCE, AND USE About 10 species of creepers are native to eastern Asia and North America (Gill and Pogge 1974d). Virginia (*Parthenocissus quinquefolia*) and Japanese (*P. tricuspidata*) creeper may ascend to about 15 to 18 m above the ground. Thicket creeper (*P. inserta*) usually lacks the adhesive disk of the other two species, and, with rare exceptions, has a low, rambling growth form. All 3 prefer soils that are moist, but otherwise they grow well in a wide variety of soil types. They are at least moderately tolerant of shading, but are most likely to occupy places such as the edges of clearings, fence rows, old walls and other structures, and stream banks. Their chief uses are as ornamentals and for wildife habitat. The creepers have attractive bluish black fruits and handsome foliage that turns scarlet, crimson, or orange in the fall. Cultivation began in the United States in 1622 for Virginia creeper, and before 1800 for thicket creeper. Japanese creeper was imported in 1862.

FLOWERING AND FRUITING The flowers are small, greenish, and borne in rather inconspicuous, long-stemmed clusters. The flowers are usually perfect, but some vines have both perfect and unisexual flowers. The periods of flowering vary with species (Gill and Pogge 1974d):

Species	Flowering dates	Fruit ripening dates	Seed dispersal dates
P. inserta	June–July	July–Aug.	Aug.–Nov.
P. quinquefolia	June–Aug.	Aug.–Oct.	Sept.–Feb.
P. tricuspidata	June–July	Sept.–Oct.	—

Ripe berries (Fig. 1) of all 3 species are bluish black in color. Thicket creeper fruits usually are 3–4 seeded and slightly larger than the 1- to 3-seeded fruits of the other species. Seeds have small embryos (Figs. 2, 3). Good seed crops are borne frequently. Seed dispersal is largely by animals.

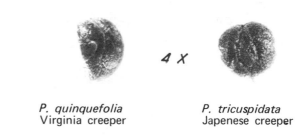

4 X

P. quinquefolia
Virginia creeper

P. tricuspidata
Japenese creeper

Figure 2. *Parthenocissus*: seeds, ×4 (Gill and Pogge 1974d).

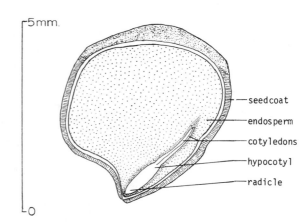

Figure 3. *Parthenocissus inserta*, thicket creeper: longitudinal section through the embryo of a seed, ×10 (Gill and Pogge 1974d).

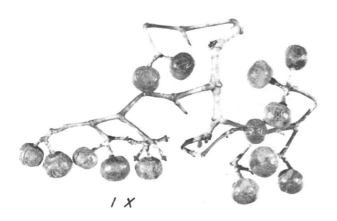

1 X

Figure 1. *Parthenocissus quinquefolia*, Virginia creeper: cluster of berries, ×1 (Gill and Pogge 1974d).

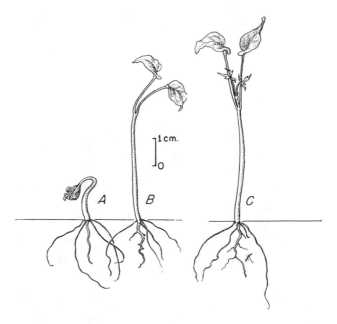

Figure 4. *Parthenocissus quinquefolia*, Virginia creeper: seedling development at 1, 3, and 22 days after germination (Gill and Pogge 1974d).

COLLECTION, EXTRACTION, AND STORAGE OF SEED After the fruits have turned color to bluish black, they can be stripped from the vines. Leaves and other debris mixed with the fruit can be removed by air screening. Seeds can be recovered by maceration of the fruits and flotation. They should be throughly dried before storage. The whole berries can also be stored. Storage should be in sealed containers at low temperatures. There are about 40 thicket and 30 Virginia creeper seeds per gram.

GERMINATION Natural germination takes place during the first , or perhaps the second, season after maturity. Germination is epigeal (Fig. 4). It is enhanced by 60 days of prechilling. Germination tests can be conducted on a sand substrate at 20/30°C using prechilled seeds.

NURSERY AND FIELD PRACTICE Untreated seeds can be sown in the fall or preferably, prechilled seeds can be sown in the spring. Seeds should be sown in drill rows and covered with 0.9 cm of soil. Optimum density is about 105 seeds per m². The creepers can also be propagated by cuttings.

BIGNONIACEAE

Paulownia tomentosa (Thunb.) Steud. — Royal paulownia

GROWTH HABIT, OCCURRENCE, AND USE Royal paulownia is a common sight along roadsides and near old house sites in eastern and southern United States (Bonner and Burton 1974). Native to China, it has been planted for its ornamental value from New York south and west to Texas. It has escaped from cultivation in many locations. This deciduous tree reaches heights of 9 to 21 m at maturity. Its rapid early growth has attracted the interest of the paper industry.

FLOWERING AND FRUITING The showy, violet or blue, perfect flowers appear in terminal panicles up to 25 cm long in April and May before the leaves emerge. The fruits are ovoid, pointed, woody capsules about 3.1 to 4.3 cm long (Fig. 1). They turn brown in fall when mature and persist on the tree through winter. The tiny, winged, flat seeds are about 0.15 to 0.3 cm long (Figs. 2, 3).

COLLECTION, EXTRACTION, AND STORAGE OF SEED Dry fruits can be collected and opened by hand any time before they disperse their seeds. It has been estimated that there are 3100 fruits per 35 liters, 2000 seeds per fruit, 4.9 kg of seeds per 35 liters of fruit, and 6200 seeds per gram (Bonner and Burton 1974).

GERMINATION Seeds exhibit no dormancy, but light is necessary for germination. Germination can be tested on creped cellulose at an incubation temperature of 20/30°C, with light during the warm period. The rapid growth rate of *Paulownia* species has sparked interest in

Figure 1. *Paulownia tomentosa*, royal paulownia: capsule, ×1 (Bonner and Burton 1974).

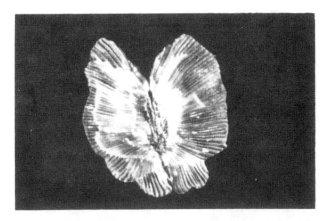

Figure 2. *Paulownia tomentosa*, royal paulownia: winged seed, ×12 (Bonner and Burton 1974).

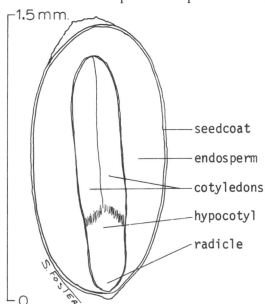

Figure 3. *Paulownia tomentosa*, royal paulownia: longitudinal section through a seed, ×50 (Bonner and Burton 1974).

the germination of their seeds as indicated by selected recent literature:

Author—Date	Subject
Barnhill et al. 1982	Germination characteristics.
Beckjord 1982	Container propagation.
Carpenter et al. 1980	Prechilling and gibberellin.
Carpenter & Smith 1979	Dry storage and prechilling.
Carpenter & Smith 1981	Light.
Cunningham & Carpenter 1980	Slurry seeding.
Donald 1987	Viability.
Grubisic et al. 1985	Changes in light sensitivity.
Wang & Hung 1979	Germination.

SCROPHULARIACEAE — FIGWORT FAMILY

Penstemon Mitch. — Penstemon

GROWTH HABIT, OCCURRENCE, AND USE The penstemons consist of about 230 species, mostly in western North America (Hylton 1974). They are herbaceous or semi-woody perennial plants, usually erect and tufted, but occasionally low and creeping. The leaves are opposite, entire or toothed, the upper ones sessile and often clasping. Included in the genus are some of the most spectacular wild flowers. Four species native to California are planted for their showy flowers (Hylton 1974):

Species	Common name	Flowering dates
P. cordifolius	Vine penstemon	May–July
P. corymbosus	Thymeleaf penstemon	June–Oct.
P. heterophyllus	Chaparral penstemon	Apr.–July
P. lemmonii	Lemmon's penstemon	June–Aug.

FLOWERING AND FRUITING The fruit is an ovoid or oblong, 2-celled capsule dehiscing along the septae, with numerous seeds that have irregularly angled, cellular coats (Figs. 1, 2). Flowering and fruiting begin 1 year after planting. Seeds of chaparral penstemon become ripe from July to September 15. Those of other species ripen soon after their flowering dates.

COLLECTION, EXTRACTION, AND STORAGE OF SEED The best time to collect seeds should be determined by frequent examination of the plants. The capsules should be gathered and placed in dry containers after the seeds have ripened in the field. After drying, the capsules can be threshed and the seeds recovered by air screening. Chaparral penstemon seeds number about 2000 per gram. Seeds may be stored in sealed containers at low temperatures.

GERMINATION Germination standards for domestic species of penstemon (ASOA 1985) are substrate, paper; incubation temperature, 15°C; duration 8 days; and additional instructions, provide light.

NURSERY AND FIELD PRACTICE Seeds should be sown on the top of flats of milled peat, covered with clear plastic film and given light. When seedlings are big enough to handle they can be transplanted to pots.

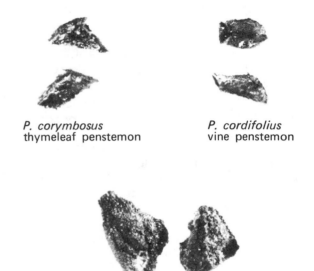

P. corymbosus
thymeleaf penstemon

P. cordifolius
vine penstemon

P. heterophyllus
chaparral penstemon

Figure 1. *Penstemon*: seeds, ×12 (Hylton 1974).

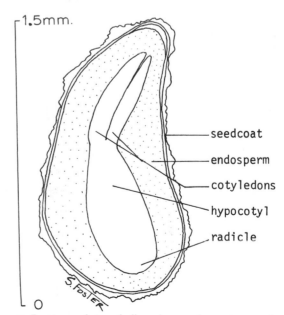

1.5mm.

seedcoat
endosperm
cotyledons
hypocotyl
radicle

0

Figure 2. *Penstemon heterophyllus*, chaparral penstemon: longitudinal section through a seed, ×50 (Hylton 1974).

NURSERY AND FIELD PRACTICE Most of the semishrubby species of penstemon can be readily grown from seed. Seeds can be sprinkled on the surface of a fine silt loam soil in a flat and covered with vermiculite. Flats should be kept continuously moist. When the seedlings are large enough to handle they can be transplanted to containers. If emergence is low an alternative procedure would be to put the seeds on top of vermiculite-covered silt loam, cover the flat with clear plastic film, and provide light.

ROSACEAE — ROSE FAMILY

Peraphyllum ramosissimum **Nutt.** — Squawapple

GROWTH HABIT, OCCURRENCE, AND USE Squawapple, the only species in the genus, is an attractive shrub with intricate and rigid branches, dark grayish bark, and a mature height of 0.5 to 2 m (Smith 1974b). Its principal habitat includes sagebrush and juniper woodlands in portions of eastern Oregon, northeastern California, and east through Idaho and Utah to Colorado. It occurs mainly on dry foothill and mountain slopes on well-drained soils. Depending on its location squawapple is considered poor-to-fair browse for domestic animals and wildlife.

FLOWERING AND FRUITING The perfect flowers with their pinkish, spreading petals appear in May and June. The fruit is a pome about 8 to 10 mm long containing several seeds (Figs. 1, 2). The fruit turns yellowish to reddish brown as it ripens in late June and July. It is usually dropped or eaten by birds by late August.

COLLECTION, EXTRACTION, AND STORAGE OF SEED The ripe fruits may be picked from the shrubs and mashed in water, and the seeds recovered by flotation. Seeds store well in dry, cool storage. Seed yield from fruit is about 7 to 10%. The average number of seeds per gram ranges from 57 to 98.

GERMINATION Apparently the seeds of squawapple require prechilling for abundant germination. Seeds prechilled for 90 days had about 50% germination when incubated at 20/30°C with light during the warm period. Germination is epigeal (Fig. 3).

NURSERY AND FIELD PRACTICE Local native plant gardeners report a lot of difficulty growing seedlings of squawapple to maturity. Seedling growth rate is slow and survival poor.

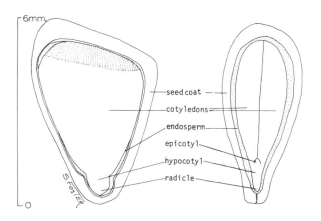

Figure 1. *Peraphyllum ramosissimum*, squawapple: seed, ×4 (Smith 1974b).

Figure 2. *Peraphyllum ramosissimum*, squawapple: longitudinal section through a seed, ×9 (Smith 1974b).

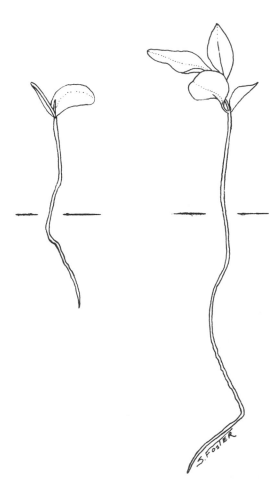

Figure 3. *Peraphyllum ramosissimum*, squawapple: seedling development at 2 and 9 days after germination (Smith 1974b).

LOASACEAE — LOASA FAMILY

Petalonyx Gray

GROWTH HABIT, OCCURRENCE, AND USE The genus *Petalonyx* consists of about 6 species native to southwestern United States and Mexico.

Petalonyx thurberi is a semiwoody species that reaches 0.6 m in height and is found in desert plant communities. The fruit is a capsule.

GERMINATION Germination of freshly collected seeds was less than 10%. After 4 years of storage germination increased to 30% (Kay et al. 1988). It is difficult to collect good seeds of this species.

COMPOSITAE — SUNFLOWER FAMILY

Peucephyllum scottii Gray — Pigmy-cedar

GROWTH HABIT, OCCURRENCE, AND USE *Peucephyllum* is monospecific genus containing a shrub, 0.5 to 2.5 m tall, native to Death Valley and adjacent deserts.

GERMINATION Initial germination of the seeds was under 10% (Kay et al. 1988). After 4 years of storage, germination increased to 40%, if storage was in the laboratory or warehouse. Germination did not increase if the seeds were stored at cold temperatures in sealed containers.

RUTACEAE — RUE FAMILY

Phellodendron amurense Rupr. — Amur corktree

GROWTH HABIT, OCCURRENCE, AND USE Amur corktree is native to China, Manchuria, Korea, and Japan (Read 1974a). This small- to medium-sized tree, 7 to 14 m in height, has long been cultivated in Asia and Europe. It was introduced to the United States about 1865. The thick, corky bark and massive, irregular branches have made this tree of special interest as an ornamental species.

FLOWERING AND FRUITING The flowers are dioecious, small, and yellowish green, in large clusters of terminal panicles that appear in May and June. The fruits are subglobose drupes about 1 cm in diameter (Fig. 1), green to yellowish green, turning black when ripe in September and October. They remain on the terminal branches long after the leaves have fallen. The fruits are very oily and aromatic. Each fruit contains 2 to 3 full-sized seeds and an equal number of aborted seeds. The minimum seed-bearing age is 7 to 13 years. The seeds are brown to black, 5 mm long, 2 mm wide, and about 1 mm thick (Figs. 2, 3); they have a moderately hard, stony coat.

COLLECTION, EXTRACTION, AND STORAGE OF SEED The fruits should be harvested with pruning shears in late September and through October. They can be run through a macerator and the seeds recovered by flotation. Seed yield from fruit is about 5% by weight. Seeds per gram have been reported to range from 59 to 105.

Figure 1. *Phellodendron amurense,* Amur corktree: fruit clusters, ×1 (Read 1974a).

Figure 2. *Phellodendron amurense,* Amur corktree: seed, ×4 (Read 1974a).

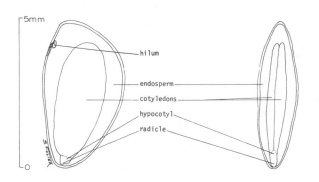

Figure 3. *Phellodendron amurense*, Amur corktree: longitudinal section through 2 planes of a seed, ×8 (Read 1974a).

GERMINATION Seeds will germinate without pretreatment, but 30 days of prechilling enhances germination. Mukai and Yokoyama (1985) compared the germination of prechilled and control seeds at constant and alternating temperatures. Nonprechilled had 38% germination at alternating temperatures and 3% germination at constant. Seeds prechilled for 8 weeks had 90% germination. All tests involved light. Lin et al. (1979) found germination on nonprechilled seeds to be zero when incubated at constant temperatures. The best alternating temperatures were 15/35 and 5/35°C.
NURSERY AND FIELD PRACTICE Untreated seeds can be sown in the fall or prechilled ones in the spring. This species can also be propagated vegetatively.

HYDRANGEACEAE — HYDRANGEA FAMILY

Philadelphus lewisii Pursh. — Mockorange

GROWTH HABIT, OCCURRENCE, AND USE The range of mockorange is from British Columbia and southwestern Alberta southward to western Montana, southern Idaho, and northern California (Stickney 1974b). It is a deciduous shrub from 1 to 3 m tall that bears showy white fragrant flowers. It was introduced into cultivation early in the 19th century.
FLOWERING AND FRUITING Flowering occurs from May through July. The fruit is a capsule, 6 to 10 mm long (Fig. 1), which matures in late summer, and the seeds are dispersed in fall. Seeds (Fig. 2) can be extracted by threshing and air screening. The seeds are very small, running about 12,000 per gram.

GERMINATION Seeds require prechilling for about 8 weeks to germinate. They may be light sensitive.

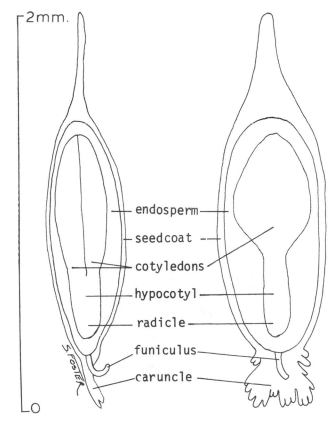

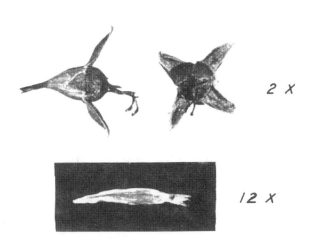

Figure 1. *Philadelphus lewisii*, mockorange: side and top views of capsules, ×2 (above), and seed, ×12 (Stickney 1974b).

Figure 2. *Philadelphus lewisii*, mockorange: longitudinal section through a seed, ×50 (Stickney 1974b).

ROSACEAE — ROSE FAMILY

Physocarpus Maxim. — Ninebark

GROWTH HABIT, OCCURRENCE, AND USE The genus *Physocarpus* consists of about 10 species native to North America and Asia. Among the important species in North America are the following:

Species	Common name	Mature tree height (m)	Location
P. capitatus	Ninebark	1–2.5	Pacific Northwest
P. malvaceus	Mallow ninebark	2–3	Northern Rocky Mountains
P. opulifolius	Common ninebark	1–3	Northeastern and Lake states

The ninebarks are important understory species in coniferous woodlands. They reach their greatest development in full sunlight, but persist in partial shade.

FLOWERING AND FRUITING The flowers are bisexual, white to pink, and borne in flat-topped clusters (Gill and Pogge 1974e). Flowering dates vary from May to June and fruit ripening occurs between late August and early October. The fruit is an inflated follicle resembling a bellows. Ripe follicles have 2 to 5 yellowish shiny seeds about 2 mm long (Figs. 1, 2). Ninebark fruits are fairly persistent on the plants. Some fruits may persist through the winter.

COLLECTION, EXTRACTION, AND STORAGE OF SEED Ripe fruits can be picked from the shrubs or shaken onto drop cloths. Dried fruits can be threshed and the seeds recovered by air screening. There are about 1670 mallow ninebark seeds per gram amd 2300 seeds of common ninebark per gram.

GERMINATION The quality of ninebark seeds is generally low, with many nonviable seeds. They apparently require prechilling for germination, but can be seeded in the fall without prechilling. The plants can be easily rooted from cuttings.

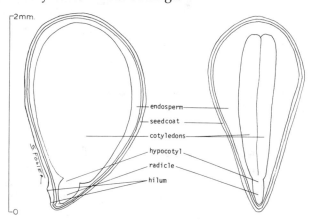

Figure 1. *Physocarpus opulifolius*, common ninebark: capsules, ×2 (above), and seeds, ×20 (below) (Gill and Pogge 1974e).

Figure 2. *Physocarpus malvaceus*, mallow ninebark: longitudinal section through a seed, ×25 (Gill and Pogge 1974e).

PINACEAE — PINE FAMILY

Picea A. Dietr. — Spruce

GROWTH HABIT, OCCURRENCE, AND USE The spruces consist of about 40 species of medium to tall evergreen conifers (Safford 1974). Their crowns are generally conical in outline. The branches are small and whorled, but internodal branches are common. The leaves, borne on peglike projections (pulvini) on the twigs, are needlelike, angled, or flattened in cross section, and persist for several years. On drying, needles fall readily from twigs. Boles are slender and gradually tapering along the entire length, sometimes with buttressed

base. The bark is thin and scaly, sometimes furrowed at the base of old trees. Roots are shallow with long, stringy, tough rootlets. Open-grown trees retain live branches to the ground, and in some species branch tips in contact with moist soil take root and produce full-sized trees.

Evolving in northeastern Asia, members of the genus *Picea* occur throughout the Northern Hemisphere. In the northernmost latitudes spruces grow on all soils and at all elevations up to tree line. In southern latitudes, they are usually restricted to wet, cold, or shallow soils of bogs, or to higher elevations on mountain slopes. Spruces are shade tolerant, often replacing stands of birch, aspen, or other species on disturbed areas.

The strong, light-weight, light-colored, fine-grained, even-textured, long-fibered wood makes most species of spruce potential timber trees. Only restricted ranges or inaccessible locations cause some species to be commercially unimportant. Most species are important watershed protectors because of their occurrence at high elevations and on steep slopes.

Spruces also provide important winter shelter for wildlife in higher latitudes. Seed-eating birds and mammals feed on spruce seeds. Red squirrels clip twigs and terminals and eat reproductive and vegetative buds. Some animals browse on spruce foliage, although it is not a highly preferred source of food for either domestic or wild animals.

Tolerance to extreme exposure to wind and cold temperatures make spruce especially well suited to shelterbelt plantings. White (*P. glauca*) and Sitka (*P. sitchensis*) spruce have been widely used for this purpose. The conical form and dense persistent branches place spruces high on the list for environmental plantings. Several species have color variants or dwarf forms that are valuable as ornamentals. Spruces are generally not tolerant to hot weather or air pollution.

GEOGRAPHICAL RACES The wide range and diverse environments to which spruces are naturally adapted provide a vast array of variation species. Natural hybridization and introgression are common among the spruces where ranges of compatible species overlap.

Seeds of Norway spruce (*P. abies*), the most intensively studied species, show latitudinal and elevational gradients. Seeds from northern latitudes and higher elevations are lighter in weight than those from southern latitudes and lower elevations.

On a worldwide basis there is a lot of research interest in ecotypic variability in the seed and germination characteristics of species of spruce (Table 1).

FLOWERING AND FRUITING Male and female strobili arise in spring (Table 2) in axils of needles of the previous year's shoot on different branches of the same tree. Cone buds may be distinguished as early as July preceding flowering in the following spring; they provide a possible means of predicting seed years. Lindgren et al. (1977) correlated Norway spruce cone crops in southern Sweden with climatic factors during the year of flower primordia initiation. High temperatures in June were the key to increased flowering.

The yellow to bright purple or crimson-colored male strobili are ovoid to cylindrical and pendant. Microsporophylls are spirally arranged on a central axis. Male strobili are usually well distributed over the crown. They dry out and fall soon after pollen shed. Female strobili arise near the end of shoots in the upper part of the tree. They are erect and cylindrical, yellowish green, crimson, or purple-colored, and 0.6 cm to 1.9 cm in diameter. Each megasporophyll bears 2 megaspores at its base. Megasphorophylls are spirally arranged on a central axis. Cones are receptive to pollen when fully open, a period which lasts for only a few days. Fertilization follows pollination within a few days or weeks and cones mature the autumn following flowering (Table 2).

The persistent cone scales may be rounded, pointed, notched, or reflexed at the ends. The pendant mature cones open on ripening to shed seeds during the autumn and winter. Cones remain on the tree for about 1 year and some may fall all year. Black spruce (*P. mariana*) is a special case. Its cones are semiserotinous, remaining on the tree and retaining viable seed for several years. Simpson and Powell (1981) studied factors influencing cone production of black spruce in

Table 1. *Picea*: phenology of flowering and fruiting, characteristics of mature trees, and cleaned seed weight (Safford 1974).

Species	Flowering dates	Fruit ripening dates	Seed dispersal dates	Ripe cone color	Mature tree height (m)	Year first cultivated	Seed-bearing Age (years)	Seed-bearing Interval (years)	Seeds/ gram
P. abies	Apr.–June	Sept.–Nov.	Sept.–Apr.	Brown	61	1548	30–50	3–13	160
P. asperata	–	–	–	Fawn gray	30	1910	–	–	160
P. breweriana	–	Sept.–Oct.	Sept.–Oct.	Brown to black	30	1893	–	–	130
P. engelmannii	June–July	Aug.–Sept.	Sept.–Oct.	Shining brown	30	1862	16–25	2–3	300
P. glauca	May	Aug.	Sept.	Pale brown	30	1700	30	2–6	500
var. *albertiana*	–	–	–	–	–	–	–	–	400
P. glehnii	–	–	–	Shining brown	30	1877	–	–	–
P. jezoensis	–	–	–	Leather brown	46	1861	–	–	400
P. koyamai	–	–	–	Shining brown	18	1915	–	–	230
P. mariana	May–June	Sept.	–[1]	Purple-brown	27	1700	10	–	890
P. omorika	May	Oct.	–	Dark brown	30	1884	–	–	300
P. orientalis	–	–	–	Brown	55	1839	–	–	170
P. polita	–	–	–	Cinnamon	30	1861	–	–	60
P. pungens	Apr.–June	Fall	Fall–Winter	Pale brown	51	1862	20	–	230
P. rubens	Apr.–May	Sept.–Oct.	Oct.–Mar.	Shining brown	30	1750	35–40	–	300
P. sitchensis	May	Aug.–Sept.	Oct.–Spring	Yellow-red	74	1831	20	–	400
P. smithiana	Apr.–May	Oct.–Nov.	Oct.–Mar.	Bright brown	61	1818	20	–	80

[1]*P. mariana* retains cones in a semiserotinous state.

New Brunswick. Trees on south slopes outproduced trees growing on other exposures. Payandeh and Haavisto (1982) determined the production of seeds of black spruce in northern Ontario in relation to tree dominance: Intermediate trees produced more seed than dominant trees and dispersed their seeds at a greater rate.

Spruce seeds are small, oblong to acute at the base, with a single well-developed wing 2 to 4 times the length of the seed (Fig. 1). Testa of mature seeds are brown to black. Cotyledon number varies from 4 to 15 (Fig. 2).

Table 2. *Picea:* selected recent literature.

Author—Date	Subject	Location
Ecotypic Variability		
Aleksandrov 1985	Norway spruce seed weight and germination.	Bulgaria
Fraser 1971	White spruce optimum temperatures for germination.	Canada
Nather and Krissl 1983	Cotyledon number of Norway spruce.	Austria
Popov 1982	Norway spruce and *P. ovata* germination in relation to temperature.	USSR
Stoehr and Farmer 1986	Black spruce germination and seed yield.	Canada
Vanselow 1981	Long-term seed size and provenance of production.	Germany
Pregermination Treatments		
Belcher 1974	*Picea* species, substrate moisture level.	USA
Chandra and Chauhan 1977	*P. smithiana*, gibberellin influence on seed germination.	India
DeCarli et al. 1987	*Picea* species, subcellular biochemistry during germination.	–
Farmer et al. 1984	*P. mariana*, light requirement for seed germination and prechilling.	Canada
Gul'binene and Murkaite 1978	*P. abaies*, ultrasonic treatment of seeds.	USSR
Haavisto and Winston 1974	*P. mariana*, germination at 0.5°C.	Canada
Kiprianov et al. 1982	*P. abies*, sulphate black liquors influence on germination of seeds.	USSR
Popov 1986	*Picea* species, prechilling.	USSR
Pulliainen and Lajunen 1984	*P. abies*, chemical composition of seeds under subarctic conditions.	Canada
Samsonova 1974	*P. abies*, gibberellin and seed soaking.	USSR
Sandberg 1988a	*P. abies*, solvents for gibberellin application to spruce seeds.	–
Sandberg 1988b	*P. abies*, growth regulators and germination.	–

Author—Date	Subject	Location
Pregermination Treatments		
Sandberg and Ernstsen 1987	*P. abies*, indole-3-acetic acid and germination.	–
Shibakusa 1980	*P. glehnii*, growth regulators and seed germination.	Japan
Tanako et al. 1986	*P. engelmannii*, prechilling of seeds.	USA
Taylor and Wareing 1979a	*P. sitchensis*, influence of light and growth regulators on germination of seeds.	USA
Wang 1974	*P. glauca*, light, KNO₃, and prechilling influence on germination.	Canada
Seed Production and Germination		
Arai 1983	*P. jezoensis*, germination.	Japan
Becwar et al. 1987	*P. abies*, tissue culture.	USA
Dolgolikov 1977	*Picea* species, seed orchards.	USSR
Galaaen and Venn 1979	*P. abies*, effect of benomyl on germination.	Norway
Gordon et al. 1979	*P. sitchensis*, prechilling and *Geniculodendron pyriforme* infection of seeds.	UK
Haavisto and Winston 1977	*P. mariana*, germination and paraquat residue.	Canada
Murasov 1976	*P. obovata*, seed production.	USSR
Mittal et al. 1982	*P. glauca*, prechilling and fungal infection of seeds.	USA
Smirnov 1979	*Picea* species, seed maturity and seedling growth.	USSR
State Seed Orchards 1975	*Picea* species, seed orchards.	France
Sutherland 1981	*P. sitchensis*, *Caloscypha* infection of seeds.	Canada
Sutherland et al. 1978	*P. sitchensis*, *Geniculodendron pyriforme* infection of seeds.	Canada
Sutherland and Woods 1978	*P. sitchensis*, *Geniculodendron pyriforme* infection of stored seeds.	Canada
Taylor and Shaw 1983	*P. engelmannii*, allelopathic effects.	Canada
Woodward and Cummins 1987	*P. engelmannii*, germination on ash seedbeds.	USA
Nursery Practices		
Fleming and Lister 1984	*P. mariana*, osmotic priming seed germination.	Canada
Fraser and Adams 1980	*Picea* species, pelleting of seeds.	Canada
Hager 1985	*Picea* species, pregerminating seeds by floating them in water and sucking them down tubes.	Sweden
Kiprianov et al., "Stimulation of Growth," 1982	*P. obovata*, growth of seedlings from seeds treated with pulp liquor.	USSR
Popov 1980	*P. obovata*, nursery sowing.	USSR

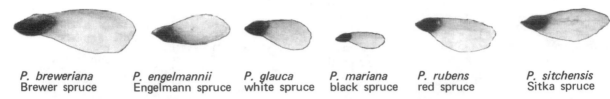

P. breweriana   P. engelmannii   P. glauca   P. mariana   P. rubens   P. sitchensis
Brewer spruce   Engelmann spruce   white spruce   black spruce   red spruce   Sitka spruce

Figure 1. *Picea*: seeds with wings, ×2. (Safford 1974).

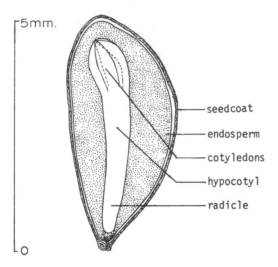

Figure 2. *Picea brewerana,* Brewer spruce: longitudinal section through the embryo of a seed, ×12 (Safford 1974).

COLLECTION, EXTRACTION, AND STORAGE OF SEED Spruce cones must be promptly harvested to avoid loss of seed. They may be collected from standing trees, slash, or animal caches. Seeds are generally mature before the cone has changed to the tan, purple, gray, or brown color characteristic of ripe cones (Table 2). In general, cones exposed to sunlight mature first. A moisture content of 30% or less for Norway spruce, or a soft spongy feel when squeezed for western white spruce (*P. glauca*) indicates a sufficient degree of ripeness for harvesting. Black spruce is an exception. Cones of this species turn purple even though they remain firm and unopened.

Color of testa and firmness of seeds give a good indication of seed ripeness. Dark brown or black testa and seeds that snap when cut with a sharp blade are characteristic of mature seed. Kochkar (1977) distinguished three stages of seed ripeness for Norway spruce in relation to scale position on cones. The Soviet literature contains numerous references to various methods of estimating the seed yield from spruce trees (Paal 1984, Voichal and Barabin 1983).

Spruce seeds are more sensitive to adverse storage conditions than those of pines and may lose viability if not extracted promptly from the cones. Cones should be air dried for a few weeks or in a simple convection kiln for 6 to 24 hours at 38 to 49°C, followed by shaking on a screen or tumbling to extract the seeds. Temperatures above 56°C were harmful to white spruce, but black spruce tolerated brief periods at 84°C and low relative humidity. Winston and Hadden (1981)

reported the influence of early cone collection and artificial ripening on the germination of white spruce seeds. Zasada (1983) studied the influence of cone storage techniques on the subsequent germination of seeds of white spruce collected in Alaska. The following special procedure has been developed for the semiserotinus cones of black spruce:

1. Soak cones in water for 3 to 4 hours, then dry slowly at room temperature for about 20 minutes.
2. Heat cones to 55°C over a 3- to 4-hour period in a kiln and maintain this temperature for 5 to 11 hours.
3. Extract seeds on a screen or tumbler.
4. Repeat these steps 2 or 3 times. The 2nd and 3rd repetitions often yield as much seed as the first.

Separation of wings and chaff from seeds is facilitated by moistening the seeds and stirring them in a round bowl with a soft plastic scraper. On drying, sound seeds are readily separated from wings, chaff, and empty seeds by air screening. An aspirator or flotation in alcohol can be used to remove most of the empty seeds, which are normally a fairly high proportion of the total number extracted from cones.

Species of spruce appear to be fairly similar in longevity characteristics and storage requirements of their seeds. Storage in sealed containers at 1 to 5°C at a moisture content of 4 to 8% proved satisfactory for Norway, western white, and black spruce. Spruce seeds have been stored for periods of 5 to 17 years or longer under similar conditions. The specific moisture content must be maintained during the entire storage period for maximum longevity. A long-term study conducted by Zveidre et al. (1984) with Norway spruce seeds in Latvia indicated that 7% moisture content was the optimum level for storage in sealed jars. Surber et al. (1973) attempted to freeze-dry spruce seeds for long-term storage.

Quality standards for certified spruce seeds are: purity 95%; foreign seeds 0%; inert matter (including wings and wing fragments) 5%. The number of cleaned seeds per gram is given in Table 2.

PREGERMINATION TREATMENT Seeds of most species of spruce germinate promptly without pretreatment, but prechilling has been used for a few species. Seeds of some species may require light for germination, but prechilling will usually overcome the light requirement. Cold water soaking or short periods of prechilling increase germination energy without influenc-

Table 3. *Picea:* standard germination procedures (ASOA 1985, Genebank Handbook 1985).

Species	Substrata	Temperature (°C)	Duration (days)	Additional directions
P. abies	TB	20/30	16	20 and 30°C temperatures may also be used.
P. engelmannii	TP, P	20/30	16	Light required, seeds sensitive to excessive moisture; use KNO$_3$ if dormant.
P. glauca	TB	20/30	21	Light required; some Canadian seed sources require prechilling 14–21 days at 3–5°C.
P. glauca var. albertiana	·TB	20/30	21	Light.
P. glauca var. glauca	TB	20/30	21	Light.
P. glehnii	TB, P	20/30	14	Prechill 21 days at 3–5°C.
P. jezoensis	TB, P	20/30	21	Prechill 21 days at 3–5°C.
P. koyamai	TB	20/30	21	Light.
P. mariana	TB	20/30	16	Light.
P. omorika	TB	20/30	16	Light.
P. orientalis	TB	20/30	21	Light.
P. polita	TB	20	21	–
P. pungens	TB, P	20/30	16	20 and 25°C constant temperatures are also satisfactory.
P. rubens	TB	20/30	28	Light.
P. sitchensis	TB, P	20/30	21	Light; more than 8 hours may be beneficial for some lots; use KNO$_3$ if dormant.

ing germination capacity. Prolonged soaking may lead to mechanical injury of soft seedcoats.

There is a very extensive literature on pregermination treatments for species of *Picea* (Table 1).

GERMINATION TEST Germination standards for *Picea* species are given in Table 3.

The spruce species are such important forest species that there is a considerable general literature concerning aspects of seed production and germination (see Table 1).

NURSERY AND FIELD PRACTICE Fall seeding is not recommended for seeds of Engelmann spruce (*P. engelmannii*), Koyama spruce (*P. koyamai*), blue spruce (*P. pungens*), and tigertail spruce (*P. polita*). Seeds of these species may germinate at low temperatures and seedlings are subject to winter frost damage. Fall sowing has been satisfactory with such species as Sakhalin spruce (*P. glehnii*), Yeddo spruce (*P. jezoensis*), Serbian spruce (*P. omorika*), oriental spruce (*P. orientalis*), and red spruce (*P. rubens*).

Seeds should be covered with soil to a depth of 2 times their diameter. Seedbeds should be firm. Spring-seeded beds are often covered with burlap until the seeds start to emerge; then the burlap is removed. Fall-seeded beds are mulched with 2.5 to 5 cm of straw,pine needles, or sawdust, which is removed in the spring prior to germination. Germination is epigeal (Fig. 3).

During the first season partial shade or overhead sprinklers are provided. At the end of the first season of growth, seedlings should be mulched to their own

height with straw for winter protection. Spruce appear to require a higher level of mineral nutrition than pine seedlings.

Newly germinated seedlings of red spruce in the greenhouse require light for 16 hours daily or they will become dormant. The dormancy requires cold treatment for 4 to 6 weeks at 0°C to induce subsequent growth.

Selected recent literature on nursery practices for species of *Picea* is given in Table 1.

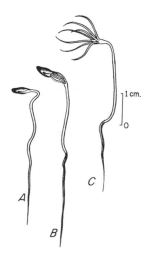

Figure 3. *Picea pungens,* blue spruce: seedling development at 2, 5, and 7 days after germination (Safford 1974).

PINACEAE—PINE FAMILY

Pinus L. — Pine

GROWTH HABIT, OCCURRENCE, AND USE The genus *Pinus*, one of the largest and most important of the coniferous genera, comprises about 95 species and numerous varieties and hybrids (Krugman and Jenkinson 1974). Pines are widely distributed, mostly in the Northern Hemisphere, from sea level to timberline. They range from Alaska to Nicaragua, from Scandinavia to North Africa, and from Siberia to Sumatra. Some species, such as Scotch pine (*P. sylvestris*), are widely distributed from Scotland to Siberia, while others have restricted ranges. Canary Island pine (*P. canariensis*) is found growing naturally only in the Canary Islands, and Torrey pine (*P. torreyana*) numbers only a few thousand individuals in 2 California localities.

There are 42 species of *Pinus* native to the United States. Artificial planting has extended the range of several of these. Many introduced pines have been planted in the United States and a few species have become naturalized.

Many pines have been successfully planted outside their native range on a worldwide basis. Monterey pine (*P. radiata*), for example, has become a commercially successful species in South Africa, New Zealand, Australia, and South America.

The pines are evergreen trees of various heights, often tall, but occasionally shrubby. Some species such as sugar pine (*P. lambertiana*), western and eastern white pine (*P. monticola* and *P. strobus*), and ponderosa pine (*P. ponderosa*), grow to more than 62 m tall. The Mexican pinyon (*P. cembroides*) and Japanese stone pine (*P. pumila*) may not exceed 10 m in height.

Pines provide some of the most valuable timber and are also used to protect watersheds, to provide habitats for wildlife, and to construct shelterbelts. Pine nuts are important food sources in some cultures and pines are widely used in environmental plantings.

GEOGRAPHIC RACES AND HYBRIDS The extensive ranges of many pine species make it important to properly identify seed sources. Krugman and Jenkinson (1974) listed extensive information on several species.

FLOWERING AND FRUITING In certain species, reproductive structures are first formed when the trees are only 5 to 10 years old (e.g., knobcone pine, *P. attenuata*); in others they do not form until the trees are much older (e.g., sugar pine) (Table 1). Pines are monoecious, with male and female strobili borne separately on the same tree. Male strobili predominate on the basal part of new shoots, mostly on older lateral branches in the lower crown. Female strobili are found most often in the upper crown, primarily at the apical end of the main branches in the position of subterminal or lateral buds. Frequent exceptions will be found to this general scheme. Several species, including jack pine (*P. banksiana*), are multinodal in the bud, and female strobili are found occasionally at a secondary whorl position. Knobcone, Monterey, and Virginia pine (*P.*

virginiana) frequently produce female strobili in all parts of the crown. In temperate climates the earliest stages of male and female strobili can be detected in the developing buds during the summer or fall; the male develops 1 to several weeks before the female.

Male and female strobili of the southern and tropical pines emerge from buds in late winter (e.g., slash pine, *P. elliottii*). Strobili of the other pines emerge from the bud in early spring, or late spring and early summer (Table 2). The male strobili are arranged in indistinct spirals in clusters 1.25 to 5 cm long. Before ripening they can be green or yellowish to reddish purple, but are light brown to brown at the time of pollen shed. In most species they fall soon after ripening. Female strobili emerge from the winter buds shortly after the male strobili and are green to red to purple. At the time of pollination they are nearly erect, and 0.9 to 3.75 cm long and sometimes longer. After pollination, scales of the female strobili close, and the strobili begin to develop slowly. At the end of the first growing season they are about one-eighth to one-fifth the length of mature cones. Where temperatures are favorable, development continues through the winter, (e.g., slash pine in Florida and ponderosa pine at low elevations in the West). Fertilization takes place in the spring and summer, some 13 months after pollination, and the cones began to grow rapidly. The growth of new shoots causes the cones to remain in a lateral position. As the cones mature they gradually turn from green, purple, or violet-purple to yellow, light brown, reddish brown or dark brown (Table 1).

Cones and seeds of most species mature rapidly during late summer and fall of the second year (Table 2). Cones of a few species mature during late winter of the second year or early spring of the third (e.g., knobcone pine). Seeds of knobcone pine are mature during the fall, about 16 to 18 months after pollination, but the cones are not fully developed until late winter. Chihuahua pine (*P. leiophylla* var. *chihuahuana*) requires 3 years for seeds and cones to ripen.

The interval between large cone crops is variable and depends on the species and environmental factors (Table 1). Some species consistently produce a large crop every year, while others show a cyclic pattern of 2 to 10 years between large cone crops.

Mature cones vary widely in size and weight (Fig. 1). Those of Swiss mountain pine (*P. mugo*) are 2.5 to 5 cm long and weigh 1.7 grams, while those of sugar pine may be 30 to 60 cm long and weigh 450 to 900 grams. The cones of digger (*P. sabiniana*) and Coulter pine (*P. coulteri*) often weigh more than a kilogram. The mature cone consists of overlapping scales, each of which bears 2 seeds at its base, on the upper surface. The cones of most species open on the tree shortly after ripening, and the seeds are rapidly dispersed (Table 2).

Table 1. *Pinus:* characteristics of mature trees (Krugman and Jenkinson 1974).

Species	Mature tree height (m)	Seed-bearing Age (years)	Interval[1] (years)	Ripe cone color
P. albicaulis	33	20–30	3–5	Purple-brown
P. aristata	15	20	102	Chocolate brown
P. armandii	37	20	–	Yellowish brown
P. attenuata	15	5–8	1	Tawny brown
P. balfouriana	19	20	5–6	Dark-red brown
P. banksiana	31	3–15	3–4	Tawny yellow
P. brutia	31	7–10	1	Yellow-red brown
P. canariensis	31	15–20	3–4	Nut brown
P. caribaea	31	–	–	Tan-light brown
P. cembra	23	25–30	6–10	Purple-brown
P. cembroides	8	–	5–8	Yellow-red brown
P. clausa	25	5	1–2	Dark yellow brown
P. contorta				
var. *contorta*	12	4–8	1	Yellow brown
var. *latifolia*	46	5–10	1	Light brown
var. *murrayana*	31	4–8	1	Clay brown
P. coulteri	23	8–20	3–6	Yellow brown
P. densiflora	37	20–30	2	Yellow brown
P. echinata	31	5–20	3–10	Green-dull brown
P. edulis	12	25–75	2–5	Light brown
P. elliottii				
var. *densa*	26	8–12	1–5	Brown
var. *elliottii*	31	7–10	3	Yellow brown
P. engelmannii	22	28–30	3–4	Light brown
P. flexilis	25	20–40	2–4	Light brown
P. gerardiana	25	–	–	Brown
P. glabra	28	10	–	Green
P. halepensis	25	15–20	1	Yellow-red brown
P. heldreichii	31	–	–	Yellow-dull brown
P. insularis	46	5–10	1	Bright-dark brown
P. jeffreyi	55	8	2–4	Purple-light brown
P. koraiensis	46	15–40	3–5	Yellow brown
P. lambertiana	69	40–80	3–5	Greenish brown
P. leiophylla	25	28–30	3–4	Light brown
P. merkusii	31	10–20	1–2	Light brown
P. monophylla	15	20–25	1–2	Russet brown
P. monticola	62	7–20	3–7	Yellow-dark brown
P. mugo	12	10	1	Yellow-dark brown
P. muricata	28	5–6	2–3	Chestnut brown
P. nigra	51	15–40	2–5	Yellow brown
P. palustris	37	20	5–7	Green-brown
P. parviflora	31	–	4–5	Red-brown
P. patula	34	12–15	–	Yellow-nut brown
P. peuce	31	12–30	3–4	Yellow brown
P. pinaster	37	10–15	3–5	Light brown
P. pinea	25	–	–	Nut brown
P. ponderosa				
var. *arizonica*	28	15–20	2–3	Green-buff brown
var. *ponderosa*	71	16–20	2–5	Green-russet brown
var. *scopulorum*	35	6–20	2–5	Purple-brown
P. pumila	3	–	–	Red-brown
P. pungens	19	5	–	Light brown
P. quadrifolia	9	–	1–5	Yellow-red brown
P. radiata	46	5–10	1	Nut brown
P. resinosa	46	20–25	3–7	Purple-nut brown
P. rigida	31	3–4	4–9	Yellow brown
P. roxburghii	55	15–40	2–4	Light brown
P. sabiniana	25	10–25	2–4	Red-chestnut brown
P. serotina	25	4–10	1	Yellow brown
P. siberica	40	25–35	3–8	Violet-brown
P. strobiformis	39	15	3–4	Green-brown
P. strobus	67	5–10	3–10	Green
P. sylvestris	40	5–15	4–6	Green
P. taeda	34	5–10	3–13	Green
P. thunbergiana	34	6–40	–	Deep purple
P. torreyana	19	12–18	1	Green-violet
P. virginiana	31	5	1	Green
P. wallichiana	46	15–20	1–2	Green

[1]Between large crops.

Table 2. *Pinus*: phenology of flowering and fruiting (Krugman and Jenkinson 1974).

Species	Flowering dates	Cone ripening dates	Seed dispersal dates
P. albicaulis	July	Aug.–Sept.	Not shed
P. aristata	July–Aug.	Sept.–Oct.	Sept.–Oct.
P. armandii	April–May	Aug.	Aug.–Sept.
P. attenuata	April	Jan.–Feb.	Closed cone
P. balfouriana	July–Aug.	Sept.–Oct.	Sept.–Oct.
P. banksiana	May–June	Sept.	Sept.
P. brutia	Mar.–May	Jan.–Mar.	Closed cone
P. canariensis	Apr.–May	Sept.	Sept.–Oct.
P. caribaea	Jan.–Feb.	July–Aug.	Sept.
P. cembra	May	Aug.–Oct.	Not shed
P. cembroides	May–June	Nov.–Dec.	Nov.–Dec.
P. clausa	Sept.–Dec.	Sept.	Sept.
P. contorta			
var. *contorta*	May–June	Sept.–Oct.	Fall
var. *latifolia*	June–July	Aug.–Sept.	Sept.–Oct.
var. *murrayana*	May–June	Sept.–Oct.	Sept.–Oct.
P. coulteri	May–June	Aug.–Sept.	Oct.
P. densiflora	May–June	Aug.–Sept.	Sept.–Oct.
P. echinata	Mar.–Apr.	Oct.–Nov.	Oct.–Nov.
P. edulis	June	Sept.	Sept.–Oct.
P. elliottii			
var. *densa*	Jan.–Apr.	Aug.–Sept.	Sept.–Nov.
var. *elliottii*	Jan.–Feb.	Sept.–Oct.	Oct.
P. engelmannii	May	Nov.–Dec.	Nov.–Feb.
P. flexilis	June–July	Aug.–Sept.	Sept.–Oct.
P. gerardiana	May–June	Sept.–Oct.	Nov.
P. glabra	Feb.–Mar.	Oct.	Oct.–Nov.
P. halepensis	May–June	Sept.	Fall
P. heldreichii	May–July	Aug.–Sept.	Sept.–Oct.
P. insularis	Jan.–Mar.	Oct.–Jan.	Feb.–Mar.
P. jeffreyi	June–July	Aug.–Sept.	Sept.–Oct.
P. koraiensis	May–June	Sept.	Oct.
P. lambertiana	June–July	Aug.–Sept.	Aug.–Oct.
P. leiophylla			
var. *chihuahuana*	May–June	Nov.	Dec.–Jan.
P. merkusii	Jan.	Apr.–June	May–July
P. monophylla	May	Aug.	Sept.–Oct.
P. monticola	June–July	Aug.	Aug.–Sept.
P. mugo	May–June	Oct.	Nov.–Dec.
P. muricata	Apr.–June	Sept.	Mid-winter
P. nigra	May–June	Sept.–Nov.	Oct.–Nov.
P. palustris	Feb.–Mar.	Sept.–Oct.	Oct.–Nov.
P. parviflora	May–June	Sept.	Nov.
P. patula	Feb.–Apr.	Dec.	Mid-winter
P. peuce	May	Fall	Fall
P. pinaster	Apr.	Nov.–Dec.	Dec.–Jan.
P. pinea	May–June	Late Summer	Late Summer
P. ponderosa			
var. *arizonica*	May	Sept.–Oct.	Oct.
var. *ponderosa*	Apr.–June	Aug.–Sept.	Aug.–Sept.
var. *scopulorum*	May–June	Aug.–Sept.	Sept.–Jan.
P. pumila	July	–	Fall
P. pungens	Mar.–Apr.	Fall	Fall
P. quadrifolia	June	Sept.	Sept.–Oct.
P. radiata	Jan.–Feb.	Nov.	Mid-winter
P. resinosa	Apr.–June	Aug.–Oct.	Oct.–Nov.
P. rigida	May	Sept.	Fall
P. roxburghii	Feb.–Apr.	Winter	Apr.–May
P. sabiniana	Mar.–Apr.	Oct.	Oct.
P. serotina	Mar.–Apr.	Sept.	Spring
P. siberica	May	Aug.–Sept.	Not shed
P. strobiformis	June	Sept.	Sept.–Oct.
P. strobus	May–June	Aug.–Sept.	Aug.–Sept.
P. sylvestris	May–June	Sept.–Oct.	Dec.–Mar.
P. taeda	Feb.–Apr.	Sept.–Oct.	Oct.–Dec.
P. thunbergiana	Apr.–May	Oct.–Nov.	Nov.–Dec.
P. torreyana	Feb.–Mar.	June–July	Sept.
P. virginiana	Mar.–May	Sept.–Nov.	Oct.–Nov.
P. wallichiana	Apr.–June	Aug.–Oct.	Sept.–Nov.

Drying causes the scales to separate, owing to differential contraction of 2 tissue systems: woody strands of short thick-walled, tracheidlike cells extending from the cone axis to the tip of the cone scale, and thick-walled sclerenchyma cells in the abaxial zone of the scale. In a few species with massive cones, the scales separate slowly, and seeds are shed over a period of several months (e.g., Coulter pine).

In some species most of or all the mature cones remain closed for several years or open only at irregular intervals (e.g., knobcone and jack pine). In addition to their serotinous cones, species such as jack pine, sand pine (*P. clausa*), pitch pine (*P. rigida*), and lodgepole pine (*P. contorta*) have forms whose cones open promptly at maturity.

The closed cone habit is the result of 3 factors: extremely strong adhesion between adjacent, overlapping cone scales beyond the winged seeds; cone structures; and the nature of the tissue system in the cone as described above. The melting point of the resin seal for Rocky mountain lodgepole pine is between 46 and 50°C. Heat, especially that from fire, melts the resin and permits the cones to open. Still other species shed partially opened cones, and seeds are dispersed only when cones have disintegrated on the ground (e.g., whitebark pine, *P. albicaulis*).

Commonly, cones that open on the tree are shed within a few months to a year after the seeds are dispersed. In some species, however, opened cones may remain on the trees for up to 5 years or indefinitely (e.g., knobcone, lodgepole, and pitch pine).

Mature seeds vary widely in size, shape, and color (Fig. 2). They range in length from 2 to 3 mm for jack pine to more than 1.9 cm for digger pine. They are ellipsoid for Monterey pine, pear-shaped for Japanese stone pine, cylindrical for chilgoza (*P. gerardiana*), more or less triangular for Table mountain pine (*P. pungens*), ovoid for Balkan pine (*P. peuce*), and convex on the inner and flattened on the outer side for Italian stone pine (*P. pinea*). The seedcoat, which may be reddish, purplish, grayish, brown, or black, and is often mottled, can be rather thin and papery to hard, and even stony.

In most species a membranous wing is attached to the seed, but in some the wings are absent or rudimentary (e.g., whitebark pine). In others the wings remain attached to the cone scales when the seeds are shed (e.g., Mexican pinyon). The wings are easily detached from the seed in most hard pine species except for Italian stone pine, Chir pine (*P. roxburghii*), and Canary Island pine. They are firmly attached to the seeds of most soft pine except for bristlecone pine (*P. aristata*) and certain sources of foxtail pine (*P. balfouriana*).

The mature seeds consist of a seedcoat that encloses an embryo embedded in storage tissue, the endosperm (female gametophyte). Attached at the micropylar end of the whitish endosperm is a brown, papery cap, the remnant of the nucleus. The endosperm and papery cap are covered by a thin, brown, membranous material, the remnant of the inner layer of the ovules integument (Fig. 3). Selected recent literature concerning flowering and seed development of *Pinus* is given in Table 3.

CONE COLLECTION Cones should be collected from trees superior in growth form characteristics. Larger cones generally contain more seeds, but normally all cones are collected except those with obvious disease and insect damage. Widely spaced, dominant trees with full crowns produce the most seeds per cone, provided adequate pollen from other trees is available. When trees are isolated and pollen is limiting, seed yield may be very low. In dense young stands most species usually produce little seed. Those species which form fire thickets are exceptions (e.g., knobcone and jack pine).

Ripe cones can be picked from standing trees, freshly fallen trees, or from animal caches. To avoid immature seeds, collections from animal caches should be delayed until late fall. For most species, cone collection begins as soon as the cones are ripe and starting to crack because seeds are shed promptly from open cones. For closed cone species it is possible and often desirable, to delay cone collection. Although the seeds of closed cone species may be mature in the fall, the cones are very hard to open then.

To avoid collecting immature seeds it is advisable to first check ripeness of seeds in a small sample of cones from individual trees. A mature seed has a firm white or cream-colored endosperm and a yellow to white embryo that nearly fills the endosperm cavity. With experience this visual test is useful, and with some species it is essential.

Ripeness for some species can be estimated by changes in color (Table 1). For example, ponderosa pine (*P. ponderosa*) cones are mature when color changes from green to yellow-green to brownish green, yellow-brown, or russet-brown. Red pine (*P. resinosa*) cones turn from green to purplish, with reddish brown on the scale tips. Eastern white pine (*P. strobus*) cones turn from green to yellow-green with brown on the scale tips, or light brown as maturity progresses. Longleaf pine (*P. palustris*) cones are green when ripe and may shed seeds before turning brown.

For species where color change may not be useful for determining maturity, flotation tests of cone specific gravity have been used. The tests are based on fresh weight of cones. For several species seed maturity has been related to cone specific gravity (Table 4). The easiest way to determine if cones have reached a specific gravity is to see if the cones will sink or float in a liquid of a given specific gravity. Ponderosa pine seeds are ripe when the cones will float in kerosene. Eastern white pine seeds are ripe when the cones will float in linseed oil, and spruce pine (*P. glabra*) seeds are ripe when the cones will float in SAE 20 motor oil. This test will only give accurate readings for fresh cones.

Collection from felled trees should only take place after seeds are mature, to avoid taking immature seeds. In some species (e.g., loblolly pine, *P. taeda*, and shortleaf pine, *P. echinata*) nearly all the cones in a given crown may be ripe at the same time, but in others cone ripeness can be variable in the same tree. In some species, immature seeds can be successfully ripened in the cone after picking (e.g., slash pine). Sugar pine seeds can be ripened in cold moist storage and those of Virginia pine seeds can be ripened in prolonged cold dry storage in closed containers. It is always better to collect fully mature seeds than to try to deal with afterripening.

Cones are often handpicked from the trees, either from ladders or by climbing. A number of tools have been developed to aid in cutting and reaching cones during harvesting. Mechanical cone harvesting is used with several of the important southern pine species. Selected recent literature on seed production of *Pinus* species is given in Table 5.

CONE PROCESSING AND SEED EXTRACTION Cones should be dried immediately after collection to avoid mold development and excessive internal heating, which lead to rapid seed deterioration. Drying can be accomplished in 2 to 60 days by immediately spreading the fresh cones in thin layers on a dry surface in the sun or on trays in a well-ventilated building, or by placing them in sacks hung from overhead racks protected from rain. The cones should be dried slowly to prevent hardening the cones, known as "hard casing." After initial drying the cones can be stored in well-ventilated bags or on trays. For many species, ripe cones open satisfactorily under these conditions, but cones of some species require additional heat in either a cone drying kiln or a heated shed. Properly air-dried cones of a few species may open satisfactory in the kiln after a few hours, but those of others may require several days. Cones of most species can be opened in a kiln at temperatures not exceeding 55°C and a humidity of about 20%; those of a few species (e.g., jack pine) require higher temperatures (Table 6).

Cones stored long enough in containers to have dried without opening or cones dried under cool conditions may not open properly during kiln drying. In such cases, the cones must be soaked in water for 12 to 24 hours, and then kiln dried.

Serotinous cones have been opened by dipping them in boiling water for 10 to 120 seconds. Immersion times up to 10 minutes, however, have been needed for some cone lots. This procedure, by melting the resin between the cone scales and wetting the woody cone, produces maximum scale reflexing.

After the cones are opened they are shaken to remove the seeds. Seeds normally are extracted by placing the cones in a large mechanical tumbler or shaker, or in a small manual shaker for small lots. Seeds are then dewinged by machines of various types, by being flailed in a sack, or by rubbing. Dewinging of a few species is simplified by first wetting the seeds, then letting them dry; the wings are loosened by this method and can be removed with an air screen. Care must be used with mechanical dewingers to not injure the seeds. Seeds of bristlecone pine, longleaf pine, and Scotch pine are especially subject to mechanical injury.

P. albicaulis
whitebark pine

P. aristata
bristlecone pine

P. attenuata
knobcone pine

P. cembroides
Mexican pinyon

P. banksiana
jack pine

P. clausa
sand pine

P. engelmanii
Apache pine

P. flexilis
limber pine

P. leiophylla var. *chihuahuana*
chihuahua pine

Figure 1. *Pinus:* mature cones collected before seed dispersal, ×0.5 (Krugman and Jenkinson 1974).

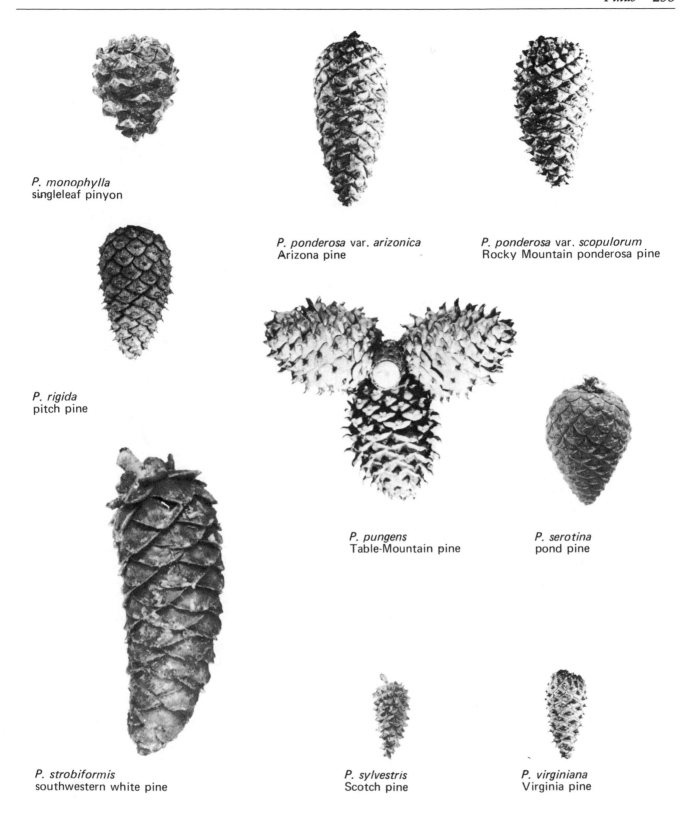

P. monophylla
singleleaf pinyon

P. ponderosa var. arizonica
Arizona pine

P. ponderosa var. scopulorum
Rocky Mountain ponderosa pine

P. rigida
pitch pine

P. pungens
Table-Mountain pine

P. serotina
pond pine

P. strobiformis
southwestern white pine

P. sylvestris
Scotch pine

P. virginiana
Virginia pine

Figure 1. Continued.

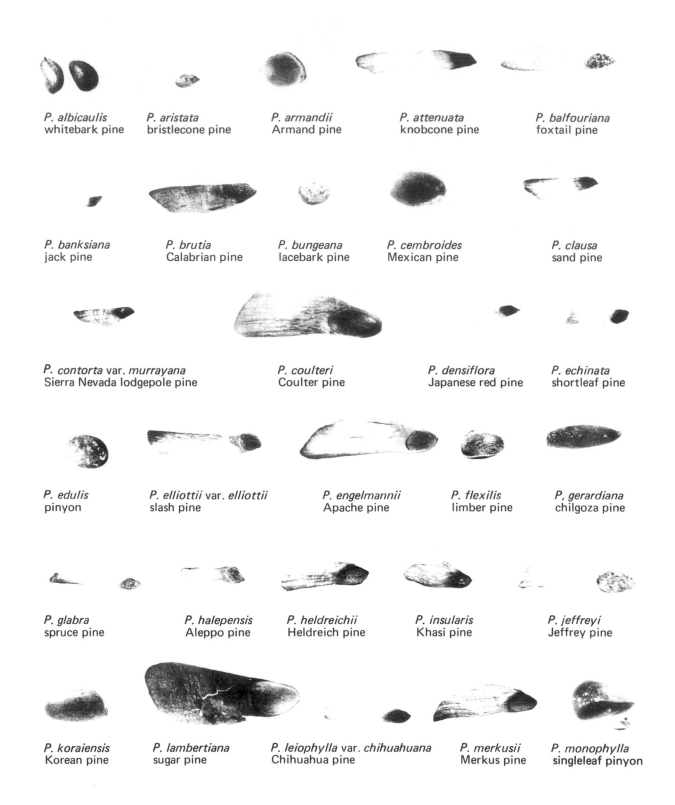

Figure 2. *Pinus:* seeds as shed naturally from their cones, ×1; some are wingless when shed (Krugman and Jenkinson 1974).

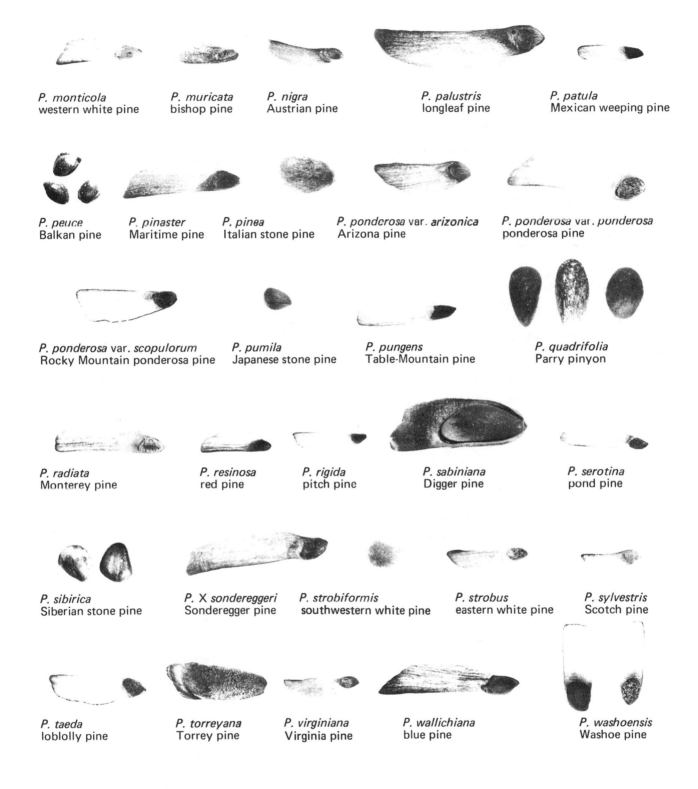

Figure 2. Continued.

Table 3. *Pinus:* selected recent literature on flowering and seed development.

Author—Date	Subject	Location
Bramlett n.d.	*P. echinata,* seed and aborted ovules.	USA
Cecich and Bauer 1987	*P. banksiana,* accelerated seed development.	USA
Cecich and Rudolph 1982	*P. banksiana,* timing of seed maturity.	USA
Dreimanis 1975	*P. sylvestris,* fall of female strobili.	USSR
Dreimanis 1978	*P. sylvestris,* heritability of cone characteristics.	USSR
Fernandes 1977	*Pinus* species, induction of flowering.	Brazil
Howcroft 1974	*P. merkusii,* ovulate strobili.	New Guinea
Jo et al. 1983	*P. densiflora,* self pollination.	China
Laura 1978	*P. sylvestris* flowering.	USSR
Matheson and Willcocks 1976	*P. radiata,* seed yield following pollarding.	New Zealand
McLemore 1977	*P. elliottii, P. echinata,* and *P. palustris,* strobili losses.	USA
McLemore 1976	*P. echinata,* seeds from grafted material.	USA
Min ct al. 1974	*P. koraiensis,* stimulating seed production.	Korea
Rim and Shidei 1974	*P. densiflora,* and *P. thunbergii,* seed production.	Japan
Ronis and Kodola 1977	*P. sylvestris,* stimulation of flowering.	USSR
Sagwal 1984	*P. roxburghii,* seed production from individual cones.	India
Sato and Yamamoto 1977	*P. densiflora,* cone scales.	Japan
Truus 1984	*P. sylvestris,* seed production stimulation with GA.	USSR

Table 4. *Pinus:* specific gravity of ripe cones and liquids used for testing ripeness by flotation (Krugman and Jenkinson 1974).

Species	Specific gravity of ripe cones	Flotation test liquid
P. aristata	0.59–0.80	kerosene
P. contortata		
var. *latifolia*	0.43–0.89	–
P. densiflora	1.10	–
P. echinata	0.88	SAE 20 motor oil
P. edulis	0.80–0.86	kerosene
P. elliottii		
var. *densa*	<0.89	SAE 20 motor oil
var. *elliottii*	<0.90	SAE 20 motor oil
P. glabra	0.88	SAE 20 motor oil
P. jeffreyi	0.81–0.86	–
P. lambertiana	0.70–0.80	–
P. palustris	0.80–0.89	
P. ponderosa		
var. *arizonica*	0.88–0.97	–
var. *ponderosa*	0.80–0.86	kerosene
var. *scopulorum*	<0.85	kerosene
P. radiata	<1	water
P. resinosa	0.80–0.94	kerosene
P. serotina	0.88	–
P. strobiformis	0.85–0.95	95% ethanol
P. strobus	0.92–0.97	linseed oil
P. sylvestris	0.88–1.00	–
P. taeda	0.88	SAE 20 motor oil
P. virginiana	<1.00	–

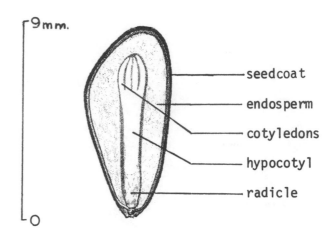

Figure 3. *Pinus resinosa,* red pine: seedling development at 1, 7, and 30 days after germination (Krugman and Jenkinson 1974).

Table 5. *Pinus:* selected recent literature on seed production.

Author—Date	Subject	Location
Alfjorden and Remrod 1975	*Pinus,* seed yield forecasting.	Sweden
Barnett 1976	*P. elliotti, P. taeda,* X-ray and biochemical analysis.	USA
Bhumibamon and Atipanumpai 1980	*P. merkusii,* X-ray analysis.	Thailand
Bhumibamon et al. 1980	*P. merkusii*	Thailand
Bobrinev 1978	*Pinus* species, harvesting time.	USSR
Bogdanov 1978	*P. strobus,* seed sources.	Bulgaria
Bramlett et al. 1977	Cone analysis of southern pines.	USA
Edwards and McConnell 1982	Netting to collect seeds.	USA
Eldridge 1982	*P. radiata,* seed orchards.	Australia
Heidmann 1986	*P. ponderosa,* seed production.	Australia
Hellum et al. 1983	*P. contorta,* seed and cone maturity.	Canada
Hodgson 1977	Southeastern U.S. pines and seed flotation techniques.	South Africa
Hoff 1981	*P. monticola,* seed production.	USA
Jeffers 1985	*P. banksiana,* seed quality.	USA
Junttlia 1974	*P. mugo,* seed morphology.	Sweden
Kamra 1982a	Closed cone pines.	Sweden
Lyuhich 1985	*Pinus,* seed quality and improvement.	Sweden
Minina and Iroshnikov 1977	*P. sylvestris* and *P. sibirica,* seed orchards.	USSR
Pederick n.d.	*P. radiata,* seed orchards.	Australia
Pishchik 1978	*P. sylvestris,* seed stands.	USSR
Rudolph and Cecich 1979	*P. banksiana,* variation in seed yield.	USA
Standnitskii 1985	*P. sylvestris,* seed orchards.	USSR
Tyystjarvi 1978	*P. sylvestris,* seed orchards.	Finland
Wilcox and Firth 1980	*P. radiata,* artificial cone ripening.	New Zealand
Wynens and Brooks 1979	*P. taeda,* seed collection.	USA
Ying et al. 1985	*P. contorta,* seed yield.	USA

Table 6. *Pinus:* cone processing schedule, viable periods for seeds in cold storage, seed weight and yield data (Krugman and Jenkinson 1974).

Species	Time in boiling water (seconds)	Air-drying time (days)	Kiln drying Time (hours)	Kiln drying Temperature (°C)	Storage period (years)	Cone wt./ 35 liters (kg)	Seeds/ 45.5 kg cones (kg)	Seeds/ gram
P. albicaulis	0	15–30	0	–	8	–	–	5.7
P. aristata	0	2–8	0	–	9	59	1.8	39.9
P. armandii	0	15	0	–	–	–	–	35.0
P. attenuata	15–30	1–3	48	48	16	–	–	55.9
P. balfouriana	0	2–8	0	–	16	88	–	37.2
P. banksiana	0	3–10	2–4	72	10	–	0.5	288.5
P. brutia	0	3–20	0	–	3	–	–	20.0
P. canariensis	0	2–10	0	–	18	–	–	9.3
P. caribaea	–	–	–	–	3	–	–	69.3
P. cembra	–	–	–	–	1+	–	–	4.4
P. cembroides	0	2–8	0	–	–	–	–	2.4
P. clausa	10–30	1	2–4	62	5	–	1.6	165.2
P. contorta								
var. contorta	0	2–20	0–96	48	17	–	–	297.3
var. latifolia	30–60	2–30	0–8	60	7+	94	0.4	207.0
var. murrayana	0	2–30	0	–	17	99	0.3	257.7
P. coulteri	0–120	3–15	0–72	48	5	99	1.0	3.1
P. densiflora	0–30	3–4	0	–	2–5	–	0.9	114.0
P. echinata	0	–	48	40	35	77	1.1	102.8
P. edulis	0	2	0	–	–	37	1.5	4.2
P. elliottii								
var. densa	0	4	8–10	48	–	–	0.3	29.7
var. elliottii	0	42	8–10	48	35	–	0.3	22.0
P. engelmannii	0	–	60	44	–	81	0.2	29.7
P. flexilis	0	15–30	–	–	5	–	–	10.8
P. gerardiana	0	15	–	–	–	–	–	2.4
P. glabra	0	–	48	48	1+	–	0.2	101.3
P. halepensis	0	–	10	10	10	–	–	61.7
P. heldreichii	0	3–10	0	–	–	–	–	46.3
P. insularis	0	5–20	0	–	–	–	–	59.5
P. jeffreyi	0	5–7	24	48	18	61	0.6	8.2
P. koraiensis	0	–	–	–	–	–	–	1.8
P. lambertiana	0	5–7	24	48	21	50	0.8	4.6
P. leiophylla	0	–	–	–	–	–	0.4	88.1
P. merkusii	0	5–7	0	–	2+	–	–	40.0
P. monophylla	0	2–3	0	–	–	–	1.5	2.4
P. monticola	0	5–7	14	43	20	61	0.3	59.5
P. mugo	0	–	48	48	5	86	0.4	152.0
P. muricata	0	–	48	48	–	99	0.1	103.1
P. nigra	0	–	24	46	10+	79	0.4	57.3
P. palustris	0	–	48	38	5–10	77	0.4	10.8
P. parviflora	0	5–15	0	–	–	–	–	8.6
P. patula	15–30	1–2	48	46	21	–	–	115.8
P. peuce	–	–	–	–	–	–	–	24.2
P. pinaster	0	4–10	–	46	11	–	2.0	22.0
P. pinea	–	–	–	–	18	–	–	1.3
P. ponderosa								
var. arizonica	0	–	60	43	–	79	0.7	25.1
var. ponderosa	0	4–12	3	48	18	89	2.0	26.4
var. scopulorum	0	4–12	2	74	15+	86	0.7	28.9
P. pumila	–	–	–	48	–	–	–	23.8
P. pungens	0	30	72	–	9	–	0.2	75.3
P. quadrifolia	0	2–8	0	–	–	–	–	2.1
P. radiata	60–120	3–7	48–72	48	–	83	0.1	29.3
P. resinosa	0	–	9	–	30	77	0.3	114.5
P. rigida	–	–	–	55	11	79	1.1	135.9
P. roxburghii	0	–	–	–	4+	–	–	12.3
P. sabiniana	0	–	48	48	5	–	–	1.3
P. serotina	–	–	48	40	–	–	–	118.9
P. siberica	0	–	–	–	2+	–	–	4.0
P. strobiformis	0	14	0	–	–	57	3.6	5.9
P. strobus	0	–	4–12	54	10	33	0.5	58.3
P. sylvestris	0	–	0–16	48	15	42	0.2	165.2
P. taeda	0	–	48	40	9+	77	0.4	40.0
P. thunbergiana	0–30	5–30	0	–	11	90	0.4	74.9
P. torreyana	0	5–20	0	–	6	–	–	1.1
P. virginiana	0	–	2	78	5+	–	0.3	122.0
P. wallichiana	–	–	–	–	–	–	–	20.0

After completing the dewinging and cleaning process, empty seeds of many species can be separated from sound seeds by flotation in a liquid having a suitable specific gravity (Table 4).

Viability may be reduced after seeds have been immersed in an organic liquid such as ethanol, pentane, or petroleum ether. The reduction, however, can be minimized by using a short immersion time and by evaporating all traces of the liquid from the seeds before they are placed in storage.

When water is used for floating off empty seeds, the remaining sound seeds should be dried to a moisture content between 5 and 10% before being placed in storage.

Numbers of cleaned seeds per gram are given for pine species in Table 6. Selected recent literature on cone and seed technology of *Pinus* is given in Table 7.

SEED STORAGE For most pines, high seed viability can be maintained for long periods with proper storage methods. Red pine seeds stored 30 years still produced vigorous seedlings in the nursery, as did seeds of shortleaf and slash pine stored 35 years. Seeds of many species are routinely stored for periods of 5 to 10 years. Storage temperature and moisture content are the two most important factors affecting the success of seed storage. As a general rule, seeds should be dried to a moisture content between 5 and 10%. Cold temperatures are preferred for long-term storage, with 2 to 5°C used most often. The viability periods for seeds stored under these conditions are listed in Table 6. Seeds of a few species such as Khasi pine (*P. insularis*) and blue pine (*P. wallichiana*) have remained viable for several years at room temperature. Some seed lots deteriorate rapidly following removal from cold storage if they are held at room temperature before sowing. Seeds should be removed from storage a week before prechilling or sowing. Selected recent literature concerning seed storage of *Pinus* species is included in Table 8.

PREGERMINATION TREATMENT Most pines of temperate climates shed their seeds in the fall and the seeds germinate promptly during the following spring. For some species, such as Swiss stone pine (*P. cembra*) and Balkan pine (*P. peuce*), germination may take place the second or even third year following dispersal. Pine seeds display highly variable germination behavior when sown following extraction or storage. The type and degree of dormancy varies among and within species, among geographic races, and in lots from the same source. Seed dormancy may result from prolonged extraction at high temperatures, and may increase with prolonged storage. Seeds of many species ordinarily germinate satisfactorily without pretreatment, but germination is greatly improved and hastened by first subjecting the seeds to prechilling, especially if they have been stored.

Prechilling is accomplished by first soaking the seeds in water for 1 or 2 days and then placing them in a moist medium or in a plastic bag and holding them at a temperature between 3 to 5°C for a specific period of time (Table 9).

Seeds of some species may exhibit extreme dormancy that requires more than 60 days of prechilling. The dormancy may be due to physical or physiological factors. A pretreatment may be needed to overcome a physiological block in the embryo (e.g.,

Table 7. *Pinus*: selected recent literature on cone and seed technology.

Author—Date	Subject	Location
Anderson 1983	Cone storage.	USA
Boado and Lasmarias 1976	*P. merkkusii*, extraction of seeds.	Philippines
Bonner 1987	*P. palustris*, cone storage and seed quality.	USA
Dale and Schenk 1978	*P. ponderosa*, insect losses.	USA
Danielson n.d.	*P. ponderosa*, air-drying seeds.	USA
Galeev and Chikizov 1976	*P. sylvestris*, cone collection and processing.	USSR
Karrfalt 1983	*P. elliotti* var. *elliottii*, fungus-damaged seeds.	USA
Maithani et al. 1986	*P. roxburghii*, artificial seed extraction.	India
Mamonov 1976a	*P. sylvestris*, processing cones.	USSR
Pauomarenko and Petrovski 1977	*P. sylvestris*, physical and mechanical properties of seeds.	USSR
Skrynuikov 1985	Airflow seed cleaning.	USSR
Sullivan and Sullivan 1982	*P. contorta*, seed predation.	USA

Table 8. *Pinus*: selected recent literature on seed storage.

Author—Date	Subject	Location
Barnett 1979	*P. glabra*, seed storage.	USA
Barnett and McGilvary 1976	*P. clausa*, seed storage.	USA
Barnett and Vozzo 1985	*P. echinata* and *P. elliottii*, 50-year seed storage.	USA
Belcher 1982	*P. taeda* and *P. elliottii*, storage of prechilled seeds.	USA
Carrillo et al. 1980	*P. montezumae*, *P. oocarpa*, *P. pseudostrobus*, *P. ayacahuite*, and *P. michoacana*, seed moisture content.	Mexico
Chen 1981	*P. koraiensis*, seed quality.	China
Danielson and Tanaka 1978	*P. ponderosa*, drying and storing prechilled seeds.	USA
Lobov 1973	*P. sylvestris*, seed storage in snow.	USSR
Mullin 1980	*P. resinosa* and *P. strobus*, water dipped and frozen seed storage.	USA
Ryynanen 1980	*P. sylvestris*, X-ray analysis of aging seeds and seed aging.	Finland
Siddiqui and Parvez 1981	*P. wallichiana*, seed storage.	Pakistan
Simak 1973	*P. sylvestris*, light and seed storage.	–
Urgenc 1973	*P. nigra* and *P. brutia*, 10-year cold storage trials.	–

sugar pine), or effect a physical change in the seedcoat to make it permeable to water (e.g., digger pine, *P. sabiniana*). The dormancy can also be more complex; an anatomically immature embryo with a physiological block may be coupled with an impermeable seedcoat as in Swiss stone pine. Acid scarification of seedcoats has been used with several species (e.g., Swiss stone pine, Balkan pine, and digger pine), but prolonged prechilling for 6 to 9 months is more effective. Acid scarification is not recommended for seeds of pines (Krugman and Jenkinson 1974).

Seeds of Swiss stone pine, Korean pine (*P. koraiensis*), Japanese pine (*P. parviflora*), and Siberian pine (*P. sibirica*) are suspected of having immature embryos at the time of collection. Germination has been increased by first placing seeds of these species in a warm moist environment (warm stratification) for several months and then prechilling the seeds for several more.

Since 1974, when the last revision of *Seeds of Woody Plants in the United States* was published, there have been several hundred publications on the germination of *Pinus* species. During this time period there may have been more published about the germination of seeds of pine species than what was published about any other wood plant and perhaps all other plants! Selected recent additions to the literature concerning the germination of seeds of pine species are listed, by species, in Table 10. Some of the material is from before 1974, but was not included in the last revision of the Handbook.

GERMINATION TEST For reliable tests of seed viability, seed is allowed to germinate under near-optimum conditions of aeration, moisture, temperature, and light. On the basis of extensive experience and experimentation, standardized seed tests for a number of pine species have been developed by the Association of Official Seed Analysis, the International Seed Testing Association, and other organizations. Because each organization has its own standards for the same species, we combined the standards to provide the most information for each species. A composite set of standards using the available published sources is given in Table 11.

The germination of pine seeds can be effectively tested in any medium or container that provides good aeration and holds adequate moisture. For a number of species, light, commonly supplied as cool white fluorescent light, is required for a reliable test. When light is necessary, the usual exposure is 8 hours in each 24-hour period. For some species the light requirement may be longer. Different temperatures are employed for testing. The standard for many species is 20/30°C, often because this fluctuating regime was the one commonly found in nonrefrigerated germinators in laboratories. The higher temperature commonly is for 8 hours in each 24-hour period and the lower for the remaining 16 hours. The duration of tests varies from 2 to 4 weeks. The number of seeds tested should be based on a statistical measurement of variability in the seed lot, which is gained from replicated experimental designs. Germination is epigeal (Fig. 4).

Table 9. *Pinus:* recommended prechilling periods for seeds (Krugman and Jenkinson 1974).

Species	Prechilling period Fresh seed (days)	Stored seed (days)
P. albicaulis	90–120	90–120
P. aristata	0	0–30
P. armandii	90	90
P. attenuata	60	60
P. balfouriana	90	90
P. banksiana	0–7	0–7
P. brutia	0	0–45
P. canariensis	0	0–20
P. caribaea	0	0
P. cembra	90–270	90–270
P. cembroides	0	0–30
P. clausa	21	21
P. contorta		
var. *contorta*	0	20–30
var. *latifolia*	0	30–56
var. *murrayana*	0	21–90
P. coulteri	0	21–90
P. densiflora	0	0–21
P. echinata	0–15	15–60
P. edulis	0	0–60
P. elliottii		
var. *densa*	30	30
var. *elliottii*	0	15–60
P. engelmannii	0	0
P. flexilis	21–90	0
P. gerardiana	0	0–30
P. glabra	28	28
P. halepensis	0	0
P. heldreichii	30–42	30–42
P. insularis	0	0
P. jeffreyi	0	0–60
P. koraiensis	90	90
P. lambertiana	60–90	60–90
P. merkusii	0	0
P. monophylla	28–90	28–90
P. monticola	30–120	30–120
P. mugo	0	0
P. muricata	0	20–30
P. nigra	0	0–60
P. palustris	0	0–30
P. parviflora	90	90
P. patula	60	60
P. peuce	0–60	60–180
P. pinaster	0	60
P. pinea	0	0
P. ponderosa		
var. *arizonica*	0	0
var. *ponderosa*	0	30–60
var. *scopulorum*	0	20–60
P. pumila	120–150	120–150
P. pungens	0	0
P. quadrifolia	0	0–30
P. radiata	0–7	7–20
P. resinosa	0	60
P. rigida	0	0–30
P. roxburghii	0	0
P. sabiniana	60–120	60–120
P. serotina	0	0–30
P. siberica	60–90	60–90
P. strobiformis	60–120	60–120
P. strobus	60	60
P. sylvestris	0	15–90
P. taeda	30–90	30–60
P. thunbergiana	0	30–60
P. torreyana	30–90	30–90
P. virginiana	0–30	30
P. wallichiana	0–15	15–90

Table 10. *Pinus:* recent additions to the literature on germination of seeds.

Author—Date	Subject	Location	Author—Date	Subject	Location
Pinus albicaulis			Mohamad and Nig 1982	Influence of shading on emergence in field. Reduced light did not produce dormancy. Suggested seedbank half-life would be short.	Malaysia
Pintel and Wang 1980	ISTA (1976) standards unsatisfactory. Acid scarification or cutting of seedcoat enhanced germination. Low germination capacity caused by lack of embryo or endosperm development.	Canada	Muselem 1975	Cone morphology related to subsequent germination. Cones collected in Costa Rica.	Mexico
			Venator 1973	GA soaking increased germination of stored seeds.	Puerto Rico
Pinus aristata			*Pinus cembra*		
Reid 1972	Seeds germinated rapidly without pretreatment.	USA	Scoz and Grossoni 1985	Warm stratification followed by prechilling 2 to 3 months gave 50% germination.	Italy
Pinus attenuata					
Conkle 1971	Isozyme specificity.	USA	*Pinus cembroides*		
Pinus ayachuite var. *veitchii*			Villagomez and Garcia 1979	Prechilling did not consistently improve germination.	Mexico
Villagomez and Garcia 1979	Prechilling 30 to 75 days increased rate of germination, not capacity.	Mexico	*Pinus contorta*		
Pinus banksiana			Cochran 1984	Symposium proceedings on species	USA
Durzan et al. 1971	Influence of tritium on seeds.	Canada	Groat 1985	Application of germination inhibitors in organic solvents.	Canada
Ramaiah et al. 1971	Changes in nitroenous compounds during germination.	Canada	Hall 1984	Dry heat influence on germination.	USA
Pinus brutia			Hellum and Dymock 1986	Prechilling did not increase germination of immature seeds. Rate of germination was hastened.	USA
Heit 1976c	Incubate seeds at 20°C on moist substrate without pretreatment.	USA			
Isik 1986	Variation study using germination.	Turkey	Hellum and Hackett 1988	Variable dormancy related to level of seed ripeness at cone collection. Mature seeds may go in and out of dormancy.	Canada
Shafiq 1978	Highest germination from seeds collected in April.	Iraq			
Shafiq 1979	Influence of light on germination.	Iraq	Helms 1987	Natural establishment in meadows.	USA
Pinus bungeana			Pitel et al. 1984	Isoenzyme patterns during germination.	Canada
Dong, Yoa, and Han 1987	Seeds very dormant due to hard seedcoat. Seedcoat broken with microorganism Trichoderma.	China	Tanaka et al. 1986	Influence of surface drying on prechilling duration.	–
Dong, Zhang, and Zhang 1987	Apparently same study, with additional information on germination characteristics. Seeds require 4 months of prechilling.	China	*Pinus densiflora*		
			Hyun n.d.	Conservation of genetic resources.	Korea
			Jo et al. 1983	No difference in germination of seeds from self-pollinated trees.	Korea
Pinus canariensis			Washitani and Saeki 1986	Gap-detection mechanism in seed germination. Seeds are light sensitive. Filtered canopy light inhibited germination.	Japan
Citharel 1979	Nitrogen fractions during germination.	Morocco			
Pinus caribaea					
Bhatnagar 1980	Presoaking in 10 ppm GA increased rate and germination capacity; higher rates of GA reduced germination.	India	*Pinus edulis*		
			Gottried and Heidmann 1986	Prechilling 30 and 60 days did not influence germination energy and capacity. Hydrogen peroxide treatments gave variable results.	USA
Madhuaraja 1982	Trials on optimum prechilling period for stored seeds.	India			

Author—Date	Subject	Location
Murphy 1985	Acetone production by fatty seeds during germination.	USA
Pinus eldarica		
Djavanshir and Reid 1975	Germination and radicle development.	Canada
Heit 1976c	Test seeds at 20°C with good moisture supply.	USA
Pinus elliottii		
Barnett 1977	Soil wetting agent, poly-oxethylene ether, reduced germination of containerized seeds.	USA
Barnett and Hall 1977	Freezing in ice decreased germination.	USA
Barnett and Vozzo 1985	Results of storing seed 50 years.	USA
Layton and Goddard 1983	No significant differences among back, half-sib, and outcrosses in germination.	USA
Mann 1979	Seed weight and seedling characteristics were related.	USA
Rowan and DeBarr 1974	Moldy seed and poor germination linked to seedbug damage.	USA
Pinus gerardina		
Sprackling 1976	Seeds stored at 4°C for 3 years failed to germinate; others stored at −10°C.	USA
Pinus halepensis		
Calamansi 1982	Germination inhibited by light and constant temperature, but not by light and alternating incubation temperatures.	Italy
Calamansi et al. 1984	Interacted prechilling duration and incubation temperature.	Italy
Thalourian 1976	HgCl₂ and KC1 enhanced germination.	Morocco
Pinus hartwegii		
Niembro et al. n.d.	Influence of color and seed size on germination.	Mexico
Pinus insularis		
Agpoa 1980	Mine tailings did not inhibit germination.	Philippines
Pinus kesiya		
Agpoa and Pulmano 1978	Acid scarification and hot water pregermination treatments had adverse effects.	Philippines
Verma and Tanden 1983	Preliminary experiments with growth regulators.	India
Verma and Tanden 1984a	Influence of imbibition temperature, pH, and water potential on germination.	India

Author—Date	Subject	Location
Verma and Tanden 1984b	Germination enhanced by light. A 16-hour photoperiod was optimum.	India
Pinus koraiensis		
Guo et al. 1981	Prechilling influenced peroxidase isoenzyme.	China
Zhang et al. 1981a	Exposure to 25°C before prechilling enhanced subsequent germination.	China
Zhang et al. 1981b	Prechilling stimulated peroxidase activity. Both peroxidase and isoperoxidase isolated from seeds.	China
Pinus lambertiana		
Baron 1978	Complete removal of seed-coat yielded rapid and complete germination. A thin inner coat layer was critical.	USA
Pinus leucodermis		
Borghetti et al. 1986	Prechilling 40 days enhanced germination.	Italy
Yamamori and Oyama 1974	Optimum temperatures for germination were 20 and 25°C.	Japan
Pinus merkusii		
Bhumibamon and Kanchanabhum 1980	Seed moisture content.	Indonesia
Sadjad and Hadi 1976	Germination test procedures.	Indonesia
Pinus monticola		
Danielson 1986	Prechilling methods using peat.	USA
Hoff 1986	Prechilling 15 weeks produced 87% germination, compared to 7% for control seeds.	USA
Hoff and Steinhoff 1986	Viability test by cutting seeds.	USA
Leadem 1986	Dormancy problems.	USA
Malone 1985	Germination.	USA
Pitel and Wang 1985	Greatest germination enhancement of stored seeds with prechill 42 days and GA enrichment. Continuous light and 20°C gave optimum germination of pretreated seeds.	Canada
Pinus morrisonicola		
Koa et al. 1980	Biochemical changes during germination.	China
Pinus nigra		
Nekrasov et al. 1974	Variation in seed quality.	USSR
Pinus oocarpa		
Kandya 1978	Seed weight and seedling development.	India

Table 10. Continued.

Author—Date	Subject	Location
Robbins 1983	Monograph of species.	Denmark
Pinus patula		
Donald 1981	Pregermination treatment with polyethylene glycol.	South Africa
Harvey 1978	Seedbed influence on germination.	Australia
Pinus pentaphylla		
Kandya and Ogino 1987	Excised embryo vigor.	India
Pinus pinaster		
Bonnet-Masimbert 1975a	Extraction and storage methods.	France
Bonnet-Masimbert 1975b	Red light stimulated germination of most sources. Seeds should be soaked 24 hours under high intensity fluorescent light before planting.	France
Donald 1987	Prechilling 6 to 7 months improved germination over 60 to 90 days.	South Africa
Pinus pinea		
Giannini et al. 1983	Variation in seed weight.	Italy
Pinus ponderosa		
Hasse 1986	Influence of prescribed burning on germination.	USA
Heidmann 1986	Acetone a poor solvent for growth regulator applied to seeds.	USA
Larson and Davault 1974	Freeze-drying of seeds.	USA
Moore and Kidd 1982	Seed sources interacted with germination under moisture stress.	USA
Pinus resinosa		
Wang 1973	Compared nursery bed emergence and laboratory germination.	USA
Pinus roxburghii		
Chauhan and Raina 1980	Seed weight and germination.	India
Ghosh and Kumar 1981	Thermal sensitivity of seeds.	India
Ghosh et al. 1976	Influence of seed grading on subsequent germination.	India
Pinus sibirica		
Kudashova and Osetreva 1976	Influence of space flight on germination.	USSR
Pinus strobus		
Hall 1984	Germination enhancement by soaking in running water.	USA
Pinus sylvestris		
Belcher and Perkins 1985	Substrate moisture content of 10% was optimum.	USA

Author—Date	Subject	Location
Inyushin et al. 1983	Laser radiation increased germination 24%.	USSR
Georgieva 1978	Seeds treated with ^{32}P had enhanced germination.	Bulgaria
Kardell 1974	Prechilling of stored seeds.	Sweden
Lehtiniemi 1977	Small doses of gamma-radiation stimulated germination; large doses inhibited it.	Finland
Loken 1977	Leaching influence on germination.	Norway
Platonova et al. 1977	Germination in space flight.	USSR
Rosochacka and Grzywacz 1980	Seedcoat color and chemical composition. Black seeds better adapted to resisting damping-off.	Poland
Rostovtsev et al. 1975	Season had no influence on germination.	USSR
Salmia et al. 1978	Proteinase activity in germinating seeds.	Finland
Samoshkin 1977	Influence of N-nitrosomethylurea on seed germination.	USSR
Pinus taeda		
Barnett 1976	Soaking seeds 30 to 60 minutes in 30% hydrogen peroxide sterilized seedcoat surface.	USA
Barnett and McLemore 1984	Speed of germination to predict seedling growth.	USA
Bonner 1983	Thermal gradient plate.	USA
Bonner 1986	Seed vigor.	USA
Boyer et al. 1985	Germination speed and seedling vigor.	USA
Campbell 1982b	Influence of surface seedcoat sterilizing treatments on nursery emergence.	USA
Campbell et al. 1983b	Light increased total germination.	USA
Carpita et al. 1983	Prechilling and radicle growth.	USA
Chaney and Kozlowski 1974	Influence of antitranspirants on seed germination.	USA
Dunlap and Barnett 1983	Influence of seed size on germination and seedling development.	USA
Dunlap and Barnett 1984	Low temperatures and/or moisture stress slowed and reduced germination.	USA
Hare 1981	Nitric acid treatment improved germination.	USA
McIntosh and Rolfe 1977	Influence of cattle waste.	USA
Zhou 1987	Light, Ga, and prechilling control germination.	–
Pinus taiwanensis		
Yang 1976	Viability estimates and germination.	China
Pinus teucodermis		
Borghetti et al. 1986	Prechilling influence on germination.	Mexico

Author—Date	Subject	Location	Author—Date	Subject	Location
Pinus thunbergii				November had highest germination.	
Chou et al. 1982	Catalase and TZ viability test.	China	*Pinus virginiana*		
Kandya and Ogino 1987	Vigor of excised embryos.	India	Bramlett et al. 1983	Genetic influences on germination.	USA
Mori 1979	Effect of light on adenosine triphosphate levels during imbibition.	Japan	*Pinus wallichiana*		
Yamamoto and Masuda 1979	Seeds collected from July to November and sown immediately; untreated seeds collected in	Japan	Siddiqui and Parvez 1981	Viability significantly reduced by short periods of storage at high temperatures.	Pakistan

Table 11. *Pinus:* germination standards (Genebank Handbook 1985, ASOA 1985).

Species	Substrata	Temperature (°C)	Duration (days)	Additional directions
P. aristata	TB	20/30	14	–
P. banksiana	TB	20/30	14	–
P. canariensis	TB	20	28	Presoak 1 day; light; sensitive to warm temperatures.
P. caribaea	TB	20/30	21	–
P. cembra	S	20/30	28	Prechill 6–9 months at 3–5°C, or excise embryo.
P. cembroides	S	20	28	Prechill 21 days at 3–5°C.
P. clausa	TB, TS	20	21	Sensitive to excessive moisture.
P. contorta	TB	20/30	21	Two tests, with and without prechilling, 21 days at 3–5°C.
P. contorta var. *latifolia*	TB	20/30	21	Light, and prechill 21 days at 3–5°C.
P. coulteri	TB	20/30	28	Prechill 60–90 days at 3–5°C, or excise embryo.
P. densiflora	TB	20/30	21	Prechill 14 days; provide light.
P. echinata	TB	20/30	28	Light 8 hrs/day, or incubate at 22°C with 16 hours light; 2 tests, with and without prechilling, 28 days at 3–5°C; sensitive to drying out.
P. edulis	TB	20/30	28	–
P. elliottii	TB	20/30, 22	28	Light 8 hrs/day; 2 tests, with and without prechilling, 28 days at 3–5°C; sensitive to drying out.
P. flexilis	TB	20/30	21	Prechill 21 days at 3–5°C.
P. glabra	TB	20/30	21	Prechill 21 days at 3–5°C; provide light.
P. halepensis	TB	20	28	Sensitive to warm temperatures.
P. heldreichii	TB	20/30	28	Prechill 42 days at 3–5°C or excise embryo.
P. jeffreyi	TB, S	20/30	28	Light, prechill 4–8 weeks at 3–5°C or excise embryo.
P. kesiya	TB	20/30	21	–
P. khasyana	TB	20/30	21	Light.
P. koraiensis	S	20/30	28	Warm stratification 2 months at 25°C followed by prechilling 3 months at 3–5°C or excise embryo.
P. lambertiana	TB, S	20/30	28	Prechill 60–90 days (8–12 weeks) at 3–5°C or excise embryo.
P. luchensis	TB	20/30	21	Light.
P. merkusii	TB	20/30	21	Light.
P. monticola	TB	20/30	28	Prechill 60–90 days (8–12 weeks) at 3–5°C or excise embryo.

Table 11. Continued.

Species	Substrata	Temperature (°C)	Duration (days)	Additional directions
P. mugo	TB	20/30	14–21	Light.
P. muricata	TB	20/30	21	Light.
P. nigra	TB	20/30	21	–
var. *poiretiana*	TB	20/30	14	Light.
P. oocarpa	TB	20/30	21	–
P. palustris	TB, S	20	21	Light, 8 hrs/day.
P. parviflora	S	20/30	28	Prechill 6–9 months at 3–5°C or incubate 10–14 days at 18–22°C on TB with excised embryos.
P. patula	TB	20	18	Light; sensitive to temperature.
P. peuce	S	20/30	28	Prechill 6 months at 3–5°C or excise embryo.
P. pinaster	TB	20	28	Light for 16 hrs/day; 2 tests, with and without prechilling, 28 days at 3–5°C; sensitive to temperature and moisture.
P. pinea	TB	20	21	Presoak 1 day; sensitive to warm temperatures.
P. ponderosa	TB	20/30	21	Light and prechill 28 days at 3–5°C.
P. pumila	S	20/30	21	Prechill 4 months at 3–5°C.
P. radiata	TB	20	28	Light more than 8 hrs/day; prechill 21 days at 3–5°C; prefers good moisture.
P. resinosa	TB	20/30, 25	14	–
P. rigida	TB	20/30	14	Light.
P. serotina	TB	22	21	–
P. strobus	TB	20/30, 22	21–28	Light 8–16 hrs/day; prechill 28–42 days at 3–5°C; sensitive to drying out.
P. sylvestris	TB	20/30	14–21	Light, prechill 21 days at 3–5°C.
P. tabulaeformis	TB	20/30	14	–
P. taeda	TB	20/30	28	Light more than 8 hrs/day; sensitive to drying out.
P. taiwanensis	TB	22	28	Light 16 hrs/day; 2 tests, with and without prechilling, 28 days at 3–5°C.
P. thunbergiana	TB	20/30	21	Light more than 8 hrs/day.
P. virginiana	TB	20/30, 22	21	Light 16 hrs/day.
P. wallichiana	TB	20/30	28	Light more than 8 hrs/day.

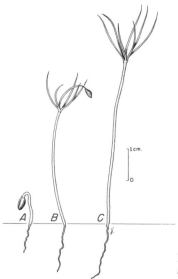

Figure 4. *Pinus resinosa,* red pine: seedling development at 1, 7, and 30 days after germination.

Table 12. *Pinus:* nursery practices (Krugman and Jenkinson 1974).

Species	Season for sowing	Seeds per m²	Sowing depth cm	Mulch	Tree percent
P. attenuata	Spring	260	0.9	None	80
P. banksiana	Either	320	0.6	None	55
P. canariensis	Spring	–	0.6	Sponge rock	43
P. clausa	Spring	–	0.9	None	70
P. contorta					
var. contorta	Spring	320	0.3	None	48
var. latifolia	Spring	510	0.3	Sawdust	60
var. murrayana	Spring	450	0.9	Peat moss	73
P. coulteri	Spring	260	1.3	None	80
P. densiflora	Spring	530	0.5	Sawdust	–
P. echinata	Spring	420	1	Pine needles	60
P. edulis	Spring	320	0.6	Sawdust	80
P. elliottii					
var. densa	Spring	370	1	Pine needles	80
var. elliottii	Spring	340	1	Pine needles	66
P. insularis	Spring	320	0.9	None	50
P. jeffreyi	Spring	290	0.7	None	69
P. lambertiana	Spring	340	1.0	None	51
P. monophylla	Spring	290	1.3	Sawdust	33
P. monticola	Spring	820	0.7	Sawdust	60
P. mugo	Spring	530	0.9	Peat moss	55
P. muricata	Spring	550	1.1	Peat moss	49
P. nigra	Fall	580	1.5	Peat moss	62
P. palustris	Spring	160	1	Pine needles	75
P. pinaster	Spring	320	0.9	–	–
P. ponderosa					
var. ponderosa	Spring	380	0.7	None	64
var. scopulorum	Spring	400	0.4	None	70
P. pungens	Either	550	1.25	None	70
P. radiata	Spring	530	0.9	Straw	24
P. resinosa	Either	420	0.7	Peat moss	73
P. rigida	Spring	340	1	Sand	–
P. roxburghii	Spring	–	0.4	–	32
P. strobus	Fall	370	1	Sawdust	70
P. sylvestris	Either	480	0.8	Peat moss	48
P. taeda	Spring	420	1	Pine needles	60
P. thunbergiana	Either	790	0.3	Straw	31
P. virginiana	Either	340	1	Pine needles	77
P. wallichiana	Spring	–	1.25	None	46

[1] Seeds pressed into soil surface making the sowing depth approximately equal to the diameter of the seeds.

Cutting tests are commonly used for rough estimates of seed quality. Such tests can also be used as emergency guides in fall sowing of fresh seeds with embryo dormancy. X-ray methods supply information on seed soundness.

NURSERY AND FIELD PRACTICE Pines are successfully grown in nurseries in most parts of the United States and in many temperate and subtropical portions of the world. The soil should be fertile and have good drainage and aeration.

In temperate regions, pine seeds can be sown in the fall or spring. It is common practice to sow non-dormant seeds in the spring. Dormant seeds can be sown then as well, but must be prechilled. The pregermination treatment for each species and seed lot should be that which achieves best germination (Table 12). Fall-sown seedlings are commonly larger and better developed after 1 season. Fall sown seedbeds must be protected against winter frost and rodent damage.

Most large nurseries drill in rows because it is economical. The amount of seed sown per unit area varies with the species (Table 13); seeds are sown at densities from 160 to 790 seedlings per m². Higher seedling survival rates are obtained with the moderate to low densities.

Large-seeded species are covered with more soil than small-seeded ones, and seeds of southern pines are pressed into the soil rather than covered with soil.

Germination is complete for most species from 10 to 50 days after spring sowing. Seeds of dormant species, even after prechilling, may continue to emerge for some time after planting. Transplanting of seedlings and final transplanting to the field depends on the species and the nature of the site. Harsh sites generally receive older transplants.

As with all aspects of the genus *Pinus* there is considerable literature concerning nursery practices (Table 13).

Allelopathic influences on germination and growth

Table 13. *Pinus:* selected literature on nursery practices.

Author—Date	Subject	Location
Belcher et al. 1984	*P. elliottii,* nursery practices.	USA
Bogdanov 1979	*P. sylvestris,* vegetative propagation.	Bulgaria
Dan'shin et al. 1975	*P. sylvestris,* chemicals to stimulate growth.	USSR
Dominx and Wood 1986	*P. banksiana,* shelter spot seeding.	Canada
Geary et al. 1971	*P. caribaea,* direct seeding.	India
Ghosh et al. 1974a, 1974b	*P. caribaea, p. patula,* nursery techniques.	India
Ghosh et al. 1976	Seed grading.	India
Girgidov and Gusev 1976	Size grading seeds.	USSR
Gomes et al. 1978	*P. caribaea,* soil fumigation.	Brazil
Gordon et al. 1979	Fluid drilling.	England
Grzywacz and Rosochacka 1980	Seed color and damping-off.	Poland
Harvey 1978	*P. patula,* seedbed cover.	Australia
Komarova 1986	Forest fires and seed germination.	USSR
Kuo 1983	*P. luchuensis,* seedling survival.	China
Mamonov and Smurova 1978	Pretreatment with growth regulators.	USSR
Migliaccio et al. n.d.	Direct seeding.	Italy
Minko 1975	*P. radiata,* soil properties and fertilization.	Australia
Polupannev et al. 1979	*P. sylvestris,* precision drilling.	USSR
Russo 1978	*P. banksiana,* development of seedlings subject to drying cycles.	USA
Shea and Armstrong 1978	*P. caribaea,* open-root plantings.	Australia
Shepperd and Noble 1976	*P. contorta,* simulated precipitation.	USA
Sims 1970	*P. banksiana,* seedling survival.	Canada
Singh et al. 1973	*P. wallichiana,* depth of sowing.	India
Tinus 1987	Seed covering.	USA
Vedenyapina and Badanov 1974	*P. sylvestris,* use of *Azotobacter.*	USSR
Woodward 1983	*P. contorta,* ash-covered seedbeds.	Canada
Woodward and Land 1984	*P. lambertiana,* suppression of reproduction.	USA
Zelenskii and Sidorova 1973	*P. sylvestris,* trace elements.	USSR

of seeds of *Pinus* species have been investigated in many parts of the world (Table 14).

Pines as a group are very difficult to root from cuttings, an exception being Monterey pine, and vegetative propagation is usually done by grafting (Dirr and Heuser 1987). Most pines are grafted in January and February. The understock should be the same or a closely related species. For example, 5-needle pines are grafted on 5-needle pine only. There are exceptions in which pines with the same number of needles per fascicle are not compatible. Dwarf pine forms are vegetatively propagated from witches broom by rooting cuttings.

Table 14. *Pinus:* selected recent literature on allelopathic influences on germination and seed growth.

Author—Date	Subject	Location
Fisher 1979	*P. banksiana,* allelopathic effect of reindeer-moss.	USA
Gilmore 1985	*P. taeda,* allelopathic effect of giant foxtail.	USA
Harrington 1987	*P. ponderosa,* allelopathic effect of Gambel oak.	USA
Hollis et al. 1982	Allelopathic effects on southern pines.	USA
Lill and McWha 1979	*P. radiata,* influence of volatile substances from litter on germination.	–
Penafiel et al. 1982	*P. kesiya,* allelopathic influence of pasture plants.	Philippines

ANACARDIACEAE — CASHEW FAMILY

Pistacea L. — Pistachio

GROWTH HABIT, OCCURRENCE, AND USE The genus *Pistacea* consists of about 20 tree species native to southwestern Asia. Pistachio (*P. vera*) is a spreading tree to 10 m in height that is widely grown on warm temperate climates for its fruit, a drupe.

GERMINATION Seeds of *Pistacea* species can show considerable dormancy (Genebank Handbook 1985).

Nondormant seeds germinate well at 21°C, poorly at 27°C, and not at all at 33°C. Dormant seeds need to have the epicarp removed and the endocarp chipped. Seeds can also be acid scarified for 10 minutes. Treatment with gibberellins can enhance germination. Seeds of *P. khinjuk* require scarification (Dahab et al. 1975).

LEGUMINOSAE — LEGUME FAMILY

Pithecellobium Mart. — Blackbead

GROWTH HABIT, OCCURRENCE, AND USE The genus *Pithecellobium* consists of about 110 species, mostly native to Asia and tropical America (Walters et al. 1974). Ebony blackbead (*P. flexicaule*) is considered the most valuable tree in the Rio Grande valley. The wood is used in cabinetry and for fence posts, and the seeds are used for food. Raintree (*P. saman*) is valued for timber production, wildlife habitat, and aesthetics. The wood is used for paneling, furniture, and specialty items, and the pods are eaten by animals. *Pithecellobium arboreum* is a tropical rain forest tree distributed from Mexico to Ecuador (Flores and Mora 1984).

FLOWERING AND FRUITING Ebony blackbead bears 3.8 cm, yellow- to cream-colored flower clusters during June through August. Raintree bears 5 cm, pink flower clusters from spring to fall. Both species fruit from fall to spring. The pods turn from green to drab or black when ripe. Ebony blackbead pods average 12.7 cm long and 2.5 cm wide when ripe (Fig. 1). Raintree pods average 15.2 cm long and 1.9 cm wide. Pods of both species are indehiscent and may remain on the trees for a year or more. The reddish brown bean-shaped seeds have no endosperm (Figs. 2, 3).

COLLECTION, EXTRACTION, AND STORAGE OF SEED Pods are picked from the tree and dried on racks. Seeds can be removed by hand flailing or by a macerator. There are between 1.5 and seeds per gram for ebony blackbead and 4.4 to 7.7 for raintree. Raintree seeds have been stored in sealed polyethylene bags at 3 to 5°C.

GERMINATION Pregermination treatments are not necessary for raintree seeds, but a 10-minute acid scarification increases both germination energy and capacity of ebony blackbead seeds. Seeds of *P. arboreum* germinate 24 to 72 hours after being shed, although they may sometimes germinate precociously (Flores and Mora 1984)

NURSERY AND FIELD PRACTICE In Hawaii, raintree seeds are generally sown in nursery beds during March so they can be outplanted the following winter. They are sown about 2.5 cm deep, and seedbeds are not mulched. About 75 to 80% shading is provided. Seedbed densities of 160 to 210 per m^2 are used. Tree percent averages 75 to 80.

Figure 1. *Pithecellobium flexicaule*, ebony blackbead: pod, ×0.5 (Walters et al. 1974).

Figure 2. *Pithecellobium flexicaule*, ebony blackbead: seeds, ×2 (Walters et al. 1974).

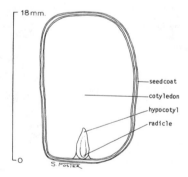

Figure 3. *Pithecellobium flexicaule*, ebony blackbead: longitudinal section through a seed, ×3 (Walters et al. 1974).

PLATANACEAE — SYCAMORE FAMILY

Platanus L. — Sycamore

GROWTH HABIT, OCCURRENCE, AND USE Sycamores are deciduous trees that range from 25 to 43 m in height at maturity (Bonner 1974i). American sycamore (*Platanus occidentalis*) is one of the largest and most valuable timber species in the eastern United States; it is widely planted in commercial plantations. An apparent hybrid between American sycamore and oriental planetree (*P. orientalis*), with the scientific name *P.* × *aceriolia*, is widely planted as an ornamental in the United States because of its tolerance to city smoke and other air pollution.

FLOWERING AND FRUITING The minute, monoecious flowers of sycamores appear in the spring (Bonner 1974i):

Species	Flowering dates	Fruit ripening dates	Seed dispersal dates
P. occidentalis	Mar.–Apr.	Nov.	Feb.–Apr.
P. orientalis	May	Sept.–Oct.	—
P. racemosa	—	June–Aug.	June–Dec.

Both staminate and pistillate flowers occur in separate, dense, globular heads. The dark red staminate flowers are usually borne along the branchlets, while the light green pistillate flowers occur at the tips. American sycamore fruiting heads are usually solitary, but oriental planetree may have 2 to 7 on a single stem (Fig. 1). They are greenish brown at maturity in the fall.

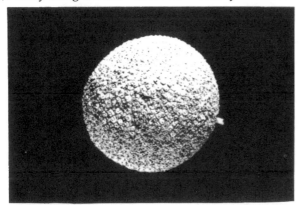

P. occidentalis
American sycamore

P. racemosa
California sycamore.

Figure 1. *Platanus*: fruiting heads, ×1 (Bonner 1974i).

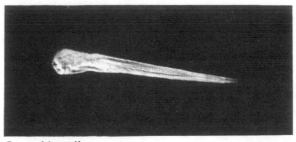

P. occidentalis
American sycamore

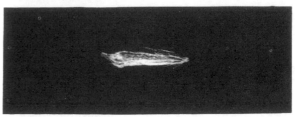

P. racemosa
California sycamore

Figure 2. *Platanus*: single achenes, ×4 (Bonner 1974i).

The fruit is an elongated, chestnut-brown, single-seeded achene with a hairy tuft at the base. The achenes, with hairs removed, are used as seeds (Fig. 2). The elongated embryo is surrounded by a thin endosperm (Fig. 3). American sycamore usually bears good seed crops every 1 or 2 years and light crops in the intervening years. Open-grown trees of this species as young as 5 years have produced good seed crops, but trees in dense natural stands are usually much older before large crops are evident.

COLLECTION, EXTRACTION, AND STORAGE OF SEED Fruiting heads of sycamore can be collected at any time after they turn brown, but the job is easiest if done after leaf drop. Since the heads are persistent, collection can be made into the next spring. Handpicking from felled trees is the most convenient method. At the northern and western limits of its range, heads of American sycamore can sometimes be collected from the ground late in the season. As they begin to fall apart in the spring, they can be shaken onto ground sheets. They should be dried on well-ventilated trays until they can be broken apart.

Studies by Delkov (1975) suggested that quality of oriental planetree seeds varied with the growth form of the parent tree. Seeds from trees with furrowed bark had higher germination than to seeds from smooth barked trees.

Seeds should be extracted by crushing the dried fruiting heads and removing the fine hairs that are attached to the individual seeds (achenes). Small lots can be rubbed through screens. Large seed lots need to be broken by mechanical treatment and the seeds

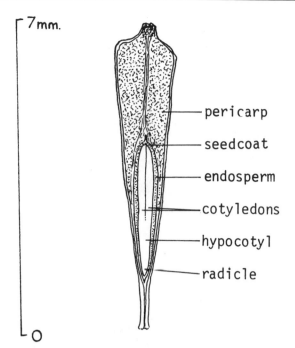

⌐7mm.

———— pericarp

———— seedcoat

———— endosperm

————cotyledons

———— hypocotyl

———— radicle

⌐O

Figure 3. *Platanus occidentalis*, American sycamore: longitudinal section through an achene, ×12 (Bonner 1974i).

recovered by air screening. The dust from such processing is dangerous, and appropriate safety equipment must be used.

Louisiana and Mississippi collections of American sycamore fruits yielded 2 to 5 kg of seed per 35 liters of fruits, and 25 to 30 kg of seed per 45 kg of fruit. American sycamore has about 330 to 440 seeds per gram and oriental planetree has 280 to 310. Bonner (1979) summarized the best practices for obtaining quality American sycamore seeds, including germination and seed storage. Karrfalt and Helmuth (1984) reported on preliminary trials with electrostatic separation of sycamore seeds.

If *Platanus* seeds are to be sown soon after collection, they may be stored in a cool, dry, well-ventilated place in open-mesh bags or spread out on shelves. For storage longer than 1 year, seeds should be dried to 10 to 15% moisture and stored in airtight containers at −8 to 3°C.

PREGERMINATION TREATMENT Prechilling 60 to 90 days at 5°C in sand, peat, or sandy loam has been reported as beneficial to seeds of California sycamore (*P. racemosa*). Pregermination treatments are not required for other species.

GERMINATION TEST Germination can be easily tested on wet paper, sand, or even in shallow dishes of water. ASOA (1985) standards for American sycamore seeds are substrata, top of blotters; incubation temperature; 20/30°C; and duration, 14 days.

NURSERY AND FIELD PRACTICE Spring is the best time to sow sycamore seeds, but fall or late winter sowings are feasible. The seeds may be broadcast or drilled in rows 15 to 20 cm apart. They should be covered with 0.6 cm of soil or mulch. Sawdust is an excellent mulch, and oat straw or pine needles are satisfactory. Fall seeding requires bird screening. Small seedlings are wanted in most cases and 260 to 370 seedlings per m² are recommended. For larger stock, 50 to 100 seedlings per m² can be used. *Platanus* species can also be rooted from cuttings.

Sigaffoos (1977) investigated the influence of seedbed quality factors on the natural regeneration of American sycamore. Highest germination occurred on bare mineral flood plain soils exposed to high light intensities. Shading by litter or herbaceous vegetation increased seedling mortality. Book and Book (1985) investigated the natural regeneration of Wright sycamore (*P. wrightii*) in stream washes in southeastern Arizona. Seedlings tended to be clumped in even-age groups near the washes, but out of the flood plain.

Sowing of American sycamore seeds by family seed lots was recommended by Johnson and Kellison (1984) to obtain uniform seed germination and seedling size.

APOCYNACEAE — DOGBANE FAMILY

Plumeria L. — Frangipani

GROWTH HABIT, OCCURRENCE, AND USE The genus *Plumeria* consists of about 7 species native to the West Indies and northeastern South America. These species were carried to the Pacific Islands and southeastern Asia by early Spanish explorers. They are widely planted in the tropics as ornamental trees.

The fruit consists of 2 leathery follicles containing winged seeds.

Plants can readily be raised from seeds, but flower colors will not be true; species are usually propagated from cuttings.

PODOCARPACEAE — PODOCARPACEAE FAMILY

Podocarpus L'Herit.

GROWTH HABIT, OCCURRENCE, AND USE The genus *Podocarpus* consists of about 60 species of widely distributed evergreen trees or shrubs. Their leaves are linear or elliptic, and sometimes scalelike.

FLOWERING AND FRUITING Male flowers are borne in catkins. The female flowers are a scale enclosing the ovule, with several bracts at its base. The fruit is a drupelike structure with the seeds borne on a fleshy receptacle.

GERMINATION Noel and Staden (1976) considered the delayed germination of *P. henkelii* seeds to be due to poor water imbibition caused by a wax coating on the seeds. Dodd and Van Staden (1982) suggest that cytokininlike compounds produce dormancy in seeds of this species.

RUTACEAE — RUE FAMILY

Poncirus trifoliata (L.) Raf. — Hardy orange

GROWTH HABIT, OCCURRENCE, AND USE The genus *Poncirus* contains 1 species, a very thorny, small tree native to China. It is used as a rootstock for citrus, and grown for hedges.

The small-diameter fruits should be collected in the fall and split to remove the seeds which can be dried and stored.

Germination without pretreatment is sporadic. Seeds prechilled for 3 to 4 weeks germinate uniformly (Dirr and Heuser 1987).

SALICACEAE — WILLOW FAMILY

Populus L. — Poplar

GROWTH HABIT, OCCURRENCE, AND USE The genus *Populus* includes about 30 species of medium to large deciduous trees native in North America—from Alaska and Labrador south to northern Mexico— Europe, North Africa, and Asia south of the Himalayas (Schreiner 1974). Some species, such as aspen, form extensive forest stands, others usually occur along stream bottoms and low-lying areas.

Poplars are important pulpwood, lumber, and veneer species. The wood is pulped by the standard chemical and mechanical processes for use in high-grade papers, corrugating paper, fiberboard, wall-board, and impregnated building board or felt. It is also used in various types of particle board. Poplar lumber is used for various structural and decorative purposes. Eastern cottonwood (*P. deltoides* var. *deltoides*) and plains cottonwood (*P. deltoides* var. *occidentalis*) have been widely planted for shelterbelts in the United States and Canada. European cultivars, particularly those of Lombardy poplar (*P. alba* 'Italica'), and natural hybrids cultivated in Europe, such as 'Canescens', 'Eugenei', 'Serotina', 'Robusta', and 'Petrowskiana', have been used quite extensively in ornamental plantings throughout the United States and Canada.

GEOGRAPHIC RACES Natural hybridization has been reported between almost all sympatric poplar species and between introduced and native poplar species both in the United States and Europe. The possibility for successful hybridization between species of different taxonomic sections as well as within such sections has been demonstrated by natural and by controlled breeding.

FLOWERING AND FRUITING Most poplar species have been classified as dioecious, but *P. lasiocarpa* has been described as a poorly known, monoecious, self-fertilizing species, and deviations from strict dioecism have been found in other species. Sex ratios in favor of male trees have been reported for quaking (*P. tremuloides*) and European (*P. tremula*) aspen.

The range in flowering and seed dates for individual species is shown in Table 1. In the aspens and balsam poplars, the flowering period and time of seed maturity appear to be quite regular within the limits of ecotypic zones; differences in flowering time from year to year apparently depend on temperatures. In the eastern cottonwood complex, seedfall may occur in May, June, July, or even August during a single season. This an adaptation of high survival value to the species. Wide variation among individual trees in date of flowering is common for many species.

Kochkar (1983) studied the fruiting of black poplar (*P. nigra*) trees 25 and 50 years old. Seed yield and quality were greater from the younger trees. In a later study the fruiting of a 40-year-old natural stand was compared to that of a 15-year-old artificial plantation (Kochkar 1984). Fruiting was more abundant in the plantation, but germination was higher in seeds collected from the natural stand.

Populus species are forced to flower in the production of artificial hybrids. Branches are placed in water in a greenhouse and the temperature and prechilling periods manipulated to induce flowering. Gladysz (1983) studied the influence of twig size, twig mass/inflorescence size, and nutrient concentration of the growth media on the quality of European × quaking aspen hybrids produced in the greenhouse. Twigs of 150 to 250 g fresh weight should be chosen, and at least 30 g twigs per inflorescence allowed, to produce seeds of suitable seed weight. Increasing nutrient content of the medium above 1.3 g/liter reduced seed weight. The age of first flowering of poplar species shows considerable inter- and intraspecific variation (Table 1). Cottonwoods and balsam populars generally reach flowering age between 10 and 15 years. Usually little seed can be collected from eastern cottonwoods less than 25 cm in diameter or less than 10 years in age.

The reported weight of poplar seeds varies between and within species, from approximately 300 per gram to 15,400 per gram (Table 1).

It has been estimated that an eastern cottonwood tree 12 m in height, with a trunk diameter of 0.6 m and a canopy diameter of 13.8 m, would bear 32,400 catkins, with 27 capsules per catkin (Fig. 1), 32 seeds per capsule, and on the average 100 seeds would weigh 0.065 grams. The tree would produce nearly 28 million seeds weighing 18 kg. European aspens in Finland and Estonia have been estimated to have the following seed production characteristics:

Age of tree (years)	Number of catkins	Number of seeds
8	9	8,700
25	1,200	1,275,000
45	10,000	3,300,000
100	40,000	54,000,000

Cottonwoods and balsam poplars produce large seed crops almost every year; aspens produce some seeds every year, but bumper crops are produced at

P; deltoides var. *occidentalis*
plains cottonwood

P. fremontii var. *fremontii*
Fremont cottonwood

P. fremontii var. *wislizeni*
Rio Grande cottonwood

Figure 1. *Populus*: catkins consisting of mature, but unopened capsules, ×0.5 (Schreiner 1974).

Table 1. *Populus:* phenology of flowering and fruiting, characteristics of mature trees, and cleaned seed weight (Schreiner 1974).

Species	Flowering dates	Seed dispersal dates	Year first cultivated	Mature tree height (m)	Seed-bearing Age (years)	Seed-bearing Interval (years)	Seeds/gram
P. acuminata	May	July	1898	15	5–10	1	–
P. alba	Apr.–May	May–June	Early	42	10–15	–	–
P. angustifolia	–	–	–	15	–	–	–
P. balsamifera	Apr.	May–July	Early	36	8–10	1	–
P. × canescens	Apr.–May	July	Early	40	8–15	–	–
P. deltoides							
var. *deltoides*	Mar.–Apr.	May–Aug.	Early	58	10	1	440
var. *occidentalis*	Apr.–May	June–Aug.	1908	31	10	1	–
P. fremontii							
var. *fremontii*	Feb.–Mar.	Mar.–Apr.	1904	31	5–10	1	–
var. *wislizenii*	Apr.–May	June–July	1894	31	5	1	–
P. grandidentata	Mar.–May	May–June	1772	31	10–20	4–5	6600
P. heterophylla	Mar.–May	Apr.–June	1656	31	10	1	330
P. laurifolia	–	–	–	15	8–10	–	–
P. maximowiczii	Apr.	July–Aug.	1830	31	10	1	–
P. nigra	Apr.	May	Early	31	8–12	1	–
P. simonii	–	–	–	12	10	1	–
P. tremula	–	–	–	39	8–10	4–5	8080
P. tremuloides	Mar.–May	May–June	1812	31	10–20	4–5	7900
P. trichocarpa	Apr.–June	June	1892	62	10	1	–

intervals of 3 to 5 years.

COLLECTION, EXTRACTION, AND STORAGE OF SEED Branches bearing nearly mature catkins can be brought into a warm room or greenhouse and placed in water to allow the capsules to open. If catkins are to be picked directly from the trees, a safe criterion for time of collection is when a small percentage of the capsules are beginning to open. For aspen, it has been suggested that catkins should be picked from the trees when the seeds are a light straw color; those collected before reaching this stage of maturity do not ripen completely and yield approximately 50% germination.

Catkins permitted to mature on tree branches can have seed collected directly with a vacuum cleaner. Catkins collected from trees can be spread in shallow layers in trays and stored at room temperature until the capsules open. Cottonwood seeds can be extracted by maceration (presumably dry) and the seeds recovered by air screening.

Removing the cotton from poplar seeds is a problem. The woolly seeds can be cleaned by tumbling them on a screen with air pressure until they are freed from the cotton and then collecting them on a smaller screen that allows small pieces of cotton to pass through. The size of the screen depends on the species being cleaned. Eastern cottonwood seeds can be cleaned in a blender operated at low speeds (Nicholson and Demeritt 1978). *Populus* seeds are shown in Figures 2 and 3.

The longevity of poplar seeds under natural conditions has been reported as from 2 weeks to a month, varying with species, the season, and local environmental conditions. With proper drying and cold storage in sealed containers, poplar seeds can be stored for several years with little loss of viability. Prestorage drying immediately after collection is an essential part of successful storage. Viability in storage is improved and germination is higher if seed moisture content is held at 5 to 8%. Cold storage at 5°C of fully matured and properly dried seeds can maintain viability for 2 years and in extreme cases as long as 6.

Comparing the germination of green and mature seeds of black poplar, Muller and Cross (1982) found that optimum germination occurred after 5 years of storage with drying to 7 or 8% moisture content and under partial vacuum in air at 5°C. Under all storage conditions the green seeds maintained higher viability than mature seeds. Hellum (1973) found that balsam poplar (*P. balsamifera*) seeds could be stored for about 140 days. Other recent literature concerning the storage of *Populus* seeds is given in Table 2.

Table 2. *Populus:* selected recent literature.

Author—Date	Subject	Location
	Seed Storage	
DenHeyer and Seymoir 1978	*P. tremuloides* and *P. balsamifera,* seed collection and storage.	USA
Kamra 1982a	*P. maximowiczii,* seed storage.	Japan
Tauer 1979	*P. deltoides,* seed tree, vacuum, and temperature effects on viability.	USA
	Germination	
Fechner et al. 1981	*P. tremuloides,* seeds from 7 clones were stored 1 to 24 months at −18°C, then germinated at water potentials from 0 to −12 bars and at 20/30 and 15/25°C. Storage did not interact with germination at reduced osmotic potentials. Cotyledon expansion occurred in at least 80% of seedlings up to −4 bars.	Canada
Gladysz and Ochlewska 1983	*P. tremula* × *P. tremuloides,* seed morphology and seedling size.	Poland
Hardin 1984	*P. deltoides,* seed weight and germination.	USA
McDonough 1979	*P. tremuloides,* temperature, substrate moisture, and seedbed quality control germination. Seeds very specific in their requirements for germination.	USA
Meyers and Fechner 1980	*P. angustifolia* and *P. fremontii* var. *wizlizenii,* germination not reduced by leaving on seed hairs.	USA
Simak 1980	*P. tremula,* germination.	–
Singh and Arya 1987	*P. ciliata,* effect of catkin parts on germination.	India
Singh and Singh 1983	*P. ciliata,* influence of moisture stress on germination.	India

Figure 2. *Populus fremontii,* Fremont cottonwood: cleaned seeds, ×4 (Schreiner 1974).

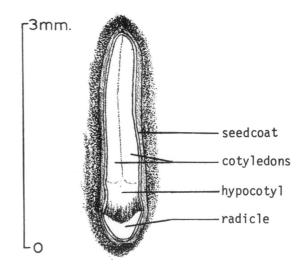

Figure 3. *Populus deltoides,* eastern cottonwood: longitudinal section through the embryo of a seed, ×20 (Schreiner 1974).

GERMINATION ASOA (1985) germination standards for *Populus* species are substrate, top of paper; incubation temperature, 23/30°C; duration, 14 days; and additional directions, provide light. It is difficult to define germination with the small seeds of *Populus* species. For most species, germinated seedlings should have well-developed hypocotyl hairs, regular growth, and a geotropic response. Abnormal seedlings would show poor development of the hypocotyl hairs, absence of firm attachment to the substrate, and imperfect geotropism. Seeds that have been dried for storage may suffer injury from rapid imbibition. Aeration with humid air after storage has been used to solve this problem.

Selected recent literature on the germination of species of *Populus* is listed in Table 2.

NURSERY AND FIELD PRACTICE The production of *Populus* nursery stock from seed requires an exacting and unique methodology. Poplar seed should not be covered nor should it be pressed into the soil of the seedbed. The seedlings are extremely susceptible to drying, to the washing action of rain or coarse irrigation, and to damping-off fungi. The critical factor for seed germination is a moisture-supplying substrate. Moisture is critical for the first month of seedling growth.

Fresh seed usually begins to germinate within a few hours, and within 12 hours the hypocotyl has begun to grow out of the seedcoat. A circular brush of delicate hairs develops rapidly around the base of the hypocotyl; the hairs become attached to the soil and, as the hypocotyl continues to grow, it straightens and lifts the seed off the ground. From 4 to 6 days after the beginning of germination, the hypocotyl has usually grown straight and upright, and the cotyledons have thrown off the seedcoat. During this time and until the primary root is established and firmly anchored to the soil, irreparable damage may be done by drying the hairs or wrenching the seedling from its attached anchorage through washing or flooding the soil surface. Beginning about the fifth day, the primary root begins to grow slowly; after 12 days the root may be only 1.5 mm long. The growth of the root system continues rather slowly for about 3 weeks to 1 month. The tap root of month-old seedlings may only be 2.5 cm. Obviously, irrigation of seedbeds must be with a fine mist, subirrigation, or soaker type applicators.

For eastern cottonwood, about 3 grams of cleaned seed (3000 seeds) per m² for broadcasting, and 325 seeds per linear meter of drill row have been suggested as seeding rates. The beds should be thinned when about 4 weeks old to approximately 210 seedlings per m².

Softwood and hardwood cuttings of most poplars root readily with the exception of gray poplars and aspens (Dirr and Heuser 1987). Cuttings are usually 25 to 30 cm long and taken from dormant one-year-old stems. Gray poplar and aspen can be propagated from root pieces. In February, pieces 5 to 10 cm long are placed in moist peat moss in the greenhouse. Cambium tissue from one-year-old terminal branches and shoot tips have been used to initiate callus tissue from which whole plants were regenerated for several species of *Populus* (Dirr and Heuser 1987).

COMPOSITAE — SUNFLOWER FAMILY

Porophyllum (Vaill.) Adans.

GROWTH HABIT, OCCURRENCE, AND USE The genus *Porophyllum* consists of about 30 species native to the U.S. Southwest and extending to South America.

The plants are multibranched perennials that grow to 0.7 m in height from a woody base. The fruit is an achene.

After 1 year of storage, germination increased from less than 2% to about 8%. Seeds remained viable over 8 years of storage (Kay et al. 1988).

ROSACEAE — ROSE FAMILY

Prinsepia sinensis D. Oliver — Cherry prinsepia

GROWTH HABIT, OCCURRENCE, AND USE Cherry prinsepias are shrubby members of the rose family. They are native to Asia and have orange-red to red drupes that ripen in July to September.

The fruit should be macerated and the seeds recovered by flotation.

Germination is enhanced by enrichment with gibberellin, and inhibited by warm temperatures (Dirr and Heuser 1987). Softwood cuttings taken in early July root readily.

LEGUMINOSAE — LEGUME FAMILY

Prosopis L. — Mesquite

GROWTH HABIT, OCCURRENCE, AND USE The genus *Prosopis* consists of about 30 to 35 species of deciduous, thorny shrubs or small trees native to subtropical and tropical regions of the Western Hemisphere, Africa, and Asia (Martin and Alexander 1974). *Prosopis juliflora* is one of the most important species and in the United States 3 varieties are recognized: Honey mesquite (var. *glandulosa*) is found throughout much of Texas, eastern New Mexico, and as far north as Kansas; western honey mesquite (var. *torreyana*) occurs mainly in New Mexico, west Texas, and southeastern Arizona; and velvet mesquite (var. *velutina*) is found in southern and central Arizona, southwestern New Mexico, and northern Sonora, Mexico.

Recently there has been considerable interest in propagation of mesquite as a fuel/forage tree for semi-arid and arid environments. Mesquite is a good fuel and charcoal species, and provides a moderately durable post for construction and fencing. It has long been used as a high-protein food source for humans and livestock. Mesquite flowers are an excellent source of honey and the plants have been used in landscaping.

FLOWERING AND FRUITING The perfect flowers of mesquite open from mid-March through May in the southwestern United States. The fruit is an indehiscent pod containing several seeds (Fig. 1) that ripen from August to September. Ripe fruits vary in color from straw to reddish brown, and are often mottled. The flat, shiny, brown seeds have no endosperm (Fig. 2).

COLLECTION, EXTRACTION, AND STORAGE OF SEED Ripe pods may be stripped from trees or picked from the ground. Seed extraction is difficult. Brown and Belcher (1979) suggest drying the pods for 18 hours at 32°C and then threshing them in a baffled seed scarifier. The problem is that the pods remain flexible and rapidly absorb moisture. In the arid Southwest, Martin and Alexander (1974) suggested allowing the pods to dry for several days at air temperatures. The pods are usually infested with insect larvae and their storage requires fumigation. (See Johnson 1983 for insects that infest seeds of *Prosopis* species.) Cleaned seeds should be fumigated if the pods are not previously treated. Seeds can be recovered after threshing by air screening. There are about 30 mesquite seeds per gram.

Felker et al. (1984) compared the pod production of *Prosopis* selections from North and South America, Hawaii, and Africa. Hawaiian and African selections were eliminated from the trials in southern California by −5°C temperatures. The most productive sources were velvet mesquite selections from southern Arizona; the most productive trees produced 7.2 kg of pods per tree. Trees that received the least irrigation produced the most pods. Pod production estimates of 3000 to 4000 kg/ha were obtained for trees that received 37 cm of rainfall the preceding season.

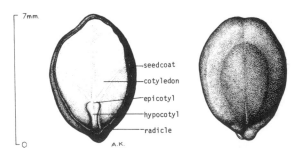

Figure 1. *Prosopis juliflora*, mesquite: pod, ×0.5 (Martin and Alexander 1974).

Figure 2. *Prosopis juliflora*, mesquite: longitudinal section through a seed and exterior view, ×5 (Martin and Alexander 1974).

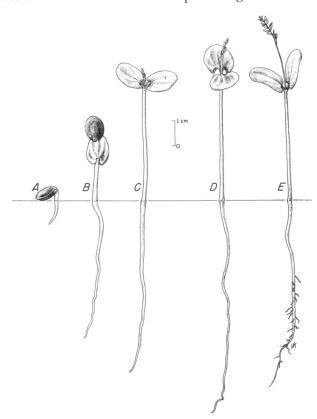

Figure 3. *Prosopis juliflora*, mesquite: seedling development at 1, 2, 5, and 25 days after germination (Martin and Alexander 1974).

PREGERMINATION TREATMENT Except for freshly harvested seeds that are not dried, seeds of mesquite have hard seedcoats that require scarification before germination will occur.

Winer (1983) compared the influence of acid scarification, hot water treatment, and passage through a goat on the germination of seeds of *P. chilensis*. Passage through a goat and acid scarification produced the greatest enhancement of germination. Acid scarification for 10 to 60 minutes enhanced germination to 100%, but control seeds had 80% germination in a study conducted by Mahmoud and El-Sheikh (1978).

Collection, handling, storage, and germination pretreatment techniques were given by Folliott and Thames (1983) for *P. tamarugo*, honey mesquite, and *P. chilensis* native to Mexico and Central America. Pregermination treatments for the Argentine species, *P. caldenia*, were reported by Arrehhini (n.d.)

Studies in India of the scarification of seeds of *P. juliflora* produced the novel idea of increasing germination by physically impacting the strophioler area of the seedcoat. This was accomplished by shaking seeds in a bottle for 15 minutes. Saxena and Khan (1974) used NaOH to clean seeds of the same species from their septa and in the process enhanced germination.

GERMINATION Germination is epigeal (Fig. 3). *Prosopis farcta* occupies a variety of habitats over a large geographic area in the Middle East (Bazzaz 1973). Several ecotypes, based on germination at reduced osmotic potentials, have been recognized. These ecotypes may be evolving and spreading in areas where improper irrigation techniques have resulted in an accumulation of salt in the soil. However, Dafni and Negbi (1978) failed to find any relationship between germination of *P. farcta* at reduced osmotic potentials and salt tolerance of mature plants.

PROTEACEAE — PROTEA FAMILY

Protea R. Br.

GROWTH HABIT, OCCURRENCE, AND USE The protea family consists of perhaps 50 or more genera and about 1000 species native to Australia, South Africa, tropical South America, and eastern Asia (Bailey 1951).

The genus *Protea* is native to South Africa. These ornamental trees and shrubs are grown in areas of the Southern Hemisphere with soils of low pH and phosphate and in shady situations for their beautiful flowers. The fruit is either winged or a hard nut depending on the species.

In New Zealand *Protea* species are propagated by sowing seeds in flats in the fall when temperatures are falling (Harre 1987). Seeds are sprinkled on the top of moist planting mix and allowed to sit for 17 to 20 days before being watered. Prechilling has not been consistently successful in enhancing germination.

ROSACEAE — ROSE FAMILY

Prunus L.

GROWTH HABIT, OCCURRENCE, AND USE The genus *Prunus* is one of the most important genera of woody plants (Grisez 1974). Its 5 well-marked subgenera include the plums and apricots (*Prunophora*), the almonds and peaches (*Amygdalus*), the umbellate cherries (*Cerasus*), the deciduous racemose cherries (*Padus*), and the evergreen, racemose, or laurel cherries (*Laurocerasus*). Nearly 200 species, ranging from prostrate shrubs to trees over 30 m tall, are found in the north temperate zone with a few species in Central and South America. By far the greatest number of species occurs in eastern Asia, but most of the long-established food-producing species originated in Europe and western Asia. Over 100 species have been brought into cultivation.

Many of the stone fruits have been cultivated since ancient times for their edible fruit and a few for their edible seeds. Wild species were a source of food for Indians and early settlers and are still collected for food. Several species are valuable ornamentals because of their attractive flowers and relatively rapid growth in a variety of soils. Trees for fruit production and many ornamentals are propagated by budding or grafting, but seed germination is necessary to grow rootstock and in breeding programs. The most important rootstock species and their scion combinations are as follows:

Rootstock	Scion combinations
Prunus americana	plum in cold climates
Prunus amygdalus	almond and plum
Prunus armeniaca	apricot and plum
Prunus avium	sweet cherry
Prunus besseyi	dwarf peach
Prunus cerasifera	almond and plum
Prunus insititia	plum
Prunus mahaleb	sweet and sour cherry
Prunus persica	peach, almond, apricot, and plum

Black cherry (*P. serotina*) is the most important timber-producing species, but several others that attain sufficient size, such as mazzard cherry (*P. avium*) and mahaleb cherry (*P. mahaleb*) in Europe and *P. serrulata* in Japan, are used for wood products. Most species are important for wildlife. The browse of some species can be poisonous to livestock. Several species are used in shelterbelt plantings. Seeds are dispersed by birds (Herrera and Jordano 1981).

FLOWERING AND FRUITING *Prunus* flowers are bisexual. They normally have 5 white or pink petals and 15 to 20 stamens. The flowers are solitary, in umbellike clusters or racemes, and usually appear before or with the leaves. They are insect pollinated. The fruit is a 1-seeded drupe that is thick and fleshy, except in the almonds, and has a bony stone or pit (Figs. 1, 2). Flowering and fruiting dates are given in Table 1.

P. americana
American plum

P. angustifolia
Chicksaw plum

P. armeniaca
apricot

P. persica
peach

Figure 1. *Prunus*: stones, ×2 (Grisez 1974).

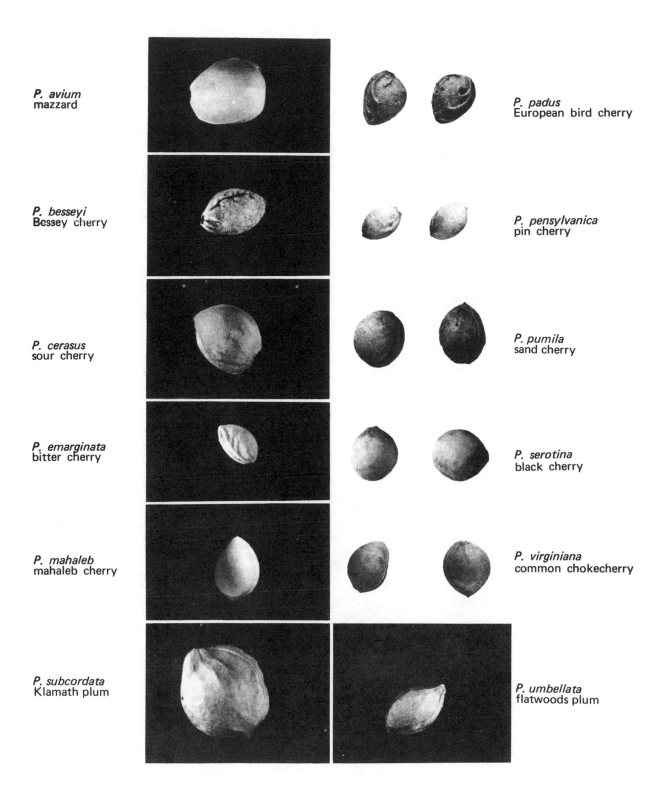

P. avium
mazzard

P. besseyi
Bessey cherry

P. cerasus
sour cherry

P. emarginata
bitter cherry

P. mahaleb
mahaleb cherry

P. subcordata
Klamath plum

P. padus
European bird cherry

P. pensylvanica
pin cherry

P. pumila
sand cherry

P. serotina
black cherry

P. virginiana
common chokecherry

P. umbellata
flatwoods plum

Figure 1. Continued.

Table 1. *Prunus*: phenology of flowering and fruiting, characteristics of mature trees, cleaned seed weight, and pregermination treatments (Grisez 1974).

Species	Flowering dates	Fruit ripening dates	Seed dispersal dates	Mature tree height (m)	Seed-bearing Year first cultivated	Seed-bearing Age (years)	Interval (years)	Ripe fruit color	Fruit diameter (mm)	Seeds/ gram	Warm stratification (days)	Prechilling (days)
P. alleghaniensis	Apr.–May	Aug.–Sept.	–	5	1889	–	1	Dark purple	10	6.5	0	150
P. americana	Mar.–May	June–Oct.	June–Oct.	9	1768	–	1–2	Red-yellow	20–30	1.9	0	90–150
P. amygdalus	Feb.–Mar.	Aug.–Oct.	–	9	Early	6–7	1	Brownish	–	0.4	0	65
P. angustifolia	Mar.–Apr.	May–July	May–July	8	1874	–	–	Red-yellow	10–20	2.3	–	–
P. armeniaca	Feb.–Mar.	May–June	–	11	Early	–	2	Yellow-red	30+	0.7	0	60–120
P. avium	Apr.–May	June–July	July–Aug.	30	Early	6–7	1	Yellow-black	20–25	5.2	0	90–125
P. besseyi	Apr.–May	July–Sept.	–	1	1892	2–3	–	Purple black	15	5.3	0	120
P. cerasifera	Apr.–May	July–Aug.	July–Sept.	8	Early	–	2–3	Red	16–25	2.2	14	189
P. cerasus	Apr.–May	June–July	–	15	Early	6–7	1	Red	8–25	6.4	0	90–150
P. domestica	May	July–Oct.	–	12	Early	–	–	Blue-purple	8–12	1.3	14	189
P. emarginata	Apr.–June	July–Sept.	Aug.–Sept.	15	1918	–	–	Red	13–17	15.5	0	90–126
P. ilicifolia	Mar.–May	Sept.–Oct.	Oct.–Dec.	9	1925	3	–	Purple-black	–	0.5	0	0
P. insititia	Apr.–May	Aug.–Sept.	–	8	Early	–	–	Black	–	3.0	0	84–112
P. mahaleb	Apr.–May	July	–	10	Early	–	1–2	Red-yellow	–	11.3	0	80–100
P. munsoniana	Mar.–May	July–Sept.	–	9	1909	–	–	Black	–	3.7	0	80–100
P. padus	Apr.–May	June–July	–	15	Early	–	2	Red	6–8	19.6	0	100–120
P. pensylvanica	Mar.–July	June–July	Aug.	12	1773	2	–	Yellow-red	5–7	31.3	60	90
P. persica	Mar.–May	July–Oct.	–	8	Early	4–8	1–2	Purple-black	30–60	0.3	0	98–105
P. pumila	May–July	July–Sept.	–	3	1756	–	–	Black	10	6.4	60	120
P. serotina	May–June	Aug.–Sept.	Aug.–Sept.	34	1629	5	1–5	Black	7–10	9.3	0	120
P. spinosa	Apr.–May	Aug.–Sept.	Sept.	4	Early	–	1–2	Blue-black	10–15	4.9	0	170
P. subcordata	Mar.–May	Aug.–Sept.	–	8	1850	–	2	Red-yellow	–	1.2	0	90
P. tomentosa	May	July	Aug.	3	1870	2–3	1–2	Red	10–31	10.4	0	60–120
P. virginiana	May	Aug.–Sept.	–	9	1724	–	–	Red-purple	8	10.6	0	120–160

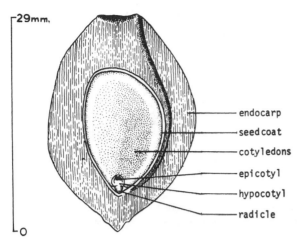

Figure 2. *Prunus persica*, peach: longitudinal section through a stone showing the embryo and no endosperm, ×2 (Grisez 1974).

COLLECTION, EXTRACTION, AND STORAGE OF SEED *Prunus* fruits should be collected when fully mature (Table 1). Color and condition of fruit indicate maturity. For species in which the ripe color is nearly black, the preripe color is red. For species where the ripe color is red, the preripe color is yellowish or partly green and red. Almonds are harvested when the husks have split.

Fruit is collected by hand stripping or spreading ground cloths to catch the naturally falling fruits. Black cherry fruits can be collected from trees in logging operations.

Seeds can be extracted by maceration and recovered by flotation. Seed weights are listed in Table 1.

For prolonged storage, seeds must be air dried and stored in sealed containers at cold temperatures. Grisez (1974) experimented with storage of black cherry seeds of high-moisture content, in plastic bags at low temperatures. After 8 years viability with this method was equal to storage at low-moisture content in sealed jars at low temperatures. Grzeskowiak and Suszka (1983) determined that seeds of mazzard cherry could be prechilled for up to 2 weeks and then dried and stored without losing the germination enhancement of the prechilling period. The partially prechilled and dried seeds could be stored without loss of viability for 16 weeks.

Wendel (1977) found that the natural germination of black cherry seeds on the forest floor occurred predominately in the first or second year after seedfall depending on the year. Most seeds germinated by the end of 3 years after seedfall. Some seeds appeared viable after 3 years, but had apparently acquired a deep dormancy.

PREGERMINATION TREATMENT *Prunus* seeds have embryo dormancy and require prolonged prechilling. Because of the presence of a hard endocarp, *Prunus* species have often been thought to have hard seedcoats. The endocarp may offer resistance to germination, but it is usually permeable to water. Removal of the endocarp may speed germination in certain species.

GERMINATION Germination standards for *Prunus* species are quite simple (ASOA 1985): substrate, paper; incubation temperature, 18 to 22°C; duration, 14 days; and additional directions, use embryo excision or TZ. Specific prechilling requirements for individual species are given in Table 1.

Selected recent literature on the germination of *Prunus* species is given in Table 2.

Table 2. *Prunus*: selected recent literature on germination.

Author—Date	Subject	Location
Auchmoody 1979	*P. pensylvanica*, influence of nitrogen fertilization on germination.	Canada
Balboa 1982	*P. avium*, bound gibberellins found in seeds.	–
Chang and Wenner 1984	*P. persica*, respiration during prechilling, and chilling requirements for peach plants.	USA
Eijsackers and Ham 1984	*P. serotina*, influence of seed production environment on germination.	Netherlands
Laidlaw 1987	*P. pensylvanica*, prechilled seeds incubated at 5/30°C for 12/12 hours to enhance germination.	USA
Lockley 1980	*P. virginiana*, wide variations in prechilling requirements (and subsequent germination) among and within 13 open-pollinated families. Correlation (poor) only between germination at high temperatures and spring emergence from fall sowing.	USA
Mehanna and Martin 1985	*P. persica*, seedcoat inhibits germination of non-prechilled seeds.	USA
Mehanna et al. 1985	*P. persica*, influence of growth hormones on prechilling requirements and germination depended on cultivar.	USA
Michalska 1983	*P. avium*, seedcoat prevents premature germination before afterripening has been completed during prechilling.	Poland
Morgenson 1986	*P. fruticosa*, *P. maacki*, require warm stratification and prechilling.	USA
Rouskas 1983	*P. persica*, seed soaked in benzyl aminopurine to replace prechilling requirement.	India
Therios 1981	*P. amygdalus*, temperature of 10°C more appropriate than usual lower temperatures for prechilling seeds of this species.	Greece

NURSERY AND FIELD PRACTICE Untreated seeds of *Prunus* species may be sown in the fall; prechilled seeds in the spring. If sowing in the fall, it is important to sow early enough so seeds can prechill before seebbeds freeze. This can be overcome by mulching the seedbeds.

Prunus seedlings reach suitable size for transplanting in 1 or 2 years. Germination is epigeal or hypogeal, depending on the species (Figs. 3, 4). Propagation techniques are given in Table 3.

Table 3. *Prunus*: propagation techniques (Dirr and Heuser 1987).

Species	Propagation techniques
P. americana	Seeds: may take 2 years to emerge unless prechilled. Fresh seeds should be warm stratified, then prechilled. Stored seeds should be soaked in aerated water for 2 weeks, then planted in July for emergence the next spring. Cuttings: not easy to root. Hardwood cuttings taken in late January have been rooted.
P. armeniaca	Seeds: prechill 1–5 months. Tissue culture: micropropagation techniques have been successful.
P. avium	Seeds: often difficult to germinate. Warm stratification for 3 weeks followed by 12 to 15 weeks of prechilling has been used. Cuttings: leaf-bud cuttings taken in mid-April will root. Tissue culture: at least 4 cultivars have been propagated in vitro.
P. besseyi	Seeds: require 3 months of prechilling. Cuttings: softwood cuttings taken in mid-August root easily.
P. caroliniana	Seeds: plant in the fall or prechill for 1 to 2 months. Cuttings: take soft to semihardwood cuttings from June to September.
P. cerasifera	Seeds: may require both warm stratification and prechilling or only 3 months of prechilling. Cuttings: propagated by hardwood cuttings in England. Tissue culture: actively growing shoot tips have been proliferated in vitro.
P. cerasus	Cuttings: cuttings taken in late June are easily rooted. Tissue culture: buds with dormancy were used for micropropagation.
P. cyclamina	Seeds: require prechilling 3 months.
P. dulcis	Seeds: prechill 2 or 3 months. Tissue culture: dormant buds collected in midwinter have been used for micropropagation.
P. glandulosa	Seeds: prechill 1 to 2 months. Fruit set rare in dwarf form. Cuttings: easily rooted from summer softwoods.
P. incisa	Cuttings: easily rooted from cuttings taken at the beginning of June.
P. laurocerasus	Seeds: prechill 2 to 3 months. Cuttings: can be rooted from midsummer cuttings.
P. maackii	Seeds: double dormancy requires 4 months of warm stratification and 4 months of prechilling. Cuttings: softwood cuttings taken in late July will root.
P. maritima	Seeds: require prechilling 2 to 3 months.
P. nigra	Seeds: require prechilling 3 months.
P. padus	Seeds: untreated seeds do not germinate. Prechilling for 3 to 4 months required.
P. pensylvanica	Seeds: require warm stratification for 2 months followed by 3 months of prechilling, or sow in early fall. Cuttings: difficult to root.
P. persica	Seeds: require prechilling 2 months. Cuttings: two-node stem cuttings of 23- to 79-day-old seedlings root readily. Several other types of cuttings are rooted. Tissue culture: actively growing shoot tips have been proliferated in vitro.
P. pumila	Seeds: prechill 3 months.
P. sargentii	Seeds: prechill 3 to 6 months.
P. serotina	Seeds: warm stratification for 2 weeks followed by 4 months of prechilling has been used. Cuttings: wide tree-to-tree variation in rooting percentages. Softwood cuttings from juvenile plants have been rooted.
P. serruls	Seeds: prechill 2 months. Cuttings: take soft tip cuttings in late July.
P. serrulata	Seeds: prechill 3 months. Cuttings: take cuttings in mid- to late summer.
P. subhirtella	Seeds: prechill 3 months. Cuttings: roots readily from softwood cuttings.
P. tenella	Seeds: prechill 3 months. Cuttings: softwood cuttings taken in mid-May will root.
P. tomentosa	Seeds: sow in the fall or prechill 2 to 3 months. Cuttings: softwood cuttings taken in mid-July will root. Tissue culture: actively growing shoot tips proliferated in vitro.
P. triloba	Cuttings: not difficult to root from softwoods.
P. virginiana	Seeds: prechill 3 months. Cuttings: terminal or basal cuttings taken in June have been rooted.

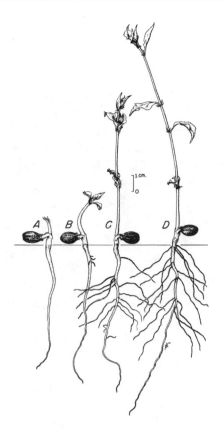

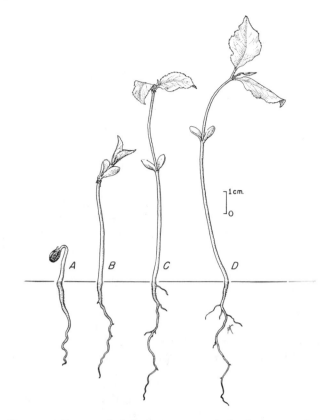

Figure 3. *Prunus americana*, American plum: seedling development at 1, 3, 5, and 9 days after hypogeal germination (Grisez 1974).

Figure 4. *Prunus virginiana*, common chokecherry: seedling development at 1, 3, 7, and 11 days after epigeal germination (Grisez 1974).

PINACEAE— PINE FAMILY

Pseudolarix kaempferi Gord. — Golden-larch

GROWTH HABIT, OCCURRENCE, AND USE *Pseudolarix* is a monospecific genus containing a deciduous tree that reaches 40 m in height and is native to eastern China. It differs from *Larix* in that the staminate flowers are clustered and pendulous, and the cone scales are deciduous. Male and female cones are borne separately on the same tree. The cones are ready to collect when they turn a bright golden brown. Seed storage at room temperature results in rapid loss of viability (Dirr and Heuser 1987). Seeds can be stored after drying at low temperatures in sealed containers. Isolated trees produce an abundance of cones, but few seeds. Trees may be self-sterile. A fair percentage of seeds germinate without pretreatment, but 1 to 2 months of prechilling improves and unifies germination (Dirr and Heuser 1987).

PINACEAE

Pseudotsuga Carr. — Douglas fir

GROWTH HABIT, OCCURRENCE, AND USE The genus *Pseudotsuga* consists of 6 species, 2 of which are native to North America (Owston and Stein 1974). *Pseudotsuga menziesii*, the major commercial species in North America, includes 2 geographic races: Coastal Douglas fir (var. *menziesii*) is fast-growing and long-lived. It sometimes becomes 90+ m tall and attains diameters of 2.5 to 3 m. Rocky Mountain Douglas fir (var. *glauca*) is slower growing, shorter lived, and seldom exceeds 40 m in height. Intermediate forms occur where the ranges overlap. The wood has exceptional strength and is widely used for structural timber. Other uses include poles, plywood, and pulp. More than a dozen ornamental cultivars are propagated. Big cone Douglas fir (*P. macrocarpa*) attains a height of 18 to 27 m and is native from Santa Barbara to San Diego counties in southern California.

FLOWERING AND FRUITING Male and female strobili burst bud during winter and spring (Owston and Stein 1974):

Species	Flowering dates	Cone ripening dates	Seed dispersal dates
P. macrocarpa	Feb.–Apr.	Aug.–Oct.	Aug.–Oct.
P. menziesii			
var. glauca	May–June	July–Aug.	Aug.–Sept.
var. menziesii	Mar.–June	Aug.–Sept.	Sept.–Mar.

Male strobili are generally borne abundantly over much of the crown on the lower half of year-old shoots, becoming somewhat pendant when mature and 2 cm long. The females, developing more distally on shoots located primarily in the upper crown, are erect at the time of pollen shedding and about 3 cm long; their appearance is dominated by large trident bracts. Color of the female ranges from deep green to deep red and that of the male from yellow to deep red. Strobili of the same sex tend to be the same color on a given tree, but colors of male amd female cones may differ. Pollination in a given location occurs over a 2- to 3-week period. Female strobili soon become pendant, and fertilization takes place about 10 weeks after pollination. Seeds develop through late spring and summer, reaching maturity in August and September. Cones generally dry and turn brown in August and September; most seed is released in September and October.

Calendar dates for these phenological events varies not only with latitude but also with elevation, between individual trees in a given locality, and in different parts of the crown. Timing also varies from year to year depending on weather conditions. Since cones open by drying, the time of seed dispersal is particularly influenced by late summer weather.

The mature pendant cones of Douglas fir are easily identified by their 3-lobed bracts that protrude beyond the cone scales (Fig. 1). Under each scale are borne 2

Figure 1. *Pseudotsuga menziesii* var. *menziesii*, coastal Douglas fir: mature, unopened cones with characteristic 3-lobed bracts, ×0.5 (Owston and Stein 1974).

seeds which have relatively large wings (Fig. 2). One side of the seed is variegated light brown; the other is more glossy and dark brown. Embryos are linear (Fig. 3). The number of seeds per cone ranges from 20 to 30 for the Rocky Mountain variety and from 26 to 50 for the coastal variety of Douglas fir. The seeds are wind disseminated; their distance of travel is quite variable. Cones may be retained on the trees for 1 or more years after seed dispersal.

Cone and seed production in Douglas fir are quite variable from tree to tree. In a good seed year, an average mature, forest-grown coastal Douglas fir produces about 0.5 kg of seed; widely spaced trees may produce 1 kg or more. Trees 100 to 200 years old are most prolific, but cones from younger trees are larger and contain more viable seeds. Seedfall was recorded in a Douglas fir stand in Washington containing trees aged 39 to 68 years (Reukema 1982). Annual seed production ranged from 0 to 3 million seeds per hectare. Thinning substantially increased seed production, but there was no lasting effect. Girdling by saw cuts has been used to increase cone production of Douglas fir (Wheeler et al. 1985).

Environmental factors make Douglas fir seed crops erratic, and abundant crops occur from 2 to 11 years apart (Owston and Stein 1974):

Species	Mature tree height (m)	Seed-bearing age (years)	Interval (years)
P. macrocarpa	27	20	Infrequent
P. menziesii			
var. glauca	40	20	3–11
var. menziesii	90	7–10	2–11

One crop failure and 2 or 3 light crops usually occur between heavy crops.

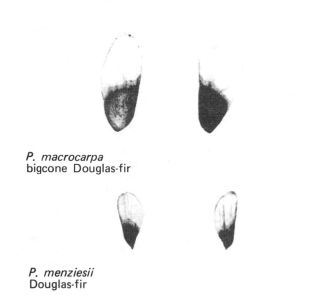

P. *macrocarpa*
bigcone Douglas-fir

P. *menziesii*
Douglas-fir

Figure 2. *Pseudotsuga*: seeds with wings intact, ×1. The 2 varieties of *P. menziesii*, coastal Douglas fir and Rocky Mountain Douglas fir, bear seeds that are similar externally and anatomically (Owston and Stein 1974).

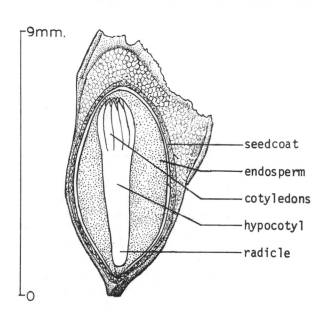

Figure 3. *Pseudotsuga menziesii*, Douglas fir: longitudinal section through a seed, ×8 (Owston and Stein 1974).

Predicting seed crops for commercial cone collection is only possible 2 months before cone maturity. Longer-range predictions can identify crop failures. Early picking of cones increases the time required for processing and produces erratic germination (Olson and Silver 1975).

Major insect pests of Douglas fir cones and seeds are the Douglas fir seed chalcid (*Megastigmus spermotrophus*), the Douglas fir cone moth (*Barbara colfaxiana*), the fir cone looper (*Eupithecia spermaphaga*), and gall midges (*Itonididae*). Leaf bugs (*Leptoglossus occidentalis*) attack developing male strobili, female cones, and seeds, as well as mature seeds within dry cones. Heavy damage to cones of the Rocky Mountain Douglas fir by a budworm (*Choristoneura occidentalis*) has been reported. Insects can destroy most of the seed in small crop years. Hedlin and Ruth (1978) reported on the influence of insect infestations on germination of Douglas fir seeds. They reported a year later on genetic variation for insect resistance in Douglas fir clones (Hedlin and Ruth 1978). Miller (1983) reported on cost/benefit ratios for the control of insects in Douglas fir seed orchards. Shearer (1984) updated the information on the influence of insects on seed production in Douglas fir.

Douglas fir strobili, cones, and seeds are depleted by frost, small mammals, and birds. Strobili and bud damage by spring frost occurs periodically. Rodents start clipping cones early in the season, but squirrels cut ripe cones in large quantities. Sullivan (1978) suggests that deer mice (*Peromyscus maniculatus*) are not a major factor in caching seeds of Douglas fir so that they would lead to natural regeneration. Alternative food sources have been offered to reduce seed predation by small mammals (Sullivan 1979).

COLLECTION, EXTRACTION, AND STORAGE OF SEED Commercial cone collection for coastal Douglas fir may begin as early as mid-August in warm, low elevation and end in October at high elevations. The collection period at any one location lasts only 2 to 3 weeks. A commonly used guide to cone maturity is a golden brown seedcoat with a wing of the same color. A firm, nonmilky endosperm enclosing a yellowish green embryo also indicates ripeness.

Douglas fir cones are often picked from standing trees. During good seed crop years, squirrel caches are also a prime source of cones. Squirrel cut cones are found scattered under trees or stored in caches. Collections from caches are feasible because the cones are cached under conditions that prevent drying and cone opening. Experiments have been conducted in Europe to develop a machine to harvest Douglas fir cones (Kofman and Workhoven 1977). Seeds can be collected from felled trees.

Cones are usually put in burlap bags for transportation and storage. Sacks should only be half-filled for storage, good ventilation must be provided around and among sacks. Under well-ventilated conditions, cones may be stored for 3 or 4 months without decreasing seed viability. Seeds not fully mature benefit from such a period of afterripening. Prolonged storage will lower seed viability.

Cones may be opened by air drying in warm dry weather, but most extractors rely on kilns. Kiln drying is done mostly between 32 and 44°C for 2 to 48 hours (Owston and Stein 1974):

Species	Air-drying time (days)	Kiln drying Time (hour)	Kiln drying Temperature (°C)
P. macrocarpa	8–10	—	—
P. menziesii			
var. *glauca*	14–60	2–10	38–44
var. *menziesii*	8–21	16–48	32–44

Drying time depends on the moisture content of the cones and rate of air flow through the kiln. Cones of coastal Douglas fir open completely when 35 to 51% of their wet weight is lost.

Extraction and cleaning usually involve (1) tumbling dried cones; (2) screening to separate seeds from cone scales, dirt, and debris; (3) dewinging; and (4) air screening. Vibratory or pneumatic separators are sometimes used for further cleaning. Care must be taken to limit seed damage during these operations. Prolonged dewinging or dewinging that involves considerable hard, sharp debris such as cone scales, is particularly damaging.

Fresh cones weigh 11 to 27 kg per 35 liters, which yields about 230 of seed. Seeds of both varieties of Douglas fir tend to become larger from northern to southern sources. In the Pacific Northwest and northern California, seeds also increase in size as elevation of the source increases: *P. macrocarpa* yields 9 seeds/gram, *P. menziesii* var. *glauca* yields 96, and *P. menziesii* var. *menziesii* yields 80 seeds/gram (Owston and Stein 1974).

Seeds of Douglas fir are generally stored at a moisture content of 6 to 9% at −18°C in sacks or plastic-lined drums. Seed viability has been maintained for 10 to 20 years under these conditions. Storage at 0°C also appears satisfactory. Viability declines rapidly at room temperatures.

PREGERMINATION TREATMENT Prechilling seed of Douglas fir often improves germination energy (rate) and capacity (total germination). Some seed lots of Douglas fir will not respond to prechilling and germinate without pretreatment.

GERMINATION Germination standards for Douglas fir are substrata, top of blotter or covered petri dish; temperature, 20/30°C; and duration, 21 days (ASOA 1985). Additional directions, however, vary (ASOA 1985): For *P. menxiesii* var. *caesia*, prechill seeds 21 days at 3 to 5°C and provide light; use vermiculite if top of blotter is not used as substrate. For *P. menziesii* var. *glauca*, provide light; central and southern Rocky Mountain sources are not sensitive to temperature; vermiculite is recommended if top of blotter is not used as substrate. For *P. menziesii* var. *menziesii* prechill seed 21 days at 3 to 5°C, and provide light; if top of blotter is not used, perlite or vermiculite is recommended.

Hydrogen peroxide treatments can be used in place of prechilling germination tests for Douglas fir seeds (Bonnet-Masimbert and Muller 1974).

Selected recent literature concerning germination of Douglas fir seeds is listed in Table 1.

Table 1. *Pseudotsuga:* selected recent literature on germination.

Author—Date	Subject	Location
	Germination	
Borno and Taylor 1975	Influence of ethylene on germination.	Canada
Campbell and Sorenson 1979	Basis for characterizing seed germination.	USA
Hedderwick 1970	Prolonged drying of prechilled seeds and subsequent germination.	New Zealand
Sorensen and Campbell 1981	Orientation of seed and germination.	–
Taylor and Waring 1979a	Influence of prechilling on growth regulator levels in seeds.	–
Vanesse 1974a	Temperature relations and dormancy.	Belgium
Vanesse 1974b	Measuring germination.	Belgium
Vanesse 1975	Prechilling to break dormancy.	Belgium
	Nursery Practices	
Guariglia and Thompson 1985	Influence of sowing depth and mulch on emergence.	USA
Minore 1984	Soils and competition influence on germination.	USA
Muhle and Hewicker 1976	Pelleted and paper strip seeding.	Germany
Noble et al. 1978	Irrigation influence.	Germany
Sorensen and Campbell 1985	Influence of seed weight on height growth.	Canada
Sutherland and Edwards 1976	Influence of seed pigments on germination.	Canada
Timmis and Worrall 1975	Cold acclimation of seedlings.	Canada

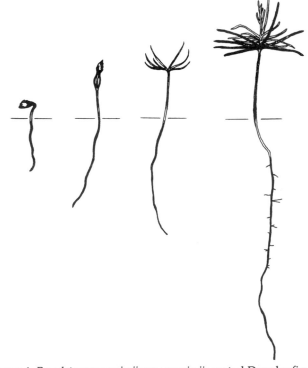

Figure 4. *Pseudotsuga menziesii* var. *menziesii*: coated Douglas fir: seedling development 2, 5, 8, and 22 days after emergence, ×0.5 (Owston and Stein 1974).

NURSERY AND FIELD PRACTICE In most western North American nurseries, seeds of Douglas fir are sown in the spring, despite the requirement for prechilling. Problems of protecting seedlings during the winter limit the use of fall seeding. Seeds are usually soaked overnight to 48 hours and prechilled in plastic bags.

Seeds are sown in drill rows about 1 cm deep. They are surface dried before drilling, but are not allowed to become dry. Sowing rate is selected to produce 190 to 500 seedlings per m². The age of transplanting seedlings (Fig. 4) to the field depends on the characteristics of the site being planted.

Douglas fir can be propagated vegetatively from stem cuttings taken from young trees.

Selected recent literature on nursery practices for Douglas fir is given in Table 1.

COMPOSITAE — SUNFLOWER FAMILY

Psilostrophe DC.

GROWTH HABIT, OCCURRENCE, AND USE The genus *Psilostrophe* consists of about 6 species found in southwestern United States and northern Mexico. *Psilostrophe cooperi* is a semi-woody perennial that reaches 0.5 m in height. The fruit is an achene.

Initial germination without pretreatment was 30% (Kay et al. 1988). Viability declined to less than 10% with all forms of storage after 1 year.

RUTACEAE — RUE FAMILY

Ptelea trifoliata L. — Common hoptree

GROWTH HABIT, OCCURRENCE, AND USE Common hoptree is a small tree or shrub up to 8 m tall with some value for wildlife, shelterbelt, and environmental plantings (Brinkman and Schlesinger 1974). The species has been cultivated since 1724, and is native from Connecticut and New York to southern Ontario, central

Figure 1. *Ptelea trifoliata*, common hoptree: fruit (samara), ×2 (Brinkman and Schlesinger 1974).

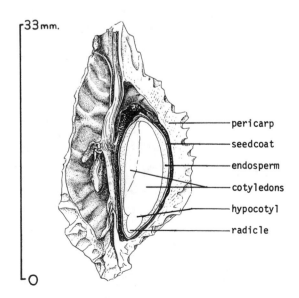

Figure 2. *Ptelea trifoliata*, common hoptree: longitudinal section through a samara, ×2 (Brinkman and Schlesinger 1974).

Michigan and eastern Kansas, south to Texas and northern Florida.

FLOWERING AND FRUITING The polygamous flowers open from April in the Carolinas to July in the north. The fruits are reddish brown samaras (Figs. 1, 2) that ripen from June to November and may persist until spring. Hoptree is an abundant seeder; the samaras are dispersed by wind.

COLLECTION, EXTRACTION, AND STORAGE OF SEED The ripe samaras may be picked by September. They may require a few days drying if they are to be stored. There are about 26 seeds per gram. Seeds stored in sealed containers at low temperatures should retain viability for 16 months.

GERMINATION Hoptree seeds germinate slowly, apparently due to embryo dormancy (Fig. 3). Prechilling for 3 to 4 months enhances germination.

NURSERY AND FIELD PRACTICE Seeds should either be sown in the fall or prechilled seeds used in the spring. Fall-sown beds need to be mulched. Propagation is possible by layering, grafting, or budding.

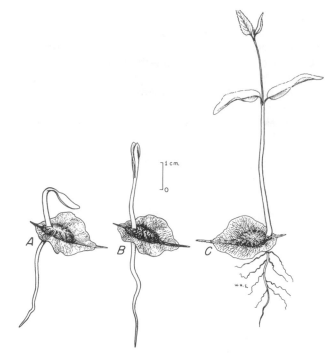

Figure 3. *Ptelea trifoliata*, common hoptree: seedling development at 1, 2, and 10 days after germination (Brinkman and Schlesinger 1974).

JUGLANDACEAE — WALNUT FAMILY

Pterocarya Kenth. — Caucasian wingnut

GROWTH HABIT, OCCURRENCE, AND USE Native to Asia from the Caucasus to China, the genus *Pterocarya* consists of about 10 species of trees that reach 30 m in height, with flowers dangling golden catkins.

GERMINATION *Pterocarya fraxinifolia* is easily grown from seed which should be collected when it turns from green to brown in September and October (Dirr and Heuser 1987). It stores satisfactorily for short periods. Fresh seeds germinate sporadically, but 3 months of prechilling produces good germination. This is a difficult species to root.

PUNICACEAE — POMEGRANATE FAMILY

Punica L. — Pomegranate

GROWTH HABIT, OCCURRENCE, AND USE In genus *Punica* there are 2 species native from the Mediterranean region to the Himalayan mountains. They are currently grown widely in tropical and subtropical regions around the world for fruit production.

The plants are deciduous shrubs or trees to 7 m tall.

The fruit is a thick-skinned, spherical, several-celled berry, with seeds that are surrounded by a fleshy pulp.

The fruits should be collected in late summer or fall and the seeds removed and cleaned. Seeds can be stored dry under cool temperatures, and germinate without pretreatment (Dirr and Heuser 1987).

ROSACEAE — ROSE FAMILY

Purshia DC. — Bitterbrush

GROWTH HABIT, OCCURRENCE, AND USE Antelope bitter-brush (*Purshia tridentata*) is found from British Columbia to western Montana and as far south as New Mexico and northern Arizona (Deitschman et al. 1974d). Desert bitterbrush (*P. glandulosa*) is limited to to the southern parts of Utah, Nevada, and California. Both grow over a wide elevational range from sea level to 3000 m in California, and both vary in appearance from low, layering shrubs to shrubs 5 m tall. They are major browse species for domestic and wild animals.

FLOWERING AND FRUITING The perfect yellow flowers are borne singly at the end of short, lateral leafy spurs. They appear in mid-April or early May. The fruit is an oblong, pubescent achene, 0.6 to 1 cm long (Figs. 1, 2). The fruit ripens and is dispersed during July or early August. Seed production can be eliminated by late spring frost, and good seed crops can be expected at widely varying intervals from 2 to 6 years.

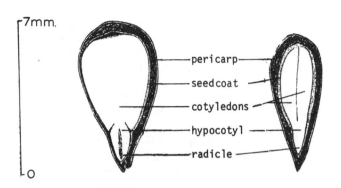

Figure 1. Hybrid achene, *Purshia tridentata* × *Cowania mexicana* var. *stanburiana*, ×2 (Deitschman et al. 1974d).

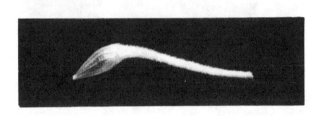

P. glandulosa
desert bitterbrush

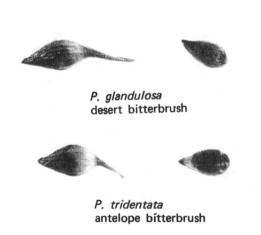

P. tridentata
antelope bitterbrush

Figure 2. *Purshia*: achenes and cleaned seeds, ×2 (Deitschman et al. 1974d).

COLLECTION, EXTRACTION AND STORAGE OF SEED Ripe seeds can be collected by shaking or stripping branches over a suitable container, or with a vacuum harvester. Seeds can shatter quite rapidly at maturity. A papery flower part remains around the seed and can be removed by any of several rubbing threshers, with the seed recovered by air screening (Fig. 3). There are about 34 antelope bitterbrush seeds per gram and 45 desert bitterbrush seeds. Bitterbrush seeds can be stored for as long as 5 years in cloth bags in warehouse conditions.

Figure 3. *Purshia tridentata*, antelope bitterbrush: longitudinal section of seeds through 2 planes. ×6 (Deitschman et al. 1974d).

GERMINATION Bitterbrush seeds are largely dormant and require prechilling before they will germinate. Prechilling temperatures from 0 to 5°C are ideal (Young and Evans 1976). Under ideal conditions 3 weeks to a month of prechilling is all that is required. Temperature or moisture stress will make the pre-chilling requirements much longer and less effective.

The prechilling requirement for bitterbrush seeds can be replaced by soaking in hydrogen peroxide solutions (Everett and Meeuwig 1975). Warm dry seed pretreatment for 4 to 8 week periods at 15 to 30°C significantly increased germination subsequent to inadequate prechilling treatments (Meyer 1989). Warm moist pretreatment markedly decreased germination. Germination standards have been recently developed for antelope bitterbrush seeds that prescribe prechilling followed by incubation at 20°C on top of paper.

NURSERY AND FIELD PRACTICE The natural regeneration of bitterbrush is extremely complex with most seeds being collected, threshed, and cached in scatter hoards by rodents. Some seeds may be consumed directly, but most seeds germinate and the rodents return to graze on the emerging seedlings. Artificial sowing of bitterbrush seeds in the field often results in rodents collecting and caching the seeds.

Both container-grown and bare root seedlings of bitterbrush are transplanted to the field often with high levels of successful plant establishment.

ROSACEAE — ROSE FAMILY

Pyracantha M. Roem. — Firethorn

GROWTH HABIT, OCCURRENCE, AND USE The genus *Pyracantha* has about 6 species of evergreen shrubs native to southeastern Europe and Asia. The fruit consists of 5 nutlets in an orange or red berrylike structure. It should be collected in fall or winter and macerated to remove the seeds, which can be recovered by flotation (Dirr and Heuser 1987).

Seeds require prechilling for 3 months before they will germinate. *Pyracantha* species can be propagated from softwood, semisoftwood, or hardwood cuttings.

ROSACEAE — ROSE FAMILY

Pyrus L. — Pear

GROWTH HABIT, OCCURRENCE, AND USE The common pear (*Pyrus communis*) is native to Europe and western Asia. It is naturalized in eastern United States (Gill and Pogge 1974f). Ussurian pear (*P. ussuriensis*), introduced from Asia, has been used on the northern Great Plains in shelterbelt and environmental plantings.

FLOWERING AND FRUITING Large bisexual flowers appear with or before the leaves during March to May. Fruits (pomes) of common pear in the wild are usually less than 5 cm long, but those of most commercial cultivars have larger fruits. Fruit color when ripe may be green, yellow, russet, red, or a combination of these. Ripe fruits contain 4 to 10 smooth, black or nearly black seeds with a thin layer of endosperm (Figs. 1, 2). Ussurian pears turn from green to yellow in ripening and are about 3.1 to 3.75 cm across. Ripening dates vary from mid-July to October.

COLLECTION, EXTRACTION, AND STORAGE OF SEED The ripe fruits may be picked from trees or shaken to the ground. Seeds can be recovered by running the fruits through a macerator, drying them, and using an air screen. The common pear has about 31 seeds per gram and Ussurian pear about 20. For storage, seeds should

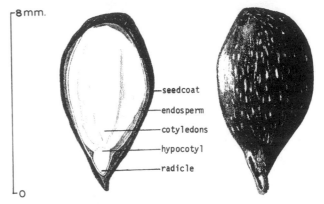

Figure 2. *Pyrus communis*, common pear: longitudinal section through a seed (left) and exterior view (right), ×6 (Gill and Pogge 1974f).

be dried to a moisture content of about 10%, sealed in containers, and stored at low temperatures.

GERMINATION Seeds of *Pyrus* species extracted from fresh mature fruits are dormant and fail to germinate unless prechilled. If partly dormant seeds of pear are tested for germination at temperatures above 20°C, a secondary dormancy will be induced. Suggested germination procedures follow (ASOA 1985, Genebank Handbook 1985):

Species	Treatment
P. amygdaliformis	Prechill 25 days at 5°C.
P. arbutifolia	Prechill 90 days at 1°C and incubate at 20°C.
P. betulaefolia	Prechill 85 days at 5°C.
P. calleryana	Prechill 80 days at 5°C and incubate at 20°C in light.
P. communis	Prechill 130 days at 5°C or use excised embryo technique.
P. dimorphophylla	Prechill 90 days at 5°C.
P. elaeagrifolia	Prechill 130 days at 5°C .
P. fauriei	Prechill 80 days at 5°C.
P. gharbiana	Prechill 80 days at 5°C.
P. mamorensis	Prechill 60 days at 5°C.
P. pashia	Prechill 40 days at 5°C.
P. pyrifolia	Prechill 160 days at 5°C.
P. syriaca	Prechill 80 days at 5°C.
P. ussuriensis	Prechill 100 days at 5°C.

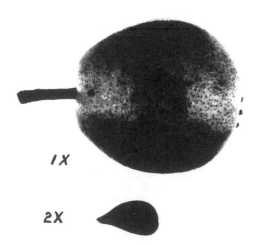

Figure 1. *Pyrus ussuriensis*, Ussurian pear: fruit, ×1, and seed, ×2 (Gill and Pogge 1974f).

NURSERY AND FIELD PRACTICE Pear seeds can be sown in the fall or in the spring if prechilled. They should be sown in drills with 32 to 48 seeds per meter of row. Germination is epigeal (Fig. 3).

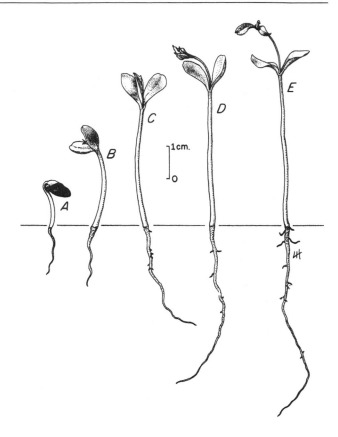

Figure 3. *Pyrus communis,* common pear: seedling development at 1, 2, 3, 6, and 12 days after germination (Gill and Pogge 1974f).

FAGACEAE — BEECH FAMILY

Quercus L. — Oak

GROWTH HABIT, OCCURRENCE, AND USE The genus *Quercus,* with many species of deciduous and evergreen trees and shrubs, is the most important aggregation of hardwoods found on the North American continent, if not the Northern Hemisphere (Olson 1974c). The oaks are widely distributed throughout temperate regions of the Northern Hemisphere, and extend southward to the mountains of Colombia in South America, and to the Indian Archipelago in the Eastern Hemisphere. There are 70 species native to the United States, and about 58 of these reach tree size. There are also about 70 recognized hybrids. The uses of oak include almost everything the human race has ever derived from trees: timber, food for people and animals, fuel, watershed protection, shade and beauty, tannin and extractives, and cork. Consequently, the oaks are widely planted for many purposes.

FLOWERING AND FRUITING The staminate flowers are borne in naked aments (catkins); the pistillate flowers are solitary, or in 2- to many-flowered spikes on the same tree, in the spring (February to May) before or with the leaves. Staminate flowers develop from leaf axils of the previous year, whereas pistillate flowers develop from leaf axils of the current year. The fruit, an acorn (nut) (Fig. 1), matures in 1 (white oaks) or 2 (red and black oaks) years. Acorns are one-seeded, or rarely 2-seeded, and occur singly or in clusters each in a scaly cup (modified involucre). They are 0.6 to 3.7 cm long, subglobose to oblong, short-pointed at the apex, and marked with a circular scar at their base which is covered by the cup. Fruit ripening and seed dispersal occur from late August to early December. The embryo has 2 fleshy cotyledons, and there is no endosperm (Fig. 2). Acorns are generally green when preripe and turn brown (sometimes black) when ripe. The oaks vary widely in initiation of seed bearing and frequency of large crops (Table 1).

COLLECTION, EXTRACTION, AND STORAGE OF SEED Ripe acorns may be collected from August to December from the ground, or flailed or shaken from branches onto canvas or plastic sheets after ripening. Acorns of the white oaks should be collected soon after they have fallen to retard early germination. Acorns of California black oak (*Q. kelloggii*) also require prompt collection because mold often infects fallen acorns and destroys the cotyledons. In addition, birds eat ripe acorns of some species while they still are on the tree, and several organisms consume acorns rapidly once they are on the ground. In years when there are light crops, acorns are sometimes infested with weevils (*Curculio* species), and collection of large quantities of sound seeds is difficult.

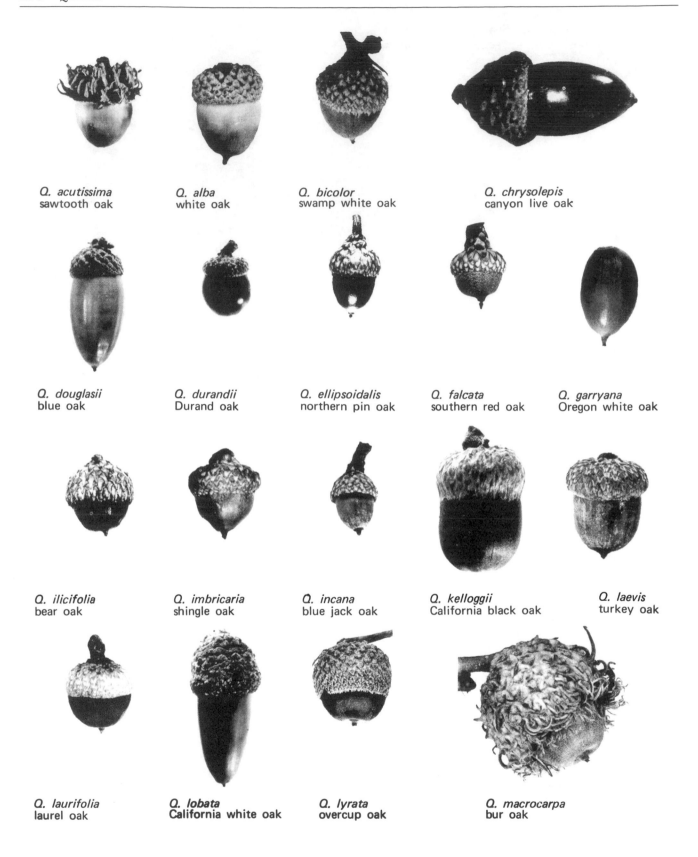

Figure 1. *Quercus*: acorns, ×1 (Olson 1974c).

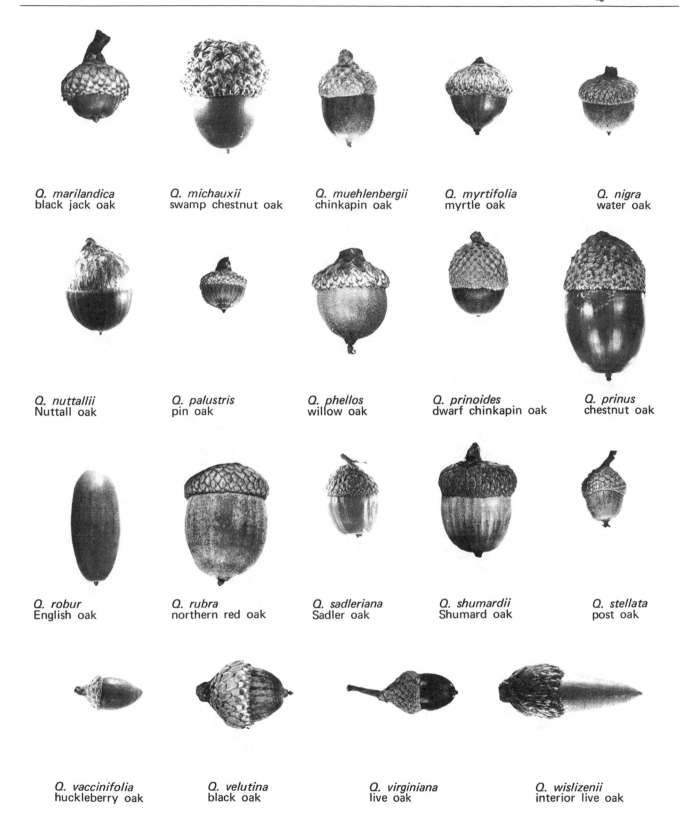

Q. marilandica
black jack oak

Q. michauxii
swamp chestnut oak

Q. muehlenbergii
chinkapin oak

Q. myrtifolia
myrtle oak

Q. nigra
water oak

Q. nuttallii
Nuttall oak

Q. palustris
pin oak

Q. phellos
willow oak

Q. prinoides
dwarf chinkapin oak

Q. prinus
chestnut oak

Q. robur
English oak

Q. rubra
northern red oak

Q. sadleriana
Sadler oak

Q. shumardii
Shumard oak

Q. stellata
post oak

Q. vaccinifolia
huckleberry oak

Q. velutina
black oak

Q. virginiana
live oak

Q. wislizenii
interior live oak

Figure 1. Continued.

Table 1. *Quercus:* characteristics of mature trees, yield data, and prechilling requirement (Olson 1974c).

Species	Mature tree height (m)	Seed-bearing Age (years)	Seed-bearing Interval (years)	Acorns/ kilogram	Oak group[1]	Prechilling (days)
Q. acutissima	15	5	1	224	W	–
Q. agrifolia	23	15	–	440	B	0
Q. alba	13	20	4–10	264	W	0
Q. bicolor	31	20	3–5	264	W	0
Q. cerris	31	–	–	242	W	0
Q. chrysolepsis	19	–	–	330	W	0–60
Q. coccinea	31	20	3–5	517	B	30–60
Q. douglasii	19	–	2–3	220	W	0
Q. dumosa	6	–	–	220	W	30–90
Q. durandii	23	–	–	638	W	0
Q. ellipsoidalis	22	–	–	539	B	60–90
Q. falcata						
var. falcata	28	25	1–2	1188	B	30–90
var. pagodaefolia	34	25	1–2	1276	B	60–120
Q. garryana	22	–	2–3	187	W	0
Q. ilicifoli	6	–	–	1540	B	60–120
Q. imbricaria	22	25	2–4	913	B	30–60
Q. kelloggii	26	30	2–3	209	B	30–45
Q. laevis	9	–	–	869	B	60–90
Q. laurifolila	28	15	1	1232	B	0
Q. lobata	31	–	2–3	286	–	–
Q. lyrata	26	25	3–4	308	W	14–90
Q. macrocarpa	31	35	2–3	165	W	30–60
Q. michauxii	31	20	3–5	187	W	0
Q. muehlenbergii	25	–	–	809	W	0
Q. nigra	25	20	1–2	809	B	30–60
Q. nuttallii	31	5	3–4	209	W	60–90
Q. palustris	26	20	1–2	902	–	–
Q. petraea	31	40	5–7	374	W	0
Q. phellos	31	20	1	1016	B	30–90
Q. prinus	25	20	2–3	220	W	0
Q. robur	34	20	2–4	286	W	0
Q. rubra	31	25	3–5	275	B	30–45
Q. shumardii	34	25	2–3	220	B	60–120
Q. stellata	19	25	2–3	836	W	0
Q. suber	22	12	2–4	165	W	0
Q. turbinella	3	–	3–5	715	W	–
Q. vaccinifolia	1	–	–	2266	W	0
Q. variabilis	25	–	2	229	–	–
Q. velutina	28	20	2–3	539	B	30–60
Q. virginiana	19	–	1	774	W	0
Q. wiselizenii	19	–	5–7	275	B	30–60

[1]B = Black; W = White.

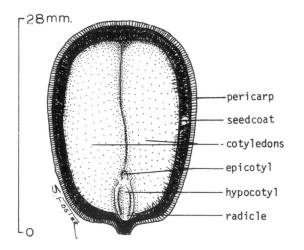

Figure 2. *Quercus rubra*, northern red oak: longitudinal section through an acorn, ×2 (Olson 1974c).

The only extraction required before storage or sowing is removal of loose cups, twigs, and other debris. However, the proportion of sound seed can be increased by removing defective, hollow, and partially consumed acorns. The sorting can be done by flotation or hand. The desired method will depend on the species of oak, moisture content of acorns, and relative cleanness of the seed lot. The species of oak vary considerably in seed yield and seed weight (Table 1). Consequently, trial procedures for sorting should be tested for specific conditions. In flotation, all acorns that float in water are discarded. Some sound seed will be lost this way, but the proportion of sound seed in the sunken acorns will be much greater than the proportion of sound seed in the floating ones.

Hand removal of obviously defective acorns (those with small holes where adult weevils have deposited eggs, cracked seedcoats, mold, etc.) will improve the planting value of an acorn collection. However, this tedious sorting may not be worth the effort in large bulk collections. Abscised acorns with the cup attached are usually defective. In southern red oak (*Q. falcata* var. *pagodaefolia*), the color of the cup scar is a good index of soundness. Sound acorns have a light, almost lemon-colored cup scar. Defective acorns have a brown cup scar. Weevil larvae inside apparently sound acorns can be killed by immersing the acorns in hot water (48°C) for 40 minutes. Temperature control is important because anything much above 50°C can kill the acorn embryos.

Acorns of most species of white oaks should not be stored, and it is impossible to store acorns of black oaks for more than 6 months, or from the time the acorns fall until sowing time in the spring. This short storage time is also the prechilling requirement for many black oak species. Therefore, they are stored under cool moist conditions. Dry storage in sealed containers has been used, but often results in rapid losses in viability. This loss occurs because the life processes within acorns are critically dependent on adequate moisture. For germination to occur, the moisture content of acorns must not drop below 30 to 50% for white oaks, and 20 to 30% for black oaks. Selected recent literature on the collection and storage of acorns is listed in Table 2.

PREGERMINATION TREATMENT With few exceptions, white oak acorns have little or no dormancy and will germinate almost immediately after falling. Black oak acorns exhibit embryo dormancy and germinate the spring following fall sowing. Black oaks require prechilling before germination testing or spring sowing. For best results, prechilling should be in moist, well-drained sand, sand and peat, or a similar material for 30 to 90 days at a temperature of 0 to 5°C. Several black oaks, notably cherrybark oak (*Q. falcata* var. *pagodaefolia*), northern red oak, and Shumard oak (*Q. shumardii*), begin germination at prechilling temperatures if prechilling is continued for more than 30 to 45 days. Epicotyl emergence, however, will not occur for at least 220 days at prechilling temperatures.

GERMINATION ASOA (1985) standards for *Quercus* spp black and red oaks include Substrate, top of creped cellulose paper; temperature, 20/30°C; duration, 14 days; and additional directions, cut ⅓ off cup scar end of acorn and remove pericarp. Substrate and temperature requirements are the same for *Q. alba*, *Q. muehlenbergii*, and *Q. virginiana*, but duration is longer—28 days. Olson (1974c) provided unofficial germination methods for a large number of oak species (Table 1). Germination is hypogeal (Fig. 3) and is generally complete in 3 to 5 weeks. Germination of live oak (*Q. virginiana*) is peculiar in that the radicle, soon after it appears, becomes enlarged just below the surface of the ground because of the transfer of food from the

Table 2. *Quercus:* selected recent literature on collection and storage of acorns.

Author—Date	Subject	Location
Anoak 1973	*Quercus* species, relative humidity and temperature induced secondary dormancy so acorns could be stored 1 year.	USSR
Arai 1982	*Q. crispula*, storage and germination.	Japan
Blanche et al. 1980	*Q. nigra*, acorns were collected at 5 dates and seed data obtained for comparison to subsequent germination.	USA
Clatenbuck 1985	*Quercus* species, utilization of food reserves during storage.	
Janson 1979	*Q. robur*, seeds stored in dry sand at low temperatures for 16 months.	Poland
Johnson 1979	*Q. nuttallii*, seeds stored submerged in water.	USA
Johnson 1983	*Q. nigra*, salt solutions for flotation of acorns.	USA
Morgan and Brohaker 1986	*Q. virginiana*, drying reduced germination of small acorns more than that of large acorns.	USA
Rink and Williams 1984	*Q. alba*, origin and storage conditions influenced storage half-life.	USA
Schroeder and Walker 1987	*Q. macrocarpa*, effects of moisture content and storage temperature on germination.	USA
Suszka and Tylkowski 1981	*Q. robra*, storage methods and viability testing.	Poland
Teclaw and Isebrands 1986	*Q. rubra*, collection procedure influence on germination.	Poland

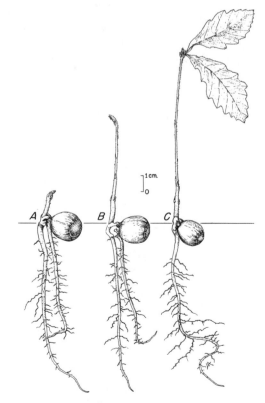

Figure 3. *Quercus macrocarpa*, bur oak: seedling development at 1, 5, and 12 days after germination (Olson 1974c).

cotyledons. Rapid tests of germination can be run by removing the pericarp of an acorn and placing it on moist blotter paper.

Selected recent literature concerning germination of *Quercus* species is given in Table 3.

NURSERY AND FIELD PRACTICE Fall seeding of oaks is preferable to spring seeding. The white oaks germinate immediately after fall sowing, and the black oaks undergo natural prechilling when fall sown, and ger-

minate promptly in the spring. If spring sowing of black oaks must be done, the acorns should be given a moist prechilling. Acorns may be drilled in rows 20 to 25 cm apart, or broadcast and covered with 0.6 cm of firmed soil. Seedbed densities of 105 to 370 per m² are recommended. Fall-sown beds should be mulched to protect the seeds and seedlings. Partial shade is beneficial for germination. Usually seedlings are transplanted after the first year (Fig. 3).

Table 3. *Quercus:* selected recent literature on germination.

Author—Date	Subject	Location
Aissa 1983	*Q. ilicifolia*, influence of origin and acorn size on germination.	France
Alam 1983	*Quercus* spp, germination for 15 species.	Bulgaria
Alam 1984	*Q. robur* (and other species), emergence and growth.	Bulgaria
Allen and Farmer 1977	*Q. ilicifolia*, no pretreatment required for germination, but prechilling or GA required for shoot elongation.	USA
Appleton and Whitcomb 1983	*Quercus* spp, effects of seed position on germination.	USA
Appleton et al. 1986	*Quercus* spp, influence of seed handling, pregermination, and planting position on seedling development.	USA
Barnett 1977a	*Q. alba*, effect of squirrel burial on acorn germination.	USA
Bonner 1974i	*Q. falcata* var. *pagodaefolia*, test for acorn vigor.	USA
Bonner 1974g	*Q. falcata* var. *pagodaefolia*, *Q. nigra*, and *Q. prinus*, maturation of acorns.	USA
Bonner 1984	*Quercus* spp, testing for acorn quality.	USA
Conner et al. 1976	*Q. alba*, correlation of flowering and acorn yield.	USA
Farmer 1974	*Q. rubar*, influence of provenance, chilling, and gibberellin on germination.	USA
Farmer 1981	*Q. alba*, variation in acorn yield.	USA
Godman and Mattson 1980	*Q. rubra*, greatest field emergence at soil temperatures near 0°C.	USA
Harif 1977	*Q. calliprinos*, heavier acorns produce larger seedlings.	Israel
Hopper et al., "Adenylate energy," 1985	*Q. rubra*, influence of prechillilng and pericarp removal on germination.	USA
Hopper et al., "Germination," 1985	*Q. rubra*, biochemical changes during prechilling and germination.	USA
Johnson and Krinard 1985	*Quercus* spp, direct seeding.	USA
Lamond 1978	*Q. pedunculata*, influence of pericarp on speed of germination.	France
Lamond and Levert 1980	*Q. rubra*, pericarp influence on water absorption.	France
Levert and Lamond 1979	*Q. rubra*, temperature and germination.	France
Luk'yanets 1978	*Q. rubra*, and other species, determined amino acid content of acorns.	USSR
Mamonov 1976b	*Q. rubra*, respiration of acorns.	USSR
Matsudaik and McBride 1987	*Quercus* spp, germination of California oaks at different elevations.	USA
McClaran 1987	*Q. douglasii*, yearly variation in emergence.	USA
Nikolova 1984	*Q. cerris*, free amino acid changes during dry and moist storage.	Bulgaria
Peterson 1983	*Q. nigra*, mechanisms involved in delayed germination.	USA
Petrov and Genov 1984	*Q. suber*, fruiting.	Bulgaria
Plumb 1982	*Quercus* spp, germination of southern California oaks.	USA
Rao 1988	*Q. floribunda*, germination and seedling growth.	–
Roberts and Smith 1982	*Q. kelloggii*, germination and survival under different watering regimes.	USA
Smith 1985	*Quercus* spp, seeding California oaks.	USA
Sopp et al. 1977	*Q. gambelii*, germination.	USA
Szczotka 1974	*Q. borealis*, amylolytic during acorn storage.	Poland
Szczotka 1975	*Q. rubra*, and *Q. borealis*, protein synthesis in the embryo axis.	Poland
Szczotka 1978	*Q. rubra*, and *Q. borealis*, respiration of acorns during storage.	Poland
Tuskan and Blanche 1980	*Q. shumardii*, mechanical shaking in water improves germination.	USA
Vogt 1970	*Q. rubra*, influence of gibberellin on germination.	USA
Vogt 1974a	*Q. rubra*, endogenous hormone changes during prechilling.	USA
Vozzo 1973	*Q. nigra*, excised embryo cylinder germination.	USA
Vozzo 1978	*Q. alba*, biochemistry of embryos.	USA
Vozzo 1985	*Q. nigra*, pericarp changes during germination.	USA
Wirges and Yeiser 1984	*Q. arkansana*, prechilling requirements.	USA

ROSACEAE — ROSE FAMILY

Raphiolepis Lindl.

GROWTH HABIT, OCCURRENCE, AND USE The genus *Raphiolepis* has 6 species, all native to Japan and China. The fruit is a drupelike pome with 1 or 2 large seeds. It should be collected in late fall or winter and the seeds removed and cleaned. Seeds germinate without pretreatment (Dirr and Heuser 1987).

RHAMNACEAE — BUCKTHORN FAMILY

Rhamnus L. — Buckthorn

GROWTH HABIT, OCCURRENCE, AND USE The genus *Rhamnus* consists of about 100 species and many varieties (Hubbard 1974b). It is chiefly native to the temperate and warm regions of the Northern Hemisphere, but a few species are found in Brazil and South Africa. Buckthorns are deciduous or evergreen shrubs or small trees that have value for wildlife food and cover as well as for environmental forestry. Some species have been cultivated for many years.

FLOWERING AND FRUITING Buckthorn flowers are bisexual or with either stamens or pistils. They are borne in axillary clusters, with 4 or 5 sepals united at the base and with a cup-shaped receptacle forming a receptocalyx, the upper part falling at maturity, the lower part remaining around the developing fruit. Four or 5 very small petals, sometimes none, are inserted on the margin of a disk lining the receptocalyx. Four or 5 stamens and 1 pistil are attached to the receptocalyx at the very base. The ovary is 3- or 4-celled. Buckthorn The buckthorn fruit is a berrylike drupe (Fig. 1) containing 2 to 4 nutlike seeds. California buckthorn (*R. californica*) and glossy buckthorn (*R. frangula*) have 2 seeds per fruit; alder buckthorn (*R. alnifolius*), 3 seeds; cascara buckthorn (*R. purshiana*) 2 to 3. Typical seeds

Figure 1. *Rhamnus purshiana,* cascara buckthorn: fruit, ×4 (Hubbard 1974b).

R. alnifolia
alder buckthorn

R. davurica
Dahurian buckthorn

R. frangula
glossy buckthorn

R. californica
California buckthorn

R. purshiana
cascara buckthorn

Figure 2. *Rhamnus:* seeds, ×4 (Hubbard 1974b).

Table 1. *Rhamnus:* phenology of flowering and fruiting, characteristics of mature trees, and cleaned seed weight (Hubbard 1974b).

Species	Flowering dates	Fruit ripening dates	Growth habit	Mature tree height (m)	Ripe fruit color	Seeds/gram
R. alnifolius	May–June	Aug.	Deciduous	1	Black	144
R. californica	May–June	Sept.–Oct.	Evergreen	2	Black	9
R. cathartica	Apr.–June	July–Nov.	Deciduous	6	Black	42
R. crocea						
var. crocea	Feb.–May	June–Sept.	Evergreen	1	Bright red	154
var. ilicifolia	Feb.–Apr.	June–Sept.	Evergreen	5	Bright red	158
R. davurica	May–June	Sept.–Oct.	Deciduous	9	Black	59
R. frangula	May–June	July–Oct.	Deciduous	6	Purple	59
R. purshiana	Apr.–July	July–Sept.	Deciduous	9	Purple	27

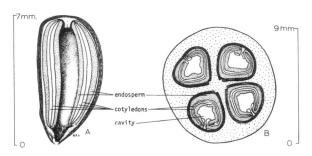

Figure 3. *Rhamnus cathartica*, European buckthorn: (A) longitudinal section through 1 seed, ×5; and (B) transverse section through the 4 seeds of a fruit showing the investing type of cotyledons, ×3.3 (Hubbard 1974b).

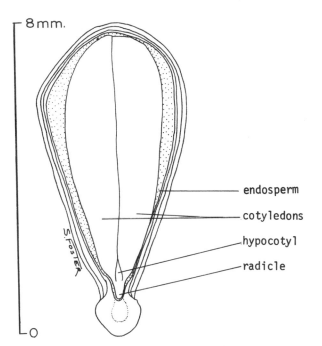

Figure 4. *Rhamnus californica*, California buckthorn: longitudinal section through a seed, ×10 (Hubbard 1974b).

are small and round, with a flat side marked by a slight central rib, and a rounded side with a small terminal knob (Figs. 2, 3, 4). Cascara buckthorn begins to fruit when 5 to 7 years old. Good seed crops of Dahurian buckthorn are common. Most other species produce abundant seed crops on alternate years.

COLLECTION, EXTRACTION, AND STORAGE OF SEED Seed can be picked from the shrubs when they ripen. Picking should be done about 2 weeks before the fruit is fully ripe. If harvesting is delayed, many seeds can be lost to birds. The fruits can be run through a macerator and the seeds recovered by flotation. Fruit and seed characteristics are given in Table 1. Seeds can be stored for several years in sealed containers at low temperatures.

PREGERMINATION TREATMENT Fresh seeds of alder, California, redberry (*R. crocea*), and hollyleaf buckthorn require no pretreatment for germination. Stored seeds of these species should be prechilled for 2 to 3 months before planting. Seeds of glossy and Dahurian buckthorn require 20 minutes acid scarification before prechilling. Apparently no formal germination standards exist for *Rhamnus* species, except that in the Genebank Handbook (1985) there is a suggestion for *R. frangula*: substrate, sand; incubation temperature, 20/30°C; duration, none given; additional instructions, provide light for 8 hours daily, and prechill at 1 to 5°C for 8 weeks. The same standards are suggested for *R. cathartica*, except the prechilling is reduced to 2 to 4 weeks.

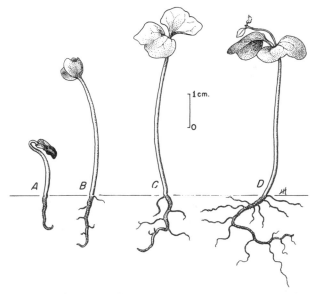

Figure 5. *Rhamnus cathartica*, European buckthorn: seedling development at 1, 4, 19, and 28 days after germination (Hubbard 1974b).

NURSERY AND FIELD PRACTICE Apparently fall sowing of most species has been successful. Germination is epigeal (Fig. 5). Most species can be propagated from cuttings.

Some species of *Rhamnus* are the alternate host for oat rust, *Puccinia coronata*.

ERICACEAE — HEATH FAMILY

Rhododendron L. — Rhododendron

GROWTH HABIT, OCCURRENCE, AND USE The rhododendrons comprise over 600 species of evergreen and deciduous shrubs or small trees (Olson 1974d). Many are cultivated for their beautiful flowers; the evergreen species are cultivated for their foliage as well. Pinxterbloom azalea (*Rhododendron nudiflorum*) is a deciduous plant native to the area from Massachusetts to southern Ontario, south to Tennessee and South Carolina. Catawba rhododendron (*R. catawbiense*) is an evergreen species native to the mountains and piedmont area from West Virginia to Georgia and Alabama. Pacific rhododendron (*R. macrophyllum*) is native to the Pacific coast of North America, from British Columbia to Monterey, California. Rosebay rhododendron (*R. maximum*) occurs from Novia Scotia to Georgia and Alabama.

FLOWERING AND FRUITING The perfect, showy flowers of rhododendron appear from March to August (Olson 1974d):

Species	Flowering dates	Fruit ripening dates	Seed dispersal dates
R. catawbiense	Apr.–June	July–Oct.	Fall
R. macrophyllum	Apr.–May	Aug.–Sept.	Late summer–fall
R. maximum	May.–Aug.	Sept.–Oct.	Oct.–Nov.
R. nudiflorum	Mar.–May	Sept.	Fall

Flower colors vary widely, with white, pink, and purple predominating. The flowers are pollinated by insects.

The fruit is an oblong capsule (Fig. 1) that ripens in the fall. It splits along the sides soon after ripening and releases minute seeds (Figs. 2, 3). Rosebay rhododendron capsules contain as many 440 sound seeds per capsule.

COLLECTION EXTRACTION AND STORAGE OF SEED Collection should begin as soon as capsules start to lose their green color and turn brown; it should be completed before the capsules open (Olson 1974d):

Species	Mature tree height (m)	Large seed crop interval (years)	Ripe capsule color
R. catawbiense	6	—	Rusty pubescent
R. macrophyllum	6	1	Rusty pubescent
R. maximum	12	1–2	Brown
R. nudiflorum	3	—	—

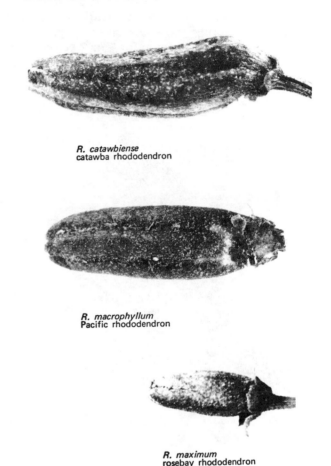

R. catawbiense
catawba rhododendron

R. macrophyllum
Pacific rhododendron

R. maximum
rosebay rhododendron

Figure 1. *Rhododendron*: capsules with styles removed, ×3 (Olson 1974d).

The fruits can be spread in thin layers for air-drying or they can be oven dried at 35°C for 12 to 24 hours. After drying the capsules can be rubbed, or otherwise threshed, and the seeds recoverd by air screening.

The seeds of *Rhododendron* are minute and few determinations of number of seeds per gram have been reported. Estimates of the number of cleaned seeds per gram range from 4400 to 12,500.

Seeds with a moisture content of 4 to 9% will remain viable for about 2 years at room temperature. They have been successfully stored in plastic vials or sealed in plastic bags at −4°C.

GERMINATION ASOA (1985) germination standards for *Rhododendron* species are substrata, crepe cellulose or top of blotters; incubation temperatures 20/30°C or

25°C; duration 21 days; and additional instructions, provide light. Germination of *Rhododendron* seeds must be conducted under light. They will germinate without pretreatment. Germination is epigeal (Fig. 4).

NURSERY AND FIELD PRACTICE Seeds are germinated in flats of sandy peat or sand and half-decayed oak leaves covered with shredded or sifted sphagnum. They are sown on the surface and covered lightly with pulverized sphagnum. Sowing can be made in a cool greenhouse during winter or in cold frames in April. Some growers cover the flats with glass or plastic. The seedlings are left in the flats in the shade the first year. The following spring the plants are transplanted to containers or beds and grown for 1 or 2 years before being transplanted to the field. Plants should be grown in acid soil with a high content of organic matter and ample soil moisture provided at all times.

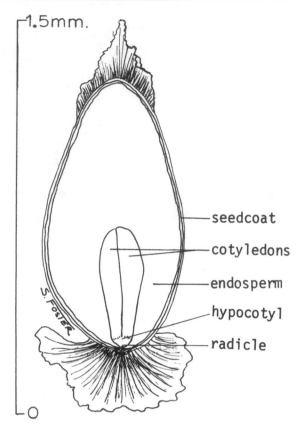

Figure 3. *Rhododendron maximum*, rosebay rhododendron: longitudinal section through a seed, ×70 (Olson 1974d).

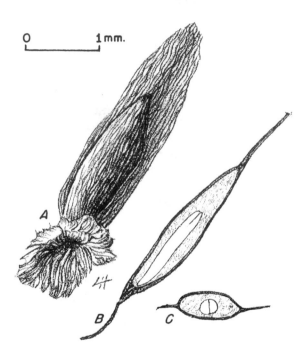

Figure 2. *Rhododendron macrophyllum*, Pacific rhododendron: seeds, ×20; (A) external view; (B) longitudinal section; (C) transverse section (Olson 1974d).

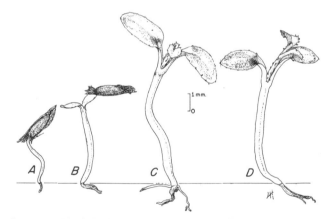

Figure 4. *Rhododendron macrophyllum*, Pacific rhododendron: seedling development at 1, 9, 40, and 60 days after germination (Olson 1974d).

ROSACEAE — ROSE FAMILY

Rhodotypos scandens (Thunb.) Mak. — Black jetbead

GROWTH HABIT, OCCURRENCE, AND USE Native to Japan and central China, black jetbead (*Rhodotypos scandens*) is an upright, spreading, deciduous shrub usually 1 to 2 m tall (Rudolf 1974n). It was introduced to the United States as an ornamental species, and has been used as a wildlife species.

FLOWERING AND FRUITING The showy, white, perfect flowers, 2.5 to 5 cm across, bloom from April to June. Black jetbead fruits are shiny, black, dry drupes, obliquely ellipsoid in shape. They ripen in October or November and persist on the plant well into the winter. Each fruit contains 1 small, stubby ellipsoidal stone about 0.6 cm long, dull tan in color, and characteristically sculptured in the manner of leaf venation with the midrib extending around the longest periphery (Fig. 1).

COLLECTION, EXTRACTION, AND STORAGE OF SEED The fruits can be collected from the bushes by hand or flailed to ground sheets from October to midwinter. Extraction of stones from the fruits may not be necessary. There are about 11 seeds per gram. Seeds can be stored in open containers at low temperatures up to 9 months without loss of viability.

GERMINATION The seeds are initially dormant and require 3 months of prechilling for germination.

NURSERY AND FIELD PRACTICE Seed should be sown in the fall in mulched or covered frames. Some germination will not take place until the second year. Slightly imma-

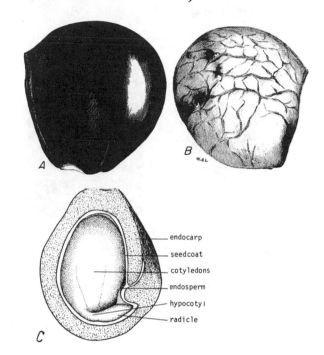

Figure 1. *Rhodotypos scandens*, black jetbead: (A) fruit; (B) stone; (C) longitudinal section through a stone, ×6 (Rudolf 1974n).

ture seeds planted immediately after collection may provide better seedling establishment.

ANACARDIACEAE — CASHEW FAMILY

Rhus L. — Sumac

GROWTH HABIT, OCCURRENCE, AND USE The sumacs include about 150 species of shrubs and trees of the temperate and subtropical regions (Brinkman 1974i). Their handsome foliage often assumes brilliant colors in the fall, and the showy red fruits provide additional ornamental value. Because of their suckering habit, edible fruit, and browse value, many sumac species are valuable for erosion control and wildlife habitat.

FLOWERING AND FRUITING The small, rather inconspicuous sumac flowers are dioecious or polygamous and are borne in terminal or axillary clusters in the spring (Brinkman 1974i):

Species	Flowering dates	Fruit ripening dates	Ripe fruit color
R. aromatica	Mar.–May	July–Aug.	Red
R. copallina	July–Aug.	Sept.–Oct.	Crimson
R. glabra	June–Aug.	Sept.–Oct.	Bright-dark red
R. integrifolia	Jan.–July	Aug.–Sept.	Reddish
R. laurina	May–July	Aug.–Sept.	Whitish
R. ovata	Mar.–May	Aug.–Sept.	Reddish
R. trilobata	Mar.–Apr.	Aug.–Sept.	Red
R. typhina	May–July	June–Sept.	Dark red

The fruit is a small, smooth to hairy drupe with a single

bony nutlet lacking endosperm (Figs. 1, 2, 3). In most species, the fruits form a dense cluster and ripen in the fall; many remain on the plant over winter. Seed dispersal is almost entirely by birds or other animals. Some seed is produced nearly every year.

R. trilobata
skunkbush sumac

R. typhina
staghorn sumac

Figure 1. *Rhus*: fruits, ×4 (Brinkman 1974j).

R. glabra
smooth sumac

R. integrifolia
lemonade sumac

R. laurina
laurel sumac

R. ovata
sugar sumac

R. trilobata
skunkbush sumac

R. typhina
staghorn sumac

Figure 2. *Rhus*: nutlets, ×4 (Brinkman 1974j).

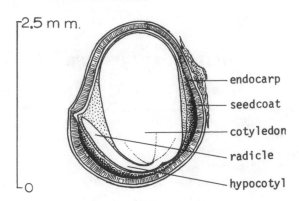

endocarp

seedcoat

cotyledon

radicle

hypocotyl

Figure 3. *Rhus typhina*, staghorn sumac: longitudinal section through a seed, ×18 (Brinkman 1974j).

COLLECTION, EXTRACTION, AND STORAGE OF SEED The fruit clusters may be picked by hand as soon as ripe, and often are available until late in the year. Fruits of smooth sumac (*R. glabra*) and staghorn sumac (*R. typhina*), which occur in very dense clusters, may require drying by spreading in shallow layers. If collected late in the fall such drying is not required.

The fruits can be run through a macerator and the seeds recovered by flotation. Such complete cleaning is seldom practiced except for fruits of skunkbush (*R. trilobata*). The seeds of other species are sown with the fruits more or less intact. Sound seeds of smooth sumac sink in water, but this test does not work with other species. The number of seeds per gram for sumacs are listed below (Brinkman 1974i):

Species	Seeds/gram	Mature tree height (m)
R. aromatica	—	2
R. copallina	125	9
R. glabra	110	8
R. integrifolia	20	3
R. laurina	290	4
R. ovata	50	3
R. trilobata	45	2
R. typhina	120	12

Seeds of *Rhus* species store over winter without special treatment. For long-term storage, seeds should be dried and stored in sealed containers at low temperatures.

PREGERMINATION TREATMENT *Rhus* seeds germinate poorly without pretreatment. Dormancy of most species is caused by hard, impervious seedcoats. Seeds of fragrant sumac (*R. aromatica*) and skunkbush also have embryo dormancy that requires prechilling. Acid scarification or hot water treatments are necessary to break the hard seedcoats.

GERMINATION Using scarified seeds, and prechilled seeds in the case of species with dormant embryos, germination can be tested with a variety of substrata ranging from sand to paper in petri dishes. Incubation temperatures that have been used are 20/30°C or a constant 20°C. Heit (1967a) recommended that light be supplied during incubation. Germination is epigeal (Fig. 4).

Brinkman (1974j) glossed over the extreme dermatitis that is caused by some species of *Rhus*. We determined that the seeds of poison oak (*R. diversiloba*) have hard seedcoats that require acid scarification before germination will occur. Various burning techniques using oak leaves and pine needles were not successful in breaking dormancy.

Selected recent literature concerning the germination of *Rhus* species includes the following:

Author—date	Subject
Baghugana & Rauat 1986	*R. parviflora*, seed germination.
Farmer et al. 1983	*R. glabra, R. coppalina*, influence of scarification, incubation temperature, and genotype on germination.
Lumley & Oliver 1987	*R. erosa*, propagation techniques.
Marks 1979	*R. typhina*, fire stimulation of seed germination.
Norton 1985	*R. typhina*, influence of gibberellin and prechilling on germination.
Rasmussen & Wright 1988	*R. copallina*, germination requirements.
Washitani 1988	*R. javanica*, influence of high temperatures on germination.

NURSERY AND FIELD PRACTICE Sumac seeds can be sown in the fall after scarification. They should be planted 1.25 cm deep in rows, at a rate of 260 viable seeds per m². *Rhus* species can be propagated by cuttings.

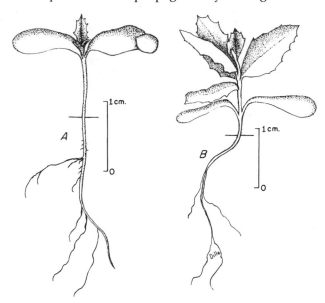

Figure 4. *Rhus ovata*, sugar sumac: seedling development of 2 age classes (Brinkman 1974j).

GROSSULARIACEAE — CURRANT OR GOOSEBERRY FAMILY

Ribes L. — Currant, Gooseberry

GROWTH HABIT, OCCURRENCE, AND USE *Ribes* includes about 150 species of deciduous or, rarely, evergreen shrubs that occur in the colder temperate parts of North America, Europe, and Asia, and in the Andes of South America (Pfister 1974). They have been placed in Saxifragaceae by most authors, but were considered part of Grossulariaceae by Pfister (1974). The unarmed species are called currants; the prickly species, gooseberries. All are valuable as food and cover for wildlife. Many species of *Ribes* serve as the alternate host to white pine blister rust, a disease that has had a severe impact on forest management.

FLOWERING AND FRUITING The flowers are usually bisexual, small and greenish, or yellow to red in some species. They are borne from April to June in few- to many-flowered racemes or are solitary (Table 1). The fruit is a green, many-seeded, glandular or smooth berry, 0.6 to 1.25 cm in diameter (Fig. 1). It ripens in early to late summer. The ripe fruit is red in some species, purple to black in others, and occasionally can be red, yellow, or black within a given species (Table 1). The mature seed (Fig. 2) contains a large endosperm in which a minute, rounded embryo is imbedded (Fig. 3).

Seeds are dispersed by birds and mammals. Depending on the species, good seed crops are borne annually to 2- to 3-year intervals.

COLLECTION, EXTRACTION, AND STORAGE OF SEED The fruits should be picked or stripped from the bushes as soon as they are ripe to minimize losses to birds. They can be macerated and the seeds recovered by flotation. Dried seeds can be stored for long periods in sealed vials at low temperatures.

GERMINATION The seeds of most species of *Ribes* are highly dormant and require prolonged prechilling, warm stratification followed by prechilling, and/or a wide range of diurnal temperatures during incubation to obtain germination (Genebank Handbook 1985). There is wide variation within and among seed lots. Prechilling for 6 months may be necessary.

NURSERY AND FIELD PRACTICE *Ribes* seeds are usually sown in the fall. They should be sown at the rate of 630 to 840 per m² and covered with 0.6 cm of soil. Seedbeds should be mulched. About 400 seedlings are produced per kilogram of seed. Most species can be readily propagated from hardwood cuttings taken in autumn. Germination is epigeal (Fig. 4).

R. cereum
wax currant

R. cynosbati
pasture gooseberry

R. lacustre
prickly currant

R. montigenum
mountain gooseberry

R. cereum
wax currant

R. hudsonianum
Hudson-Bay currant

R. irriguum
Idaho gooseberry

R. lacustre
prickly currant

R. montigenum
mountain gooseberry

R. nevadense
Sierra currant

R. roezli
Sierra gooseberry

R. sanguineum
winter currant

R. viscosissimum
sticky currant

Figure 2. *Ribes*: seeds, ×12 (Pfister 1974).

R. sanguineum
winter currant

R. viscosissimum
sticky currant

Figure 1. *Ribes*: berries, ×2 (Pfister 1974).

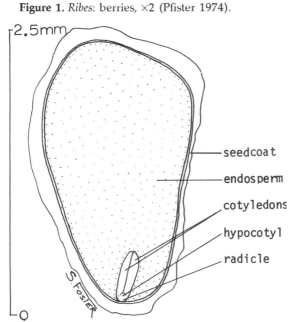

2.5mm

seedcoat

endosperm

cotyledons

hypocotyl

radicle

0

Figure 3. *Ribes missouriense*, Missouri gooseberry: longitudinal section through the embryo of a seed, ×30 (Pfister 1974).

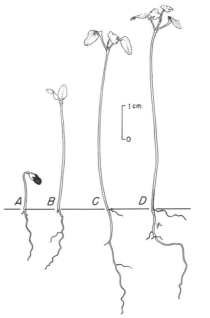

1 cm.

0

A   B   C   D

Figure 4. *Ribes missouriense*, Missouri gooseberry: seedling development at 2, 7, 23, and 44 days after germination (Pfister 1974).

Table 1. *Ribes:* phenology of flowering and fruiting, fruit characteristics, and cleaned seed weight (Pfister 1974).

Species	Flowering dates	Fruit ripening dates	Mature tree height (m)	Prickles	Fruit characteristics Surface	Ripe color	Seeds/gram
R. alpinum	Apr.–May	July–Aug.	2.5	No	Glabrous	Scarlet	–
R. americanum	Apr.–June	June–Sept.	2.0	No	Glabrous	Black	690
R. aureum	Apr.–May	June–July	3.0	No	Glabrous	Red, black	510
R. cereum	Apr.–June	Aug.	1.5	No	Glandular	Red	550
R. cynosbati	Apr.–June	July–Sept.	1.5	Yes	Glandular	Purple	450
R. hudsonianum	May–July	–	2.0	No	Smooth	Black	2130
R. inerme	May–July	–	2.3	Yes	Smooth	Purple	810
R. irriguum	Apr.–June	–	2.5	Yes	Smooth	Purple	–
R. lacustre	Apr.–July	Aug.	2.0	Yes	Glandular	Black	1130
R. missouriense	Apr.–May	June–Sept.	–	Yes	Smooth	Black	360
R. montigenum	June–July	Aug.–Sept.	1.0	Yes	Glandular	Red	310
R. nevadense	May–July	–	2.0	No	Glandular	Blue-black	860
R. odoratum	May–June	June–Aug.	3.0	No	Smooth	Black-red	370
R. roezlii	May–June	July–Sept.	2.5	Yes	Glandular	Purple-brown	520
R. rotundifolium	Apr.–May	July–Aug.	1.0	Yes	Smooth	Purple	–
R. sanguineum	Apr.–May	July–Aug.	3.5	No	Glandular	Blue-black	630
R. viscosissimum	May–June	Aug.–Sept.	2.0	No	Glandular	Black	660

LEGUMINOSAE — LEGUME FAMILY

Robinia L. — Locust

GROWTH HABIT, OCCURRENCE, AND USE *Robinia* includes about 20 species which are native to the United States and Mexico (Olson 1974e). Three or 4 species are deciduous trees and the others are shrubs. Black locust (*R. pseudoacacia*) is a medium-sized tree 12 to 30 m high. It reaches its best development in the Appalachian region, and has been widely planted in the Western Hemisphere and in Europe. The rapid growth of black locust on good sites, its nitrogen-fixing ability, and the durability of its wood makes it the most valuable species in the genus. Rose acacia locust (*R. hispida*) and bristly locust (*R. fertilis*) are low shrubs that are useful for erosion control because of their prolific root sprouting. Both will grow on strip mine spoils with low soil pH values. New Mexico locust (*R. neomexicana*) is a small tree to 8 m in height.

FLOWERING AND FRUITING The perfect flowers occur in racemes originating in the axils of leaves of the current year, and appear in the spring and early summer. Pollination is by insects. The fruit, a legume (Fig. 1), ripens in the fall and contains 4 to 10 dark brown to black seeds about 4 to 6 mm long (Fig. 2). When they ripen, the fruits become brown and open on the tree, releasing the seeds. Black locust begins seed bearing at about 6 years of age, and produces good crops at 1- to 2-year intervals. The seeds contain no endosperm (Fig. 3).

COLLECTION, EXTRACTION AND STORAGE OF SEED Collection of the ripe seeds must begin before the legumes open. The legumes can be picked from the trees by hand or flailed to ground cloths from late August throughout the winter (Olson 1974e):

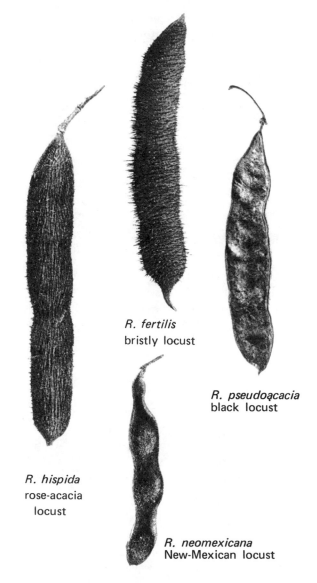

R. fertilis
bristly locust

R. pseudoacacia
black locust

R. hispida
rose-acacia
locust

R. neomexicana
New-Mexican locust

Figure 1. *Robinia:* legumes (pods), ×1 (Olson 1974e).

Species	Flowering dates	Fruit ripening dates	Seed dispersal dates	Seeds/ gram
R. fertilis	Early June	Sept.	Oct.–Nov.	50
R. hispida	May–June	July–Sept.	—	60
R. neomexicana	May–June	Sept.	Sept.–Oct.	50
R. pseudoacacia	May–June	Sept.–Oct.	Sept.–Apr.	50

The dried legumes can be threshed with a rubbing thresher and the seeds recovered by air screening. Seeds retain their viability as long as 10 years in storage in closed containers under cool conditions.

PREGERMINATION TREATMENT The seedcoats of *Robinia* seeds are hard and require scarification before they will germinate. Mechanical, acid, or hot water scarification can be used; acid scarification is most effective. The duration of acid soaking, which can range from 10 to 120 minutes, must be determined for each seed lot.

GERMINATION ASOA (1985) standards for black locust are substrate, between blotters; incubation temperature, 20°C; duration, 21 days; and additional directions, soak in concentrated H_2SO_4 for 1 hour.

Selected recent literature on the germination of locust seeds is given in Table 1.

NURSERY AND FIELD PRACTICE *Robinia* seeds may be drilled from March to May in rows 15 to 20 cm apart at a rate of 65 to 100 seeds per meter of row, or broadcast in fertile soil. They should be covered with 0.6 cm of soil, sand, or sand and sawdust and should be treated with a nitrogen inoculant. Mulching with light straw is helpful. Germination is epigeal (Fig. 4). Seedlings can be outplanted after 1 year.

Table 1. *Robinia:* selected recent literature on germination.

Author—Date	Subject	Location
Bogoroditskii and Sholokhov 1974	R. pseudoacacia, vacuum water saturation of seeds.	USSR
Hamilton 1972	R. pseudoacacia, breaking dormancy with 2-chloroethyl phosphonic acid.	–
Kochkar 1978	R. pseudoacacia, calculations of seed yield.	USSR
Kulygin 1974	R. pseudoacacia, seed production.	USSR
Kulygin 1977	R. pseudoacacia, influence of drought on seed quality and seed quality as index of drought resistance.	USSR
Marjal 1972	R. pseudoacacia, soil seed banks.	Hungary
Pathak and Misra 1978	R. pseudoacacia, seed quality and germination.	–

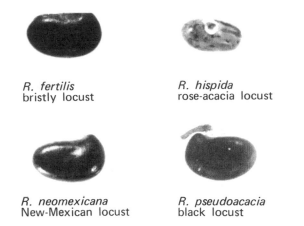

R. fertilis
bristly locust

R. hispida
rose-acacia locust

R. neomexicana
New-Mexican locust

R. pseudoacacia
black locust

Figure 2. *Robinia:* seeds, ×4. The speckled and mottled appearance of the seed of *R. hispida* also occurs in other species of *Robinia* (Olson 1974e).

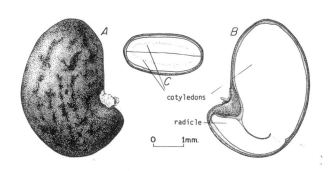

cotyledons

radicle

0   1mm.

Figure 3. *Robinia pseudoacacia,* black locust: seeds, ×8. (A) exterior view; (B) longitudinal section; (C) cross section (Olson 1974e).

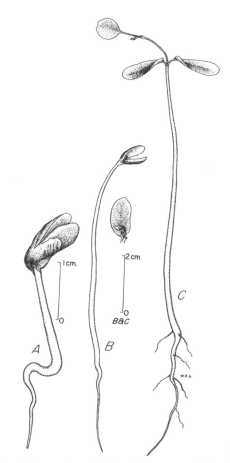

Figure 4. *Robinia pseudoacacia,* black locust: seedling development at 1, 3, and 8 days after germination (Olson 1974e).

ROSACEAE — ROSE FAMILY

Rosa L. — Rose

GROWTH HABIT, OCCURRENCE, AND USE The genus *Rosa* consists of about 100 species, many native to temperate portions of the Northern Hemisphere. The native rose species of North America provide valuable cover and food for wildlife (Gill and Pogge 1974g).

FLOWERING AND FRUITING The flowers are bisexual and are borne singly or in groups of 2 or 3. The exceptions are Japanese (*R. multiflora*), Prairie (*R. setigera*), and wichura (*R. wichuraiana*) roses, which bear flat-topped clusters of few to many flowers. Blossom color is usually white in Japanese and wichura rose; pink in baldhip (*R. gymnocarpa*), meadow (*R. blanda*), and sweetbrier (*R. eglantaria*) roses; and ranges from purple to white in the other species. Colors of ripe fruit range from orange-red to scarlet, and those of Nootka rose (*R. nutkana*) and prairie rose are sometimes purplish.

Flowering and fruiting dates vary among species and location, but hips of all species remain on the plants after ripening (Gill and Pogge 1974g):

Species	Flowering dates	Fruit ripening dates	Seeds/gram
R. blanda	May–June	Sept.–Oct.	100
R. canina	June	Oct.	70
R. eglanteria	June–July	Sept.–Oct.	70
R. gymnocarpa	June–July	Aug.–Sept.	60
R. multiflora	—	Sept.–Oct.	150
R. nutkana	May–July	Aug.	100
R. rugosa	May–Sept.	—	140
R. setigera	May–July	Aug.	110
R. wichuraiana	July–Sept.	—	190
R. woodsi	June–Aug.	July–Aug.	110

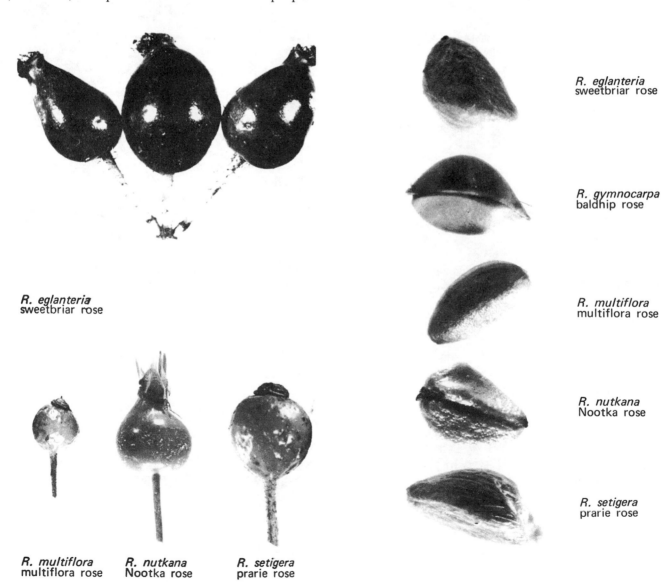

R. eglanteria
sweetbriar rose

R. gymnocarpa
baldhip rose

R. multiflora
multiflora rose

R. nutkana
Nootka rose

R. setigera
prarie rose

R. eglanteria
sweetbriar rose

R. multiflora
multiflora rose

R. nutkana
Nootka rose

R. setigera
prarie rose

Figure 1. *Rosa*: fruits (hips), ×2 (Gill and Pogge 1974g).

Figure 2. *Rosa*: seeds (achenes), ×8 (Gill and Pogge 1974g).

The seeds are achenes borne within a fleshy, berrylike hip (Figs. 1, 2, 3). Plants flower and produce seeds at a young age and some seeds are produced almost every year. Seed dispersal is by birds and mammals. Bird digestion may enhance germination.

COLLECTION, EXTRACTION, AND STORAGE OF SEED The hips can be handpicked soon after the dark green color fades to reddish and any time thereafter. Seeds collected shortly after ripening and not allowed to dry will be less dormant than fully dried seeds. The seeds can be recovered by macerating the hips and recovering the seeds by flotation. Seeds stored dry in sealed vials will maintain viability for 2 to 4 years.

GERMINATION ASOA (1985) standards for multiflora rose (*R. multiflora*) are substrate, top of blotters in closed petri dishes; incubation temperature, 10/20°C; duration, 28 days; and additional instructions, prechill 28 days. Both the excised embryo and TZ methods have been used to test seeds of various species. The most effective treatment to enhance germination of all rose species is prolonged prechilling. Germination is epigeal (Fig. 4).

NURSERY AND FIELD PRACTICE One of the most effective methods of propagating roses is to sow freshly cleaned seeds before they have a chance to dry. They can be either broadcast or drilled, and covered with a shallow layer of firm soil and mulched.

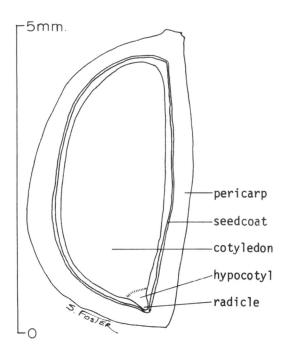

Figure 3. *Rosa setigera*, prairie rose: longitudinal section through an achene, ×16 (Gill and Pogge 1974g).

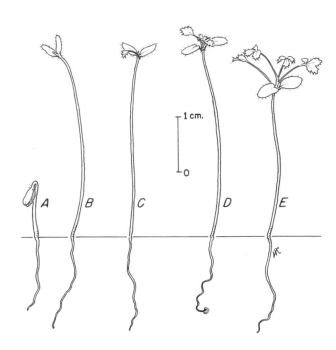

Figure 4. *Rosa blanda*, meadow rose: seedling development at 1, 3, 6, 26, and 41 days after germination (Gill and Pogge 1974g).

LABIATAE—MINT FAMILY

Rosmarinus officinalis L. — Rosemary

GROWTH HABIT, OCCURRENCE, AND USE The genus *Rosmarinus* consists of a single species native to the Middle East. A garden plant, its aromatic leaves are used as a seasoning. The fruit is a nutlet. The seeds require no pretreatment for germination (Dirr and Heuser 1987).

ROSACEAE — ROSE FAMILY

Rubus L. — Blackberry, Raspberry

GROWTH HABIT, OCCURRENCE, AND USE *Rubus* includes about 400 species of deciduous or evergreen, often prickly, erect or trailing shrubs or vines (Brinkman 1974k). Most are native to the cool temperate regions of the Northern Hemisphere; a few are found in the tropics and the Southern Hemisphere. Many species are cultivated for their fruit, flowers, and foliage, and nearly all provide food and shelter for wildlife. Some are valuable for erosion control.

FLOWERING AND FRUITING The perfect flowers bloom in the spring or summer (Table 1). The fruit, which ripens unevenly in the summer or early fall, is an aggregate of small, usually succulent drupes (Fig. 1), each containing a single hard-pitted nutlet (seed) (Fig. 2) that has a negligible amount of endosperm (Fig. 3). Natural dispersal is mostly by birds and mammals. Good seed crops occur nearly every year. Although most fruits are either red or black at maturity, varieties of some species produce yellow to orange fruit (Table 1).

COLLECTION, EXTRACTION, AND STORAGE OF SEED When ripe the fruits of *Rubus* should be picked to reduce loss to birds. The fruits of most species mature over an

Table 1. *Rubus:* phenology of flowering and fruiting, ripe fruit color, and cleaned seed weight (Brinkman 1974k).

Species	Flowering dates	Fruit ripening dates	Seed dispersal dates	Ripe fruit color	Seeds/gram
R. allegheniensis	May–July	Aug.–Sept.	Aug.–Sept.	Black-purple	580
R. canadensis	June–July	July–Sept.	July–Sept.	Black	480
R. flagellaris	May–July	June–Sept.	June–Sept.	Black	290
R. hispidus	June–Sept.	Aug.–Oct.	Aug.–Oct.	Red-black	410
R. idaeus	May–July	June–Oct.	July–Oct.	Red	720
R. laciniatus	June–Aug.	July–Aug.	Sept.–Oct.	Black	300
R. macropetalus	Apr.–June	Aug.–Sept.	Oct.–Nov.	Black	850
R. occidentalis	Apr.–July	June–Aug.	June–Aug.	Purple	740
R. odoratus	June–Sept.	July–Sept.	July–Sept.	Red	1,090
R. procerus	June	Aug.–Sept.	–	Black	320
R. spectabilis	May–June	June–Aug.	June–Aug.	Orange-red	310

R. allegheniensis
Allegheny blackberry

R. canadensis
thornless blackberry

R. hispidus
swamp blackberry

R. procerus
Himalaya blackberry

R. parviflorus
western thimbleberry

R. macropetalus
trailing blackberry

Figure 1. *Rubus:* fruits, ×1 (Brinkman 1974k).

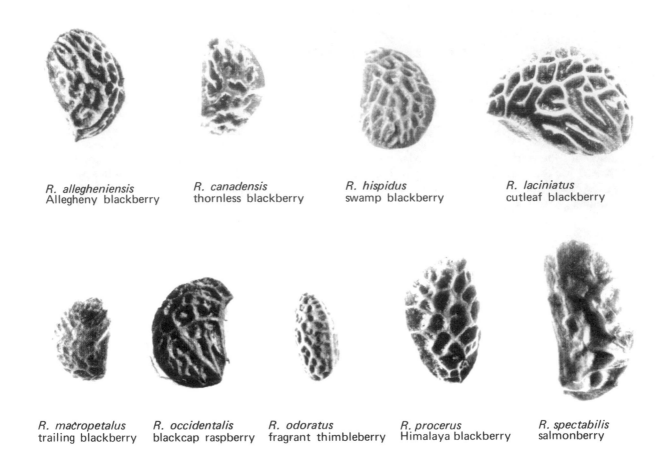

R. allegheniensis
Allegheny blackberry

R. canadensis
thornless blackberry

R. hispidus
swamp blackberry

R. laciniatus
cutleaf blackberry

R. macropetalus
trailing blackberry

R. occidentalis
blackcap raspberry

R. odoratus
fragrant thimbleberry

R. procerus
Himalaya blackberry

R. spectabilis
salmonberry

Figure 2. *Rubus*: nutlets (seeds), ×12 (Brinkman 1974k).

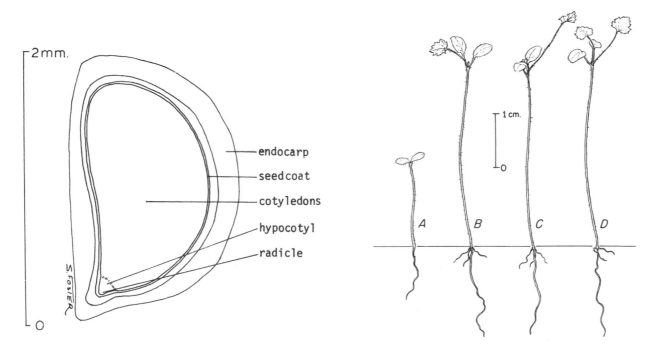

2mm.

endocarp

seedcoat

cotyledons

hypocotyl

radicle

Figure 3. *Rubus canadensis*, smooth blackberry: longitudinal section through a seed, ×36 (Brinkman 1974k).

Figure 4. *Rubus occidentalis*, blackcap raspberry: seedling development at 1, 13, 22, and 36 days after germination (Brinkman 1974m).

extended period of time. Seeds may be extracted by maceration and flotation. Cleaned seeds should be dried before storage. They can be stored at low temperatures.

PREGERMINATION TREATMENT Seeds of most species of *Rubus* have low germination because of hard seedcoats. Red raspberry (*R. idaeus*) seeds will germinate after prolonged prechilling (120 days or more), but the seeds of other species require warm stratification followed by prolonged prechilling. Heit (1967c) recommended soaking them 20 to 60 minutes in sulfuric acid or a 1% solution of sodium hyperchlorite for 7 days prior to subjecting them to both warm stratification and prechilling.

GERMINATION Pretreated seeds can be tested on sand in germinators set at 20/30C with 8 hours of light daily. Campbell et al. (1988) found seeds of sand blackberry (*R. cuneifolius*) had their endocarps broken by 18 hours

of soaking in 15% sodium hyperchlorite. Dale and Jarvis (1983) suggested raspberry seeds treated with calcium hypochlorite could be germinated without prechilling. Nesme (1985) obtained 100% germination of naked embryos of raspberry seeds and considered embryo dormancy nonexistent in this species. Warr et al. (1979) determined that seeds of bakeapple (*R. chamaemorus*) also had hard seedcoats that prevented germination. Lundergan and Carlisi (1984) germinated freshly harvested seeds of blackberry (*Rubus*. spp.) without prechilling by oxygenating them in water and gibberellin plus benzylamino purine. Ke et al. (1985) used tissue culture techniques to grow *Rubus* seeds and embryos.

NURSERY AND FIELD PRACTICE Best emergence usually follows late summer or early fall sowing of scarified seeds. They should be sown in drill rows, covered with a thin layer of soil and the beds mulched. Germination is epigeal (Fig. 4).

PALMA — PALM FAMILY

Sabal Adans. — Palmetto

GROWTH HABIT, OCCURRENCE, AND USE *Sabal* is native to the Western Hemisphere from the Bermuda islands and the South Atlantic and Gulf states through the West Indies to Venezuela and Mexico (Olson and Barnes 1974c). Cabbage palmetto (*Sabal palmetto*) is found from North Carolina to Florida, in low, flat woods and on offshore islands in the north, becoming common throughout the lower part of the Florida peninsula. It is used by rural residents for food, timber, and craft weaving; it has also been widely planted as an ornamental species. Etonia palmetto (*S. etonia*) is a low, spreading species compared to the 30 m height

obtained by cabbage palmetto. It has a restricted range in the dry pinelands and scrub of central Florida.

FLOWERING AND FRUITING The perfect flowers of cabbage palmetto are about 0.6 cm in diameter, white, and borne in drooping clusters up to 2 m long from June to August, depending on the latitude. The fruit is a dark brown to black berry, subglobose or slightly obovoid, about 0.8 cm in diameter, that ripens in late autumn or winter. Each fruit contains one light brown seed about 0.6 cm in diameter. The fruits and seeds of etonia palmetto are slightly larger (Fig. 1). Embryos are minute (Fig. 2).

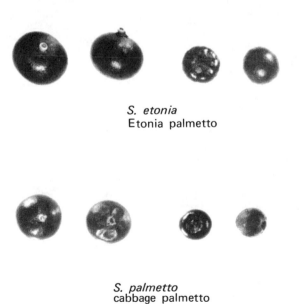

S. etonia
Etonia palmetto

S. palmetto
cabbage palmetto

Figure 1. *Sabal*: left, fruits, and right, seeds, ×1 (Olson and Barnes 1974e).

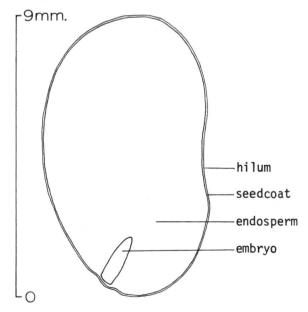

Figure 2. *Sabal etonia*, etonia palmetto: longitudinal section through the embryo of a seed, ×8 (Olson and Barnes 1974e).

COLLECTION, EXTRACTION, AND STORAGE OF SEED The fruits of palmettos may be picked from the plants when ripe. The seeds may be separated by maceration and recovered by flotation. Cabbage palmetto has about 2.8 seeds per gram.

GERMINATION The seeds of *Sabal* species native to North America apparently require no pretreatment for germination, but prechilling for 30 days makes it more rapid and uniform. Removal of the micropyle cap enhances germination.

The Genebank Handbook (1985) offers germination information for several species of *Sabal*:

Species	Germination
S. bermudana	88 to 137 days after sowing.
S. blackburniana	100 to 120 days after sowing.
S. causiarum	43 to 131 days after sowing.
S. domingensis	Delay of up to 48 days.
S. glaucescens	2-month delay.
S. jamaicensis	37-day delay.
S. mexicana	48- to 120-day delay.
S. minor	Considerable dormancy; requires 7 to 24 months of afterripening at room temperature. Treatment with gibberellin prompts germination. The optimum incubation temperature is 25°C.
S. palmetto	3- to 4-month delay possible. Optimum test incubation temperatures are 25 or 30°C, with a sand substrate.
S. parviflora	22- to 42-day delay possible.
S. texana	70- to 169-day delay possible.
S. yapa	82- to 220-day delay possible.

Becwar et al. (1983) used differential thermal analysis to determine threshold moisture levels below which seed tissue water was in an unfreezable condition. *Sabal parviflora* seeds survived dehydration as low as 2 to 12%. Brown (1976) determined that a soil temperature of 20°C must be reached before cabbage palmetto seeds can germinate in the field, and a maximum of 35 to 40°C must not be exceeded. Sunlight retards radicle elongation.

NURSERY AND FIELD PRACTICE *Sabal* seeds should be planted 1.25 to 2.5 cm deep in light textured soil, soon after collection, and the seeds should not be permitted to dry.

LABIATAE — MINT FAMILY

Salazaria Torr. — Bladder sage

GROWTH HABIT, OCCURRENCE, AND USE *Salazaria* is a monotypic genus. The single species, *S. mexicana*, is a low, dense, and divaricately branched shrub with spinescent branchlets. It is common in dry washes and canyons in the American Southwest from Texas to Utah, California, and adjacent Mexico.

GERMINATION Initial germination of untreated seed was greater than 80% (Kay et al. 1988). Seeds stored in sealed containers maintained viability for 11 years, but those stored in the open rapidly lost viability.

SALICACEAE — WILLOW FAMILY

Salix L. — Willow

GROWTH HABIT, OCCURRENCE, AND USE The willows consist of about 300 species of deciduous trees and shrubs widely distributed in both hemispheres from the Arctic region to South Africa and southern Chile (Brinkman 1974l). There are numerous hybrids. Of the some 70 North American species, about 30 attain tree size and form, and many are valuable for wood products. Nearly all species provide browse for animals and enhance riparian habitats.

FLOWERING AND FRUITING Staminate and pistillate flowers are borne in catkins on separate trees, usually appearing before or with the leaves (Table 1). The fruit, a capsule (Fig. 1) occurring in elongated clusters, contains many minute, hairy seeds (Figs. 2, 3). These usually ripen in early summer, but seeds of some species mature in the fall. Seed are disseminated by wind and water.

COLLECTION, EXTRACTION, AND STORAGE OF SEED Willow seeds must be collected as soon as the fruits ripen, as indicated by the capsule turning from green to yellowish (Table 1). Frequent observations are necessary to determine maturity, at which time the capsules can be collected by picking from the trees. Seeds from trees growing near the water can often be collected from the surface of the water or from drifts along the shore.

It is unnecessary to separate the seeds from the open capsules. Because of the short time seeds of willow are viable, there is usually no commercial seed available. The maximum period of storage is from 4 to 6 weeks (see more modern information below), but germination rates drop off rapidly after 10 days for seeds stored

Table 1. *Salix:* phenology of flowering and fruiting, fruit ripeness criteria, and cleaned seed wieght (Brinkman 1974L).

Species	Flowering dates	Fruit ripening dates	Seed dispersal dates	Fruit color		Seeds/gram
				Preripe	Ripe	
S. amygdaloides	May–June	–	–	–	–	5,700
S. bebbiana	Apr.–May	May–June	May–June	Green	Yellowish	5,500
S. caroliniana	Apr.–June	Mar.–Apr.	–	–	Pale yellow	18,300
S. discolor	Mar.–Apr.	–	–	Green	Yellowish	–
S. exigua	May	Apr.–May	Apr.–May	Green	Yellowish	22,000
S. fragilis	Mar.–Apr.	June–July	June–July	–	–	7,000
S. interior	–	Aug.	May–June	–	–	–
S. lasiandra	Apr.–May	May–June	June–Aug.	Green	Yellowish	25,300
S. nigra	Feb.–June	Apr.–May	Apr.–May	Green	Yellowish	–
S. petiolaris	May–June	June–July	June–July	–	–	1,100
S. rigida	Apr.–June	June	–	Green	Yellowish	–
S. scoulerana	Apr.–June	May–July	May–July	–	–	14,300

Figure 1. *Salix:* three stages of opening capsule, ×2 (Brinkman 1974l).

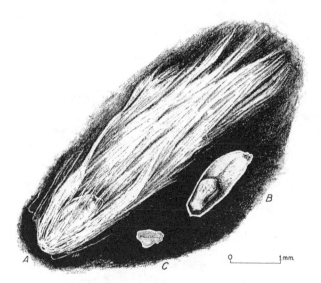

Figure 2. *Salix fragilis,* crack willow: (A) exterior view of seed with cotton, showing corona; (B) longitudinal section of seed showing embryo; (C) cross section of seed, all at ×14 (Brinkman 1974l).

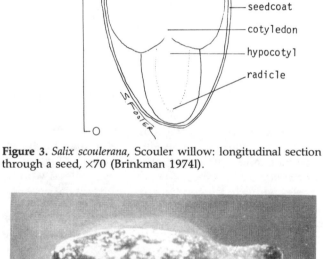

Figure 3. *Salix scoulerana,* Scouler willow: longitudinal section through a seed, ×70 (Brinkman 1974l).

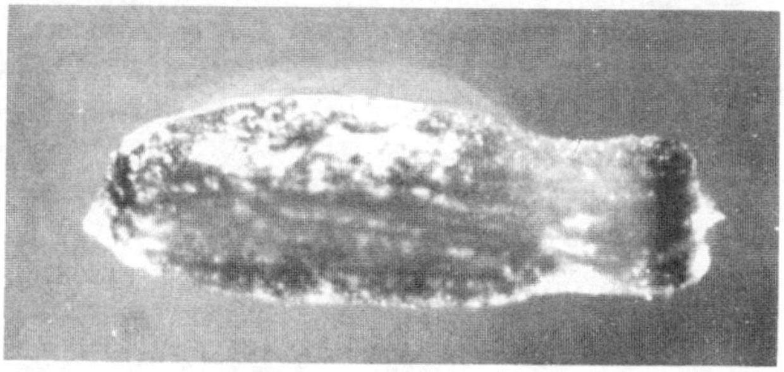

Figure 4. *Salix bebbiana,* Bebb willow: imbibed seed (cotton removed), ×20 (Brinkman 1974l).

at room temperature. Moistened seeds (Fig. 4) may be stored up to a month if refrigerated in sealed containers.

GERMINATION Under natural conditions, willow seeds will usually germinate in 12 to 24 hours on moist sand or alluvium. Germination is epigeal. It may be tested on moist sand or on paper in petri dishes.

Despite their widespread distribution and impor-

tance in riparian habitats, there has been relatively little recent research on the germination of *Salix* species. Gorobets (1978) determined that 2 species of willow native to the USSR were capable of germinating under a wide range of temperatures. In a study of germination of willows in Norway, Junttila (1976) found the optimum temperature for germination of seeds of *S. herbacea, S. pentandra, S. polaris,* and *S. reticulata* was a warm 26 to 32°C. Seed germination was greatly reduced below 16–20°C, but moist chilling at 4°C for 2 to 3 weeks lowered optimum temperatures for all but *S. herbacea*. The optimum temperature for germination of *S. caprea* seed was 20 to 28°C but germination was good even at 12°C without pretreatment.

Zasada and Densmore (1980) compared the longevity of summer- and winter-dispersed seeds of willow species in Alaska. Seeds of *S. glauca*, a winter-dispersed species, had to be prechilled for 2 weeks before testing. The seeds were stored at −10°C for 3 years and then tested at 5, 10, 15, 20, and 25°C. After 3 years all species had 70% germination at one or more of the temperatures tested. The prechilling requirement of *S. glauca* persisted during storage.

NURSERY AND FIELD PRACTICE Seeds must be sown immediately after collection. The opened capsules and seeds are broadcast on well-prepared beds, followed by light packing with a roller. The seedbeds must be kept moist until the seedlings are well established. Shading is usually applied to keep relative humidity high near the surface. If the initial stand density is too great, seedlings can be transplanted after 3 to 4 weeks.

CHENOPODIACEAE — GOOSEFOOT FAMILY

Salsola L.

GROWTH HABIT, OCCURRENCE, AND USE The only woody species of *Salsola* that has been introduced to the United States is *S. vermiculata*. This shrub is considered a valuable browse species. It has become naturalized along the west side of the San Joaquin valley in California but is under an eradication program because it serves as an alternate host for leaf hoppers that transmit diseases to crop species.

GERMINATION Without pretreatment, *S. vermiculata* seeds had 70% germination (Kay et al. 1988). Viability was maintained for 7 years under low temperature storage.

LABIATAE — MINT FAMILY

Salvia L. — Sage

GROWTH HABIT, OCCURRENCE, AND USE The genus *Salvia* consists of over 500 species widely distributed in temperate and warmer regions. Some are cultivated as ornamentals and others for flavoring. Several woody species native to western North America have promise as revegetation species. Creeping sage (*S. sonomensis*) is an aromatic, semiprostrate, suffrutescent plant found in the chaparral zone along the Sierra Nevada and Coast Range in California at elevations below 2000 m (Nord and Gunter 1974). It has value for soil protection and wildfire hazard abatement.

FLOWERING AND FRUITING The bisexual flowers occur in dense, globose whorls along erect spicate inflorescences. Flowering occurs in late spring, and good seed crops are borne nearly every year. The fruit consisting of nutlets is enclosed in persistent, papery calyx parts (Figs. 1, 2).

COLLECTION, EXTRACTION, AND STORAGE OF SEED Seeds should be collected soon after maturity by stripping the inflorescence. After drying, the seeds can be recovered by air screening. They store well in sealed containers at low temperatures.

GERMINATION *Salvia* seeds are usually initially dormant and require scarification and prechilling before they will germinate. The Genebank Handbook (1985) suggests that gibberellin can be substituted for the pre-chilling and provides germination suggestions for the following species:

Species	Germination
S. coccinea	Prechill.
S. farinacea	Prechill and provide light.
S. glutinosa	Remove seed covering structure, apply gibberellin, and incubate in light.
S. officinalis	Prechill.
S. patens	Prechill.
S. pratensis	Prechill.
S. reflexa	Prechill.
S. sclarea	Prechill and provide light. Can be pretreated at 40°C.
S. splendens	Prechill and provide light.

Figure 1. *Salvia sonomensis*, creeping sage: nutlet, ×12 (Nord and Gunter 1974).

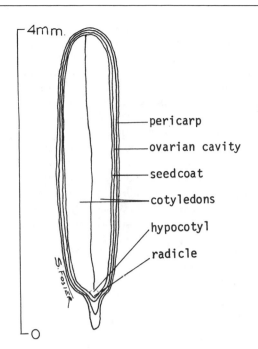

Capon et al. (1978) stored seeds of 10 collections of *S. columbariae* under room temperature, 50°C preheating conditions and at 2°C. Cold storage did not improve germination, but preheating did enhance germination of the seeds of some collections. *Salvia mellifera* seeds from chaparral communities in southern California failed to germinate in the dark unless exposed to charcoal (Keeley 1968). It was suggested this was an adaptation to germination after wildfires. Verzar-Petri and Then (1974) investigated the volatile oils of *Salvia* seeds. Weerakeen and Levett (1986) found that germination of seeds of *S. reflexa* increased as osmotic potentials of the germination substrate became more negative.

NURSERY AND FIELD PRACTICE Pretreated seeds of creeping sage can be seeded directly in the spring.

Figure 2. *Salvia sonomensis*, creeping sage: longitudinal section through a nutlet, ×20 (Nord and Gunter 1974).

CAPRIFOLIACEAE — HONEYSUCKLE FAMILY

Sambucus L. — Elder

GROWTH HABIT, OCCURRENCE, AND USE The elders include about 20 species of deciduous shrubs or small trees, rarely herbs, native to temperate and subtropical regions of both hemispheres (Brinkman 1974m). The fruit of most species is used by wild animals and humans.

FLOWERING AND FRUITING The large clusters of small, white or yellowish white, perfect flowers bloom in the spring or summer (Brinkman 1974m):

Species	Flowering dates	Fruit ripening dates	Seed dispersal dates
S. callicarpa	June	July–Aug.	Aug.–Sept.
S. canadensis	June–July	June–Sept.	Aug.–Oct.
S. glauca	May–July	Aug.–Sept.	Aug.–Oct.
S. pubens	Apr.–July	June–Aug.	June–Nov.

The fruit is a berrylike drupe (Fig. 1) containing 3 to 5 one-seeded nutlets or stones (Figs. 2, 3). When ripe, the fruits vary from red to nearly black, depending on the species (Brinkman 1974m):

Species	Mature height (m)	Ripe fruit color	Seeds/gram
S. callicarpa	6	Red	470
S. canadensis	3	Purplish-black	510
S. glauca	9	Blue-black	450
S. pubens	3	Scarlet red	630

Seed dispersal is chiefly effected by birds and other animals.

COLLECTION, EXTRACTION, AND STORAGE OF SEED Elder fruits are collected by stripping or cutting the clusters from branches. Collections should be made as soon as the fruits ripen to avoid losses to birds. If the seeds are not extracted at once, care must be taken to avoid heating. The fruits can be run through a macerator and the seeds recovered by flotation. Commercial seed may consist of dried fruits or clean nutlets. Seeds may be stored in closed containers at low temperatures for several years.

PREGERMINATION TREATMENT Elder seeds are difficult to germinate because of their dormant embryos and hard seedcoats. Pretreatment usually consists of 90 days of warm stratification followed by 90 days of prechilling. Heit (1967c) suggests 10 to 15 minutes soaking in acid followed by 2 months of prechilling.

GERMINATION The Genebank Handbook (1985) offers the following suggestions:

Species	Temperature (°C)	Additional directions
S. glauca	20/30	Prechill 16 weeks; provide light 8 hrs/day.
S. nigra	20/30	Warm stratify 10 weeks at 25°C, then prechill 12 weeks; provide light 8 hrs/day.
S. racemosa	20/30	Warm stratify 10 weeks at 25°C, then prechill 12 weeks; provide light 8 hrs/day.

Tylkowski (1982) determined that a thermal treatment in the form of warm stratification at 15 to 25°C for

Figure 1. *Sambucus glauca*, blueberry elder: fruit cluster, ×0.75, and single fruit, ×4 (Brinkman 1974m).

3 weeks greatly improved the germination of European elder (*S. nigra*) and red elder (*S. racemosa*), compared to prechilling alone. Norton (1986) used prechilling and gibberellin to stimulate the germination of *S. caerulea*. The germination of members of the honeysuckle family (which includes elder) was investigated by Cran (1982).

NURSERY AND FIELD PRACTICE Elder seeds can be sown in the fall soon after collection or prechilled and sown in the spring. In either case, germination is not complete until the spring of the second year after sowing. A seedling density of 370 plants per m² is desired. It may be desirable to sow seeds as soon as they are collected, without allowing them to dry. Germination is epigeal (Fig. 4).

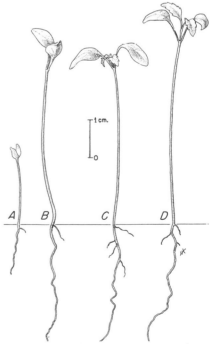

Figure 3. *Sambucus pubens*, scarlet elder: longitudinal section through a nutlet, ×24 (Brinkman 1974m).

S. callicarpa
Pacific red elder

S. canadensis
American elder

S. glauca
blueberry elder

Figure 2. *Sambucus*: seeds, ×12 (Brinkman 1974m).

Figure 4. *Sambucus canadensis*, American elder: seedling development at 2, 20, 33, and 45 days after germination (Brinkman 1974m).

SANTALACEAE — SANDALWOOD FAMILY

Santalum album L. — Sandalwood

GROWTH HABIT, OCCURRENCE, AND USE Native to tropical Asia, sandalwood is a tree with aromatic, sweet-scented wood prized for use in cabinet making and in perfumery.

GERMINATION Nagaveni and Srimathi (1980, 1981) investigated the germination of sandalwood seeds. Removal of the seedcoat gave the best germination of these rather dormant seeds. Acid scarification and gibberellin enrichment were considered practical for large-scale treatments.

COMPOSITAE — SUN FLOWER FAMILY

Santolina L. — Lavender cotton

GROWTH HABIT, OCCURRENCE, AND USE The genus *Santolina* consists of about 8 species of shrubs (or rarely, herbs) that have aromatic leaves. The fruit is an achene, and seeds germinate without pretreatment (Dirr and Heuser 1987).

SAPINDACEAE — SOAPBERRY FAMILY

Sapindus drummondii Hook. & Arn. — Western soapberry

GROWTH HABIT, OCCURRENCE, AND USE Western soapberry grows on clay soils and dry limestone uplands from southwestern Missouri to Louisiana and westward through Oklahoma and Texas to southern Colorado, New Mexico, southern Arizona, and northern Mexico (Read 1974b). It is a small to medium deciduous tree, 7 to 15 m tall, first introduced into cultivation in 1900. The glossy yellow fruit and long pinnate leaves make this species attractive for environmental plantings. The heavy, strong, close-grained wood is split into strips and used in basketry.

FLOWERING AND FRUITING The small, white, polygamo-dioecious flowers, borne in rather large clusters of terminal or axillary panicles, open during May to July. The fruit—a yellow, translucent, globular drupe, 10 to 14 mm in diameter, usually contains a single, dark brown, hard-coated seed (Figs. 1, 2), but occasionally 2 or 3 seeds are present. The fruit ripens during September and October and persists on the tree until late winter or spring. Seed crops are usually abundant each year.

COLLECTION, EXTRACTION, AND STORAGE OF SEED Fruits may be collected any time during late fall or winter by hand-picking or by flailing onto ground sheets. They can be macerated and the seeds recovered by flotation. There are about 1.2 seeds per gram.

PREGERMINATION TREATMENT Germination of the seeds is usually slow and delayed. Scarification in acid for 2 to 2.5 hours followed by 90 days of prechilling is a recommended treatment. Freshly collected seeds that have not been allowed to dry before planting often give better germination than stored pretreated seeds.

Figure 1. *Sapindus drummondii*, western soapberry: fruit and seed, ×2 (Read 1974b).

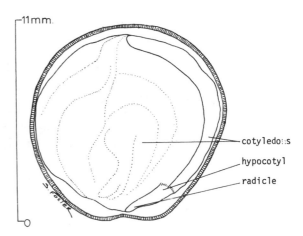

Figure 2. *Sapindus drummondii*, western soapberry: longitudinal section through a seed showing folded cotyledons, ×5 (Read 1974b).

Sheikh (1979) and Munson (1984) both determined that acid scarification was the most suitable means of breaking the hard seedcoats of *Sapindus* seeds.

NURSERY AND FIELD PRACTICE Fresh, moist seeds can be sown in the fall without pretreatment. Sowing densities of about 200 viable seeds per m² are satisfactory. Seedlings develop a strong taproot and exhibit slow shoot growth.

EUPHORBIACEAE — EUPHORBIA FAMILY

Sapium sebiferum (L.) Roxb. — Tallowtree

GROWTH HABIT, OCCURRENCE, AND USE Tallowtree is a small, deciduous tree that attains a height of about 10 m at maturity (Bonner 1974j). A native of China, this species has been widely planted in the coastal plain from South Carolina and Florida to Texas, Oklahoma, and Arkansas. The bright red fall foliage makes the tree a popular ornamental, and the seeds have value for wildlife. In China, the wax that coats the seeds is extracted and used in soaps, candles, and cloth dressings. Tallowtree readily escapes from cultivation and is common along roadsides of the Gulf Coast.

FLOWERING AND FRUITING Both pistillate and staminate flowers are borne on the same yellowish green spike in the spring. The fruit, ripening in October to November, is a rounded, 3-lobed capsule, 0.7 to 1.25 cm in diameter. Its greenish color changes to brownish purple at maturity. There are usually 3, sometimes 2, waxy seeds per capsule (Figs. 1, 2).

COLLECTION, EXTRACTION, AND STORAGE OF SEED The dry fruits can be collected from the trees by hand after fruit splitting (dehiscence) has started. On a sample of fruits from a tree in central Mississippi, Bonner (1974j) recorded 10,700 capsules per 35 liters, 6 seeds per gram, and a moisture content of 6%.

GERMINATION Fresh seeds germinate about 38% on creped cellulose when incubated at 20/30°C for 30 days. Tallowtree is easily propagated by cuttings.

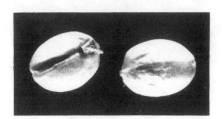

Figure 1. *Sapium sebiferum*, tallowtree: seeds, ×2 (Bonner 1974j).

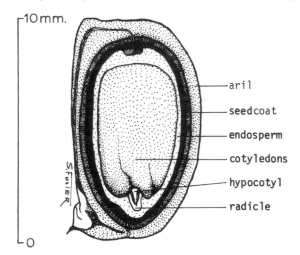

Figure 2. *Sapium sebiferum*, tallowtree: longitudinal section through a seed, ×6 (Bonner 1974j).

BUXACEAE — BOX FAMILY

Sarcococca Lindl. — Sweet Box

GROWTH HABIT, OCCURRENCE, AND USE The genus *Sarcococca* consists of about 16 species of evergreen shrubs native to China. Many are planted in the southern United States as ornamentals. The fruit is drupelike, fleshy or leathery, with 1 or 2 seeds.

GERMINATION The pulp should be removed by maceration and the seeds recovered by flotation. They germinate without pretreatment (Dirr and Heuser 1987).

LAURACEAE — LAUREL FAMILY

Sassafras albidum (Nutt.) Nees — Sassafras

GROWTH HABIT, OCCURRENCE, AND USE Sassafras is a short to medium-tall, deciduous tree native from southwestern Maine to central Michigan and southeastern Iowa, south to eastern Texas and central Florida (Bonner and Maisenhelder 1974b). On more fertile sites the trees may reach 30 m in height, but such heights are exceptional. Sassafras is valuable for timber and wildlife. The lightweight wood is soft but very durable. Bark has been used for tea, oil, and soaps perfume. *Sassafras albidum* has been cultivated since 1630.

FLOWERING AND FRUITING The dioecious, greenish yellow flowers are borne in 5 cm axillary racemes in March and April as the leaves appear. The drupaceous fruits are ovoid, dark blue, and about 0.7 to 1.25 cm long (Fig. 1); they mature in August and September and are dispersed within a month. Dispersal is aided by birds, which often eat the fruits before they fall. The fruit is borne on a thickened red pedicel, and the pulpy flesh covers a hard, thin endocarp that encloses the seeds (Figs. 1, 2). The minimum seed-bearing age is 10 years, and good seed crops are produced every 1 or 2 years.

COLLECTION, EXTRACTION, AND STORAGE OF SEED Fruits may be picked from the trees or knocked to ground cloths. They are green before maturity, and a change to dark blue indicates they are ready for collection. The pulpy flesh can be removed by rubbing the fruits over hardware cloth or by mechanical maceration. In the northern half of the range, seeds average 13 per gram. To prevent deterioration, the cleaned seeds should be placed in storage soon after cleaning. Seeds in sealed containers stored at cool temperatures maintain viability for a few seasons.

GERMINATION The seeds exhibit strong embryo dormancy, which can be overcome with prechilling for 120 days.

NURSERY AND FIELD PRACTICE Best results are obtained by sowing cleaned seeds. They should be sown late in the fall to prevent fall germination in drill rows 20 to 30 cm apart and 0.6 to 1.25 cm deep. The beds should be mulched until after late frost.

Figure 1. *Sassafras albidum*, sassafras: fruit and seed, ×2 (Bonner and Maisenhelder 1974b).

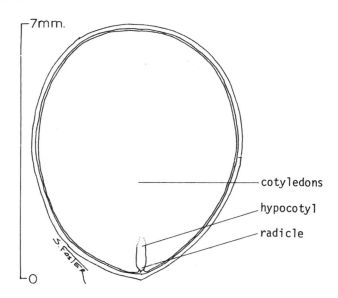

Figure 2. *Sassafras albidum*, sassafras: longitudinal section through a seed, ×8 (Bonner and Maisenhelder 1974b).

MELASTOMATACEAE — MELASTOME FAMILY

Schizocentron Meissn.

GROWTH HABIT, OCCURRENCE, AND USE *Schizocentron* is a creeping vine similar to *Heterocentron*. The fruit is a 4-valved capsule. The seeds germinate without pretreatment, but 2 to 3 months of prechilling improves the speed and uniformity of germination.

TAXODIACEAE — TAXODIUM FAMILY

Sciadopitys verticillata (Thunb.) Sieb. & Zucc. — Umbrella pine

GROWTH HABIT, OCCURRENCE, AND USE Native to central Japan at elevations of 215 to 1690 m, the umbrella pine is an evergreen tree from 22 to 40 m tall, most commonly grown for ornamental purposes but also planted for erosion control (Rudolf 1974i). In Japan the species provides lumber and oakum for caulking boats. Umbrella pine is the only species in the genus.

FLOWERING AND FRUITING Flowers of both sexes occur at the ends of branches in the spring. The male flowers are in clusters, and the female flowers, which develop into cones, are solitary. When the cones ripen at the end of the second season, they become gray brown and are about 7.5 to 12.5 cm long and 3.8 to 5 cm wide. Each cone bears several ovoid, compressed, narrowly winged seeds about 1.25 cm long.

COLLECTION, EXTRACTION, AND STORAGE OF SEED Ripe cones may be picked in the fall and placed in a warm, dry place to open; seeds are removed by shaking and then dewinged. The number of cleaned seeds per gram ranges from 32 to 42, averaging 38. Seeds dried to 10% moisture content and stored in closed containers at low temperatures retain viability for a couple of years.

PREGERMINATION TREATMENT Apparently both warm stratification for 100 days and prechilling for 90 days is required for germination.

GERMINATION Germination of pretreated seeds can be tested in sand at a 20/30°C incubation temperature for 75 days. Light may be required for germination. More than 8 hours of light daily, however inhibits germination.

NURSERY AND FIELD PRACTICE Seeds should be sown in the fall or prechilled for spring sowing. Umbrella pine has the tendency to form several leaders and is slow growing in the nursery. It can be propagated from cuttings taken in the summer.

TAXODIACEAE — TAXODIUM FAMILY

Sequoia sempervirens (D. Don) Endl. — Redwood

GROWTH HABIT, OCCURRENCE, AND USE Redwood is one of the largest of the forest trees (Boe 1974a). Its natural range is in the summer fog belt of the Coast Range from Little Redwood Creek on the Chetco River in southwestern Oregon to Salmon Creek in the Santa Lucia mountains of southern Monterey County, California. The redwood belt is an irregular coastal strip about 750 km long and 8 to 60 km wide. Redwood thrives in cool, moist places. Elevations of native stands range from 30 to 770 m. The beautifully colored and grained wood is valued for its durability and weathering characteristics.

FLOWERING AND FRUITING Tiny, inconspicuous male and female flowers are borne separately on different branches of the same tree. The ovulate conelets grow into broadly oblong cones with scales that are closely packed, woody, persistent, and thick. The ovules are usually borne in 1 concentric row on each cone scale. Ripe seeds have brown wings and slightly darker seedcoats; each wing is about equal in width to the seed and is part of the seedcoat (Fig. 1). Embryos have two cotyledons (Fig. 2). Many open cones persist through the next growing season. Trees start to bear at 5 to 15 years of age, and good seed crops occur frequently.

COLLECTION, EXTRACTION, AND STORAGE OF SEED In the northern redwoods, cone collection should begin in late September and October, since seed dispersal from

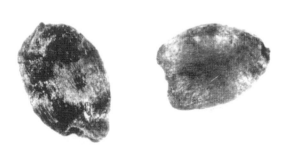

Figure 1. *Sequoia sempervirens*, redwood: seed, ×8 (Boe 1974a).

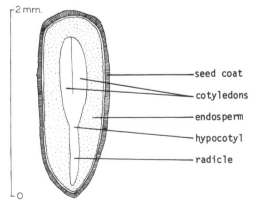

Figure 2. *Sequoia sempervirens*, redwood: longitudinal section through a seed, ×25 (Boe 1974a).

cones proceeds rapidly after October, reaching a peak from November to February. Seeds are mature when the cone color changes from green to greenish yellow or when the cone scales separate slightly .

Cones can be air dried at room temperatures of 22 to 24°C in 14 days. They should be stirred daily and a small fan used to circulate the air. In large nurseries, the cones are kiln dried at 78°C for 24 hours. The seeds are extracted in tumbling screens and recovered by pneumatic cleaners. Seed yield from cones, on a weight basis, is about 5 to 11%.

Viability of seeds in storage has been maintained longer at freezing temperatures than with any other procedure.

GERMINATION Redwood seeds require no pretreatment for germination. ASOA (1985) standards for germination are substrata, the top of paper or rolled in paper; incubation temperature, 20/30°C; and duration, 21 days. The soundness of seed is usually low.

NURSERY AND FIELD PRACTICE Seeds are sown in the spring when frost is not likely and soil temperatures are warm. They are sown in drill rows at a depth of 0.5 cm. Optimum seedling densities are 315 plants per m² of seedbed. Half-shade should be provided the first growing season.

TAXODIACEAE — TAXODIUM FAMILY

Sequoiadendron giganteum (Lindl.) Buchholz — Giant sequoia

GROWTH HABIT, OCCURRENCE, AND USE The giant sequoia grows to heights exceeding 77 m in central California on the western slopes of the Sierra Nevada in more or less isolated groves at 1385 to 2310 m elevations (Boe 1974b). Its north-south range is about 433 km. It has been cultivated rather widely since 1853.

FLOWERING AND FRUITING Small male and female flowers grow separately on the branches of the same tree. Although the small, enclosed terminal buds are present the previous summer, flowering and pollination usually occur in mid-April to mid-May, when the conelets are quite small. Conelets are about half size in July and reach full size in August, when fertilization takes place. At the start of winter, the embryos have only a few cells, and they remain this way over winter. They develop rapidly the following summer, and by late August of the second year following pollination, they are morphologically mature. Young trees start to bear cones at the age of about 20 years.

Cones may remain attached to the trees for many years, and much of the seed will be retained. During late summer, however, when cone scales shrink, some seed is shed. As soon as cones become detached, they dry out, and seeds are liberated within a few days. This fruiting characteristic provides seeds every year. Boe (1974b) reported there were 25 to 40 scales per cone, 3 to 9 seeds per scale, 230 seeds per cone, and a cone length of 5 to 8.75 cm.

The seeds are compressed, 3 to 6 mm long, and surrounded by laterally united wings broader than the body of the seed (Fig. 1). The embryos have 3 to 5 cotyledons.

COLLECTION, EXTRACTION, AND STORAGE OF SEED The old, persistent cones can be collected at any time, but collection of fresh cones should be made in August and later. Squirrels cut and cache cones that furnish considerable numbers of seeds for collections.

Cones have been air dried at 30°C for 7 days and the seed then extracted in a screened tumbler and separated from fine material with a pneumatic separator. Seed yield on a weight basis from cones is about 1.6%; there are about 180 seeds per gram. Stored seeds of the giant sequoia retain moderate viability for many years. They can be stored in plastic bags at low temperatures.

GERMINATION ASOA (1985) standards for germination are substrata, on top of blotters or on paper in covered petri dishes; incubation temperature, 20/30°C; duration, 28 days; and additional directions, paired tests can be conducted with seeds prechilled for 30 days. Studies by Fins (1981) suggest that giant sequoia seeds should be soaked overnight in distilled water and then prechilled for 60 days.

NURSERY AND FIELD PRACTICE Nursery workers often prechill seeds and then sow in the spring. The desired seedling density is 460 to 650 per m². There is about 75% success in seedling production per seeds sown.

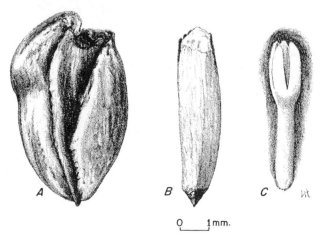

Figure 1. *Sequoiadendron giganteum*, giant sequoia: (A) seed with wings; (B) seed with outer coat removed; (C) excised embryo; all ×8 (Boe 1974b).

PALMACEAE — PALM FAMILY

Serenoa repens (Bartr.) Small — Saw-palmetto

GROWTH HABIT, OCCURRENCE, AND USE Saw-palmetto usually is an evergreen shrub, 0.7 to 2.2 m tall, with creeping horizontal stems. Occasionally the species attains the size of a small tree, reaching 6 to 8 m, with an erect to oblique stem (Olson and Barnes 1974f). The common name is derived from the ascending, palm-shaped leaves, which are rather stiff, with long petioles heavily armed with sharp, rigid, recurved teeth. Saw-palmetto occurs from coastal South Carolina southward to Florida and westward to eastern Louisiana. It reaches its most extensive development in the pine flatwoods of the lower coastal plain of Georgia and Florida.

Saw-palmetto provides wildlife habitat, and several species of wildlife eat the fruit. Large quantities of saw-palmetto leaves are shipped north for Christmas decorations. The flowers are a significant source of honey and the stems a source of tannic acid extract.

FLOWERING AND FRUITING The small white flowers are borne in panicles from April to early June, depending on the latitude. They appear on branches that are shorter than the leaves and are usually numerous.

The fruit is a drupe about 1.5 to 2.5 cm long, ovoid-oblong, green or yellow before ripening, and bluish to black when ripe (Fig. 1). It contains a single globose seed (Figs. 1, 2). Fruiting panicles sometimes weigh as much as 4 kg. Fruits ripen during September and October.

COLLECTION, EXTRACTION, AND STORAGE OF SEED The fruiting panicles should be collected by hand, by picking them from the shrubs when ripe, or by cutting the fruit-bearing branches and allowing them to drop onto ground cloths.

Seeds must be extracted from the fruits or germination will either be delayed or will not occur, even after 7 months. Seeds may be extracted by maceration and recovered by flotation. Dried saw-palmetto fruits average about 700 per kilogram, and the dry seeds average 2.4 per gram. Seed stored dry at room temperature for 3 months retained its viability.

GERMINATION Seeds with the micropyle cap removed that are incubated on moist filter paper germinate fairly quickly. Seeds with the cap in place have delayed germination. In planting trials, 65 to 85% emergence has been obtained over an 8-month period.

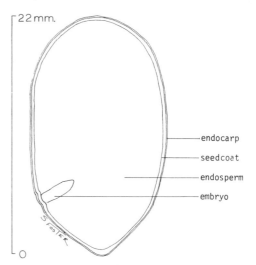

Figure 1. *Serenoa repens*, saw-palmetto: fruit and seed, ×2 (Olson and Barnes 1974f).

Figure 2. *Serenoa repens*, saw-palmetto: longitudinal section through a seed, ×3 (Olson and Barnes 1974f).

LEGUMINOSAE — LEGUME FAMILY

Sesbania Scop.

GROWTH HABIT, OCCURRENCE, AND USE *Sesbania* is a genus of short-lived, woody species that are 6 to 12 m tall. Native to tropical areas, some are naturalized in the West Indies and southern Florida. The fruit is a pendulous legume up to 30 cm long. The seeds are separated by ingrowths from the legume wall. *Sesbania* species are grown as ornamentals and for erosion control and have recently received much interest as fodder trees.

GERMINATION Recent literature on the germination of *Sesbania* species includes the following:

Author—Date	Subject	Location
Ghai et al. 1985	*S. grandiflora, S. glabra, S. aegyptiaca, S. sesban,* germination in relation to osmotic potentials reduced with NaCl, Na$_2$SO$_4$, and CaCl$_2$. *Sesbania grandiflora* is least tolerant to salts, and *S. glabra* most tolerant.	India
Graaff & Staden 1984	Germination in relation to salt concentration.	South Africa
Misra & Singh 1985	*S. grandiflora,* germination with osmotic potential reduced with NaCl.	India
Pathak et al. 1976	*S. grandiflora, S. microcarpa, S. bispinosa, S. sesban,* influence of storage duration and soil moisture stress on germination.	India

ELAEAGNACEAE — ELAEAGNUS FAMILY

Shepherdia Nutt. — Buffaloberry

GROWTH HABIT, OCCURRENCE, AND USE Silver buffaloberry (*Shepherdia argentea*) and russet buffaloberry (*S. canadensis*) are deciduous shrubs that may reach 3 m in height (Thilenius et al. 1974). Silver buffaloberry is a thorny species, russet buffaloberry is smaller and thornless. Both species are nitrogen fixers and have been used in shelterbelt plantings.

FLOWERING AND FRUITING The small, yellowish male and female flowers are borne on different plants, either solitary or in clusters on the branchlets. Drupelike, ovoid fruits, about 0.3 to 0.6 cm long, develop during the summer. The fruit is an achene enveloped in a fleshy perianth. Cleaned achenes are used as seed (Fig. 1). Seeds are dispersed chiefly by animals. Flowering phenologies for the species follow (Thilenius et al. 1974):

Species	Flowering dates	Fruit ripening dates	Seed dispersal dates
S. argentea	Apr.–June	June–Sept.	Through winter
S. canadensis	Apr.–June	June–Aug.	June–Sept.

Both species usually produce good seed crops every year. The minimum seed-bearing age is 4 to 6 years.

COLLECTION, EXTRACTION AND STORAGE OF SEEDS Fruits are ripe when they turn yellow or red. They may be gathered by stripping or flailing the branches onto ground cloths. Heavy gloves are required when harvesting silver buffaloberry because of the thorns. There are about 20 fruits per gram (apparently for both species).

The fruits are macerated and the seeds recovered by flotation. For short-term storage, seed extraction is not necessary. Care should be taken to prevent heating of freshly harvested fruits. There are about 870 seeds of silver buffaloberry per gram.

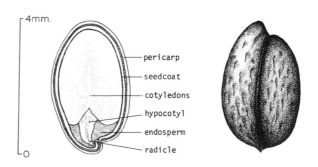

Figure 1. *Shepherdia argentea*, silver buffaloberry: (right) exterior view of cleaned achene; (left) longitudinal section through the embryo of an achene, ×9 (Thilenius et al. 1974).

GERMINATION Both hard seedcoats and embryo dormancy occur in buffaloberry seeds. Acid scarification followed by 60 to 90 days of prechilling is required. Pretreated seeds can be tested for germination by incubation in petri dishes at 20/30°C. Fung (1984) has given the most recent review of *Shepherdia* seed germination.

NURSERY AND FIELD PRACTICE Commercial nurseries usually do not acid scarify seeds before planting. Either fall planting or spring planting with prechilled seeds can be done. Seeds should be planted 0.6 cm deep and the beds mulched. Successful seedling percentage is about 50%. Germination is epigeal (Fig. 2).

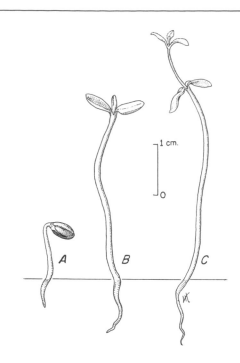

Figure 2. *Shepherdia argentea*, silver buffaloberry: seedling development at 1, 9, and 38 days after germination (Thilenius et al. 1974).

DIPTEROCARPACEAE — DIPTEROCARP FAMILY

Shorea Roxb. ex Gaertn. — Mahogany

GROWTH HABIT, OCCURRENCE, AND USE *Shorea* species are tropical trees. They are among the most important sources of mahogany lumber, often referred to as Philippine mahogany.

PREGERMINATION Yap (1981) determined that seeds of *Shorea* stored best at 22 to 28°C and that temperatures of 10°C inhibited subsequent germination. Tompsett (1985) determined that temperatures of 16°C were harmful to fruits of *S. alnoa*, *S. robusta*, and *S. roxburghii*. Maury-Lechen et al. (1981) successfully stored *S. parvifolia* seeds for 2 weeks by using a nitrogen atmosphere and silica gels.

Seeds of *S. talura* retained over 50% germination when stored 5 months in a sealed bag at 4°C (Sasaki 1980). Obviously there is a lot of interest in storing *Shorea* seeds for short time periods; equally obvious are the conflicting reports from different researchers and from different species.

Big seeds of *S. contorta* appear to produce seedlings with a higher chance of survival compared to those from small seeds. Sharma and Purohit (1980) investigated the natural germination and survival of seedlings of *S. robusta*.

BUXACEAE — BOX FAMILY

Simmondsia chinensis (Link) C.K. Schneid. — Jojoba

GROWTH HABIT, OCCURRENCE, AND USE Jojoba is a grayish green, rounded-to-erect shrub generally up to 2.5 m high (Nord and Kadish 1974). In desert habitats it is prostrate, but in moist habitats it may reach 15 m. Jojoba occurs in southern California and Arizona, as well as Sonora and Baja California, Mexico, from sea level to 1230 m. It is climatically adapted both to mesic, equable coastal climates and the continental inland desert, where diurnal temperature extremes may be 5 to 30°C. Optimum development occurs where annual

precipitation is about 25 cm for coastal populations and 40 to 45 cm for inland populations.

Jojoba is useful for food, and the foliage is highly palatable and nutritious for livestock and wildlife. The edible seed has a slightly bitter, nutty flavor and contains about 50% liquid wax, made up of esters of long chain alcohols and fatty acids. Steroidal alkaloids and related compounds also occur in the seed and vegetative parts.

FLOWERING AND FRUITING The flowering period is usually

from December to April, but in tropical Baja California, it may be almost any time. In California and Arizona, flowering appears to be triggered by cool periods of 2 weeks or longer, followed by warmer conditions. Some off-season flowering occurs occasionally. Viable seed may develop regardless of flowering date. The male and female flowers occur on separate plants. Instances of hermaphroditic flowers on male plants were found in peripheral populations of jojoba in Arizona.

The female flowers are axillary, usually solitary, greenish in color, and about 1.25 cm long on short pedicels. In certain populations they may occur in fascicles of up to 20 flowers. The ovary contains 3 ovules. The yellowish male flowers, about 0.6 cm long, are grouped in dense, rounded axillary clusters.

The fruit is a capsule containing 1 to 3 seeds that mature between July and October. Most capsules split at maturity and release their seeds. Occasionally, capsules drop from the bush without opening. Some capsules remain on the bush through the winter. Seed colors vary from light brown to black, and seed length is about 1.25 cm (Fig. 1). Plants grown under irrigation may bear fruit in 3 years; otherwise, longer periods are required.

COLLECTION, EXTRACTION, AND STORAGE OF SEED Fruit collection can begin when 10% are in the hard-dough stage. If properly dried, seeds in this stage will be viable. The most satisfactory harvesting technique is to rake or vacuum ripe seeds from the ground. Capsules can be threshed after drying and seeds recovered by air screening. There are about 2 seeds per gram.

GERMINATION Jojoba seeds germinate without pretreat-

ment. Optimum germination occurs at 15 to 25°C. Germination is hypogeal (Fig. 2). Rao-Vasudeva and Ivensar (1982) reported on detailed studies of the morphology of jojoba seeds. Aragua et al. (1980) investigated the influence of gibberellin on seed germination.

NURSERY AND FIELD PRACTICE Direct seeding in the field is possible if protection from rodents is provided. Seeding should be carried out in the spring after the danger of frost. Seeds are planted relatively deeply at 4 to 5 cm in coarse-textured soils. Jojoba can be propagated from softwood cuttings taken in late spring or early summer.

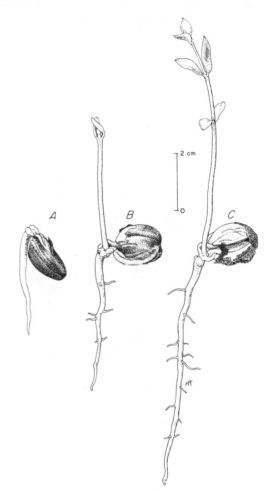

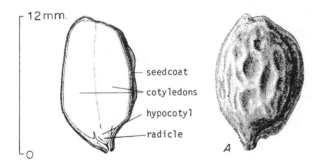

Figure 1. *Simmondsia chinensis*, jojoba: (left) longitudinal section through a seed, and (right) exterior view, ×3 (Nord and Kadish 1974).

Figure 2. *Simmondsia chinensis*, jojoba: seedling development at 3, 7, and 14 days after germination (Nord and Kadish 1974).

RUTACEAE — RUE FAMILY

Skimmia Thunb.

GROWTH HABIT, OCCURRENCE, AND USE The genus *Skimmia* consists of about 9 species of evergreen shrubs native to the region from the Himalayas to Japan. They are grown for their handsome foliage and red berries. The fruit is a small drupe with 2 to 5 stones. Seeds should be cleaned and sown immediately (Dirr and Heuser 1987).

SOLANACEAE — NIGHTSHADE FAMILY

Solanum dulcamara L. — Bitter nightshade

GROWTH HABIT, OCCURRENCE, AND USE Bitter nightshade is a climbing perennial vine, somewhat woody at the base, that grows to a height of 1.8 to 3.6 m (Crossley 1974b). It is native to Europe, northern Africa, and Asia. Naturalized in North America, it is often found in moist thickets from Nova Scotia to Minnesota, south to North Carolina and Missouri, and in the Pacific Northwest. Bitter nightshade has long been cultivated as an ornamental; it also has cover and food value for wildlife. The fresh berries are poisonous to humans, but some species of wildlife apparently eat them with impunity.

FLOWERING AND FRUITING The violet flowers, occurring in long peduncled cymes, bloom from July to August. The ovoid scarlet berries, about 1.25 cm long, ripen from August to October. The seeds are small (about 0.3 cm long), flesh-colored, irregular disks, dully glistening, as if coated with fine sugar (Fig. 1). Good seed crops are borne almost annually.

COLLECTION, EXTRACTION, AND STORAGE OF SEED Fruits can be handpicked from July through September. The fruits can be macerated and the seeds recovered by flotation. There are about 770 seeds per gram. Dried seeds can be stored at least a year at low temperatures.

GERMINATION Freshly collected seeds with no pretreatment have a high germination capacity. In some seed lots, germination capacity and uniformity are increased by prechilling the seeds for 30 days. For seed testing, 20/30°C incubation temperatures with light for 8 hours daily is recommended. Germination is epigeal (Fig. 2).

NURSERY AND FIELD PRACTICE Seeds should be sown in the fall. *Solanum* can also be propagated by stem cuttings.

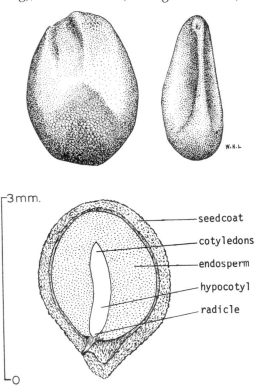

Figure 1. *Solanum dulcamara*, bitter nightshade: exterior view of seed (above) and longitudinal section through the embryo of a seed (below), ×16 (Crossley 1974b).

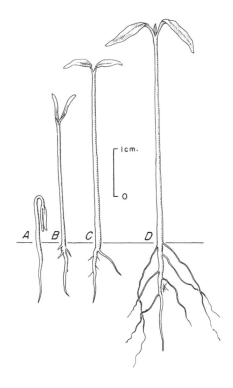

Figure 2. *Solanum dulcamara*, bitter nightshade: seedling development at 1, 2, 6, and 12 days after germination (Crossley 1974b).

LEGUMINOSAE — LEGUME FAMILY

Sophora L.

GROWTH HABIT, OCCURRENCE, AND USE The genus *Sophora* consists of about 20 species of ornamental, woody, and only rarely herbaceous species native to tropical and temperate regions of both hemispheres. The legume is stalked, almost terete or 4-winged, rarely compressed, and indehiscent or tardily dehiscent.

GERMINATION The seeds have hard seedcoats that require scarification. Acid or mechanical scarification have given better results than hot water treatments with mamane (*S. chrysophylla*) (Scowcroft 1978, 1981). Mamane builds large reserves of viable seeds in forest soils of Hawaii (Scowcroft 1982).

ROSACEAE — ROSE FAMILY

Sorbaria sorbifolia (L.) A. Br. — Ural false spiraea

GROWTH HABIT, OCCURRENCE, AND USE The Ural false spiraea is native to northern Asia from the Urals to Kamchatka, Sakhalin, and Japan (Rudolf 1974p). It is a deciduous shrub, 1 to 2 m tall, usually grown as an ornamental for its bright green foliage and conspicuous panicles of white flowers. It is also useful for watershed protection and wildlife habitat. One of about 8 species native to northern and eastern Asia, the Ural false spiraea often escapes from cultivation in the eastern United States.

FLOWERING AND FRUITING The shiny, white, bisexual flowers bloom in May, June, and July in the northern United States. The fruits are small, shiny follicles that ripen in August in Minnesota. Good seed crops are borne almost every year. The seeds are small and fusiform (Fig. 1).

COLLECTION, EXTRACTION, AND STORAGE OF SEED The ripe fruits should be picked from the bushes by hand and separated from the panicles. After threshing, the seeds can be recovered by air screening. There are about 40 dried follicles and 1700 seeds per gram. Dried seeds can be stored in sealed containers at low temperatures.

GERMINATION Apparently some of the seeds have embryo dormancy that requires prechilling for 30 to 60 days. Germination can be tested on sand at a 20/30°C incubation temperature.

Seeds should be sown immediately after collection in the late summer or prechilled seeds can be sown in the spring. The small seeds should be covered with only a thin layer of soil.

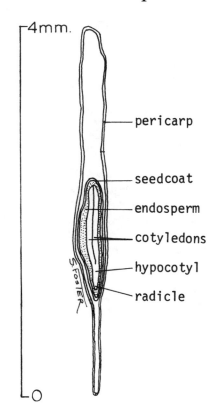

Figure 1. *Sorbaria sorbifolia*, Ural false spiraea: longitudinal section through a seed, ×24 (Rudolf 1974p).

ROSACEAE—ROSE FAMILY

Sorbus L.—Mountain ash

GROWTH HABIT, OCCURRENCE, AND USE The mountain ashes include more than 80 species of deciduous trees and shrubs distributed through the Northern Hemisphere (Harris and Stein 1974). Their graceful foliage, showy flowers, and brightly colored fruits make them especially sought after for ornamental plantings. The fruits are an important food for birds and rodents, and twigs furnish browse for deer and moose. The strong, close-grained wood is sometimes used for specialty products. European mountain ash (*Sorbus aucuparia*) is among the most widely planted species, and in some areas of North America it has escaped from cultivation.

Sorbus, like other genera in the rose family, is a plastic genus that shows extensive introgression where ranges overlap. Geographic races undoubtedly exist.

FLOWERING AND FRUITING The white, perfect flowers are borne in large, rather flattened clusters from April until July, depending on species and location, with fruit ripening from August until October (Harris and Stein 1974):

Species	Flowering dates	Fruit ripening dates	Seed dispersal dates
S. americana	May–July	Aug.–Oct.	Aug.–Mar.
S. aucuparia	Apr.–July	Aug.–Oct.	Aug.–Winter
S. decora	May–July	Aug.–Sept.	Aug.–June
S. sitchensis	June–July	Aug.–Oct.	Sept.–Winter

The showy fruits are orange-red to bright red when ripe. They are 2-to 5-celled, berrylike pomes (Fig. 1), with each cell containing 1 or 2 small brown seeds (Figs. 2, 3). Fruits may remain on trees until late winter and are thus available for birds during critical periods. Seed is chiefly dispersed by birds.

S. decora
showy mountain-ash

Figure 1. *Sorbus decora*, showy mountain ash: cluster of fruits, X1 (Harris and Stein 1974).

S. americana
American mountain-ash

S. sitchensis
Sitka mountain-ash

Figure 2. *Sorbus*: seeds, ×8 (Harris and Stein 1974).

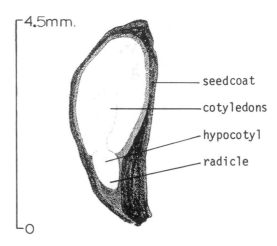

4.5mm.

— seedcoat

— cotyledons

— hypocotyl

— radicle

Figure 3. *Sorbus aucuparia*, European mountain ash: longitudinal section through a seed, ×12 (Harris and Stein 1974).

European mountain ash begins bearing seed at about 15 years of age and good seed crops occur almost annually.

COLLECTION, EXTRACTION, AND STORAGE OF SEED The fruit must be picked or shaken from the tree as soon as it is ripe to prevent losses to birds. It may be picked earlier, as soon as it begins to color.

Seeds may be extracted by macerating the fruits and recovering the seeds through flotation. After drying, the seeds can be cleaned with an air screen.

Cleaned seeds can be stored for 2 to 8 years with little loss of viability. For best results, storage in sealed containers at 6 to 8% moisture content and low temperatures is recommended. The yield and size of cleaned seeds vary among species (Harris and Stein 1974):

Species	Ripe fruit color	Mature height (m)	Seeds/gram
S. americana	Bright red	9	350
S. aucuparia	Orange-red	20	280
S. decora	Vermilion	12	280
S. sitchensis	Bright red	6	310

On a weight basis, seed yield from fruit is only 1 to 7%.

PREGERMINATION TREATMENT *Sorbus* seeds require 60 days or more prechilling for germination. Warm stratification before prechilling does not enhance germination.

GERMINATION Viability checks can be quickly run by using excised embryos. Using pretreated seeds germination test can be conducted on a variety of substrata at 15 to 20°C incubation temperatures.

NURSERY AND FIELD PRACTICE Seeds can be sown in the fall at shallow depths and the beds mulched. Germination is epigeal (Fig. 4). Mountain ash seedlings are quite hardy and easily grown. They are subject to browsing by deer.

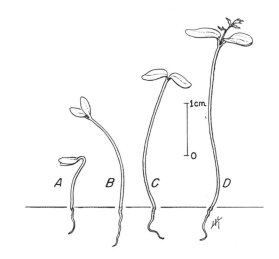

Figure 4. *Sorbus americana*, American mountain ash: seedling development at 1, 3, 7, and 24 days after germination.

ROSACEAE — ROSE FAMILY

Spiraea betulifolia var. *lucida* (Dougl.) C. L. Hitchc. — Birchleaf spirea

GROWTH HABIT, OCCURRENCE, AND USE Birchleaf spirea is a low, deciduous shrub 0.1 to 0.5 m in height (Stickney 1974c). In North America, the variety *lucida* is native from British Columbia to northern Oregon and eastward to Saskatchewan, South Dakota, and Wyoming. Another variety, *corymbosa*, occurs naturally in the eastern United States. Birchleaf spirea grows on a wide range of forest sites, from sea level to subalpine, but its best development is in recently opened habitats in mesic forest types at moderate elevations. This strongly rhizomatous shrub is potentially useful for watershed rehabilitation, particularly on road fills. It has been cultivated since 1885 and the showy, flat-topped inflorescences of white flowers make it an desirable ornamental.

FLOWERING AND FRUITING The perfect, very small (1 to 1.5 mm) flowers bloom from June through early August. At elevations around 980 m, in the northern Rocky Mountains, the beginning of the flowering period may vary from early June to early July. The period of fruit ripening ranges from mid-July to early September, and seed dissemination occurs in October.

Seeds, borne in a follicle 2 to 3 mm long, are shed when the fruit becomes straw colored or light brown and split down one side. The seeds are very small (Fig. 1). Shrubs that grow in full sun are the ones that bear seeds.

GERMINATION Seeds require no pretreatment and will germinate at very low incubation temperatures. Seeds sown in the fall and overwintering under snow are ready to germinate in early spring.

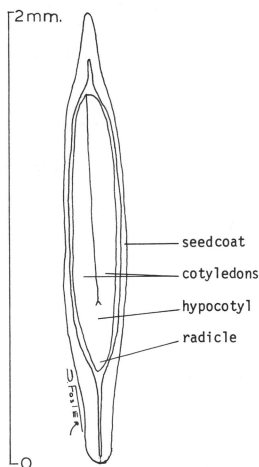

Figure 1. *Spiraea betulifolia* var. *lucida*, birchleaf spirea: longitudinal section through a seed, ×60 (Stickney 1974c).

CRUCIFERAE — MUSTARD FAMILY

Stanleya Nutt. — Prince's plume

GROWTH HABIT, OCCURRENCE, AND USE The genus *Stanleya* consists of about 6 species native to western North America. These plants are semiwoody and die back to a woody base annually. *Stanleya pinnata* is a concentrator of selenium when it grows on seleniferous soils. In flower, it is a very attractive plant. Seleniferous plants do not require selenium to grow, so *S. pinnata* can be safely grown in nonseleniferous soils.

Without pretreatment, seeds germinate at about 75% (Kay et al. 1988). They store well for 3 or 4 years.

STAPHYLEACEAE — BLADDERNUT FAMILY

Staphylea L. — Bladdernut

GROWTH HABIT, OCCURRENCE, AND USE The genus *Staphylea* consists of about 11 species of deciduous shrubs or small trees native to the Northern Hemisphere.

The fruit is a capsule with 2 to 3 inflated lobes containing 1 to 4 large, bony seeds. The white or white-tinged-with-pink flowers occur in panicles.

Capsules should be collected when brown and the seeds extracted by threshing.

GERMINATION The seedcoats are extremely hard and require acid scarification. The seeds should not be allowed to dry. They require 3 months of warm stratification and 3 months of prechilling for germination (Dirr and Heuser 1974).

ROSACEAE — ROSE FAMILY

Stephanandra Sieb. & Zucc.

GROWTH HABIT, OCCURRENCE, AND USE The genus *Stephanandra* consists of 4 or 5 species of deciduous shrubs, native to Japan and China, that are grown for ornamentals in the United States. Cutleaf stephanandra (*S. incisa*) is an example. The fruit is a follicle.

GERMINATION Seeds require acid scarification for 15 minutes followed by 3 months of prechilling (Dirr and Heuser 1987). Seeds may require light for final germination. Propagation by cuttings is easy.

COMPOSITAE — SUNFLOWER FAMILY

Stephanomeria Nutt.

GROWTH HABIT, OCCURRENCE, AND USE The genus *Stephanomeria* consists of 12 or more species native to western North America. *Stephanomeria pauciflora* is a common semishrub found in washes and open places from Kansas to California. The fruit is an achene.

GERMINATION Germination is initially below 10%, but increases with storage to 70% at the end of 3 years (Kay et al. 1988). After 3 years, viability declines rapidly.

THEACEAE — TEA FAMILY

Stewartia L.

GROWTH HABIT, OCCURRENCE, AND USE The genus *Stewartia* consists of about 8 species of deciduous trees and shrubs native to eastern North America and eastern Asia. Lovely white flowers and rich autumn colors make these woody plants desirable ornamental species. The fruit is a woody capsule that dehisces into 5 valves.

COLLECTION, EXTRACTION, AND STORAGE OF SEED Seeds are dispersed from the capsule at maturity, and the capsules should be collected while still green. After drying, the seeds can be threshed from the capsules. They are difficult to recover from the chaff (Dirr and Heuser 1987).

GERMINATION Fall-planted seeds will not emerge until the second year. Seeds lose viability rapidly in storage. Either sow seeds as soon as collected and wait for emergence, or warm stratify for 5 months and then prechill for 3 months before planting.

ROSACEAE — ROSE FAMILY

Stranvaesia Lindl.

GROWTH HABIT, OCCURRENCE, AND USE The genus *Stranvaesia* consists of 4 or 5 species of evergreen shrubs and small trees native to China and the Himalayas. Several species are planted as ornamentals.

The fruit is a bright red pome that persists into early winter. It should be macerated to remove the pulp; the seeds can be recovered by flotation.

GERMINATION Seeds display embryo dormancy and require 2 months of prechilling (Dirr and Heuser 1987). They can be sown in the fall or prechilled and sown in the spring. Cuttings taken from September through November can be rooted.

STYRACACEAE — STORAX FAMILY

Styrax L. — Snowbell

GROWTH HABIT, OCCURRENCE, AND USE The genus *Styrax* consists of about 100 species of deciduous or evergreen shrubs or trees native to tropical and subtropical regions and often grown for their attractive flowers.

FLOWERING AND FRUITING The fruits are globose or oblong drupes that can be collected as early as August but will drop from the plant by November. At maturity the outer covering of the fruit splits from the single stony, brown, hard seed.

GERMINATION *Styrax* seeds are difficult to germinate. Treatments that have been tried include warm stratification, acid scarification, prechilling, and repeated warm-cold treatments. Probably the best approach is to sow the seeds as soon as they are mature without letting them dry.

Germination techniques for individual species follow (Dirr and Heuser 1987):

Species	Germination techniques
S. americanus	Prechill 3 months or fall plant.
S. dasyanthus	Warm stratification 3 months and prechilling 3 months.
S. grandifolius	Prechill 3 months or fall plant.
S. hemsleyana	Warm stratification 3 to 5 months and prechilling 3 months.
S. japonicus	Warm stratification 3 to 5 months and prechilling 3 months.
S. obassia	Best to start with fresh seeds; warm stratification 3 to 5 months followed by prechilling 3 months.

The 6 species mentioned above are easily rooted from softwood cuttings (Dirr and Heuser 1987).

CHENOPODIACEAE — GOOSEFOOT FAMILY

Suaeda Forsk. ex Scop. — Seep weed

GROWTH HABIT, OCCURRENCE, AND USE The genus *Suaeda* consists of about 50 widely distributed species. *Suaeda torreyana* occurs in saline coastal marshes and saline/alkaline sinks in interior deserts.

The fruit is an utricle enclosed in the calyx. The seeds consist of a coiled embryo without endosperm.

Seed germination without pretreatment is below 10% and highly variable (Kay et al. 1988).

CAPRIFOLIACEAE — HONEYSUCKLE FAMILY

Symphoricarpos Duham. — Snowberry

GROWTH HABIT, OCCURRENCE, AND USE Snowberries occur in North America from Alaska to Mexico. One species is native to China (Evans 1974). In North America the genus includes about 15 closely related dwarf or medium, thicket-forming deciduous shrubs. Snowberries have been used in wildlife plantings and have long been grown as ornamentals.

FLOWERING AND FRUITING The pinkish to yellow-white, perfect flowers are borne in dense axillary or terminal clusters in midsummer (Evans 1974):

Species	Flowering dates	Fruit ripening dates	Ripe fruit color
S. albus			
var. *albus*	June–Sept.	Aug.–Oct.	Waxy white
var. *laevigatus*	June–Sept.	Aug.–Oct.	White
S. occidentalis	June–July	Sept.–Oct.	White-black
S. orbiculatus	July–Aug.	Sept.–Oct.	Purplish red

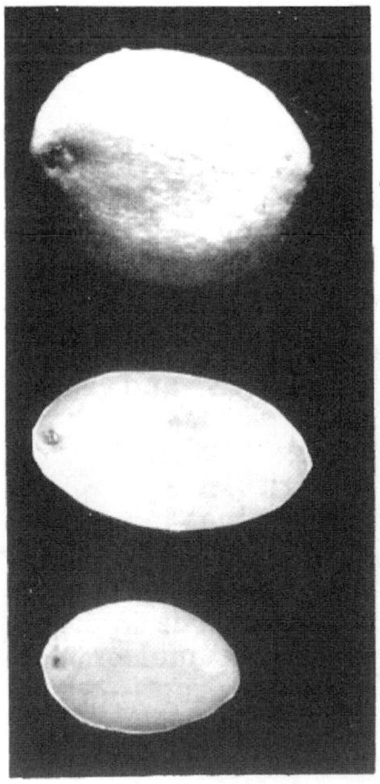

S. albus var. *laevigatus*
garden snowberry

S. occidentalis
western snowberry

S. orbiculatus
Indian-currant

Figure 1. *Symphoricarpos:* nutlets, ×8 (Evans 1974).

The fruit is a berrylike drupe, white in most species, but darker colors do occur. Fruits ripen in the late summer or early fall. Each contains 2 nutlets (pyrenes). These are flattened on one side and are composed of a tough, bony endocarp, a seedcoat, a fleshy endosperm, and a small embryo (Figs. 1, 2). The nutlets are used as seeds. They are dispersed from late fall to the following spring, largely by birds and mammals. Normally a good seed crop is produced each year.

COLLECTION, EXTRACTION, AND STORAGE OF SEED Ripe fruits can be collected any time during the fall and winter by stripping or flailing the shrubs. The fruits collected in the early fall contain more moisture and must be handled with care to prevent heating.

The fruits can be macerated and the seeds recovered by flotation. Dried fruits should be soaked for several days before macerating. The numbers of seeds per gram and pregermination treatments follow (Evans 1974):

Species	Seeds/gram	Pregermination treatments	
		Warm stratification (days)	Prechilling (days)
S. albus			
var. *albus*	170	20–60	180
var. *laevigatus*	120	0–112	120–182
S. occidentalis	160	—	
S. orbiculatus	90	0–30	120–180

Dried seeds can be stored in sealed containers at low temperatures for at least 2 years.

PREGERMINATION TREATMENT Nutlets of *Symphoricarpos* are very difficult to germinate. The endocarp is hard and impermeable and requires from 20 to 75 minutes of acid scarification, depending on the species and seed lot. The seeds also have embryo dormancy that requires both warm stratification and prechilling.

GERMINATION Pretreated seeds can be incubated at 20/30°C using a suitable substrate. The Genebank Handbook (1985) suggests for *S. rivularis* that a 20/30°C incubation temperature be used with light for 8 hours daily. The seeds should be pretreated by warm stratification at 25°C for 12 to 16 weeks, followed by prechilling at 1 to 5°C for 16 to 26 weeks. Germination is epigeal (Fig. 3).

NURSERY AND FIELD PRACTICE The warm stratification requirement means that emergence is not going to be appreciable with fall sowing until the second year after sowing. Pretreating by warm stratification for 90 to 120

days has been used with fall sowing. Early shade is beneficial for seedlings of some species.

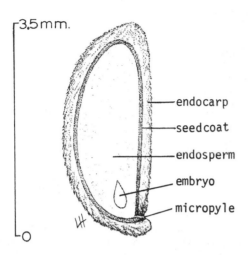

Figure 2. *Symphoricarpos albus* var. *albus*, common snowberry: longitudinal section through a nutlet, ×20 (Evans 1974).

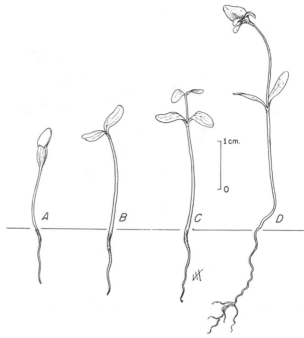

Figure 3. *Symphoricarpos albus* var. *albus*, common snowberry: seedling development at 5, 7, 13, and 20 days after germination (Evans 1974).

OLEACEAE — OLIVE FAMILY

Syringa L. — Lilac

GROWTH HABIT, OCCURRENCE, AND USE The lilacs include about 28 species of deciduous shrubs or small trees native to Asia and southeastern Europe (Rudolf and Slabaugh 1974). They are grown primarily for ornament because of their large, showy, often fragrant panicles of flowers.

FLOWERING AND FRUITING The perfect flowers, ranging from white to violet, purple, and deep reddish purple, bloom in the spring or early summer (Rudolf and Slabaugh 1974):

Species	Flowering dates	Fruit ripening dates
S. amurensis	June	Sept.–Oct.
S. persica	May–June	—
S. villosa	May–June	—
S. vulgaris	Apr.–June	Aug.–Oct.

The fruit is an oblong, smooth, leathery, brown, 2-celled capsule (Fig. 1) that ripens in late summer or fall. Each capsule contains 4 thin, flat, bright brown, lozenge-shaped seeds about 1.25 cm long and 0.5 cm wide (Fig. 2). Fair to good seed crops are produced annually.

COLLECTION, EXTRACTION, AND STORAGE OF SEED The ripe capsules may be picked from the shrubs by hand in the

fall. Amur lilac (*S. amurensis*) capsules collected in October yield better quality seeds than earlier-collected material. After drying, the capsules can be threshed and the seeds recovered by air screening.

For common lilac (*S. vulgaris*) there are about 190 seeds per gram and for late lilac (*S. villosa*) there are 90.

Seeds can be stored dry for 2 years with little loss of viability. An interesting study conducted by Junttila (1974) suggests that the temperature under which lilac seeds are produced influences subsequent germination. The influence is more through mechanical changes related to osmotic potential changes induced by moisture stress than on embryo dormancy.

PREGERMINATION TREATMENT It is common to prechill lilac seeds for 30 to 90 days, but the treatment may not be necessary for all species or seed lots. Heit (1974) determined that prechilling was not necessary for germination if proper moisture, incubation temperatures, and light were provided during germination testing.

GERMINATION The Genebank Handbook (1985) suggests the following germination test for *S. reflexa*, *S. villosa*, and *S. vulgaris*: substrate, top of paper; temperature, 20°C (20/30°C for *S. villosa*); duration, 21 days; and additional directions for *S. reflexa*, 2 tests, with and without prechilling, for 21 days. Heit (1974) suggests

20/30°C incubation temperatures, a good supply of moisture, and light with a 21- to 25-day test.

NURSERY AND FIELD PRACTICE Lilac seed should be sown at the rate of 260 to 420 seeds per m²; fall-seeded beds should be mulched. The seed should be covered with 0.6 cm of soil. The beds need to receive half-shade and be protected from late frost. Lilac can be propagated from cuttings.

Figure 1. *Syringa amurensis*, Amur lilac: fruits (capsules), ×2 (Rudolf and Slabaugh 1974).

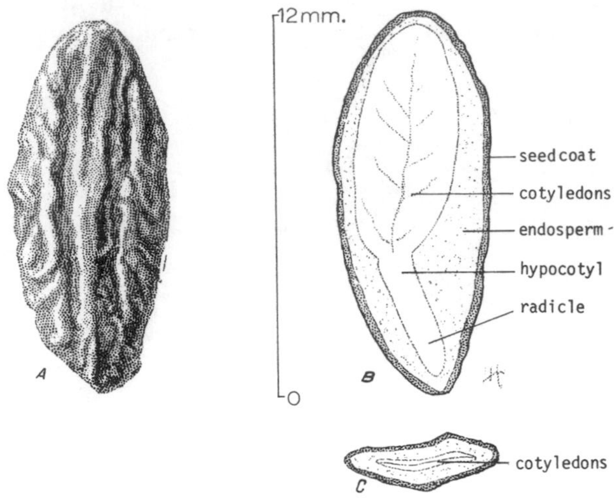

Figure 2. *Syringa vulgaris*, common lilac: (A) exterior view of seed; (B) longitudinal section through a seed; (C) transverse section; all at ×4 (Rudolf and Slabaugh 1974).

BIGNONIACEAE — BIGNONIA FAMILY

Tabebuia Gomes ex DC. — Trumpet tree

GROWTH HABIT, OCCURRENCE, AND USE The genus *Tabebuia* consists of about 100 species of evergreen trees or shrubs native to tropical America and planted as ornamentals in warm areas. The fruit is a linear dehiscent capsule that contains many winged seeds. The seeds do not retain viability for long periods in storage. Maeda and Matthes (1984) determined that storage in sealed flasks at 10°C maintained seed viability the longest. No information was found on seed germination.

TAMARICACEAE — TAMARISK FAMILY

Tamarix pentandra Pall. — Tamarisk

GROWTH HABIT, OCCURRENCE, AND USE The natural range of tamarisk, a finely branched, deciduous shrub to small tree, is from southeastern Europe to central Asia (Reynolds and Alexander 1974b). Introduced into the eastern United States in the 1820s, tamarisk escaped cultivation, and it now grows along major river drainages at lower elevations throughout most of the western United States.

Tamarisk is widely cultivated as an ornamental because of its showy flowers and graceful foliage. In many places these plants are also used as windbreaks and for erosion control. Tamarisks are an important source of honey. The naturalized groves are important nesting habitat for white-winged doves.

FLOWERING AND FRUITING The pink to white flowers, borne in terminal panicles, bloom from March through September. A succession of small capsular fruits ripen and split open during the period from late April through October. The seeds are minute and have an apical tuft of hairs (Figs. 1, 2) that facilitates dispersal by wind.

COLLECTION, EXTRACTION, AND STORAGE OF SEED The fruits can be collected by hand during the spring, summer, and fall. It is not practical to extract the seeds from the small fruits. Seeds stored at low temperatures will retain viability for nearly 2 years. Seeds stored at room temperature rapidly lose viability.

GERMINATION Seeds germinate within 24 hours of

imbibing water without pretreatment.

NURSERY AND FIELD PRACTICE Early seedling growth is very slow. Seedling height after 30 days is only 2.5 cm and 11.5 cm after 60. The soil must be kept continually moist during the seedling development period. After the seedlings become established, they can withstand severe drought. Tamarisk is readily propagated by cuttings.

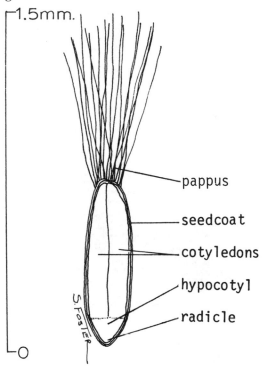

Figure 1. *Tamarix pentandra,* tamarisk: longitudinal section through a seed, ×60 (Reynolds and Alexander 1974b).

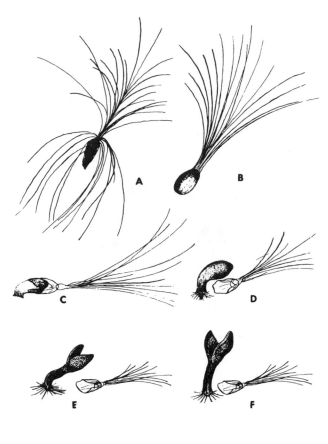

Figure 2. *Tamarix pentandra,* tamarisk: (A) dry seed and seedling development at intervals after moistening the seed; (B) several hours; (C) 8 hours; (D) 24 hours; (E) 40 hours; all approximately ×10 (Reynolds and Alexander 1974b).

TAXODIACEAE — TAXODIUM FAMILY

Taxodium distichum (L.) Rich. — Baldcypress

GROWTH HABIT, OCCURRENCE, AND USE Baldcypress is a large, deciduous tree that often reaches heights of 40 m at maturity (Bonner 1974k). Two varieties of this important timber species are found in the United States: baldcypress (*Taxodium distichum* var. *distichum*) and pondcypress (*T. distichum* var. *nutans*). Very similar in their botanical and silvical characteristics, they are distinguished by their foliage. Widely grown as ornamental species, these trees also have value as wildlife habitat.

FLOWERING AND FRUITING The monoecious flowers of baldcypress appear in March and early April. The male flowers are borne at the end of the previous year's growth in slender, purplish, tassellike clusters 7.5 to 12.5 cm long. Female conelets are found singly or in clusters of 2 or 3 in leaf axils near the ends of the branchlets. The globose cones turn from green to brownish purple as they mature in October to December. The cones are 1.25 to 3.1 cm in diameter (Fig. 1) and consist of a few 4-sided scales that break away irregularly after maturity. Each scale bears 2 irregularly shaped seeds that have thick, horny, warty coats and projecting flanges (Figs. 2, 3). Each cone contains 18 to 30 seeds. Some seeds are borne every year, and good crops occur at 3- to 5-year intervals. Few seeds mature in the northern range of the species.

COLLECTION, EXTRACTION, AND STORAGE OF SEED Mature dry cones can be picked by hand from standing or felled trees and spread in a thin layer for air drying. The dried cones should be broken apart by flailing or trampling. Separation of cone fragments from the seeds is very difficult, and they usually are sown together. Thirty-five liters of cones weigh 18 to 22 kilograms. Seed yield from cones on a weight basis is as

T. distichum var. **distichum**
baldcypress

T. distichum var. **nutans**
pondcypress

Figure 1. *Taxodium*: cones, ×1 (Bonner 1974k).

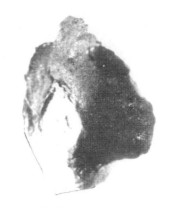

T. distichum var. **distichum**
baldcypress

T. distichum var. **nutans**
pondcypress

Figure 2. *Taxodium*: seeds, ×4 (Bonner 1974k).

high as 50%. There are 2600 to 3500 cones per 35 liters. For baldcypress there are about 11.5 seeds per gram and for pondcypress about 8.9. Baldcyress seeds store well at low temperatures after drying.

GERMINATION The Genebank Handbook (1985) lists the following standards for *Taxodium distichum* based on International Seed Testing Association standards: substrate, sand; incubation temperatures, 20/30°C or 20°C; duration, 28 days. Additional instructions suggest prechilling for 30 days. Bonner (1974k) suggested a 90-day prechilling period preceded by a 5-minute soak in ethyl alcohol. Gibberellin and potassium nitrate improved the germination of non-prechilled seeds. None of the materials tested improved the germination of prechilled baldcypress seeds. Murphy and Stanley (1975) determined that removal of the seedcoat of baldcypress and pond-cypress seeds resulted in prompt germination. They interpreted this as evidence that embryo dormancy was not involved. Scarification treatments did not increase germination capacity but did increase the germination energy or rate.

NURSERY AND FIELD PRACTICE Spring sowing of pre-treated seeds and fall sowing of untreated seeds are both practiced. The latter method has proved practical in northern nurseries. Seeds and cone fragments are broadcast together and covered with a shallow layer of sand or peat moss. Fall-sown beds should be mulched with leaves to protect the seedlings. Shade is required

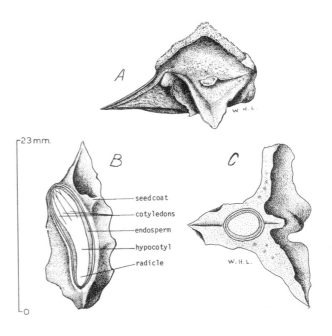

Figure 3. *Taxodium distichum* var. *distichum*, baldcypress: (A) exterior view of seed; (B) longitudinal section; and (C) transverse section, ×2 (Bonner 1974k).

in the South from June to September, and the beds must be kept well watered. The resinous seeds are not eaten by any extent to birds and rodents. Germination is epigeal (Fig. 4).

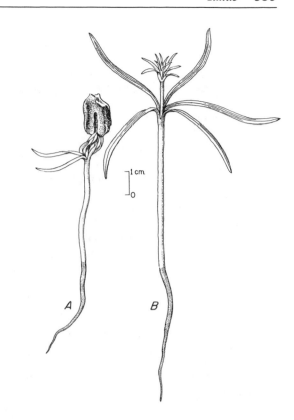

Figure 4. *Taxodium distichum* var. *distichum*, baldcypress: seedling development at 3 and 8 days after germination (Bonner 1974k).

TAXACEAE — YEW FAMILY

Taxus L. — Yew

GROWTH HABIT, OCCURRENCE, AND USE About 8 species of yew, which are sometimes considered geographical varieties of the same species, occur in the temperate zone of the Northern Hemisphere (Rudolf 1974q). They are nonresinous, evergreen trees or shrubs used for ornamental purposes. The wood is prized for making bows and arrows and is used in cabinet making and turnery. The fruits are eaten by birds, but the foliage is poisonous to livestock.

The most frequently planted species in North America are listed below (Rudolf 1974q):

Species	Flowering dates	Fruit ripening dates	Seed dispersal dates	Mature height (m)
T. baccata	Mar.–May	Aug.–Oct.	Aug.–Oct.	28
T. brevifolia	June	Aug.–Oct.	Oct.	12
T. canadensis	Apr.–May	July–Sept.	—	1
T. cuspidata	Mar.–Apr.	Oct.–Nov.	—	20

English yew (*Taxus baccata*) is the most widely cultivated.

FLOWERING AND FRUITING Both male and female flowers are rather inconspicuous and are borne on the same plant (monoecious) or different plants (dioecious). The fruit, which ripens in late summer or autumn, consists of a scarlet, fleshy, cuplike aril (Fig. 1) bearing a single hard seed (Figs. 2, 3). The seeds have a large, oily, white endosperm and a minute embryo.

T. canadensis
Canada yew

Figure 1. *Taxus canadensis*, Canada yew: fruit and needles, ×1 (Rudolf 1974q).

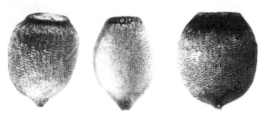

T. baccata   *T. brevifolia*   *T. canadensis*
English yew   **Pacific yew**   **Canada yew**

Figure 2. *Taxus*: seeds, ×5 (Rudolf 1974q).

Little information is available on the frequency of good seed crops among the yews, but most of them produce some seeds almost every year. For Japanese yew (*T. cuspidata*), good crops are reported every 6 to 7 years. English yew begins to bear seed at about 30 years of age.

COLLECTION, EXTRACTION, AND STORAGE OF SEED To prevent losses to birds, yew fruits should be picked from the branches by hand as soon as they are ripe. For the dioecious species, good seed is produced only where there is a good intermixture of male and female plants. The seeds may be extracted by maceration and recovered by flotation. They should not be allowed to dry after flotation but should directly put into pre-chilling treatments. Yew seeds can be stored for several years at prechilling temperatures in a moist medium. Dried seeds can be stored even longer but will be highly dormant. Cleaned seed yields are as follows (Rudolf 1974q): *T. baccata* yields 17 seeds/gram; *T. brevifolia*, 34; *T. canadensis*, 46; and *T. cuspidata*, 16.

PREGERMINATION TREATMENT Yews are slow to germinate, natural germination takes place the second year after maturity (Fig. 4). Most of the germinating seeds in natural stands of English yew are from seeds that passed through the digestive system of birds such as nutcrackers. Yews have a strong but variable dormancy that can be broken by warm stratification followed by prechilling. Rudolf (1974q) suggested 90 to 120 days at 15°C followed by 60 to 120 days at 3 to 5°C. Another suggestion for *T. baccata* and *T. cuspidata* is 120 days of warm stratification followed by 365 days of prechilling. International Seed Testing Association rules suggest prechilling 270 days.

GERMINATION Laboratory germination involves prolonged warm stratification and prechilling. Light apparently does not play a role. There has been a lot of very recent interest in the germination of seeds of *Taxus* species because of the discovery of a chemical in extracts from the trees that has medical uses. This research is not yet published, so the most recent germination information is from LePage-Degiury (1973) and LePage-Degiury and Garello (1973) on research conducted in France. These researchers succeeded in leaching germination inhibitors from *Taxus* embryos by culturing them in a liquid nutritive medium. Embryo germination on agar depended on the composition of the medium. Sucrose had to be present, and the addition of a source of Ca^{2+} ions increased germination.

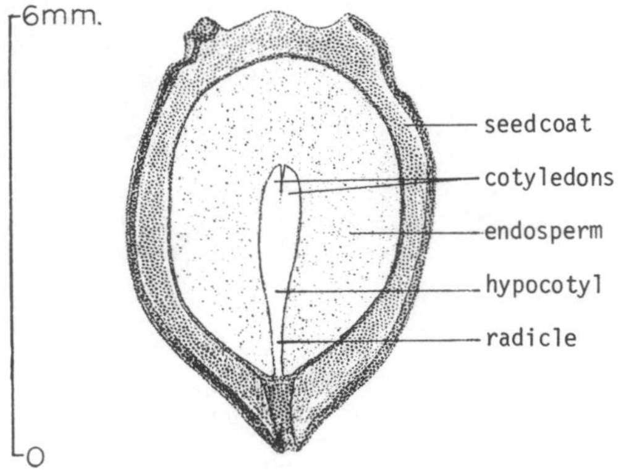

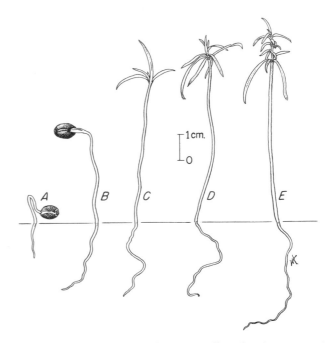

Figure 3. *Taxus brevifolia*, Pacific yew: longitudinal section showing small embryo, ×8 (Rudolf 1974q).

Figure 4. *Taxus baccata*, English yew: seedling development at 1, 8, 12, 22, and 39 days after germination (Rudolf 1974q).

NURSERY AND FIELD PRACTICE Freshly collected seeds can be sown in July. Stored seeds can be pretreated and seeded in the spring. The seeds should be planted shallowly and the beds mulched with pine needles or wood shavings. The beds should be shaded the first summer, mulched the second winter, and protected with shade cloth.

VERBENACEAE — VERBENA FAMILY

Tectona grandis L. — Teak

GROWTH HABIT, OCCURRENCE, AND USE Native to south-eastern Asia, teak is the only important tree species in the genus *Tectona* (Schubert 1974). It is a large, deciduous tree that grows best in a warm, moist, tropical climate with a marked dry season of several months. Teak wood is famous the world over for its strength, durability, dimensional stability, working qualities, and the fact that it does not cause corrosion when it comes in contact with metal. It is used for building, in making fine furniture, and as a veneer for fine plywood.

FLOWERING AND FRUITING The small white perfect flowers of teak are borne on short pedicels in large erect terminal panicles, about 2 months after the dry season has ended and the large obovate leaves have emerged. The dates vary depending on the climate regime, but flowering generally takes place for several months between June and September, and the fruits ripen 2.5 to 3 months later. They gradually fall to the ground during the dry season.

The fruit consists of a subglobose, 4-lobed, hard, bony stone about 1.25 cm in diameter surrounded by a thick, felty, light-brown covering (Fig. 1), the whole enclosed in an inflated bladderlike papery involucre. The stone (often called a nut) contains 1 to 3, or, rarely 4 seeds (Fig. 2) and has a central cavity giving the appearance of a fifth cell.

COLLECTION, EXTRACTION, AND STORAGE OF SEED Teak has borne viable seeds when 3 years old, and good seed crops are produced by plantations less than 20 years old. The bladder like involucre turns from green to brown when the seeds are ripe. The fruits can be swept up as they fall on the ground beneath the trees or can be shaken from the branches. Drying can be completed by spreading the fruits on racks in the sun. For convenience of handling, the dry involucre can be removed. Teak fruits in Honduras average 700 per kilogram with involucres intact and 880 per kilogram with involucres removed. In other parts of the world, the number of cleaned fruits per kilogram ranges from 880 to 3,080. The fruits have been successfully stored for at least a year in a dry warehouse. Longer periods of storage have not been necessary because teak produces good seed crops almost every year.

GERMINATION TEST The Genebank Handbook (1985) provides International Seed Testing Association standards for *Tectona grandis*: substrate, sand; incubation temperature, 30°C; duration, 28 days. Additional instructions suggest presoaking fruit and predrying in a 3- day cycle repeated 6 times. Germination is epigeal.

Selected recent literature on teak seed germination includes the following:

Author—date	Subject	Location
Bagchi & Emmanuel 1983	Germination studies.	India
Bhumibhamon & Kanchanabhum 1980	Germination reported for fruits with 1, 2, 3, or 4 seeds. Extracts from fruits inhibited germination of seeds of other species.	Thailand
Banik 1978	Grading of teak seeds.	—
Dabral 1976	Extraction of teak seeds from fruits. Presoaking seeds reduced germination.	India
Gupta & Pattanath 1975	Soaking in water increased germination.	India
Gupta & Kumar 1976	Estimating potential germination.	India
Keiding 1985	Germination.	Denmark
Keiding & Knudsen 1974	International provenance testing of germination.	—
Khanal 1975	Nursery techniques.	Nepal
Kumar 1979	Influence of fruit size on germination.	—
Madsen 1975	Determining moisture content of fruits.	Denmark
Muttiah 1975	Germination—soaking and drying gave best results.	Sri Lanka

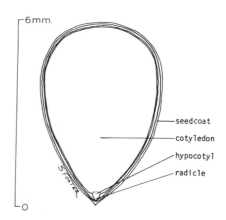

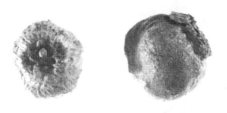

Figure 1. *Tectona grandis*, teak: top and side views of fruits with their bladderlike involucres removed, ×2 (Schubert 1974).

Figure 2. *Tectona grandis*, teak: longitudinal section through a seed (Schubert 1974).

NURSERY AND FIELD PRACTICE Teak fruits are usually broadcast in nursery beds and covered with about 1.25 cm of sand, sawdust, or soil. A tree percentage of 25 can be expected from good seed. Field planting is generally done with plants 1.25 to 2.5 cm in diameter at the soil surface that have the top cut back to about 2.5 cm and the taproot pruned to 17 to 20 cm. Outplanting is done at the start of the rainy season. Teak plantations are renewed by coppicing.

EUPHORBIACEAE — SPURGE FAMILY

Tetracoccus Engelm. ex Parry

GROWTH HABIT, OCCURRENCE, AND USE The genus *Tetracoccus* consists of 4 species native to California, Arizona, and Baja California, Mexico. *Tetracoccus hallii* is a shrub, 2 m in height, with stiff, spinescent branches.

The fruit is a capsule containing 3 smooth seeds. Germination without pretreatment is about 50% (Kay et al. 1988). The seeds can be stored for a couple of years without loss of viability.

COMPOSITAE — SUNFLOWER FAMILY

Tetradymia DC.

GROWTH HABIT, OCCURRENCE, AND USE The genus *Tetradymia* consists of about 6 species native to semiarid and arid areas of the western United States. *Tetradymia canescens* is an important member of many *Artemisia tridentata* communities. The fruit is an achene that ranges from glabrous to densely hairy, depending on the species. Some germination will occur without pretreatment but is greatly enhanced by prechilling for 4 to 6 weeks.

LABIATAE — MINT FAMILY

Teucrium chamaedrys L. — Wall germander

GROWTH HABIT, OCCURRENCE, AND USE Wall germander is a shrubby plant, 0.3 m in height, that can be decumbent or ascending at the base. It is often grown as an ornamental border plant.

The flowers are bright red-purple or rose-red. The fruit is a nutlet.

Seeds germinate without pretreatment (Dirr and Heuser 1987). It is easily rooted from softwood cuttings.

CUPRESSACEAE — CYPRESS FAMILY

Thuja L. — Arborvitae

GROWTH HABIT, OCCURRENCE, AND USE The genus *Thuja* includes 2 species native to North America and 4 species native to Asia (Schopmeyer 1974c). All are aromatic, evergreen trees, but some also have shrubby forms. Three species are commercially important in the United States. Northern white cedar (*T. occidentalis*) grows both in swamp and uplands and does well in extremely wet or dry situations. Western red cedar (*T. plicata*) obtains its greatest size in stream bottoms, moist flats, and north-facing slopes at low elevations in the Pacific Northwest. Both are valuable timber species because their heartwood is lightweight and resists decay. The wood is used extensively for shingles, shakes, siding, and poles. Oriental arborvitae (*T. orientalis*) is widely cultivated as an ornamental species. The foliage of young trees is browsed by deer.

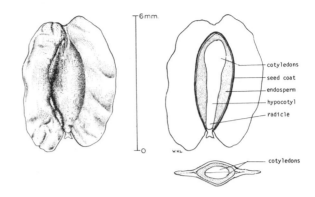

Figure 2. *Thuja occidentalis*, eastern white cedar: exterior view of seed, longitudinal section with labels and transverse section showing 2 cotyledons; all at ×6 (Schopmeyer 1974c).

Figure 1. *Thuja plicata*, western red cedar: cone, ×6 (Schopmeyer 1974c).

FLOWERING AND FRUITING Male and female flowers are borne on the same tree but usually on separate twigs or branchlets. Flower buds are formed during the fall season and develop into small, erect cones during the following summer (Fig. 1). Cone size and number of scales vary with the species. During the ripening period, cones change in color from green to yellow and finally to a pale cinnamon-brown. Their light chestnut seeds are 0.6 cm long and have lateral wings about as wide as the body (Fig. 2). Cones of eastern arborvitae have thick scales with conspicuous hooks on their tips, and seeds are dark red-purple and wingless. Embryos of all 3 species have 3 cotyledons. Phenology of flowering for the North American species follows (Schopmeyer 1974c):

Species	Flowering dates	Cone ripening dates	Seed dispersal dates
T. occidentalis	Apr.–May	Aug.–Sept.	Sept.–Oct.
T. plicata	Apr.–June	Aug.–Sept.	Aug.–Sept.

Cone descriptions and cone bearing information for the 3 most important species are listed below (Schopmeyer 1974c):

Species	Mature height (m)	Seed bearing Age (years)	Seed bearing Interval (years)	Cone length (m)
T. occidentalis	25	20–30	3–5	0.8–1.25
T. orientalis	12	—	—	1.5–2.5
T. plicata	62	15–25	3–4	0.8–1.25

COLLECTION, EXTRACTION, AND STORAGE OF SEED Cones may be picked by hand from standing or recently felled trees. A good time to collect is when the seeds have become firm and the cones have turned from green to brown. For eastern white cedar the period between

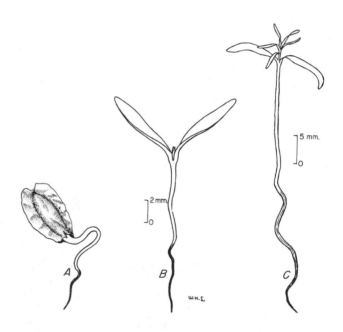

Figure 3. *Thuja occidentalis*, eastern white cedar: seedling development at 1, 5, and 25 days after germination (Schopmeyer 1974c).

cone ripening and the start of cone opening is only 7 to 10 days. Cones of western red cedar also start to open soon after they ripen. The peak rate of seedfall occurs 4 to 6 weeks after the cones first open. Mature trees of both species produce abundant crops of cones every 3 to 5 years. All cones do not open at the same time: seed release progresses slowly, and the collection period can last a month after cones on the first tree open.

Kiln drying is needed to extract seeds from large quantities of cones. Kiln temperatures of 45°C are preferred to prevent damage to the seeds. Cones of western red cedar can be opened in 24 to 36 hours at 32°C, but cones are often dried in the sun. Seeds should not be dewinged.

The percentage of empty seeds is often high, but they can be removed with an air screen. Seeds of eastern white cedar with a moisture content of 6 to 8%

have been stored at low temperatures for as long as 5 years. Similar results have been obtained with seeds of western red cedar.

GERMINATION ASOA (1985) standards follow:

Species	Substrata	Temperature (°C)	Duration (days)
T. occidentalis	TB, P	20/30	21
T. orientalis	TB	20	21
T. plicata	TB, P	20/30	21

Occasionally, dormant seed lots are encountered that require prechilling or the addition of potassium nitrate. Germination is epigeal (Fig. 3).

NURSERY AND FIELD PRACTICE Practices vary among nurseries. In general, fall sowing is preferred for eastern white cedar and spring sowing for western red cedar. Average seedbed densities are 530 per m². Half-shade over the seedbeds is recommended during the first growing season. Many horticultural cultivars are propagated by cuttings.

CUPRESSACEAE — CYPRESS FAMILY

Thujopsis dolabrata (L.) Sieb. & Zucc. — Hiba arborvitae

GROWTH HABIT, OCCURRENCE, AND USE The genus *Thujopsis* consists of a single species of evergreen tree native to central Japan that can reach 15 m in height. The leaves (needles) are green and glossy above with a white band below. The cones are subglobose, with 6 to 10 hard, woody scales and 3 to 5 seeds. Plants grown from seed show great variability.

GERMINATION There is limited knowledge on the germination of seeds of this species, but reportedly, freshly harvested seeds that are planted immediately readily germinate (Dirr and Heuser 1987). This species is one of the easiest conifers to root. Cuttings can be taken from November to January.

TILIACEAE — LINDEN FAMILY

Tilia L. — Basswood

GROWTH HABIT, OCCURRENCE, AND USE The basswoods include about 30 species of small- to medium-sized trees native to temperate regions in eastern North America south to Mexico, and in Asia to south central China and Japan (Brinkman 1974n). They are valuable as timber species for lumber and veneer, and as ornamental species. American basswood (*Tilia americana*) and littleleaf linden (*T. cordata*) have been widely planted in the United States. Both are relatively tolerant of smog.

FLOWERING AND FRUITING The perfect, yellowish flowers of both species open in June and July and are borne in drooping clusters attached to large bracts. The fruits, which ripen in September and October, are round to egg-shaped, indehiscent capsules. Each consists of a crustaceous or woody pericarp usually enclosing a single seed but sometimes 2 to 4 (Figs. 1, 2). The seeds have a crustaceous seedcoat, a fleshy yellowish endosperm, and a well-developed embryo (Figs. 2, 3). Natural dispersion of seeds is by wind and animals. Seeding information for the 2 species follows (Brinkman 1974n):

T. americana
American basswood

T. cordata
littleleaf linden

Figure 1. *Tilia:* fruits, ×1 (Brinkman 1974n).

Species	Mature height (m)	Seed bearing Age (years)	Seed bearing Interval (years)	Fruit color Preripe	Fruit color Ripe
T. americana	40	15	1+	Green	Brown
T. cordata	30	32–30	1+	Gray	Brown

Figure 2. *Tilia cordata*, littleleaf linden: seed, ×4 (Brinkman 1974n).

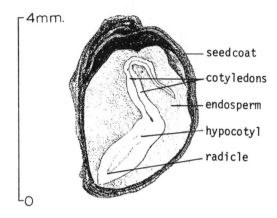

Figure 3. *Tilia americana*, American basswood: longitudinal section through a seed, ×12 (Brinkman 1974n).

COLLECTION, EXTRACTION AND STORAGE OF SEED After frost, basswood fruits can be shaken from the branches onto ground sheets and spread to dry. If the fruits are collected early, when they first turn brown, they must be picked from the trees.

When dry, the persistent bracts can be removed by flailing or threshing. Air screening will remove the debris. The pericarp of littleleaf linden also can be removed by threshing, but fruits of American basswood must be run through a hammer mill or acid treated to remove the pericarp. Seed yield from fruits on a weight basis is about 75 to 80%. There are about 8.8 seeds per gram for American basswood and 30.4 seeds per gram for littleleaf linden. Basswood seeds should be dried to 10 to 12% moisture content and stored at low temperatures. Under these conditions seeds will retain viability for 2 or 3 years.

PREGERMINATION Basswood seeds show delayed germination because of impermeable seedcoats, embryo dormancy, and a hard pericarp. Scarification to break both the pericarp and seedcoat has been used to enhance germination. The planting of early collected fruits before they have a chance to dry has sometimes given good results.

GERMINATION The Genebank Handbook (1985) suggest the following germination standards for *T. cordata* and *T. platyphyllos*: substrata, sand or soil; temperature, 20/30°C; duration, 28 days; additional instructions, prechill 6 to 9 months. For *Tilia* species in general, it suggests the following additional instruc-

tions: scarify in concentrated sulphuric acid or soak in 70°C water and dry 5 times. Germination is epigeal (Fig 4). For American basswood germination capacity is about 30%.

The dormancy problems of *Tilia* species have attracted considerable research attention. Selected recent literature includes the following:

Author—date	Subject	Location
Nagy & Keri 1984	*T. platyphyllos*, study builds on previous studies by senior author concerning plant growth regulator active in embryo during germination.	Hungary
Szalai & Nagy 1974	*T. platyphyllos*, isolation of inhibitors from seeds.	Hungary
Valase et al. 1973	*T. tomentosa*, moisture content and storage conditions.	—
Vanstone 1978	*T. americana*, monitor pericarp color and collect when they turn from green to grayish brown. Plant seeds at once.	Canada
Vanstone & Ronald 1982	*T. americana*, excised embryos germinated at all stages of maturity, but germination of intact seeds declined after earliest collection.	Canada

NURSERY AND FIELD PRACTICE Both fall and spring sowing of prechilled seeds result in seedling emergence over a 2- to 3-year period. Increased production of first-year seedlings can be obtained if the seeds are treated with acid 5 months before sowing. For American basswood, early collection of fruit when it starts to turn brown is recommended, followed by immediate sowing. Sowing of 2100 seeds per m² results in about 315 seedlings per m².

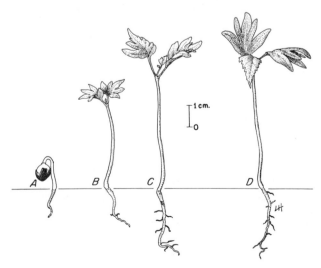

Figure 4. *Tilia americana*, American basswood: seedling development at 1, 3, 16, and 19 days after germination (Brinkman 1974n).

MELIACEAE — MAHOGANY FAMILY

Toona australis Harms — Australian toon

GROWTH HABIT, OCCURRENCE, AND USE Australian toon is important timber species native to Australia (Walters 1974c). It was introduced to Hawaii from the coastal rain forest of Australia in 1914. Australian toon is a deciduous tree 30 to 33 m tall. It keeps its leaves longer in moist sites than dry. The wood is valued for cabinet making and furniture. The red, often highly figured wood is durable and seasons rapidly.

FLOWERING AND FRUITING In Hawaii, Australian toon flowers from April to June. The flowers are bisexual. The 5-valved, tear-shaped capsules, 1.8 to 2.5 cm long in pendulous clusters, ripen from July to September. Seeds are dispersed from August to October. Trees begin to seed as early as 5 years of age but generally do not produce seed regularly until 10 to 15 years old.

COLLECTION, EXTRACTION, AND STORAGE OF SEED The capsules turn from green to brown or reddish brown when ripe. When the first capsule in a cluster opens, the entire cluster should be picked. The harvested fruits should be spread on trays to dry. The light brown, membranous winged seeds (Figs. 1, 2) fall from the capsules as they open. Shaking aids in separating the seeds from the capsules, and an air screen can be used to recover them. There are about 310 seeds per gram. They can be stored in plastic bags at low temperatures.

GERMINATION Australian toon seeds germinate without pretreatment, but 30 days of prechilling greatly increases the speed of germination. Presoaking in water also enhances the speed of germination. Germination is epigeal. Investigations in China indicate that 25°C is the optimum temperature for the germination of seeds of *Toona sinensi.*

NURSERY AND FIELD PRACTICE Australian toon seeds can be sown in Hawaii in any month of the year, but the best results are obtained by sowing from March to November. No mulch or shade is used on the beds. Seedling densities of 105 to 150 per m² are used.

Figure 1. *Toona australis,* Australian toon: seed, ×2 (Walters 1974c).

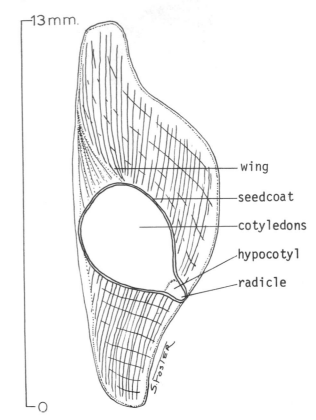

Figure 2. *Toona australis,* Australian toon: longitudinal section through a seed with wing attached, ×8 (Walters 1974c).

TAXACEAE — YEW FAMILY

Torreya Arn.

GROWTH HABIT, OCCURRENCE, AND USE The genus *Torreya* includes 6 closely related species of trees and shrubs found in North America, and central and northern China (Roy 1974d). These species are isolated relics of a pre-Pleistocene circumpolar flora growing at high latitudes. California (*T. californica*) and Florida (*T. taxifolia*) torreya are native to the United States.

The wood of both California torreya and Florida torreya is excellent for cabinet making, but large trees are rare. The wood of California torreya is offered for sale as turning blocks under the name of California nutmeg. California torreya has long been cultivated as a rare specimen tree.

FLOWERING AND FRUITING *Torreya* species are nominally dioecious but occasionally monoecious. Male flowers are small, budlike, and numerous on the undersides of twigs at the base of leaves produced the previous year. Stamens are clustered in 6 to 8 whorls with 4 stamens in a whorl. Each filament supports 4 globose yellow pollen-sacs.

Scattered single female flowers generally appear on the lower sides of shoots of the current year's growth. The ovule develops in a sessile, fleshy, arillike structure. By the end of the second season, the fertilized ovule forms a thinly fleshed, green-to-purple drupe-like fruit 2.5 to 4.5 cm long (Fig. 1). The fleshy layer containing numerous thin, flat fibers splits longitudinally after maturity to expose the single, oblong-ovate, yellow brown seed (Fig. 2). The thick, woody inner seedcoat is irregularly folded into the endosperm, and the embryo is minute (Fig. 3).

Flowering of both species occurs in March and April and extends into May for California torreya. Fruits ripen from August to October.

GERMINATION Seeds germinate without prechilling, but very slowly. Seeds of California torreya require a long afterripening. Germination is usually delayed several months after planting.

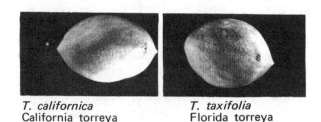

T. californica
California torreya

T. taxifolia
Florida torreya

Figure 2. *Torreya* : seeds, ×1 (Roy 1974d).

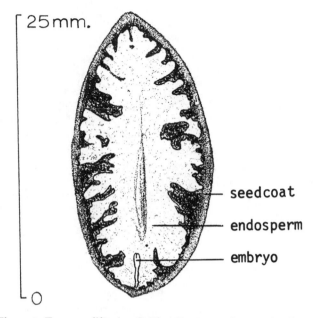

Figure 3. *Torreya californica*, California torreya: longitudinal section through a seed showing the folds of the inner seedcoat extending into the endosperm, ×3 (Roy 1974d).

Figure 1. *Torreya taxifolia*, Florida torreya: fruit, ×1 (Roy 1974d).

CELASTRACEAE — STAFF-TREE FAMILY

Tripterygium regelii Sprague and Takeda

GROWTH HABIT, OCCURRENCE, AND USE *Tripterygium regelii* is a deciduous, polygamous shrub native to Asia. The fruit is a lime-green, bladderlike samara.

Fresh seeds have low germination but respond to 3 months of prechilling (Dirr and Heuser 1987). The species can be rooted from softwood or hardwood cuttings.

MYRTACEAE — MYRTLE FAMILY

Tristania conferta R. Br. — Brushbox

GROWTH HABIT, OCCURRENCE, AND USE Brushbox is native to Australia (Pettys 1974). This species has become naturalized in Asia and Africa, as well as California and Hawaii. Brushbox is an evergreen tree that attains heights of 37 to 46 m. It is generally planted for timber but also is used for environmental plantings. The wood is hard, heavy, and durable and is fairly resistant to termites and shipworms. It is used for decking, general construction, and specialty items.

FLOWERING AND FRUITING Brushbox produces seeds moderately well in 15 to 20 years. The bell-shaped, 3-celled capsules, up to 0.75 cm in diameter, turn from green to brown when ripe. In Hawaii the trees can be found in all stages of reproduction at any time of the year.

COLLECTION, EXTRACTION, AND STORAGE OF SEED The capsules can be picked from standing or felled trees. The fruit can be spread on trays for ripening. The seed is released as the capsules dry. The seed is very small, about 2 mm in diameter (Figs. 1, 2). There are about 4800 seeds per gram. Seeds have been stored at low temperatures.

Figure 1. *Tristania conferta*, brushbox: seed, ×12 (Pettys 1974).

GERMINATION Seeds germinate without pretreatment. NURSERY AND FIELD PRACTICE Brushbox seeds are mixed with fine soil and the mixture spread on seedbeds. Germination usually occurs in 10 to 14 days. No mulch or shading is provided. In Hawaii seeds are sown from November to March, and seedlings are outplanted the following winter. Seedling densities of 210 to 315 per

m² are common. Bare root transplants must be handled with care.

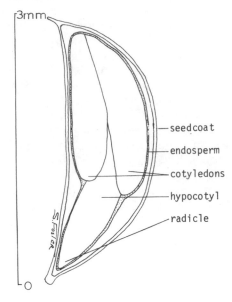

Figure 2. *Tristania conferta*, brushbox: longitudinal section through seed, ×24 (Pettys 1974).

PINACEAE — PINE FAMILY

Tsuga Carr. — Hemlock

GROWTH HABIT, OCCURRENCE, AND USE The hemlocks are tall, straight, evergreen trees with conical crowns and slender, horizontal to pendulous branches (Ruth 1974). The genus consists of about 14 species; 4 are native to North America, and the remainder are native to Asia. Of the 4 species in the United States, eastern hemlock (*Tsuga canadensis*) and western hemlock (*T. heterophylla*) are used commercially for lumber and plywood. The bark of eastern hemlock used to be an important source of tannin. Carolina hemlock (*T. caroliniana*) overlaps with the southern range of eastern hemlock but is a smaller tree, with longer needles and cones. Mountain hemlock (*T. mertensiana*) is important for watershed protection and is one of the most beautiful trees in the Sierra Nevada.

FLOWERING AND FRUITING Seed production of hemlock usually begins when trees are 20 to 30 years of age or a little later if they are shaded. Male and female strobili develop in clusters at the end of lateral branches; each consists of a central axis with spirally arranged microsporophylls. Ovulate strobili are erect with nearly orbicular scales, each with 2 basal ovules and subtended by a membranous bract about the length of the scale. They occur terminally on the lateral shoots of the previous year.

Cones mature in one year and are small, pendant, globose to ovoid or oblong, with scales longer than the bracts (Fig. 1). Eastern hemlock has the smallest cones,

Figure 1. *Tsuga*: cones, ×1 (Ruth 1974).

typically 1.25 to 1.9 cm long, followed by western hemlock with cones 2.5 cm long, and Carolina hemlock with cones 2.5 to 3.8 cm. Mountain hemlock cones often reach 8.9 cm long. The period of dissemination for western hemlock seeds can last over a full year. Cones often remain on hemlocks well into the second year, being especially conspicuous on the tops of mountain hemlocks. All species of hemlock bear some cones almost every year, and large crops are frequent (Ruth 1974):

Species	Flowering dates	Fruit ripening dates	Seed dispersal dates
T. canadensis	Apr.–June	Sept.–Oct.	Sept.–Winter
T. caroliniana	Mar.–Apr.	Aug.–Sept.	—
T. heterophylla	Apr.–June	Aug.–Oct.	Sept.–Winter
T. mertensiana	June–July	Aug.–Oct.	—

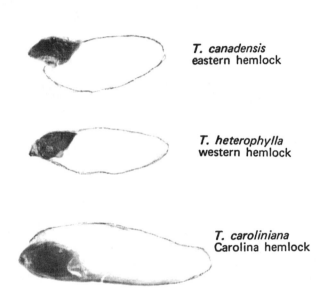

Figure 2. *Tsuga*: seeds, ×4 (Ruth 1974).

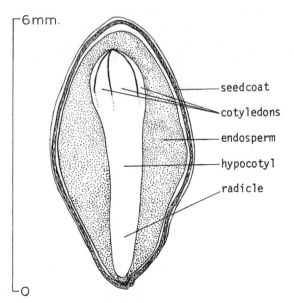

Figure 3. *Tsuga mertensiana*, mountain hemlock: longitudinal section through a seed, ×12 (Ruth 1974).

Seeds are nearly surrounded by their wings (Fig. 2). Embryos have 3 to 6 cotyledon (Fig. 3).

COLLECTION, EXTRACTION, AND STORAGE OF SEED Hemlock cones are small and more difficult to harvest than those of most other conifers. They are easily collected from the tops of recently felled trees. Usually cone collection is delayed until shortly before seed dispersal to ensure seed maturity. Seed viability increases gradually up to dispersal time (Ruth 1974):

Species	Mature height (m)	Seed bearing		Ripe cone color
		Age (years)	Interval (years)	
T. canadensis	49	15–30	2–3	Purple-brown
T. caroliniana	22	—	—	Brown
T. heterophylla	77	20–30	2–8	Reddish brown
T. mertensiana	46	20–30	1–5	Brown

Cones should be stored in permeable sacks in open-sided sheds. Kiln drying temperatures are 30 to 45°C, with a drying time of about 48 hours. Seeds are extracted by tumbling after kiln drying. Eastern hemlock seed yield from cones on a weight basis is about 4%.

Several years can elapse between hemlock seed crops, so annual planting is dependent on seed storage. Western hemlock seeds are stored at −18°C. Eastern hemlock seeds can be stored at cool or subfreezing temperatures. Moisture content of hemlock seeds in storage should be maintained at 6 to 9%.

PREGERMINATION TREATMENT Not all lots of hemlock seeds require prechilling, but most lots require some prechilling (Ruth 1974):

Species	Prechilling duration (days)	Seeds/gram
T. canadensis	30–120	410
T. caroliniana	30–90	190
T. heterophylla	21–90	570
T. mertensiana	90	250

Prechilling clearly improves and accelerates the germination of both eastern and western hemlock seeds. Care must be taken not to prechill the seed too long or damage will result.

GERMINATION ASOA (1985) standards for *Tsuga* seeds follow:

Species	Substrata	Temperature (°C)	Duration (days)	Additional directions
T. canadensis	TB,P	15	28	Prechill 28 days.
T. heterophylla	TB,P	20	28	2 tests, with and without prechilling, for 21 days.

International Seed Testing Association rules suggest providing light for the incubation of both species of hemlock.

Selected recent literature on germination of seeds of *Tsuga* includes the following:

Author—date	Subject
Campbell & Ritland 1982	*T. heterophylla*, regulation of germination timing by moist prechilling.
Coffman 1978	*T. canadensis*, field germination in Lake states. Results indicate that conditions must remain continually wet.
Edwards et al. 1973	*T. heterophylla*, photoperiod response in germination.

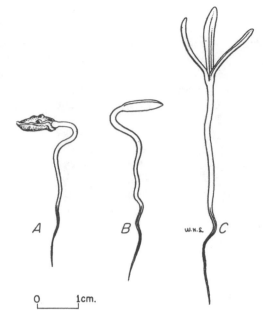

Figure 4. *Tsuga canadensis*, eastern hemlock: seedling development at 2, 4, and 7 days after germination (Ruth 1974).

NURSERY AND FIELD PRACTICE Hemlock seedlings are difficult to grow in the nursery. They are easily damaged by the hot sun, and their small size makes them susceptible to frost heaving (Fig. 4). Most nurseries prechill western hemlock seeds and sow them in the spring. One practice is to sow seeds on the surface of beds and cover them with burlap, which is kept damp until germination occurs.

Eastern hemlock seeds are also sown in the spring. Desirable seedbed densities are 320 to 530 per m². Seedling percentages are 15 to 50%.

LEGUMINOSAE — LEGUME FAMILY

Ulex europaeus L. — Gorse

GROWTH HABIT, OCCURRENCE, AND USE Gorse is a spiny, dense, deciduous shrub usually between 1 and 2 m tall (Rudolf 1974r). Native to central and western Europe, it has become naturalized in the middle Atlantic States and on Vancouver Island as well as coastal Oregon. It is one of 20 species of *Ulex*, all shrubs, native to western Europe. Common gorse has long been cultivated as a soil-binding species as well as for protecting beaches.

FLOWERING AND FRUITING The perfect flowers are solitary, golden yellow, and bloom from April to June and sometimes in the fall. The legumes are about 1.25 to 1.9 cm long and turn brown when they are ripe. They ripen irregularly from May to July in western Europe and into September in the USSR. Natural seed dispersal takes place during the summer, and natural germination takes place the next spring.

COLLECTION, EXTRACTION, AND STORAGE OF SEED When pods begin to turn brown they can be picked from the bushes. They should be spread in a thin layer for drying. During the winter the pods will open and spill the seeds. There are about 150 seeds per gram. They can be kept in dry storage for 2 years.

PREGERMINATION TREATMENT Gorse seeds have hard seedcoats and about a third will not germinate without pretreatment. Mild acid scarification or hot water treatment breaks their dormancy.

GERMINATION The Genebank Handbook (1985) suggests testing the seeds of gorse at an incubation temperature of 20°C with light 8 hours daily.

NURSERY AND FIELD PRACTICE Gorse seeds are usually sown in the spring. Seedlings can be container grown for a year before being transplanted to the field. Plants can also be established by direct seeding in the field. Gorse is often considered a terrible weed.

ULMACEAE — ELM FAMILY

Ulmus L. — Elm

GROWTH HABIT, OCCURRENCE, AND USE About 20 species of elm are native to the Northern Hemisphere (Brinkman 1974o). None occur in western North America, but some are found in northern Mexico. Most elms are valued for their tough, hard wood, and many have been used in environmental plantings. In recent years many elms have been killed by Dutch elm disease, caused by the fungus *Ceratocystis ulmi*, and by phloem necrosis, caused by the virus *Morus ulmis*. Only Japanese elm (*Ulmus japonica*), Siberian elm (*U.*

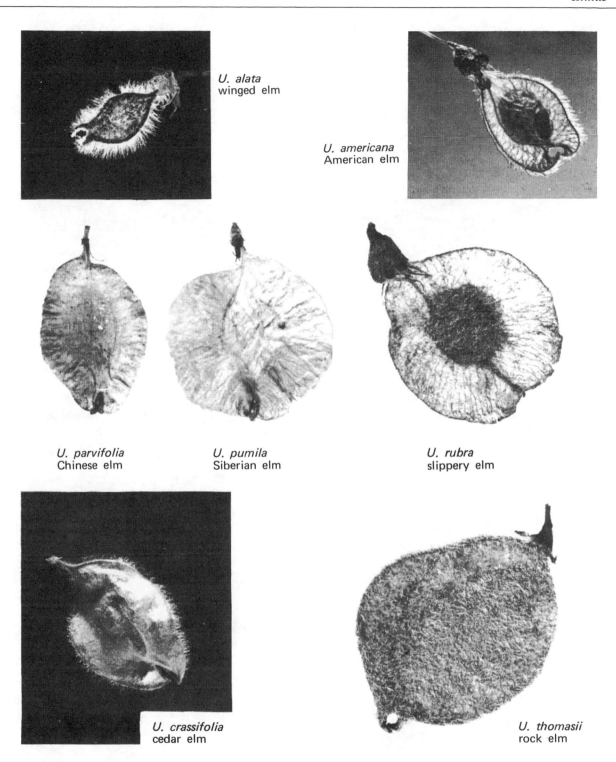

U. alata
winged elm

U. americana
American elm

U. parvifolia
Chinese elm

U. pumila
Siberian elm

U. rubra
slippery elm

U. crassifolia
cedar elm

U. thomasii
rock elm

Figure 1. *Ulmus*: samaras, ×4 (Brinkman 1974o).

pumila), and Chinese elm (*U. parvifolia*) are resistant to these diseases. Park and street plantings of the other species have practically ceased in some areas, but elms are still abundant in natural stands.

FLOWERING AND FRUITING The perfect, rather inconspicuous flowers are usually borne in the spring before the leaves appear. In a few species, they appear in the late summer or early fall. The fruit, a samara, ripens a few weeks later and consists of a compressed nutlet surrounded by a membranous wing (Figs. 1, 2). Dispersal is by wind. In most species, good seed crops are produced at 2- to 3-year intervals (Table 1).

COLLECTION, EXTRACTION, AND STORAGE OF SEED Elm seeds can be collected by sweeping them up from the ground soon after they fall or by flailing branches and collecting seeds that fall onto ground cloths. The large seeds of rock elm (*U. thomasii*) are relished by rodents, but they usually can be picked from the tree. Freshly col-

Table 1. *Ulmus:* phenology of flowering and fruiting, characteristics of mature trees, and cleaned seed weight (Brinkman 1974o).

Species	Flowering dates	Fruit ripening dates	Seed dispersal dates	Mature tree height (m)	Seed-bearing Age (years)	Interval (years)	Ripe fruit color	Seeds/gram
U. alata	Feb.–Mar.	Apr.	Apr.	15	–	–	Reddish green	250
U. americana	Feb.–May	Feb.–June	Mar.–June	37	15	–	Greenish brown	160
U. crassifolia	Aug.–Sept.	Sept.–Oct.	Oct.	31	–	–	Green	150
U. glabra	Mar.–May	May–June	May–June	40	30–40	2–3	Yellow-brown	90
U. japonica	Apr.–May	June	–	31	–	2	–	10
U. laevis	Apr.–May	May–June	May–June	31	30–40	2–3	Yellow-brown	140
U. parvifolia	Aug.–Sept.	Sept.–Oct.	Sept.–Oct.	25	–	–	–	270
U. pumila	Mar.–Apr.	Apr.–May	Apr.–May	25	8	–	–	160
U. rubra	Feb.–May	Apr.–June	Apr.–June	22	15	2–4	Green	90
U. serotina	Sept.	Nov.	Nov.	19	–	–	Greenish brown	330
U. thomasii	Mar.–May	May–June	May–June	31	20	3–4	Yellow-brown	20

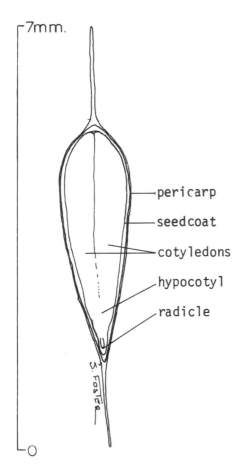

Figure 2. *Ulmus alata*, winged elm: longitudinal section through the embryo of a samara, ×16 (Brinkman 1974o).

lected seeds should be air dried a few days, but excessive drying will reduce germination (Brinkman 1974o):

Seeds should not be dewinged because it damages them. Fruits can be sown with the wings attached. Elm seeds store best at low moisture content in sealed containers at cool temperatures. The number of seeds per gram varies widely, even within species (Table 1).

PREGERMINATION TREATMENT Under natural conditions, seeds of species of elm that ripen in the spring germinate the same season, and those that ripen in the fall

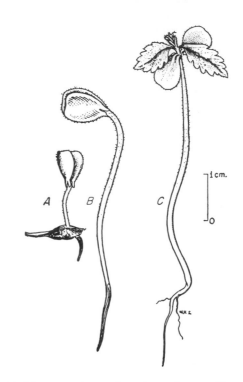

Figure 3. *Ulmus americana*, American elm: seedling development at 1, 3, and 21 days after germination (Brinkman 1974o).

germinate the next spring. Although the seeds of most species require no pretreatment, those of American elm (*U. americana*) always remain dormant until the second season, and seeds of slippery elm (*U. rubra*) from northern sources are often dormant. Prechilling for 60 to 90 days improves the germination of cedar elm (*U. crassifolia*) and September elm (*U. sertina*).

GERMINATION ASOA (1985) standards for germination of elms are as follows:

Species	Substrate	Temperature (°C)	Duration (days)
U. americana	TB	20/30	14
U. parvifolia	TB	20	10
U. pumila	TB	20	10

Apparently because of their general lack of deep seed dormancy, there is not a lot of literature concerning *Ulmus* seeds. The Soviet Chernik (1983) investigated the apparent lack of endosperm. His fellow Soviet Olovyannikova (1981) reported on the use of elms in plantations in the dark soils of Soviet deserts. Tompsett (1986) of Kew Gardens investigated the role of moisture content in the storage of seeds of *U. carpinifolia*. For ultra-low temperature storage the moisture content of the seeds had to be lowered below

optimum levels for higher temperatures.

NURSERY AND FIELD PRACTICE For species whose seeds ripen in the spring, the seeds are sown as soon as they are mature. The fall maturing species are usually planted the following spring, after prechilling. Only 5 to 12% of the viable seeds planted can be expected to make transplantable seedlings (Fig. 3). Seedlings are often container grown.

LAURACEAE — LAUREL FAMILY

Umbellularia Nutt. — Myrtlewood

GROWTH HABIT, OCCURRENCE, AND USE *Umbellularia californica* is the sole member of the genus (Stein 1974b). Over much of its range, this species is a tree from 12 to 24 m tall with a diameter from 45 to 75 cm, but near the ocean and the bordering chaparral it can be a prostrate shrub. In protected bottomlands of southern Oregon or northwestern California, the trees may reach 31 m in height.

The range of myrtlewood spans more than 11 degrees of latitude from the Umpqua Valley of Douglas County, Oregon, south to San Diego County, California. It is most abundant near the coast but occurs also in the interior mountains to the Sierra Nevada. The wood is valued for furniture making, gunstocks,

and specialty products.

FLOWERING AND FRUITING The small, yellow, perfect blossoms grow on short-stemmed umbels that originate from leaf axils or near the terminal bud (Fig. 1). Flower buds develop early; those for the following year develop while the current year's fruits are maturing. Flowering occurs from late spring into early summer. Seeds are produced in abundance when trees are 30 to 40 years old.

COLLECTION, EXTRACTION, AND STORAGE OF SEED The fruits, acrid drupes each containing a single, large, thin-shelled seed, ripen in the first autumn following

Figure 1. *Umbellularia californica*, myrtlewood: flowers develop in umbels that originate in leaf axils, ×2 (Stein 1974b).

Figure 2. *Umbellularia californica*, myrtlewood: yellow-green drupe suspended from conical cupule, ×1.5 (Stein 1974b).

flowering. At maturity, the fruits turn from medium green to speckled yellow and through various hues to purple (Fig. 2).

The fruits can be collected from under trees in late autumn. Macerating them frees the seeds, which can then be recovered by flotation. Seeds lose viability rapidly in storage, even at low temperatures.

GERMINATION Natural germination apparently takes place in the autumn after the fruits fall to the ground. The seeds germinate without pretreatment, but several months are required for emergence to occur. Germination is hypogeal (Figs. 3, 4). Myrtlewood can also be propagated by cuttings.

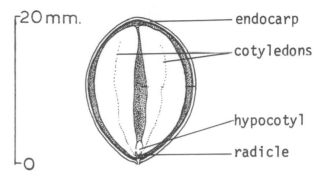

Figure 4. *Umbellularia californica*, myrtlewood: longitudinal section through a seed, ×2 (Stein 1974b).

Figure 3. *Umbellularia californica*, myrtlewood: exterior views of cleaned seeds, ×2 (Stein 1974b).

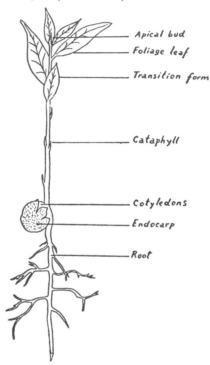

Figure 5. *Umbellularia californica*, myrtlewood: 4-month-old seedling, ×0.5 (Stein 1974b).

ERICACEAE — HEATH FAMILY

Vaccinium L. — Blueberry

GROWTH HABIT, OCCURRENCE, AND USE The blueberries include about 130 species of deciduous or evergreen shrubs (rarely trees) found in the Northern Hemisphere from the Arctic to high mountains in the tropics (Crossley 1974c). The genus contains several commercial berry-producers as well as a number of valuable ornamentals and wildlife species.

FLOWERING AND FRUITING The perfect flowers are either axillary or terminal and can be solitary or in racemes. The blossoms appear in spring or early summer, and the fruit ripens in late summer or early fall (Crossley 1974c):

Species	Flowering dates	Fruit ripening dates	Ripe fruit color
V. angustifolium	May–June	July–Aug.	Bluish black
V. caespitosum	June–July	July–Sept.	Black, bloomy
V. corymbosum	May	June–Aug.	Blue-black
V. ovalifolium	Mar.–July	July–Aug.	Bluish purple
V. ovatum	May–June	Aug.–Sept.	Purplish black
V. parvifolium	Apr.–June	July–Sept.	Bright red

Many cultivars of blueberry have been developed for commercial production. The fruits of naturally occurring species are usually less than 1.2 cm in diameter (Fig. 1).

COLLECTION, EXTRACTION, AND STORAGE OF SEED   Blueberries are easily collected by handpicking the ripe berries or by shaking bushes over containers. After collection, the berries should be chilled at 10°C for several days. After chilling, they can be shredded in a food blender and the seeds recovered by flotation. After drying for 48 hours at room temperature, the seeds (Figs, 2, 3) may be stored at cool temperatures for years. The number of seeds per gram follow (Crossley 1974c):

Species	Seeds/gram	Mature height (m)
V. angustifolium	4,350	0.6
V. caespitosum	11,700	0.6
V. corymbosum	2,150	4.5
V. ovalifolium	3,500	3.4
V. ovatum	4,300	3.7
V. parvifolium	5,500	4.5

GERMINATION   Light is essential for the germination of seeds of *Vaccinium* species (Genebank Handbook 1985), but high light intensities may inhibit germination. Seeds of most species also require fluctuating temperature regimes. Germination is often better in flats in the greenhouse than in the laboratory. Seeds should be tested for germination on top of paper in closed petri dishes, with incubation temperatures of 20/30°C and 8 hours of light daily. Gibberellin enrichment may be substituted for light requirements.

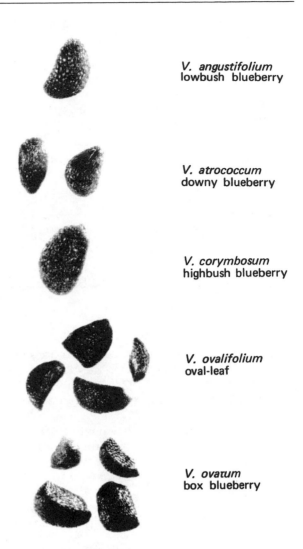

V. angustifolium
lowbush blueberry

V. atrococcum
downy blueberry

V. corymbosum
highbush blueberry

V. ovalifolium
oval-leaf

V. ovatum
box blueberry

Figure 2. *Vaccinium*: seeds, ×12 (Crossley 1974c).

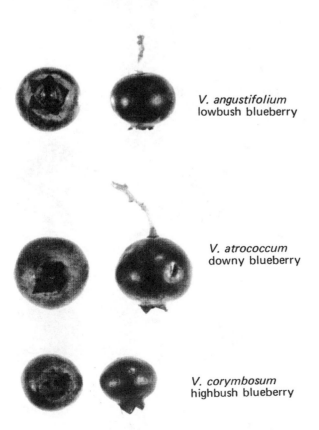

V. angustifolium
lowbush blueberry

V. atrococcum
downy blueberry

V. corymbosum
highbush blueberry

Figure 1. *Vaccinium*: berries, ×2 (Crossley 1974c).

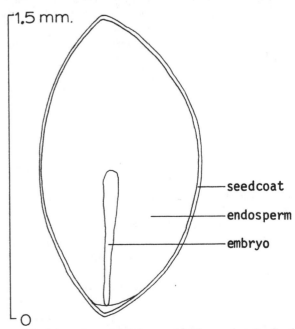

1.5 mm.

seedcoat

endosperm

embryo

0

Figure 3. *Vaccinium atrococcum*, downy blueberry: longitudinal section through a seed, ×40 (Crossley 1974c).

Selected recent literature on the germination of seeds of *Vaccinium* species includes the following:

Author—Date	Subject	Location
Hall & Biel 1970	*V. vitis-idaea,* fresh seeds germinated without pretreatment.	Canada
Targonskii & Bogdanova 1984	*V. vitis-idaea,* seed germination.	USSR
Vander-Kloet 1983	*Vaccinium* species, seed and seedling characteristics.	USA
Vander-Kloet & Austin-Smith 1986	*Vaccinium* species, fruit production.	USA

NURSERY AND FIELD PRACTICE Seeds should be planted on the surface of a fine, moist substrate in flats and the flats covered with clear plastic film until emergence occurs. Seedlings can be transplanted to containers and then to the field.

CAPRIFOLIACEAE — HONEYSUCKLE FAMILY

Viburnum L.

GROWTH HABIT, OCCURRENCE, AND USE There are about 120 species of *Viburnum* that occur in the Northern Hemisphere (Gill and Pogge 1974h). Many species are important for environmental plantings and for ornamentals.

FLOWERING AND FRUITING The small white or sometimes pinkish flowers are arranged in flat, rounded, or pyramidal cymes (Fig. 1). The flowers are typically bisexual, but the marginal blossoms in hobblebush (*V. alnifolium*) and cranberry bush (*V. opulus*) are sterile. Flowers and fruits ripen mainly in May to June and September to October, respectively, but vary among species (Table 1).

The fruit is a one-seeded drupe, 0.6 to 1.25 cm long, with a soft pulp and a thin stone (Figs. 2, 3, 4). During the ripening process, fruits of most species change in color from green to various shades of red. Most species produce fruit every year (Table 1). Much of the ornamental and wildlife value of viburnums is due to their persistent fruit.

COLLECTION, EXTRACTION, AND STORAGE OF SEED The fruits can be handpicked when ripe. Care should be taken that they do not heat. They should be macerated and the seeds recovered by flotation. Dried seeds can be stored at low temperatures for several years. Seeds per gram are given in Table 1.

GERMINATION Seeds of *Viburnum* species are difficult to germinate because of embryo dormancy and hard seedcoats. The dormancy of species native to more southern habitats is less severe. The southern species germinate in 1 year while the more northern species require 2 years. The Genebank Handbook (1985) lists the following germination standards for *V. lantana, V. lentago, V. opulus,* and *V. trilobum*: substrate, TP; temperatures, 20/30°C; and additional directions, provide light 8 hours per day. For *V. lantana* warm stratification at 25°C for 10 weeks followed by pre-chilling for 12 weeks is suggested also.

Figure 1. *Viburnum lentago,* nannyberry: cluster of fruits typical of the genus, ×1 (Gill and Pogge 1974h).

V. *cassinoides*
witherod viburnum

V. *lentago*
nannyberry

V. *rafinesquianum*
downy arrowwood

V. *opulus*
cranberrybush

Figure 2. *Viburnum*: single fruits (drupes), ×2 (Gill and Pogge 1974h).

Table 1. *Viburnum:* phenology of flowering and fruiting, characteristics of mature trees, and cleaned seed weight (Gill and Pogge 1974h).

Species	Flowering dates	Fruit ripening dates	Seed dispersal dates	Mature height (m)	Growth habit	Seed-bearing Age (years)	Seed-bearing Interval (years)	Seeds/gram
V. acerifolium	Apr.–Aug.	July–Oct.	Fall	2	Shrub	2–3	1	30
V. alnifolium	May–June	Aug.–Sept.	Fall	3	Shrub	–	3–4	25
V. cassinoides	May–July	July–Oct.	Oct.–Nov.	3	Shrub	–	1	60
V. dentatum	May–Aug.	July–Nov.	Fall–Winter	5	Shrub	3–4	–	45
V. lantana	May–June	Aug.–Sept.	Fall–Winter	5	Tree	–	–	20
V. lentago	Apr.–June	July–Oct.	Fall–Spring	–	–	–	–	15
V. opulus	May–July	Aug.–Sept.	Fall–Spring	10	Tree	8	1	30
V. prunifolium	Apr.–June	July–Oct.	Fall–Spring	4	Shrub	3–5	–	10
V. rafinesquianum	May–June	July–Oct.	Fall	2	Shrub	8–10	1	–
V. recognitum	Apr.–June	July–Sept.	Fall–Winter	3	Shrub	5–6	–	–

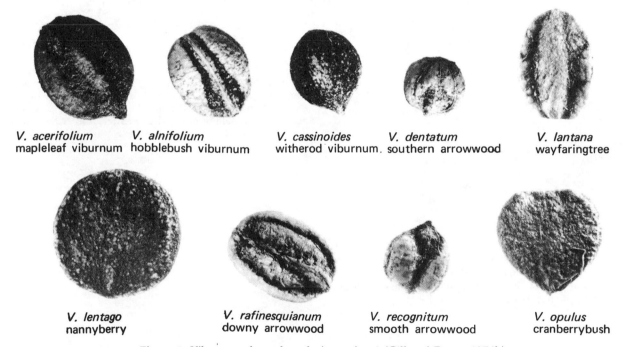

V. acerifolium
mapleleaf viburnum

V. alnifolium
hobblebush viburnum

V. cassinoides
witherod viburnum.

V. dentatum
southern arrowwood

V. lantana
wayfaringtree

V. lentago
nannyberry

V. rafinesquianum
downy arrowwood

V. recognitum
smooth arrowwood

V. opulus
cranberrybush

Figure 3. *Viburnum:* cleaned seeds (stones), ×4 (Gill and Pogge 1974h).

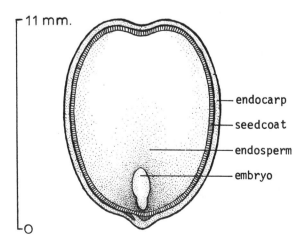

Figure 4. *Viburnum lentago,* nannyberry: longitudinal section through a stone, ×5 (Gill and Pogge 1974h).

NURSERY AND FIELD PRACTICE Seeds can be sown in the spring in nursery beds for warm stratification. The following winter they will be prechilled, and emergence will occur the next spring. Seedlings of some species require shading. Seedling development is shown in Figures 5 and 6. Most species can be propagated by hard or softwood cuttings or by air layering.

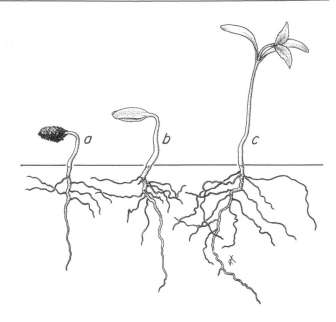

Figure 6. *Viburnum lentago,* nannyberry: seedling development from stratified seed; (A) root development during warm stratification, (B) very little development during ensuing prechilling for breaking epicotyl dormancy; and (C) subsequent development at germination temperature (Gill and Pogge 1974h).

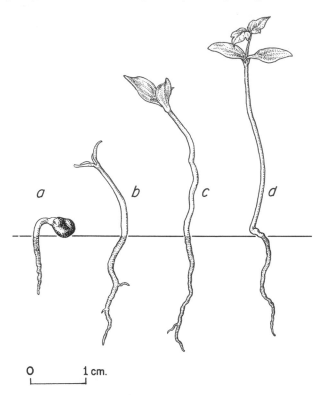

O   1 cm.

Figure 5. *Viburnum dentatum,* southern arrowwood: seedling development at 1, 2, 11, and 29 days after germination. Roots and shoots develop concurrently (Gill and Pogge 1974h).

VERBENACEAE — VERBENA FAMILY

Vitex agnus-castus L. — Lilac chastetree

GROWTH HABIT, OCCURRENCE, AND USE Native to southern Europe, lilac chastetree is a deciduous, strongly aromatic shrub or small tree (Schopmeyer 1974d). The genus consists of about 100 species, largely native to the tropics and subtropics. On the Pacific coast, *Vitex agnus-castus* is grown from Washington to southern California. Its height ranges from 2 m in the north to 8 m in the low land deserts of the south. In the eastern United States, it is hardy as far north as New York State. It is grown as an ornamental but has potential in shelterbelts.

FLOWERING AND FRUITING The fragrant flowers occur in dense spikes about 20 cm long during the late summer and autumn. The pungent fruits are small drupes about 0.3 to 0.4 cm in diameter that ripen in late summer and fall. Good seed crops occur almost every year. Each drupe contains a rounded, brownish purple, 4-celled stone about 0.3 cm long, frequently covered with a lighter colored, membranous cap. Each stone contains 1 to 4 seeds (Fig. 1).

COLLECTION, EXTRACTION, AND STORAGE OF SEED Seeds may be handpicked or recovered from ground sheets spread under the trees. The fruits should be threshed and the seeds recovered by air screening. Seed yield from fruits on a weight basis is about 75%, and there are about 100 seeds per gram. They can be stored in prechilling conditions for 1 year.

GERMINATION Stored seeds require 90 days of prechilling. Fresh seeds planted immediately may germinate without pretreatment.

NURSERY AND FIELD PRACTICE Prechilled seeds can be sown in the spring and covered with 0.6 cm of soil. Seeds planted result in about 16% successful seedlings (Fig. 2). This species can be rooted from softwood cuttings.

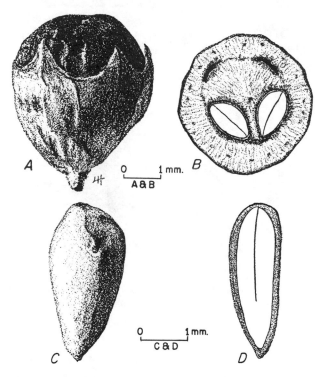

Figure 1. *Vitex agnus-castus*, lilac chastetree: (A) fruit, ×10; (B) tranverse section through 2 seeds within a fruit, ×10; (C) cleaned seed, ×12; (D) longitudinal section through a seed. Embryo occupies entire seed cavity (Schopmeyer 1974d).

Figure 2. *Vitus agnus-castus*, lilac chastetree: seedling showing cotyledons and first leaves (Schopmeyer 1974d).

VITACEAE — GRAPE FAMILY

Vitis labrusca L. — Fox grape

GROWTH HABIT, OCCURRENCE, AND USE Fox grape is a deciduous, woody vine found from New England to Illinois and south to Georgia and Arkansas (Bonner and Crossley 1974). It may climb to heights of 12 m. It hybridizes readily with other *Vitis* species and has been the most important grape in the development of U.S. viticulture. It is also an important food source for wildlife.

FLOWERING AND FRUITING The dioecious flowers are borne in short panicles, 5 to 10 cm long, in May and June. The fruit clusters usually have fewer than 20 globose berries, 0.8 to 2.5 cm in diameter, that mature in August to October and drop singly. Mature berries are brownish purple to dull black and contain 2 to 6 brownish, angled seeds 0.5 to 0.8 cm long (Figs. 1, 2).

COLLECTION, EXTRACTION, AND STORAGE OF SEED Ripe fruits can be stripped from the vines by hand or shaken onto canvas sheets. The fruits can be washed through a screen and the seeds recovered. There are about 30 seeds per gram. They can be stored for a couple of years at low temperatures.

PREGERMINATION TREATMENT *Vitis* seeds require prechilling before they will germinate. Fox grapes require from 120 to 220 days of prechilling.

GERMINATION Many investigations have shown that gibberellin enrichment enhances the germination of grape seeds. Pal et al. (1976) suggested that gibberellin enrichment improved germination and subsequent seedling growth of some grape cultivars. Patil and Patil (1984) reported on detailed studies of the seed morphology of numerous grape species. Ellis et al. (1983) developed the following procedure for germination of grape seeds: (a) soak for 24 hours in 0.5M H_2O_2, (b) soak an additional 24 hours in 1000 ppm GA_3, and (c) prechill for 21 days at 3 to 5°C. The optimum temperature regime for testing germination was 20/30°C, and 25°C was the best constant temperature. Rajasekaran et al. (1982) attributed the dormancy of grape seeds to abscissic acid concentrations in the embryo.

NURSERY AND FIELD PRACTICE Propagation by cuttings is the common method of increasing plants of *Vitis labrusca*.

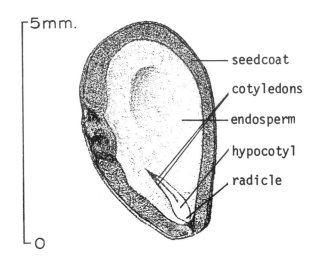

Figure 1. *Vitis labrusca*, fox grape: seeds, ×4 (Bonner and Crossley 1974).

Figure 2. *Vitis riparia*, riverbank grape: longitudinal section through a seed, ×12 (Bonner and Crossley 1974).

PALMACEAE — PALM FAMILY

Washingtonia filifera (Lindl.) Wendl. —
California Washington palm

GROWTH HABIT, OCCURRENCE, AND USE The California Washington palm is the only palm native to California; it is also the largest native palm in the United States (Krugman 1974e). The sturdy, massive, cylindrical trunk, growing to 18 to 23 m in height, tapers gradually from a diameter of 50 to 90 cm at the base to slightly less at the top. It has a broad crown of as many as 50

fan-shaped leaves with petioles as long as 1.5 m. Dead leaves may remain around the trunk for many years, forming a dense, thatchlike shroud that reaches almost to the ground. *Washingtonia filifera* is native to rocky stream beds or other nearby water sources bordering the Colorado Desert of southeastern California; Yuma County, Arizona; and Baja California, Mexico. It is widely planted in the Southwest and Gulf states as an ornamental.

FLOWERING AND FRUITING In August, small but showy clusters of white, vase-shaped flowers appear, initially enclosed by a spathe. The mature flower stalks may average 3.7 m in length and extend almost horizontally in the crown. The flowers are perfect and occur annually in great abundance once the trees reach reproductive maturity. The calyx is tubular, and the corolla is funnel-shaped with the stamens inserted in the tube.

The fruit and seeds mature during December and January. The ripe fruit is a spherical or elongated black berry about 0.9 to 1.25 cm long, with a thin, fleshy layer surrounding a single seed. The seed is pale chestnut in color and about 0.6 to 0.8 cm long by 0.3 cm thick. There are 12 seeds per gram (Fig. 1). They are mature by the time the fruits drop.

COLLECTION, EXTRACTION AND STORAGE OF SEED The fleshy covering of the seeds should be removed in a macerator and the seeds recovered by flotation. The cleaned seeds may be stored or sown immediately. Seeds should not be permitted to dry. Long-term storage is not recommended.

GERMINATION Fresh seeds will germinate in 4 to 15 weeks without pretreatment.

The coyote (*Canis latrans*) is considered to be the prime dispersal agent of California Washington palm seeds (Connett 1985). Khan (1982) investigated the allelopathic potential of the fruits. Dry fruits were found to be highly allelopathic to germination of seeds of other species.

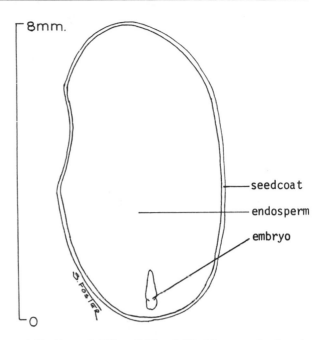

Figure 1. *Washingtonia filifera*, California Washington palm: longitudinal section through the embryo of a seed, ×10 (Krugman 1974e).

NURSERY AND FIELD PRACTICE Seeds can be sown in sand or peat-sand mixtures. Bottom heat is recommended to speed germination. Germination is hypogeal. With the appearance of the elongated second leaf, the seedlings should be transplanted to containers and grown in partial shade.

CAPRIFOLIACEAE — HONEYSUCKLE FAMILY

Weigela florida (Bunge.) DC. — Old-fashioned weigela

GROWTH HABIT, OCCURRENCE, AND USE *Weigela florida* is a spreading shrub up to 2 m in height that is native to northern China and Korea. The fruit is a long, narrow capsule that splits into 2 valves. The seeds germinate without pretreatment (Dirr and Heuser 1987), and the species is very easy to root.

LEGUMINOSAE — LEGUME FAMILY

Wisteria Nutt.

GROWTH HABIT, OCCURRENCE, AND USE The genus *Wisteria* consists of about 9 species of woody twiners that are native to the eastern United States and eastern Asia. They are planted as ornamentals. Individual plants attain a great age and develop thick trunks. The fruit is a flattened legume.

The seeds germinate readily without pretreatment. Dry seeds should be soaked overnight before planting (Dirr and Heuser 1987). They should be planted quite shallowly in the seedbed.

SAPINDACEAE — SOAPBERRY FAMILY

Xanthoceras scorbifolia **Bunge.** — Yellowthorn

GROWTH HABIT, OCCURRENCE, AND USE *Xanthoceras* consists of a single species that is a deciduous shrub native to northern China. It is planted in North America for its attractive flowers.

The fruit is a hard, thick-walled capsule that splits into 3 valves. The seeds are dark brown and require no pretreatment for germination, but germination speed and uniformity are enhanced by 2 to 3 months of prechilling (Dirr and Heuser 1987).

AGAUACEAE — AGUAVE FAMILY

Yucca **L.**

GROWTH HABIT, OCCURRENCE, AND USE There are about 30 species of *Yucca* native to North America and the West Indies (Alexander and Pond 1974). Most grow in the semiarid southwestern United States and adjacent Mexico. Great Plains yucca (*Y. glauca*) is a small, acaulescent shrub 1 to 2 m tall, with narrow, sword-shaped, spine-tipped, upright leaves 0.6 to 1.25 cm wide. Soaptree yucca (*Y. elata*) is a medium to large caulescent shrub up to 10 m tall, with leaves that are similar to *Y. glauca* but wider (5 cm) and longer. Natural reproduction by seed is apparently limited. Most reproduction occurs through sprouts from rhizomes. Early growth of seedlings is very slow. Soaptree yucca seedlings average 20 cm tall at 16 years of age. The yucca species are important components of the native vegetation and have been used as ornamental species.

FLOWERING AND FRUITING The greenish to creamy white flowers are perfect. They appear on terminal panicles from mid-May to mid-July (Alexander and Pond 1974):

Species	Flowering dates	Fruit ripening dates	Seed dispersal dates
Y. elata	May–July	Aug.–Sept.	Sept.–Oct.
Y. glauca	May–June	July–Aug.	Sept.

Plants can begin to bear flowers at 5 to 6 years of age.

The fruit is a capsule containing 120 to 150 flat, round to ovoid, black seeds that are wind dispersed (Figs. 1, 2). Pollination seldom occurs without the aid of the yucca moth, *Pronuba yuccasella*, or *Prodoxus quinquepunctellus*. These moths gather pollen, place it on the stigmatic tube, and lay their eggs. The larvae feed exclusively on the maturing seeds but usually consume only 20% of the total.

COLLECTION, EXTRACTION, AND STORAGE OF SEED Since the capsules are dehiscent, fruits should be collected just before or at the time they open. Seeds are easily extracted when the capsules are dry, and they can be cleaned with an air screen. There are 50 seeds per gram. They can be stored dry at room temperature.

PREGERMINATION TREATMENT There is some evidence that seedcoats of the Yucca species may be indurate. Presoaking for 24 hours or mild scarification have been considered to enhance germination.

GERMINATION McCleary and Wagner (1973) investigated the germination of *Y. angustissima*, *Y. baccata*, *Y, brevifolia*, *Y. elata*, *Y. glauca*, and *Y. kanabensis* at 10, 15, 20, and 25°C. All species germinated readily at 20 or 25°C without pretreatment. Keeley and Tufenkian (1984) determined that seeds of *Y. whipplei* germinated readily at moderate temperatures, but were inhibited from germinating at high incubation temperatures.

NURSERY AND FIELD PRACTICE Seedlings are container grown and protected from frost the first winter.

Figure 1. *Yucca elata*, soaptree yucca: seed ×4 (Alexander and Pond 1974).

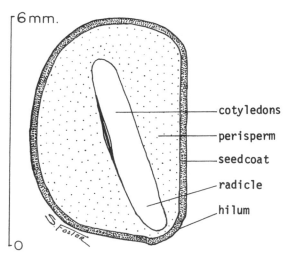

Figure 2. *Yucca elata*, soaptree yucca: longitudinal section through the embryo of a seed, ×10 (Alexander and Pond 1974).

RUTACEAE — RUE FAMILY

Zanthoxylum L. — Prickly ash

GROWTH HABIT, OCCURRENCE, AND USE Most species of prickly ash are large shrubs or small trees (Bonner 1974m). Their deciduous foliage is very aromatic, and the bark and fruit were once used for medicinal purposes.

FLOWERING AND FRUITING The greenish white dioecious flowers are borne in inconspicuous axillary cymes in common prickly ash (*Zanthoxylum americanum*) and in large terminal cymes on Hercules club (*Z. clava-herculis*) (Fig. 1). Flowering occurs in the spring, with fruit ripening in late summer. The fruits are single seeded capsules about 0.5 to 0.6 cm in diameter. During ripening, they turn from green to reddish brown. At maturity, the round, black, shiny seeds hang from the capsules (Figs. 1, 2, and 3).

COLLECTION, EXTRACTION, AND STORAGE OF SEED Seeds may be stripped from mature fruit clusters by hand as the capsules open, or entire clusters of Hercules club may be picked. Unopened capsules open with drying and gentle threshing. Seeds can be cleaned by air screening. Common prickly ash seeds can be stored in sealed containers for 25 months at low temperatures.

GERMINATION *Zanthoxylum* seeds exhibit strong dormancy, which apparently is partially imposed by the seedcoat. Scarification with concentrated sulfuric acid for 2 hours has given fair germination. Prechilling for 120 days has helped the germination of seeds of common prickly ash. A germination test can be conducted at 20/30°C.

NURSERY AND FIELD PRACTICE Fall sowing of seeds immediately after collection is recommended. Germination is epigeal (Fig. 4). Propagation by root cuttings and suckers is also practiced.

Figure 2. *Zanthoxylum clava-herculis,* Hercules club: single carpel and seed, ×4 (Bonner 1974m).

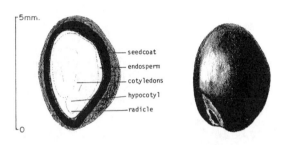

Figure 3. *Zanthoxylum americanum,* common prickly ash: longitudinal section of a seed (left) and exterior view (right), ×6 (Bonner 1974m).

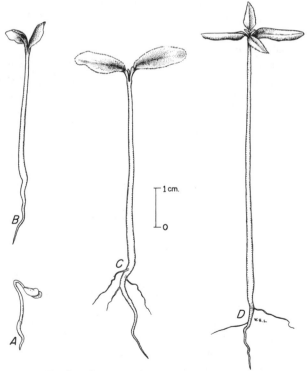

Figure 4. *Zanthoxylum americanum,* common prickly ash: seedling development at 1, 3, 13, and 18 days after germination (Bonner 1974m).

Figure 1. *Zanthoxylum clava-herculis,* Hercules club: cluster of mature fruits, ×1 (Bonner 1974m).

ULMACEAE — ELM FAMILY

Zelkova Spach.

GROWTH HABIT, OCCURRENCE, AND USE *Zelkova* consists of 5 species native to the region from the Caucasus to eastern Asia. *Zelkova serrata* is grown in the United States as a street tree. It reaches 31 m in height and has a broad crown. The fruit is a one-seeded drupe.

GERMINATION Seeds do not require pretreatment for germination, but germination speed and uniformity are enhanced by prechilling for 2 to 3 months. Seeds should not be allowed to dry or viability is rapidly lost (Dirr and Heuser 1987). Ishii (1979) determined that increasing the prechilling time for *Z. serrata* seeds increased germination capacity and energy.

ERICACEAE — HEATH FAMILY

Zenobia pulverulenta (Bartr. ex Willd.) Pollard — Dusty zenobia

GROWTH HABIT, OCCURRENCE, AND USE Dusty zenobia is a shrub native to eastern North America that is occasionally grown as an ornamental. The fruit is a loculicidally dehiscent capsule. Fresh seeds require no pretreatment and germinate immediately (Dirr and Heuser 1987).

RHAMNACEAE — BUCKTHORN FAMILY

Ziziphus Mill. — Jujube

GROWTH HABIT, OCCURRENCE, AND USE There are about 40 species of *Ziziphus*, chiefly found in tropical and subtropical regions of the world (Bonner and Rudolf 1974). Common jujube (*Z. jujuba*) and Christ thorn (*Z. spina-christi*) are grown in the United States as ornamentals.

FLOWERING AND FRUITING The perfect yellow flowers of common jujube appear in the spring, and the fruits mature from July to November. The fruits are globose to slender, fleshy drupes that turn from green to brown at maturity. If left on the tree, the fruits turn black. Common jujube fruits are about 2.5 to 5 cm long and contain a 2-celled, 2-seeded, deeply furrowed, reddish brown to gray, oblong, pointed stone about 1.9 to 2.5 cm long (Fig. 1). Good fruit crops are borne annually. The fruits are eaten in Europe.

COLLECTION, EXTRACTION, AND STORAGE OF SEED Jujube fruits may be picked by hand or flailed from trees. Stones can be cleaned in a macerator and recovered by flotation. The cleaned stones are used as seeds. There are 1.7 seeds per gram for common jujube and 1.5 for Christ thorn. They should be stored dry at low temperatures.

PREGERMINATION TREATMENT *Ziziphus* seeds require pretreatment for germination. Both prechilling and prolonged hot water treatments have been used to enhance germination.

GERMINATION The Genebank Handbook (1985) recommends complete removal of the cover structure of the stone and presoaking for 24 hours.

Lotebush (*Z. obtusifolia* var. *obtusifolia*) is a spiny shrub that occurs in the U.S. Southwest. Speer and Wright (1981) found optimum germination of lotebush seeds occurred at 20 to 30°C. Both light and prechilling were necessary for substantial germination. Zietsman and Botha (1987) found that seeds of *Ziziphus mucronata* subsp. *mucronata*, which occurs in South Africa, were light-neutral in germination trials. They also found that the pericarp restricted germination, and the seeds were very sensitive to moisture stress. Kim and Kim (1983) determined that growth regulators did not enhance germination of *Ziziphus* seeds.

NURSERY AND FIELD PRACTICE Seeds can be soaked in warm water for 2 days and then seeded in the fall. Germination is epigeal.

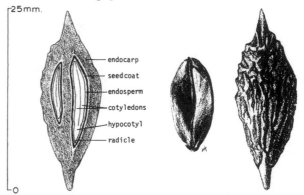

Figure 1. *Ziziphus jujuba*, common jujube: (left) longitudinal section through 2 seeds in a stone; (center) exterior view of seed after removal from a stone; (right) exterior view of stone; all at ×2 (Bonner and Rudolf 1974).

Appendix A

COMMON AND SCIENTIFIC NAMES OF PLANTS

Common Name	Scientific Name	Common Name	Scientific Name
Abelialeaf	*Abeliophyllum*	Green	*F. pennsylvanica*
Acacia	*Acacia*	Mountain	*Eucalyptus regnans*
Actinidia	*Actinidia*	Pumpkin	*Fraxinus profunda*
African locust bean	*Parkia clapertoniana*	Shamel	*F. uhdei*
Ailanthus	*Ailanthus*	Two-petal	*F. dipetala*
Akebia	*Akebia*	Velvet	*F. velutina*
Alamo	*Populus*	White	*F. americana*
Alamodela sierra	*Abies*	Aspen	*Populus*
Albizia	*Albizia*	Bigtooth	*P. grandidentata*
Alder	*Alnus*	European	*P. tremula*
American green	*A. crispa*	Quaking	*P. tremuloides*
European	*A. glutinosa*	Siebold	*P. sieboldiana*
European speckled	*A. incana*	Aucuba	*Aucuba*
Green	*A. viridis* ssp. *crispa*	Australian pine	*Casuarina*
Hazel	*A. serrulata*	Australian toon	*Toona australis*
Red	*A. rubra*	Baccharis	*Baccharis*
Sitka	*A. sinuata*	Dwarf	*B. pilularis* var. *pilularis*
Speckled	*A. rugosa*	Kidneywort	*B. pilularis* var. *consanguinea*
Thinleaf	*A. tenuifolia*	Mulefat	*B. viminea*
White	*A. rhombifolia*	Narrowleaf	*B. angustifolia*
Aliso	*Platanus*	Baldcypress	*Taxodium distichum* var. *distichum*
Alternate leaf butterfly bush	*Buddleia alternifolia*	Balsam	*Abies*
American beauty bush	*Callicarpa americana*	Baobab	*Adansonia*
American bittersweet	*Celastrus scandens*	Barberry	*Berberis*
American smoketree	*Cotinus obovatus*	Black	*B. gagnepanii*
American snowbell	*Styrax americanus*	Boxleaf	*B. boxifolia*
Amorpha	*Amorpha*	Cascades	*B. nervosa*
California	*A. californica*	creeping	*B. repens*
Dwarfindigo	*A. nana*	Cutleaf	*B. circumserrata*
Indigobush	*A. fruticosa*	Darwin	*B. darwinii*
Leadplant	*A. canescens*	European	*B. vulgatis*
Ampelopsis	*Ampelopsis*	Fremont	*B. fremontii*
Amur corktree	*Phellodendron amurense*	Japanese	*B. thunbergii*
Amur honeysuckle	*Lonicera maackii*	Korean	*B. koreana*
Antelope bitterbrush	*Purshia tridentata*	Nevin	*B. nevinii*
Anacua	*Ehretia accuminata*	Oregon-grape	*B. aquifolium*
Appalachian tea	*Ilex*	Red	*B. haematocarpus*
Apple	*Malus pumila*	Sargent	*B. sargentii*
Japanese flowering crab	*M. floribunda*	Threesine	*B. thriacanthopora*
Oregon crab	*M. diversifolia*	Verna	*B. vernae*
Prairie crab	*M. ioensis*	Warty	*B. verruculosa*
Siberian crab	*M. baccata*	Wildfire	*B. gilgiana*
Sweet crab	*M. coronaria*	Wintergreen	*B. julianae*
Apple gum	*Angophora*	Bargras	*Eucalyptus deglupta*
Apricot	*Prunus armeniaca*	Basswood	*Tilia*
Aralia	*Acanthopana*	Bay	*Umbellularia*
Arborvitae	*Thuja*	Bayberry	*Myrica*
Arizona ironwood	*Olneya*	Northern	*M. pensylvanica*
Arrowleaf	*Hofmeisteria pluriseta*	Pacific	*M. californica*
Arrowwood	*Holodiscus*	Southern	*M. cerifera*
Ash	*Fraxinus*	Bead tree	*Adenanthera*
Alpine	*Eucalyptus delagatensis*	Bearberry	*Rhamnus*
Black	*Fraxinus nigra*	Bearberry	*Arctostaphylos uva-ursi*
Blue	*F. quadrangulata*	Beardtongue	*Penstemon*
Carolina	*F. caroliniana*	Beargrass	*Yucca*
European	*F. excelsior*		
Flowering	*F. ornus*		

Common Name	Scientific Name	Common Name	Scientific Name
Bearmat	*Chamaebatia foliolosa*	Bottle tree	*Brachychiton*
Beauty bush	*Callicarpa*	Bower actinidia	*Actinidia arguta*
Beech	*Fagus*	Bowwood	*Maclura*
American	*F. grandifolia*	Box	*Buxus*
European	*F. sylvatica*	Boxthorn	*Lycium*
Beefwood	*Casuarina*	Breadfruit	*Artocarpus altilis*
Bent enkianthus	*Enkianthus deflexus*	Brickle bush	*Brickellia*
Big tree	*Sequoiadendron*	Bristly aralia	*Aralia hispida*
Bigleaf snowbell	*Styrax grandifolius*	Broom	*Genista*
Birch	*Betula*	Broom snakeweed	*Gutierrezia sarothrae*
Bog	*B. glandulosa*	Brushbox	*Tristania conferta*
Chinese paper	*B. albo-sinensis*	Buckeye	*Aesculus*
Dahurian	*B. davurica*	Arizona	*A. × bushii*
Erman	*B. etmanii*	California	*A. californica*
European white	*B. pendula*	Ohio	*A. glabra* var. *glabra*
Hairy	*B. pubescens*	Painted	*A. sylvatica*
Jacquemont	*B. jacquemontii*	Texas	*A. glabra* var. *arguta*
Japanese cherry	*B. grossa*	Red	*A. pavia*
Japanese white	*B. platyphylla* var. *japonica*	Yellow	*A. octandra*
Low	*B. pumila* var. *glandulifera*	Buckthorn	*Bumelia*
Monarch	*B. maximowiczina*	Buckthorn	*Rhamnus*
Paper	*B. payrifera*	Alder	*R. alnifolius*
River	*B. nigra*	California	*R. californica*
Sweet	*B. lenta*	Cascara	*R. purshiana*
Szechuanica white	*B. platyphylla* var. *szechuanica*	Dahurian	*R. davurica*
		European	*R. cathartica*
White	*B. pendula*	Glossy	*R. frangula*
Yellow	*B. alleghaniensis*	Hollyleaf	*R. crocea* var. *ilicifolia*
Birchleaf mountain mahogany	*Cercocarpus betuloides*	Redberry	*R. crocea*
		Buckwheat	*Eriogonum*
Birchleaf spirea	*Spiraea betulifolia* var. *lucida*	Budsage	*Artemisia spinescens*
Bitter nightshade	*Solanum duilcamara*	Buffaloberry	*Shepherdia*
Bitterbrush	*Purshia*	Bunchberry	*Cornus canadensis*
Bittersweet	*Celastrus*	Bunya-bunya	*Araucaria bidwilli*
Black alder	*Ilex*	Burning bush	*Euonymus*
Black butt	*Eucalyptus pilularis*	Bush honeysuckle	*Diervilla*
Black huckleberry	*Gaylussacia baccata*	Butterfly bush	*Buddleia*
Black snap	*Gaylussacia*	Alternate leaf	*B. alternifolia*
Black twinberry	*Lonicera involucrata*	Gum	*B. langinosa*
Blackbead	*Pithecellobium*	Butternut	*Juglans cinerea*
Blackberry	*Rubus*	Buttonball tree	*Platanus*
Allegheny	*R. allegheniensis*	Cabbage palmetto	*Sabal palmetto*
Cutleaf	*R. laciniatus*	California brickle bush	*Brickellia californica*
Himalaya	*R. procerus*	California buckwheat	*Eriogonum fasciculatum*
Smooth	*R. canadensis*	California fremontia	*Fremontodendron californicum*
Trailing	*R. macropetalis*	California torreya	*Torreya californica*
Blackbrush	*Coleogyne ramosissima*	California Washington palm	*Washingtonia filifera*
Blackhaw	*Viburnum prunifolium*	Camellia	*Camellia sinensis*
Black jetbead	*Rhodotypos scandens*	Camphor tree	*Cinnomomum camphor*
Blackthorn	*Prunus*	Candlenut	*Aleurites moluccana*
Blackwood	*Acacia melanoxylon*	Carob	*Ceratonia siliqua*
Bladder sage	*Salazaria*	Carolina allspice	*Calycanthus floridus*
Bladdernut	*Staphylea*	Carolina silverbell	*Halesia carolina* var. *carolina*
Blue gum	*Eucalyptus globulus*	Carpenteria	*Carpenteria californica*
Bluebeard	*Caryopteris incana*	Cashew	*Anacardium occidentalis*
Bluebeech	*Carpinus*	Catalpa	*Catalpa*
Blueberry	*Vaccinium*	Northern	*C. speciosa*
Box	*V. ovatum*	Southern	*C. bignoniodes*
Dwarf	*V. caespitosum*	Catawba rhododendron	*Rhododendron catawbiense*
Early	*V. ovalifolium*	Caucasian wingnut	*Pterocarya fraxinifolia*
Highbush	*V. corymbosum*	Ceanothus	*Ceanothus*
Lowbush	*V. angustifolium*	Buckbrush	*C. cuneatus*
Blueblossom	*Ceanothus thyrsiflorus*	Deerbrush	*C. integerrimus*
Bluebush	*Ceanothus*	Desert	*C. greggii*
Bluemist shrub	*Caryopteris × clandonensis*	Feltleaf	*C. arboreanus*
Bog kalmia	*Kalmia poliifolia*	Fendler	*C. fendleri*
Bog rosemary	*Andromeda polifolia*	Hoaryleaf	*C. crassifolius*

Common Name	Scientific Name
Inland	*C. ovatus*
Monterey	*C. rigidus*
Redstem	*C. sanguineus*
Santa Barbara	*C. impressus*
Snowbrush	*C. velutinus*
Trailing	*C. diversifolius*
Cedar	*Cedrus*
Alaska	*Chamaecyparis nootkatensis*
Atlantic white	*Chamaecyparis thyoides*
Atlas	*Cedrus atlantica*
Australian red	*Toona*
Cyprian	*Cedrus brevifolia*
Deodar	*Cedrus deodara*
Eastern red	*Juniperus virginiana*
Eastern white	*Thuja occidentalis*
Northern white	*Thuja occidentalis*
Port Orford	*Chamaecyparis lausoniana*
Southern red	*Juniperus silicicola*
West Indian	*Cederalla odorta*
Western red	*Thuja plicata*
Cedar	*Juniperus*
Chamise	*Adenostoma fasiculatum*
Chamiza	*Atriplex*
Checkerberry	*Gautheria procumbens*
Cherry	*Prunus*
Allegheny	*P. alleghaniensis*
Almond	*P. amygdalus*
American	*P. americana*
Bessey	*P. besseyi*
Bitter	*P. emarginata*
Black	*P. serotina*
Bullace	*P. insititia*
Cornealian	*Cornus mas*
European bird	*Prunus padus*
Hollyleaf	*P. ilicifolia*
Mahaleb	*P. mahaleb*
Manchu	*P. tomentosa*
Mazzard	*P. avium*
Pin	*P. pensylvanica*
Sand	*P. pumila*
Sour	*P. cerasus*
Cherry prinsepia	*Prinsepia sinensis*
Chestnut	*Castenea*
American	*C. dentata*
Chinese	*C. mollissima*
European	*C. sativa*
Japanese	*C. crenata*
China fir	*Cunninghamia*
Chinaberry	*Melia azedarach*
Chinese quince	*Chaenomeles sinensis*
Chinese peashrub	*Caragana sinica*
Chinese privet	*Ligustrum sinense*
Chinese redbud	*Cercis chinensis*
Chinese yew	*Cephalotaxus fortunii*
Chinkapin	*Castanopsis*
Chokeberry	*Aronia*
Black	*A. melanocarpa*
Purple	*A. prunifolia*
Red	*A. arbutifolia*
Choparosa	*Beloperone californica*
Christ thorn	*Ziziphus spina-christi*
Christmas berry	*Heteromeles arbutifolia*
Cinnamon	*Cinnamomum zeylanium*
Cinnamon clethra	*Clethra acuminata*
Cistus	*Cistus incanus*
Clematis	*Clematis*
Drummond	*C. drummondii*

Common Name	Scientific Name
Italian	*C. viticella*
Plume	*C. flammula*
Rock	*C. verticilaris*
Clethra	*Clethra*
Cliffrose	*Cowania mexicana* var. *stansburiana*
Coastal tea tree	*Leptospermum laevigatum*
Colonial pine	*Arzucaria*
Columnar araucaria	*Araucaria columnaris*
Common bladder senna	*Colutea arborescens*
Common bluebeard	*Caryopteris incuna*
Common boxwood	*Buxus sempervirens*
Common buttonbush	*Cephalanthus occidentalis*
Common crape myrtle	*Lagerstoemia indica*
Common custard apple	*Annona reticulata*
Common fig	*Ficus carica*
Common hoptree	*Ptelea trifoliata*
Common jujube	*Ziziphus jujba*
Common lavendar	*Lavandula angustifolia*
Common moonseed	*Menispermum canadense*
Common ninebark	*Physocarpus opulifolius*
Common olive	*Olea europaea*
Common pearl bush	*Exochorda racemosa*
Common persimmon	*Diospyros virginiana*
Common prickly ash	*Zanthoxylum americanum*
Common sea buckthorn	*Hippophae rhamnoides*
Common smoketree	*Cotinus coggygrie*
Common trumpet creeper	*Campsis radicans*
Common winterberry	*Ilex verticillata*
Cook's araucaria	*Araucaria columnaris*
Coralbean	*Erythrina flabelliformis*
Coral tree	*Erythrina*
Cornel	*Cornus*
Cotoneaster	*Cotoneaster*
Bearberry	*C. dammeri*
Black	*C. melanocarpa*
Brighthead	*C. glaucophyllus*
Cranberry	*C. apiculata*
Hedge	*C. lucida*
Many flowered	*C. multiflorus*
Peking	*C. acutifolia*
Pyrenees	*C. congestus*
Redbead	*C. recemiflorus*
Rock	*C. horizontalis*
Willowleaf	*C. salicifolius*
Wintergreen	*C. conspicuus*
Cottongum	*Nyssa*
Cottonwood	*Populus*
Black	*P. trichocarpa*
Eastern	*P. deltoides* var. *deltoides*
Fremont	*P. fremontii*
Lanceleaf	*P. acuminata*
Narrowleaf	*P. angustifolia*
Plains	*P. deltoides* var. *occidentalis*
Rio Grande	*P. fremontii* var. *wislizenii*
Swamp	*P. heterophylla*
Coyote bush	*Baccharis pilularis* var. *consanguinea*
Cranberry bush	*Viburnum opulus*
Creambush	*Holodiscus*
Creeper	*Parthenocissus*
Japanese	*P. tricuspidata*
Thicket	*P. inserta*
Virginia	*P. quinquefolia*
Creeping sage	*Salvia sonomensis*
Creeping snowberry	*Gaultheria hispidula*
Creosote bush	*Larrea tridentata*

Common Name	Scientific Name	Common Name	Scientific Name
Crescent tree	*Aporusa benthamina*	Roundleaf	*C. rugosa*
Cross vine	*Anisostichus capreolata*	Silky	*C. amomum*
Crossosoma	*Crossosoma*	Walter	*C. walteri*
Croton	*Croton*	Douglas-fir	*Pseudotsuga*
Cucumber tree	*Magnolia acuminata*	Bigcone	*P. macrocarpa*
Curlleaf mountain		Coast	*P. menziesii* var. *menziesii*
mahogany	*Cercocarpus ledifolius*	Gray	*P. menziesii* var. *caesia*
Currant	*Ribes*	Rocky Mountain	*P. menziesii* var. *glauca*
Alpine	*R. alpinum*	Dove tree	*Davidar involucarta*
American black	*R. americanum*	Downy andromeda	*Andromeda glaucophylla*
Clove	*R. odoratum*	Downy arrowwood	*Viburnum rafinesquianum*
Golden	*R. aureum*	Downy blueberry	*Vaccinium atrocaccum*
Gooseberry currant	*R. montigenum*	Downy rosemary	*Andromeda glaucophylla*
Hudson Bay	*R. hudsonianum*	Dunkeld larch	*Larix eurolepis*
Indian	*Symphoricarpos orbiculatus*	Dusty zenobia	*Zenobia pulverulenta*
Prickly	*Ribes lacustre*	Dutchman's pipe	*Aristolochia durior*
Sticky	*R. viscosissimum*	Dwarf apple gum	*Angophora cordifolia*
Wax	*R. cereum*	Dwarf bush honeysuckle	*Diervilla lonicera*
Winter	*R. sanguineum*	Dwarf mistletoe	*Arecuthobium*
Custard apple	*Annona*	Eaglewood	*Aquilaria agallocha*
Cutleaf stephanandra	*Stephanandra incisa*	Eastern hophornbean	*Ostrya virginiana*
Cypress	*Cupressus*	Eastern virgin-bower	*Clematis virginiana*
Arizona	*C. arizonica* var. *arizonica*	Eatern wahoo	*Euonymus atropurpureus*
Arizona smooth	*C. arizonica* var. *glabra*	Ebony blackbead	*Pithecellobium flexicaule*
Cuyamaca	*C. arizonica* var. *stephensonii*	Elder	*Sambucus*
		American	*S. canadensis*
Gowen	*C. goveniana* var. *goveniana*	Blueberry	*S. glauca*
Guadalupe	*C. guadalupensis* var. *guadalupensis*	European	*S. nigra*
		Pacific red	*S. callicarpa*
Italian	*C. sempervirens* var. *sempervirens*	Red	*S. racemosa*
		Scarlet	*S. pubens*
MacNab	*C. macnabiana*	Elm	*Ulmus*
Mendocino	*C. goveniana* var. *pygmaea*	American	*U. americana*
Mexican	*C. lusitanica*	Cedar	*U. crassifolia*
Modoc	*C. bakeri*	Chinese	*U. parvifolia*
Monterey	*C. macrocarpa*	Japanese	*U. japonica*
Piute	*C. arizonica* var. *nevadanensis*	Rock	*U. thomasii*
		Russian	*U. laevis*
San Pedro Martir	*C. arizonica* var. *montana*	Scotch	*U. glabra*
Santa Cruz	*C. goveniana* var. *abramsiana*	September	*U. serotina*
		Siberian	*U. pumila*
Spreading Italian	*C. sempervirens* var. *horizontalis*	Slippery	*U. rubra*
		Smooth-leaf	*U. carpinifolia*
Dake-momi	*Abies*	Winged	*U. alata*
Daminar pine	*Agathis*	Etonia palmetto	*Sabal etonia*
Daphne	*Daphne*	Eucalyptus	*Eucalyptus*
Dawn redwood	*Metasequoia glyptostroboides*	Alpine-ash	*E. delegatensis*
Deernut	*Simmondsia*	Brown barrel eucalyptus	*E. fastigata*
Desert bitterbrush	*Purshia glandulosa*	Mountain-ash	*E. regnans*
Desert eucalyptus	*Eucalyptus rudis*	Euonymus	*Euonymus*
Desert palm	*Washingtonia*	Brook	*E. americanus*
Desert trumpet	*Eriogonum inflatum*	European	*E. europaeus*
Desert willow	*Chilopsis linearis*	Maack	*E. maackii*
Devil's walkingstick	*Aralia spinosa*	Running	*E. obovatus*
Dogwood	*Cornus*	Warty-bark	*E. verrucosus*
Alternate-leaf	*C. alternifolia*	Winged	*E. alatus*
Bigleaf	*C. macrophylla*	Eurasian winterfat	*Ceratoides latens*
Blood twig	*C. sanguinea*	European filbert	*Corylus avellana*
California	*C. californica*	Fairy duster	*Calhandra eriophylla*
Flowering	*C. florida*	False indigo	*Amorpha*
Giant	*C. controversa*	Fanpalm	*Washingtonia*
Gray	*C. racemosa*	Fat hen	*Atriplex hasta*
Japanese	*C. kousa*	Fern bush	*Chamaebatiaria*
Japanese cornel	*C. officinalis*	Fever tree	*Acacia xanthophoea*
Pacific	*C. nuttallii*	Fig	*Ficus*
Red-osier	*C. stolonifera*	Fiji tennis-ball tree	*Agathis vintiensis*
Roughleaf	*C. drummondii*	Filbert	*Corylus*

Common Name	Scientific Name
Fir	*Abies*
Balsam	*A. balsamea*
Corkbark	*A. lasiocarpa* var. *arizonica*
European silver fir	*A. alba*
Japanese	*A. firma*
Maries'	*A. mariesii*
Noble	*A. procera*
Nordmann	*A. nordmanniana*
Pacific silver	*A. amabilis*
Red	*A. magnifica* var. *magnifica*
Sakhalin	*A. sachalinensis*
Shasta red	*A. magnifica* var. *shastensis*
Subalpine	*A. lasiocarpa*
Veitch	*A. veitchii*
White	*A. concolor*
Firethorn	*Pyracantha*
Fish poison tree	*Barringtonia asiatica*
Fish tail palm	*Caryota urens*
Fiveleaf aralia	*Acanthopanax sieboldianus*
Five-corner fruit	*Averrhoa*
Five-leaf akebia	*Akebia quinata*
Florida leucothoe	*Leucothoe populifolia*
Florida torreya	*Torreya taxifolia*
Flowering quince	*Chaenomeles*
Flowering willow	*Clilopsis*
Fly honeysuckle	*Lonicera canadensis*
Forsythia	*Forsythia*
Fortune fantanesia	*Fontanesia fortunei*
Fox grape	*Vitis labrusca*
Fragrant snowbell	*Styrax obassia*
Fragrant wintersweet	*Chimomonanthus praeco*
Frangipani	*Plumeria*
Fraser magnolia	*Magnolia fraseri*
Fremont silktassel	*Garrya fremontii*
Fremontia	*Fremontodendron*
Fringetree	*Chionanthus virginicus*
Fushia	*Fushia*
Gebang palm	*Corypha elata*
Georgia bush honeysuckle	*Diervilla rivularis*
Giant cactus	*Cereus*
Giant redwood	*Sequoiadendron giganteum*
Ginkgo	*Ginkgo biloba*
Glorybower	*Clerodendrum*
Glossy privet	*Ligustrum lucidum*
Goldenchain laburum	*Laburnum anagyroides*
Golden chinkapin	*Castanopsis chrysophylla*
Goldenlarch	*Pseudolarix kaempferi*
Goldenshower	*Cassia fistula*
Goldenhead	*Acamptopappas sphaerocephalus*
Gooseberry	*Ribes*
Idaho	*R. irriguum*
Missouri	*R. missouriense*
Pasture	*R. cynosbati*
Roundleaf	*R. rotundifolium*
Sierra	*R. nevadense*
White-stem	*R. inerme*
Gopher-wood tree	*Torreya*
Gorse	*Ulex*
Gray ironbark	*Eucalyptus paniculata*
Great Plains yucca	*Yucca glauca*
Ground hemlock	*Taxus*
Groundselbush	*Baccharis halimifolia*
Guayule	*Parthenium argentatum*
Gulf cypress	*Taxodium*
Hackberry	*Celtis*
Hackberry	*C. laevigata*

Common Name	Scientific Name
Jessco	*C. jessoenis*
Netleaf	*C. reticulata*
Hairy ceanothus	*Ceanothus oliganthus*
Hardy orange	*Poncirus trifoliata*
Harland boxwood	*Buxus harlandii*
Hawthorn	*Crataegus*
Arnold	*C. arnoldiana*
Black	*C. douglasii*
Cockspur	*C. crus-galli*
Dotted	*C. punctata*
English	*C. laevigata*
Fireberry	*C. chrysocarpa*
Fleshy	*C. succulenta*
Plum leaf	*C. prunifolia*
Redhaw	*C. sanguinea*
Singleleaf	*C. monogyna*
Washington	*C. phaenopyrum*
Hazel	*Corylus*
American	*C. americana*
Beaked	*C. cornuta* var. *cornuta*
California	*C. cornuta* var. *californica*
He-huckleberry	*Lyonia ligustrina*
Heather	*Calluna*
Hemlock	*Tsuga*
Carolina	*T. caroliniana*
Eastern	*T. canadensis*
Mountain	*T. mertensiana*
Western	*T. heterophylla*
Hemsley snowbell	*Styrax hemsleyana*
Henry's aralia	*Acanthopanax henryi*
Hercules club	*Zanthoxylum clava-herculis*
Hiba arborvitae	*Thujopsis dolabrata*
Hickory	*Carya*
Bitternut	*C. cordiformis*
Mockernut	*C. tomentosa*
Nutmeg	*C. myristicaeformis*
Pignut	*C. glabra*
Red	*C. ovalis*
Shagbark	*C. ovata*
Shellbark	*C. laciniosa*
Water	*C. aquatica*
Highbush cranberry	*Viburnum trilobum*
Hinds walnut	*Juglans hindsii*
Hobblebush	*Viburnum alnifolium*
Holly	*Ilex*
American	*I. opaca*
English	*I. aquifolium*
Honey mangrove	*Avicennia nitida*
Honey locust	*Gleditsia*
Honeysuckle	*Lonicera*
Amur	*L. maackii*
Coralline	*L. chrysantha*
Donald	*L. glaucescens*
Fly	*L. canadensis*
Hairy	*L. hirsuta*
Morrow	*L. morrowii*
Southern bush	*Diervilla sessifolia*
Swamp fly	*Lonicera oblongifolia*
Tatarian	*L. tatarica*
Hoop pine	*Araucaria cunninghamii*
Hopsage	*Grayia*
Horsetail tree	*Casuarina equisetifolia*
Hornbeam	*Carpinus*
American	*C. caroliniana*
European	*C. betulus*
Horsechestnut	*Aesculus*
Huckleberry	*Vaccinium*

Common Name	Scientific Name	Common Name	Scientific Name
Red	*V. parvifolium*	Lambkill	*Kalmia augustifolia*
Hydrangea	*Hydrangea*	Larch	*Larix*
Bigleaf	*H. macrophylla*	Dahurian	*L. gmelini*
Climbing	*H. anomala*	European	*L. decidua*
Oakleaf	*H. quercifolia*	Japanese	*L. leptolepis*
Panicle	*H. paniculata*	Siberian	*L. sibirica*
Illawaru flame	*Brachychiton acerifolius*	Subalpine	*L. lyallii*
Incense cedar	*Calocedrus decurrens*	Western	*L. occidentalis*
Indian arrowweed	*Philadelphus*	Late bush	*Ziziphus obtusifolia* var.
Indian bead	*Catalpa*		*obtusifolia*
Indigo bush	*Dalea*	Lavender	*Lavandula*
Inkberry	*Ilex glabra*	Lavender cotton	*Santolina*
Ipil-ipil	*Leucaena*	Leatherwood	*Cyrilla*
Ironwood	*Olneya*	Leatherwood	*Dirca*
Ironwood	*Carpinus*	Lemmon	*Citrus limon*
Ironwood	*Ostrya*	Lemon gum	*Eucalyptus citriodora*
Ironwood	*Metrosideros*	Lespedeza	*Lespedeza*
Ironwood	*Casuarina*	Japan	*L. japonica*
Jamaican plum	*Chrysophyllum oliviforme*	Shrub	*L. bicolor*
Japan wood-oil tree	*Aleurites cordata*	Thunberg	*L. thunbergii*
Japanese ardisia	*Ardisia japonica*	Leucothoe	*Leucothoe*
Japanese aucuba	*Aucuba japonica*	Lilac	*Syringa*
Japanese clethra	*Clethra barbinervis*	Amur	*S. amurensis*
Japanese cryptomeria	*Cryptomeria japonica*	Common	*S. vulgaris*
Japanese flowering quince	*Chaenomeles japonica*	Late	*S. villosa*
Japanese pachysandra	*Pachysandra terminalis*	Persian	*S. persica*
Japanese persimmon	*Diospyros kaki*	Lilac chastetree	*Vitex agnus-castus*
Japanese raisin tree	*Hovenia dulcis*	Lillypilly	*Acmena smithii*
Japanese snowbell	*Styrax japonicus*	Limber honeysuckle	*Lonicera dioica*
Japanese tree peony	*Paeonia suffruticosa*	Lime	*Citrus auratifolia*
Jasmine	*Jasminum*	Litchi	*Litchi chinesis*
Jelutong	*Dyera costulata*	Littleleaf boxwood	*Buxus mircophylla*
Jimbush	*Ceanothus sorediatus*	Littleleaf linden	*Tilia cordota*
Jojoba	*Simmondsia chinensis*	Littleleaf peashrub	*Caragana microphylla*
Judas tree	*Cercis siliquastrum*	Littleleaf snakeweed	*Gutierrezia mircocephala*
Jujube	*Ziziphus*	Lobiolly bay	*Gordoria*
Juniper	*Juniperus*	Locust	*Robinia*
Alligator	*J. deppeana*	Black	*R. pseudoacacia*
Ashe	*J. ashei*	Bristly	*R. fertilis*
California	*J. californica*	New-Mexican	*R. neomexicana*
Common	*J. communis*	Rose-acacia	*R. hispida*
Creeping	*J. horizontalis*	Longleaf casuarina	*Casuarina glauca*
Dahurian	*J. davuricaa*	Lupine	*Lupinus*
Greek	*J. excelsa*	Pauma	*L. longifolius*
Japanese garden	*J. procumbens*	Whiteface	*L. albifrons*
One-seeded	*J. monosperma*	Magnolia	*Magnolia*
Pinchot	*J. pinchotii*	Mahala mat	*Ceanothus*
Rocky Mountain	*J. scopulorum*	Malayan eaglewood	*Aquilaria maloccensis*
Sabin	*J. occidentalis*	Mallow ninebark	*Physocarpus malvaceus*
Western	*J. occidentalis*	Mamane	*Sophora chrysophylla*
Utah	*J. osteosperma*	Mango	*Manifera indica*
Kahili flower	*Grevillea banksii*	Mangrove	*Avicennia*
Kalopanax	*Kalopanax pictus*	Manna	*Eucalyptus viminalis*
Kapok tree	*Ceiba pentandra*	Manuka	*Leptospermum scoparium*
Karna orange	*Citrus karna*	Manzanita	*Arctostaphylos*
Karri	*Eucalyptus diversicolor*	Bigberry	*A. glauca*
Katsura tree	*Cercidiphyllum*	Eastwood	*A. glandulosa*
Kauri	*Agathis*	Greenleaf	*A. patula*
Kentucky coffee tree	*Gymnocladus dioicus*	Maple	*Acer*
Kidneywood	*Eysenhardtia*	Amur	*A. ginnala*
Kiwi	*Actinidia chinensis*	Bigleaf	*A. macrophyllum*
Koa	*Acacia koa*	Bigtooth	*A. grandidentatum*
Kobus magnolia	*Magnolia kobus*	Black	*A. nigrum*
Kochia	*Kochia*	Boxelder	*A. negundo*
Korean abelialeaf	*Abeliophyllum*	Chalkbark	*A. leucoderme*
Koyama spruce	*Picea koyamai*	David	*A. davidii*
Labrador tea	*Ledum groenlandicum*	Florida	*A. barbatum*

Common Name	Scientific Name	Common Name	Scientific Name
Fullmoon	*A. japonicum*	Bur oak	*Q. macrocarpa*
Hedge	*A. campestre*	California black	*Q. kelloggii*
Henry	*A. henryi*	California live	*Q. agrifolia*
Ivy-leaved	*A. cissifolium*	California scrub	*Q. dumosa*
Japanese	*A. palmatum*	Canyon live	*Q. chrysolepis*
Mancharian	*A. mandshurian*	Cherrybark	*Q. falcata*
Manchu-striped	*Tegmon rosum*	Chinkapin	*Q. muehlenbergii*
Miyabe	*Acer miyabei*	Cork	*Q. suber*
Mountain	*A. spicatum*	Durand	*Q. durandii*
Nikko	*A. maximowiczianum*	Durmast	*Q. petraea*
Norway	*A. platanoides*	English	*Q. robur*
Oregon vine	*A. circinatum*	European turkey	*Q. cerris*
Paperbark	*A. griseum*	Huckleberry	*Q. vaccinifolia*
Planetree	*A. pseudoplatanus*	Interior live	*Q. wislizenii*
Purplebloom	*A. pseudoseiboldianum*	Laurel	*Q. laurifolia*
Purpleblow	*A. troncatum* subsp. *mono*	Live	*Q. virginiana*
Red	*A. rubrum* var. *rubrum*	Northern pin	*Q. ellipsoidalis*
Red vein	*A. rufinerue*	Northern red	*Q. rubra*
Rocky Mountain	*A. glabrum*	Nuttal	*Q. nuttallii*
Seibold	*A. sieboldoanum*	Oregon white	*Q. garryana*
Silver	*A. saccarinum*	Oriental	*Q. variabilis*
Striped	*A. pensylvanicum*	Overcup	*Q. lyrata*
Sugar	*A. saccharum*	Post	*Q. stellata*
Three flower	*A. trifolium*	Pin	*Q. palustris*
Trident	*A. buergeranum*	Sawtooth	*Q. acutissima*
Vine	*A. circinatum*	Scarlet	*Q. coccinea*
Maple leaf viburnum	*Viburnum acerifolium*	Shingle	*Q. imbricaria*
Matrimony vine	*Lycium halimifolium*	Shrub live	*Q. turbinella*
Maximowicz peashrub	*Caragana maximowicziana*	Shumard	*Q. shumardii*
Mesquite	*Prosopis*	Southern red	*Q. falcata* var. *falcata*
Honey	*P. julifora* var. *glandulosa*	Swamp chestnut	*Q. michauxii*
Velvet	*P. julifora* var. *velutina*	Turkey	*Q. laevis*
Messmate eucalyptus	*Eucalyptus obliqua*	Valley	*Q. labata*
Mexican fremontia	*Fremontodendron mexicanum*	Water	*Q. nigra*
Mockorange	*Philadelphus lewisii*	White	*Q. alba*
Monkhead vine	*Ampelopsis acontifolia*	Willow	*Q. phellos*
Mountain ash	*Sorbus*	Oceanaspray	*Holodiscus discolor*
American	*S. americana*	Ohia	*Metrosideros polymorpha*
European	*S. aucuparia*	Oil palm	*Elaeis*
Showy	*S. decora*	Old-fashioned weigela	*Weigela florida*
Sitka	*S. sitchensis*	Olive	*Olea*
Mountain big sagebrush	*Artemisia tridentata* ssp. *vaseyana*	Orange	*Citrus sinensis*
		Orange-eye butterfly bush	*Buddleia davidii*
Mountain holly	*Nemopanthus mucronatus*	Oregon grape	*Berberis aquifolium*
Mountain laurel	*Kalmia*	Oriental arborvitae	*Thuja orientalis*
Mountain mahogany	*Cercocarpus montanus*	Oriental bittersweet	*Cedlastrus orbiculatus*
Mountain misery	*Chamaebatia*	Oriental plane tree	*Platanus orientalis*
Mountain whitethorn	*Ceanothus cordulatus*	Osage orange	*Maclura pomifera*
Mountain winterberry	*Ilex montana*	Osoberry	*Oemleria*
Mu-oil tree	*Aleurites montana*	Pachysandra	*Pachysandra*
Mulberry	*Morus*	Pacific madrone	*Arbutus menziesii*
Red	*M. rubra*	Pacific rhododendron	*Rhododendron macrophyllum*
Russian	*M. alba forma tatarica*	Paeonia	*Paeonia*
White	*M. alba*	Palmetto	*Sabal*
Multiflora rose	*Rosa multiflora*	Palo verde	*Cercidium floridum*
Myrtle beech	*Nothofagus cunninghamii*	Panicled golden raintree	*Koelreuteria paniculata*
Myrtlewood	*Umbellularia*	Papaya	*Carica*
Nannyberry	*Viburnum lentago*	Paper mulberry	*Broussonetia papyrifera*
New Jersey tea	*Ceanothus americanus*	Parana pine	*Araucaria angustifolia*
New Zealand kauri	*Agathis moeri*	Parish goldenweed	*Haplopappus parishii*
Ninebark	*Physocarpus*	Partridge pea	*Cassia fasciculata*
Norfolk Island pine	*Araucaria heterophylla*	Partridgeberry	*Mitchella repens*
Northern dewberry	*Rubus flagellaris*	Pawpaw	*Asimina triloba*
Oak	*Quercus*	Peach	*Prunus persica*
Bear	*Q. ilicifolia*	Peacock flower	*Adenanthera pavonina*
Black	*Q. velutina*	Pear	*Pyrus*
Blue	*Q. douglasii*	Common	*P. communis*
		Ussurian	*P. ussuriensis*

Common Name	Scientific Name	Common Name	Scientific Name
Pearl bush	*Exochorda*	South Florida slash	*P. elliottii* var. *densa*
Pecan	*Carya illinoensis*	Southwestern white	*P. strobiformis*
Penstemon	*Penstemon*	Spruce	*P. glabra*
Chaparral	*P. heterophyllus*	Sugar	*P. lambertiana*
Lemmons	*P. lemmonii*	Swiss Mountain	*P. mugo*
Thymeleaf	*P. corymbosus*	Swiss stone	*P. cembra*
Vine	*P. cordifolius*	Table-Mountain	*P. pungens*
Peperume	*Ampelopsis arborea*	Torrey	*P. torreyana*
Peppergrass	*Lepidium fremontii*	Virginia	*P. virginiana*
Persimmon	*Diospyros*	Western white	*P. monticola*
Phellodendron	*Phellodendron*	Whitebark	*P. albicaulis*
Pigmy cedar	*Peucephyllum scottii*	Pineapple guava	*Feijoa sellowiana*
Pincushion tree	*Hakea*	Pine-mat manzanita	*Arctostaphylos nevadensis*
Pine	*Pinus*	Pink lacebark	*Brachyctiton bidwilli*
Aleppo	*P. halepensis*	Pinxterbloom azalea	*Rhododendron nudiflorum*
Apache	*P. engelmannii*	Pistachio	*Pistacia*
Arizona	*P. ponderosa* var. *arizonica*	Plum	*Prunus*
Armand	*P. armandii*	Allegheny	*P. alleghaniensis*
Austrian	*P. nigra*	American	*P. americana*
Balkan	*P. peuce*	Bullace	*P. insititia*
Bishop	*P. muricata*	Chickasaw	*P. angustifolia*
Blue	*P. wallichiana*	Garden	*P. domestica*
Calabrian	*P. brutia*	Klamath	*P. subcordata*
Canary Island	*P. canariensis*	Myrobalan	*P. cerasifera*
Chihuahua	*P. leiophylla* var. *chihuahuana*	Plum yew	*Cephalotaxus*
		Podocarpus	*Podocarpus*
Chir	*P. roxburghii*	Poison oak	*Rhus diversiloba*
Coulter	*P. coulteri*	Pomegranate	*Punica*
Chilgoza	*P. gerardiana*	Pond apple	*Annona glabra*
Digger	*P. sabiniana*	Pondcypress	*Taxodium distichum* var. *nutans*
Eastern white	*P. strobus*	Poplar	*Populus*
Foxtail	*P. balfouriana*	Balsam	*P. balsamifera*
Heldreich	*P. heldreichii*	Black	*P. nigra*
Italian stone	*P. pinea*	Euphrates	*P. euphratica*
Jack	*P. banksiana*	Gray	*P. canescens*
Japanese black	*P. thunbergiana*	Japanese	*P. maximowiczii*
Japanese red	*P. densiflora*	Laurel	*P. laurifolia*
Japanese stone	*P. pumila*	Petrowsky	*P. petrowskyana*
Japanese white	*P. parviflora*	Simon	*P. simonii*
Jeffrey	*P. jeffreyi*	White	*P. alba*
Khasi	*P. insularis*	Porcelain ampelopsis	*Ampelopsis brevipedunculata*
Knobcone	*P. attenuata*	Prickly ash	*Zanthoxylum*
Korean	*P. koraiensis*	Prince's plume	*Stanleya*
Limber	*P. flexilis*	Privet	*Ligustrum*
Loblolly	*P. taeda*	Amur	*L. amurense*
Longleaf	*P. palustris*	European	*L. vulgare*
Merkus	*P. merkusii*	Glossy	*L. lucidum*
Mexican pinyon	*P. cembroides*	Japanese	*L. japonicum*
Mexican weeping	*P. patula*	Puna	*Ehretia acuminata*
Monterey	*P. radiata*	Pygmy peashrub	*Caragana pygmaea*
Parry pinyon	*P. quadrifolia*	Quailbush	*Atriplex lentiformis*
Pinyon	*P. edulis*	Queensland auri	*Agathis robusta*
Pitch	*P. rigida*	Queensland bottle tree	*Brachychiton rupestre*
Pond	*P. serotina*	Queensland maple	*Flindersia brayleyana*
Ponderosa	*P. ponderosa*	Rabbitbrush	*Chrysothamnus*
Rocky Mountain lodgepole	*P. contorta* var. *scopalorum*	Gray	*C. viscidiflorus*
		Green	*C. nauseosus*
Rocky Mountain ponderosa	*P. ponderosa* var. *scopulorum*	Raintree blackbead	*Pithecellobium samam*
		Rangpur lime	*Citrus limonia*
Sand	*P. clausa*	Raspberry	*Rubus*
Scotch	*P. sylvestris*	Blackcap	*R. occidentalis*
Shore	*P. contorta*	Red	*R. idaeus*
Shortleaf	*P. echinata*	Rattan	*Calamus*
Siberian	*P. sibirica*	Red ironbark	*Eucalyptus siderozylen*
Sierra Nevada lodgepole	*P. contorta* var. *murrayana*	Red molly	*Kochia americana*
Singleleaf pinyon	*P. monophylla*	Red sandlewood	*Adenanthera pavonia*
Slash	*P. elliottii*	Red shank	*Adenostoma sparsifolium*

Common Name	Scientific Name
Red silk cotton tree	*Bombax malabaricum*
Red vein enkianthus	*Enkianthus campanulatus*
Redbud	*Cercis*
California	*C. occidentalis*
Eastern	*C. canadensis*
Redbud pearl bush	*Exochorda giraldi*
Redwood	*Sequoia sempervirens*
Rhododendron	*Rhododendron*
River bank grape	*Vitis riparia*
River gum	*Eucalyptus camaldulensis*
River oak casuarina	*Casuarina cunninghamiana*
Ropevine	*Clematis panciflora*
Rose	*Rosa*
Baldhip	*R. gymnocarpa*
Dog	*R. canina*
Japanese	*R. multiflora*
Meadow	*R. blanda*
Nootka	*R. nutkana*
Prairie	*R. setigera*
Sweetbrier	*R. eglantaria*
Wichura	*R. wichuraiana*
Woods	*R. woodsii*
Rose gum	*Eucalyptus grandis*
Rose of Sharon	*Hibiscus syriacus*
Rosebay rhododendron	*Rhododendron maximum*
Rose mallow	*Hibiscus*
Rosemary	*Rosmarinus officinalis*
Rough menodora	*Menodora scabra*
Royal paulowina	*Paulownia tomentosa*
Rubber	*Hevea brasiliensis*
Russet buffaloberry	*Shepherdia canadensis*
Russian olive	*Elaeagnus angustifolia*
Russian peashrub	*Caragana frutex*
Sage	*Salvia*
Sagebrush	*Artemisia*
Basin big	*A. tridentata* subsp. *tridentata*
Black	*A. arbuscula* var. *nova*
Pygmy	*A. pygmeae*
Saguaro	*Cereus giganteus*
Salal	*Gaultheria shallon*
Salmon berry	*Rhus spectabilis*
Saltbush	*Atriplex*
Cattle	*A. polycarpa*
Fourwing	*A. canescens*
Gardner	*A. gardnerii*
Nuttall	*A. nuttallii*
Trailing	*A. semibaccata*
Sand myrtle	*Leiophyllum*
Sandalwood	*Santalum*
Sandhill kalmia	*Kalmia hirsuta*
Sassafras	*Sassafras albidum*
Saw-palmetto	*Serenoa repens*
Scotch broom	*Cytisus scoparius*
Sea grape	*Coccoloba uvifera*
Seepweed	*Suaeda*
Senna	*Cassia*
Serviceberry	*Amelanchier*
Allegheny	*A. laevis*
Downy	*A. arborea*
Pacific	*A. florida*
Roundleaf	*A. sanguinea*
Saskatoon	*A. alnifolia*
Thicket	*A. canadensis*
Sesbania	*Sesbania*
Shadscale	*Atriplex confertifolia*
Shining eucalyptus	*Eucalyptus nitens*

Common Name	Scientific Name
Siberian peashrub	*Caragana arborescens*
Sierra chinkapin	*Castanopsis sempervirens*
Silk cotton tree	*Bombax*
Silktree	*Albizia julibrissin*
Silk-oak	*Grevillea robusta*
Silktassel	*Garrya*
Silver vine	*Actinidia polygama*
Silverberry	*Elaeagnus commutata*
Siris	*Albizia lebbek*
Skunkbush	*Rhus trilobata*
Sloe	*Prunus spinosa*
Small flower pawpaw	*Asimina parviflora*
Smokebush	*Dalea polyadenia*
Smoketree	*Cotinus*
Smooth arrowwood	*Viburnum recognitum*
Smoothbarked apple gum	*Angophora costata*
Snakeweed	*Gutierrezia*
Snowbell	*Styrax*
Snowberry	*Symphoricarpos*
Common	*S. albus* var. *albus*
Garden	*S. albus* var. *laevigatus*
Indian	*S. orbiculatus*
Western	*S. occidentalis*
Snowbush ceanothus	*Ceanothus velutinus*
Soaptree yucca	*Yucca elata*
Sour orange	*Citrus aurantin*
Sourwood	*Oxydendrum arboreum*
Southern arrowwood	*Viburnum dentatum*
Southern beech	*Nothofagus*
Southern magnolia	*Magnolia grandiflora*
Spicebush	*Lindera benzoin*
Spineless hopsage	*Grayia brandegei*
Spiny hackberry	*Celtis pallida*
Spiny hopsage	*Grayia spinosa*
Spruce	*Picea*
Black	*P. mariana*
Black Hills white	*P. glauca* var. *glauca*
Blue	*P. pungens*
Brewer	*P. breweriana*
Dragon	*P. asperata*
Engelmann	*P. engelmannii*
Himalayan	*P. smithiana*
Koyama	*P. koyamai*
Norway	*P. abies*
Oriental	*P. orientalis*
Red	*P. rubens*
Sakhalin	*P. glehnii*
Serbian	*P. omorika*
Sitka	*P. sitchensis*
Tigertail	*P. polita*
Western white	*P. glauca* var. *albertiana*
white	*P. glauca*
Yeddo	*P. jezoensis*
Squaw carpet	*Ceanothus prostratus*
Squawapple	*Peraphyllum ramosissimum*
Staghorn sumac	*Rhus parviflora*
Star apple	*Chrysophyllum cainito*
Stiff bushpoppy	*Dendromecon involucrata*
Sugar apple	*Annona squamosa*
Sugarberry	*Celtis laevigata*
Suina	*Erythrina speciosa*
Sulphur flower	*Eriogonum umbellatum*
Sumac	*Rhus*
Fragrant	*R. aromatica*
Laurel	*R. laurina*
Lemonade	*R. integrifolia*
Shining	*R. copalina*

Common Name	Scientific Name	Common Name	Scientific Name
Smooth	*R. glabra*	Siebold	*J. ailantifolia*
Staghorn	*R. typhina*	Water locust	*Gleditsia aquatica*
Sugar	*R. ovata*	Wattle	*Acacia*
Summer sweet	*Clethra alnifolia*	Black	*A. mearnsii*
Swamp dewberry	*Rubus hispidus*	Golden	*A. longifolia*
Swamp white oak	*Quercus bicolor*	Green	*A. decurrens*
Sweet box	*Sarcococca*	Wayfaring tree	*Viburnum lantana*
Sweet fern	*Comptonia peregrina*	Western honey mesquite	*Prosopis julifora* var. *torreyana*
Sweet-scented shrub	*Calycanthus*	Western soapberry	*Spindus drummondii*
Sweetbay	*Magnolia virginiana*	Western virgin-bower	*Clematis ligusticifolia*
Sweetgale	*Myrica gale*	White cedar	*Chamaecyparis*
Sweetgum	*Liquidambar styraciflua*	White cypress	*Callitris glauca*
Sycamore	*Platanus*	White enkianthus	*Enkianthus perulatus*
American	*P. occidentalis*	White lacebark	*Brachychiton discolor*
California	*P. racemosa*	White mangrove	*Avicennia marina*
Wright	*P. wrightii*	White sage	*Ceratoides*
Talipot palm	*Corypha umbraculifera*	White sandwood	*Santalum albium*
Tallowtree	*Sapium sebiferum*	Whitewicky kalmia	*Kalmia cuneata*
Tallow wood	*Eucalyptus microcorys*	Wild allspice	*Lindera*
Tamarack	*Larix laricina*	Wild sarsaparilla	*Aralia nudicaulis*
Tamarisk	*Tamarix pentandra*	Wild senna	*Cassia marilandica*
Tanoak	*Lithocarpus densiflorus*	Willow	*Salix*
Tatarian	*Cornus alba*	Bebb	*S. bebbiana*
Tea	*Adinandra acuminata*	Black	*S. nigra*
Tea tree	*Leptospermum*	Coastal Plain	*S. caroliniana*
Teak	*Tectona grandis*	Coyote	*S. exigua*
Tesota	*Olneya tesota*	Crack	*S. fragilis*
Texas honey locust	*Gleditsia* × *texana*	Meadow	*S. petiolaris*
Texas locust	*Gleditsia*	Missouri River	*S. rigida*
Texas persimmon	*Diospyros texana*	Pacific	*S. lasiandra*
Threadleaf snakeweed	*Gutierrezia microcephala*	Peachleaf	*S. amygdaloides*
Three-leaf akebia	*Akebia trifoliata*	Pussy	*S. discolor*
Tingiringy gum	*Eucalyptus glaucescens*	Sandbar	*S. interior*
Toddy palm	*Caryota urens*	Scouler	*S. scouleriana*
Toon	*Cederalla sinenis*	Winterberry	*Euonymus bungeanus*
Trailing arbutus	*Epigaea*	Winterfat	*Ceratoides lanata*
Travellersjoy	*Clematis vitalba*	Wintergreen	*Gaultheria*
Tree of heaven	*Ailanthus altissima*	Winter hazel	*Corylopsis*
Trifoliate orange	*Citrus trifoliate*	Wintersweet	*Chimonanthus*
True cedar	*Cedrus*	Wisteria	*Wisteria*
Trumpet tree	*tabebebuia*	Witch-hazel	*Hamamelis virginiana*
Tukas	*Caryota mitis*	Witherod viburnum	*Viburnum cassinoides*
Tung-oil tree	*Aleurites fordii*	Wolfberry	*Lycium*
Tupelo	*Nyssa*	Anderson	*L. andersoni*
Black	*N. sylvatica* var. *sylvatica*	Chinese	*L. chinense*
Ogeechee	*N. ogeche*	Rich	*L. richii*
Swamp	*N. sylvatica* var. *biflora*	Wooly nama	*Nama lobbii*
Water	*N. aquatica*	Wright silktassel	*Garrya wrightii*
Umbrella pine	*Sciadopitys verticillata*	Yaupon	*Ilex vomitoria*
Ural falsespiraea	*Sorbaria sorbifolia*	Yellow poplar	*Liriodendron tulipifera*
Viburnum	*Viburnum*	Yellow silktassel	*Garrya flavenscens*
Virgin river encelia	*Encelia virginensis*	Yellowthorn	*Xanthoceras scorbifolia*
Virginia creeper	*Parthenocissus quinquefolia*	Yellow wood	*Clandrastis lutea*
Virginia sweetspire	*Itea illicifolia*	Yew	*Taxus*
Wall germander	*Teucrium chamaedrys*	Canada	*T. canadensis*
Walnut	*Juglans*	English	*T. baccata*
Arizona	*J. major*	Japanese	*T. cuspidata*
Black	*J. nigra*	Pacific	*T. brevifolia*
California	*J. californica*	Yucca	*Yucca*
Little	*J. microcarpa*	Zigzag bush	*Eriogonum heermannii*
Persian	*J. regia*		

Appendix B
GERMINATION CHARACTERISTICS BY GENERA

As noted in the Introduction, certain types of germination characteristics are associated with each plant family. Appendix B provides an alphabetical listing of plant families, the genera within those families, and a generalized statement on the germination characteristics of each genus. Germination characteristics can be summarized as follows:

Afterripening required = Period of time after harvest, relatively independent of external environment, must elapse before seeds will germinate.

Dormant = Seeds will not germinate, for unknown reasons.

Double dormant = Two types of dormancy present in seeds, such as hard seed coat and embryo dormancy.

Germinable = Seeds require no pretreatment for germination greater than 25%.

Light required = Illuminance from cool white fluorescent source required for seeds to germinate. Sometimes the duration of light in each 24-hour period is critical (photoperiod) and sometimes the quality of light is important.

Mechanical restriction = Seed or fruit coat restricts emergence of radicle.

Moderately germinable = Seeds require no pretreatment for germination greater than 5%, but not more than 25%.

Prechilling required = Period of cool-moist incubation required before more than incidental germination occurs.

Recalcitrant = Seeds cannot be dried and they have short storage life.

Scarification required = Mechanical, thermal, or chemical rupturing of seedcoat required for germination.

Family	Genus	Germination characteristics
Acanthaceae	Beloperone	Germinable
Actinidiaceae	Actinidia	Prechilling required
Anacardiaceae	Anacardium	Slow moisture imbibition
	Bouea	Germinable
	Cotinus	Prechilling and scarification required
	Mangifera	Germinable, recalcitrant
	Pistacea	Scarification required
	Rhus	Scarification required
Annonaceae	Annona	Prechilling required
	Asimina	Scarification required
Apocynaceae	Dyera	Germinable
	Plumeria	Germinable
Aquifoliaceae	Ilex	Double dormant
	Nemopanthus	Afterripening required
Araliaceae	Aralia	Double dormant
	Fatsia	Germinable
	Kalopanax	Double dormant
Araucariaceae	Agathis	Germinable, recalcitrant
	Araucaria	Germinable, recalcitrant
Aristolochiaceae	Aristolochia	Prechilling required
Agauaceae	Yucca	Germinable
Barringtoniaceae	Barringtonia	Germinable
Betulaceae	Alnus	Mild prechilling required, light sensitive?
	Betula	Prechilling or light required
	Carpinus	Prechilling required
	Corylus	Fresh seeds germinable
	Ostrya	Double dormant

Family	Genus	Germination characteristics
Bignoniaceae	Anisostichus	Germinable
	Campsis	Prechilling required
	Catalpa	Germinable
	Chilopsis	Germinable
	Paulownia	Prechilling required, light?
	Tabebuia	—
Bombacaceae	Bombax	Germinable
	Ceiba	—
Boraginaceae	Cordia	Prechilling required
	Ehretia	Scarification required
Burseraceae	Commiphera	Germinable, dimorphic
	Ehretia	Scarification required
Buxaceae	Buxus	Germinable
	Pachysandra	Scarification enhances germination
	Sarcococca	Germinable
	Simmondsia	Germinable
Cactaceae	Cereus	Light required
Calycanthaceae	Calycanthus	Mature seeds have hard seed coats
	Chimonanthus	Scarification required for mature seeds
Caprifoliaceae	Diervilla	Germinable
	Lonicera	Prechilling required
	Sambucus	Double dormant
	Symphoricarpus	Double dormant
	Vibernum	Double dormant
	Weigelia	Germinable
Caricaceae	Carica	—
Casuarinaceae	Casuaria	Germinable
Celastraceae	Celastrus	Germinable
	Euonymus	Prechilling required
	Tripterygium	Prechilling required
Cephalotaxaceae	Cephalotaxus	Prolonged prechilling required

Family	Genus	Germination characteristics
Cercidiphyllaceae	*Cercidiphyllum*	Germinable
Chenopodiaceae	*Atriplex*	Prechilling required
	Ceratoides	Germinable
	Grayia	Germinable
	Haloxylon	Moderately germinable
	Kochia	Germinable
	Salsola	Germinable
	Suaeda	Dormant
Cistaceae	*Cistus*	Hot water treatment, fire related
Compositae	*Acamptopappus*	Dormant
	Artemisia	Germinable
	Baccharis	Germinable?
	Brickellia	Germinable
	Callistris	—
	Chrysothamnus	Moderately germinable
	Dyssodia	Afterripening required
	Eastwoodia	Germinable
	Encelia	Germinable
	Gutierrezia	Afterripening required
	Haplopappus	Moderately germinable
	Hofmeisteria	Afterripening required
	Parthenium	Afterripening required
	Peucephyllum	Afterripening required
	Porophyllum	Dormant
	Psilostrophe	Germinable
	Santolina	Germinable
	Stephanomeria	Afterripening required
	Tetradymia	Prechilling required
Cornaceae	*Aucuba*	Prechilling required
	Clethra	Germinable
	Cornus	Scarification and prechilling required
	Nyssa	Prechilling required
Crossosomataceae	*Crossosoma*	Germinable
Cruciferae	*Lepidium*	Dormant
	Stanleya	Germinable
Cupressaceae	*Callitris*	Moderately germinable
	Calocedrus	Germinable
	Chamaecyparis	Germinable
	Cupressus	Prechilling required
	Juniperus	Prolonged prechilling required
	Thuja	Germinable
	Thujopsis	Germinable
Cyrillaceae	*Cyrilla*	Germinable
Dipterocarpaceae	*Anisoptera*	Germinable
	Dipterocarpus	Germinable, recalcitrant
	Hopea	Germinable
	Shorea	Germinable, recalcitrant
Ebenaceae	*Diospyros*	Mechanical restriction
Elaeagncaeae	*Elaeagnus*	Prechilling required
	Hippophae	Germinable
	Shepherdia	Double dormant
Ephedraceae	*Ephedra*	Germinable
Ericaceae	*Andromeda*	Germinable, very small
	Arbutus	Prechilling required
	Arctostaphylos	Heat treatment, scarification required

Family	Genus	Germination characteristics
[Ericaceae]	*Calluna*	Germinable, very small
	Enkianthus	Germinable, very small
	Epigaea	Prechilling enhances germination
	Erica	Germinable, very small
	Gaultheria	Prechilling required?
	Gaylussacia	Prechilling enhances germination
	Kalmia	Prechilling enhances germination
	Ledum	Germinable, very small
	Leiophyllum	Germinable, light required
	Leucothoe	Germinable, light required
	Lyonia	Germinable
	Oxydendrum	Germinable, light required?
	Rhododendron	Germinable, light required
	Vaccinium	Light and fluctuating temperatures required
	Zenobia	Germinable
Eucommiaceae	*Eucommia*	Dry seeds require prechilling
Eupherbraceae	*Aleurites*	Sacrification required
	Aporusa	Germinable
	Baccaurea	Germinable
	Croton	Germinable
	Hevea	Germinable, recalcitrant
	Sapium	Germinable
	Tetracoccus	Germinable
Fagaceae	*Castanea*	Prechilling required
	Castanopsis	Moderately germinable
	Fagus	Prechilling required
	Lithocarpus	Germinable
	Nothofagus	Prechilling required?
	Quercus	White oak germinable, black oak requires prechilling
Flacourtiaceae	*Idesia*	Germinable
Garryaceae	*Garrya*	Prechilling required
Ginkgoaceae	*Ginkgo*	Germinable
Grossulariaceae	*Ribes*	Double dormant
Guttiferae	*Calophyllum*	Seeds very slowly germinable
Hamamelidaceae	*Corylopsis*	Double dormant
	Fothergilla	Double dormant
	Hamamelis	Double dormant
	Liquidambar	Prechilling enhances germination
	Loropetalum	Double dormant
Hippocastanaceae	*Aesculus*	Prechilling required, recalcitrant
Hydrangeaceae	*Philadelphus*	Prechilling required
Hydrophyllaceae	*Nama*	Leaching and gibberellin enrichment required
Illiciaceae	*Illicium*	Germinable
Juglandaceae	*Carya*	Prechilling required
	Juglans	Prechilling required
	Pterocarya	Prechilling required
Labiatae	*Elsholtzia*	Germinable

Family	Genus	Germination characteristics	Family	Genus	Germination characteristics
[Labiatae]	*Lavandula*	Germinable	Magnoliaceae	*Liriodendron*	Prechilling required
	Rosmarinus	Germinable		*Magnolia*	Variable dormancy
	Salazaria	Germinable	Malvaceae	*Hibiscus*	Germinable
	Salvia	Prechilling required	Melastomataceae	*Schizocentron*	Prechilling enhances germination
	Teucrium	Germinable			
Lardizabalaceae	*Akebia*	Prechilling required	Meliaceae	*Aglaia*	Germinable, recalcitrant?
Lauraceae	*Cinnamomum*	Light required?			
	Lindera	Double dormant		*Cederella*	Germinable
	Sassafras	Prolonged prechilling required		*Melia*	Germinable
				Toona	Germinable
	Umbellularia	Germinable	Menispermaceae	*Menispermum*	Prechilling required
Lecythidaceae	*Gustavia*	Germinable	Moraceae	*Broussonetia*	Germinable
Leguminosae	*Acacia*	Scarification required		*Cecropia*	Light required
	Adenanthera	Stored seeds have hard coats		*Ficus*	Light required?
				Maclura	Prechilling required
	Albrizia	Scarification required		*Morus*	Light required
	Amorpha	Fresh seeds germinable	Myriaceae	*Comptonia*	Scarification required
				Myrica	Prechilling required
	Calandra	Scarification required	Myrtaceae	*Amomyrtus*	Photoperiod sensitive
	Caragana	Prechilling enhances germination		*Eucalyptus*	Most germinable, light required, occasionally prechilling enhances germination
	Cassia	Scarification required			
	Ceratonia	Scarification required for fresh seeds			
	Cercis	Scarification required		*Feijoa*	Germinable
	Cercidium	Germinable		*Leptospermum*	Germinable
	Cladrastis	Scarification and prechilling required		*Metrosideros*	Light enhances germination
	Colutea	Scarification required	Myrsinaceae	*Ardisia*	Germinable
	Copaifera	Scarification required	Nyssaceae	*Davidia*	Mechanical restriction, embryo dormancy
	Cytisus	Scarification required			
	Dalea	Scarification required			
	Erythrina	Scarification required	Oleaceae	*Abeliophyllum*	Afterripening required
	Erythrophleum	Scarification required		*Chionanthus*	Double dormant
	Eucommia	Fresh seeds germinable		*Fontanesia*	Dry seeds require prechilling, fresh seeds germinable
	Eysenhardtia	Germinable			
	Genista	Scarification required		*Forsythia*	Prechilling required?
	Gleditsia	Scarification required		*Fraxinus*	Double dormant
	Gymnocladus	Scarification required		*Jasmimum*	Germinable
	Hartwickia	Germinable, dimorphic		*Ligustrum*	Prechilling required
				Menodora	Germinable
	Indigofera	Scarification required		*Olea*	Scarification required, light
	Laburnum	Scarification required			
	Lespedeza	Scarification required		*Syringa*	Prechilling required
	Leucaena	Scarification required	Onagraceae	*Fuchsia*	Germinable, light required
	Lupinus	Scarification required			
	Maackia	Scarification required	Palmaceae	*Butia*	Germination very slow
	Olneya	Fresh seeds germinable		*Calamus*	Germination very slow
				Caryota	Difficult to germinate
	Parkia	Scarification required, dimorphic		*Corypha*	Seeds slowly germinable
	Parkinsonia	Scarification required		*Elaeis*	Dormant
	Pithecellobium	Scarification enhances germination		*Sabal*	Germination very slow or seeds dormant
	Prosopis	Scarification required		*Serenoa*	Mechanical restriction
	Robinia	Scarification required		*Washingtonia*	Germinable
	Sesbania	Germinable	Papaveraceae	*Dendromecon*	Moderately germinable
	Sophora	Scarification required			
	Ulex	Scarification required	Pinaceae	*Cedrus*	Germinable
	Wisteria	Germinable		*Larix*	Germinable
Loganiaceae	*Buddleia*	Germinable		*Picea*	Germinable
Loasaceae	*Petalonyx*	Moderately germinable		*Pinus*	Germinable
				Pseudolarix	Prechilling enhances germination
Loranthaceae	*Arceuthobium*	Prechilling required?			
Lythraceae	*Lagerstroemia*	Germinable			

Family	Genus	Germination characteristics	Family	Genus	Germination characteristics
[Pinaceae]	*Pseudotsuga*	Prechilling enhances germination, light?	Rutaceae	*Citrus*	Scarification required
				Flindersia	Germinable
	Tsuga	Prechilling enhances germination		*Phellodendron*	Prechilling enhances germination
Platanaceae	*Platanus*	Germinable		*Pomcirus*	Prechilling required
Podocarpaceae	*Podocarpus*	Dormant		*Ptelea*	Prechilling enhances germination
Polygonaceae	*Coccoloba*	—		*Skimmia*	Fresh seeds germinable
	Erigonoum	Germinable?			
Proteaceae	*Grevillea*	Prechilling required		*Zanthoxylum*	Double dormant
	Hakea	Germinable, fire related	Salicaceae	*Populus*	Germinable, light required
Punicaceae	*Punica*	Germinable		*Salix*	Germinable, recalcitrant
Ranunculaceae	*Clematis*	Prechilling required	Santalaceae	*Santalum*	Scarification required
	Paeonia	Double dormant	Sapindaceae	*Koelreuteria*	Scarification required
Rhamnaceae	*Ceanothus*	Hot water treatment required		*Litchi*	Germinable, recalcitrant
	Hovenia	Scarification required		*Sapindus*	Scarification required
Rhamnus	*Double*	dormant?		*Xanthoceras*	Germinable
	Ziziphus	Double dormant	Sapotaceae	*Bumelia*	Scarification required
Rhizophoraceae	*Carallia*	Germinable, recalcitrant		*Chrysephyllum*	—
			Santalaceae	*Santalon*	—
Rosaceae	*Adenostoma*	Scarification required, fire related	Saxifragaceae	*Carpenteria*	Germinable
	Amelanchier	Prechilling required		*Deutzia*	Germinable
	Aronia	Prechilling required		*Itea*	Germinable
	Cerocarpus	Prolonged prechilling or hydrogen peroxide required		*Hydrangea*	Germinable
			Scrophulariaceae	*Penstemon*	Germinable, light required
	Chaenomeles	Prechilling required			
	Chamaebatia	Germinable	Simarubaceae	*Ailanthus*	Germinable
	Chamaehatatia	Prechilling required		*Brucea*	Seeds slowly germinable
	Coleogyne	Prechilling required			
	Cotoneaster	Scarification and prechilling required	Solanaceae	*Lycium*	Prechilling required?
				Solanum	Germinable
	Cowania	Prechilling required	Staphyleaceae	*Staphylea*	Scarification required
	Cratgaeus	Scarification and prechilling required	Sterculiaceae	*Brachychiton*	Germinable
				Fremontodendron	Hot water and prechilling required
	Eriobotrya	Germinable			
	Exochorda	Prechilling required	Styraceae	*Halesia*	Double dormant
	Fallugia	Germinable		*Styrax*	Double dormant
	Heteromeles	Germinable	Tamaricaceae	*Tamarix*	Germinable
	Holodiscus	Prechilling required	Taxaceae	*Taxus*	Double dormant
	Malus	Prechilling required	Taxodiaceae	*Cryptomeria*	Prechilling required
	Oemleria	Prolonged prechilling required		*Cunninghamia*	Germinable
				Metasequoia	Germinable
	Peraphyllum	Prechilling required		*Sciadopitys*	Double dormant
	Physocarpus	Prechilling required		*Sequoia*	Germinable
	Prinsepia	Prechilling required?		*Sequoiadendron*	Germinable
	Prunus	Prechilling required		*Taxodium*	Prechilling required
	Purshia	Prechilling required	Theaceae	*Adinandra*	Germinable
	Pyracantha	Prechilling required		*Camellia*	Fresh seeds germinable
	Pyrus	Prechilling required			
	Raphiolepis	Germinable		*Gordonia*	Fresh seeds germinable
	Rhodotypos	Prechilling required			
	Rosa	Prechilling required		*Stewartia*	Double dormant
	Rubus	Double dormant	Thymelaeaceae	*Aquilaria*	Seeds very slowly germinable
	Sorbaria	Prechilling required			
	Sorbus	Prechilling required		*Daphne*	Double dormant
	Spiraea	Germinable		*Dirca*	Moderately germinable
	Stephanandra	Double dormant plus light required	Tiliaceae	*Belotia*	Afterripening required
				Brownlowia	Germinable
	Stranvaesia	Prechilling required		*Helicarpus*	Heat treatment and light required
Rubiaceae	*Anthocephalus*	Seeds sensitive to light quality		*Tilia*	Double dormant
	Cephalanthus	Germinable			
	Mitchella	Prechilling required			

Family	Genus	Germination characteristics	Family	Genus	Germination characteristics
Ulmaceae	*Celtis*	Prechilling required	[Verbenaceae]	*Gmelina*	Germinable
	Ulmus	Germinable		*Tectona*	Germinable
	Zelkova	Prechilling required		*Vitex*	Prechilling required
Verbenaceae	*Avicennia*	Germinable, recalcitrant	Vitaceae	*Ampelopsis*	Prechilling required
				Parthenocissus	Prechilling required
	Callicarpa	Germinable		*Vitus*	Prechilling required
	Caryopteris	Germinable	Zygophyllaceae	*Larrea*	Germinable
	Clerodendrum	Prolonged prechilling required			

GLOSSARY

Abortive. Imperfectly or incompletely developed (e.g., abortive seed).

Abscission. The natural separation of leaves, flowers, and fruit from plants, generally associated with deterioration of a specialized layer of thin-walled cells.

Achene. A small, dry, nonsplitting, one-seeded fruit with seed attached to the ovary wall at only one point, as in *Cowania* and *Eriogonum*; or a pericarp fused with calyx tube and embryo completely filling the ovarian cavity, as in *Artemisia* and *Chrysothamnus*.

Afterripening. A period of time after harvesting in which embryo maturity must occur before a seed will germinate. We restrict this term to seeds that require time for embryo maturity, independent of environmental influences. However, the term is commonly applied to a wide range of types of dormancy, such as prechilling requirements.

Aggregate fruit. Fruit formed from several separate or fused ovaries of a single flower, as in *Rubus* and *Magnolia*. Compare **Multiple fruit** and **Simple fruit**.

Ament. See catkin.

Angiosperm. A member of the group of plants having seeds that develop in a carpellary ovary. Compare **Gymnosperm**.

Anthesis. The bursting of anthers with subsequent release of pollen.

Apomixis. Reproduction from seeds or seedlike organs but without fertilization.

Apophysis. The visible portion of a cone scale when the cone is closed.

Aril. The exterior covering or appendage of certain seeds that develops after fertilization as an outgrowth from the point of attachment of the ovule.

Asexual reproduction. Reproduction by purely vegetative means.

Berry. A pulpy, indehiscent fruit developed from a single pistil and containing one or more seeds but not true stones.

Bisexual. Having both stamens and at least 1 pistil in the same flower.

Bract. A modified leaf subtending a flower or a female cone scale.

Broadcast sowing. Scattering seeds uniformly over an area.

Browse. The shoots, twigs, and leaves of woody plants eaten by animals.

Bur. The prickly or spiny casing around a fruit.

Calyx. The outermost group of floral parts; the whorl of sepals.

Capsule. A dry, usually many-seeded fruit composed of 2 or more fused carpels that split at maturity to release their seeds.

Carpel. A simple pistil, or a single member of a compound pistil.

Catkin. A spike of unisexual flowers or fruits with imbricated scaly bracts as in *Alnus* and *Betula*. Compare **Strobile**. Synonym **Ament**.

Cone. One of the reproductive structures of conifers. A female cone consists of a central axis supporting imbricated bracts, each of which subtends a scale bearing naked seeds. A male cone consists of a central axis supporting spirally arranged microsporophylls, each of which bears pollen sacs containing pollen grains. Also used for any seed-bearing structure having a conical shape, as in *Magnolia*.

Conelet. The immature female cone of gymnosperms.

Cool-moist stratification. See **Prechilling**.

Corolla. The inner set of floral leaves consisting of separate or fused petals that surround the carpels.

Corymb. A flat-topped flower cluster.

Cotyledons. The first leaf (or one of the first pair of leaves) developed in the embryo of a seed.

Cyme. A flower cluster having main and secondary axes each terminating in a single flower.

Deciduous. Falling off at the end of the growing season, such as deciduous leaves, or at certain levels or stages of development, such as flower petals.

Dehiscent. Splitting open at maturity to discharge contents, as a capsule discharging seeds. Compare **Indehiscent**.

Determinate flowering. The terminal flower blooms slightly in advance of of its closest associates. Compare **Indeterminate flowering**.

Dichogamy. The maturation of male and female flowers on the same plant at different times.

Dioecious. Having male and female flowers borne on different individual plants. Compare **Monoecious**.

Dormancy. A general term for instances when a living tissue that is predisposed to grow does not do so. In seeds, a state that prevents germination.

Double dormancy. The state in which embryo dormancy is combined with seed coat dormancy.

Drupe. A fleshy, usually one-seeded, indehiscent fruit whose seed is completely enclosed in a hard, bony endocarp.

Embryo dormancy. A state in which a seed fails to germinate because the embryo is not in a physiological condition in which germination is possible.

Endocarp. The inner layer of the pericarp, such as the hard, bony pit in *Prunus*.

Endosperm. The storage tissue surrounding the embryo in a seed and consisting of thin-walled cells rich in carbohydrates; usually triploid from separate fertilization in angiosperms, and haploid from female tissue in seeds of gymnosperms.

Epicotyl. The portion of the axis of a plant embryo or seedling stem above the cotyledons.

Exocarp. The outermost layer of the pericarp; the skin of fleshy fruits such as apples.

Extraction factor. The weight of cleaned seeds expressed as a percentage of the weight of the fresh fruits from which they were extracted.

Fertilization. The penetration of the pollen tube through the embryo sac into the ovule (egg cell), discharge of the male nucleus into the ovule, and the union of the male nucleus with the ovule.

Follicle. A dry, one-celled fruit splitting open on only one side, as in individual fruits of a Magnolia cone.

Fruit. The reproductive unit of a seed-bearing plant developed after fertilization by a sperm cell from a pollen grain. Fruit includes the ripened ovary and its associated protective covers, appendages, and supporting structures.

Fruit wall. The outer layer of fruits in which the pericarp is not distinguished from the seed coat, as in the case of many achenes.

Full seed. Seed filled with tissue having a normal appearance as distinguished from empty or partially empty seed.

Germination. The resumption of active growth in the embryo of a seed, as demonstrated by the protrusion of the radicle.

Germination percentage. The percentage of seeds tested that germinate. See **Germinative capacity**. In the commercial seed trade, germination capacity is not used as frequently as germination percentage.

Germinative capacity. The percentage of seeds that germinate during a period of time ending when germination is practically finished.

Germinative energy. The percentage of seeds that germinate during a specific interval that is determined by peak germination rate; essentially a measure of speed of germination.

Glabrous. Smooth; without hairs or down.

Globose. Approximately or completely spherical; globular.

Gymnosperm. A member of the group of plants having naked seeds borne on the scales of a cone, on the megasporophylls of other types of strobili, or singly with arils. Compare **Angiosperm**.

Head. A densely packed cluster of stalkless flowers.

Hilum. The scar on a seed marking the point of attachment to the ovary in angiosperms, or to the megasporophyll in gymnosperms.

Husk. The outside envelope of a fruit, especially if coarse, harsh, or rough.

Hypocotyl. The portion of the axis of an embryo or stem of a seedling between the cotyledons and the radicle.

Inflorescence. A floral axis with its appendages.

Indehiscent. Remaining closed. Use of dry fruits that normally do not split open at maturity. Compare **Dehiscent**.

Indeterminate flowering. Flowers open progressively from the base of an inflorescence. Compare **Determinate flowering**.

Integuments. In angiosperms, a layer of tissue that encloses the nucleus of an ovule and develops after fertilization into the seed coats; sometimes there are 2 layers of tissue, often fused, both of which develop into seed coats. In gymnosperms, a single layer of tissue that encloses the nucleus of an ovule. In *Pinus*, this layer of tissue develops after fertilization into 3 seed coats, but the outer one is usually not distinct in harvested seeds.

Internal dormancy. Dormancy maintained by conditions within the embryo.

Involucre. One or more whorls of bracts situated below and close to a flower or flower cluster; sometimes enclosing the carpels.

Legume. A dry, dehiscent, one-celled fruit developed from a simple superior ovary and usually splitting into 2 equal parts with the seeds attached to the lower edge of each part. The fruit of all genera of Leguminosae.

Maceration. A process for removing the soft, pulpy tissue from fruits.

Mesocarp. The middle layer of the pericarp; the pulp of drupes and berries.

Micropyle. A minute opening in the integument of an ovule through which the pollen tube normally passes to reach the embryo sac; usually closed at maturity to form a superficial scar.

Monecious. Having male and female flowers on the same plant. Compare **Dioecious**.

Multiple fruit. The coalesced ripened ovaries of several distinct flowers. Compare **Aggregate fruit** and **Single fruit**.

Nut. The dry, indehiscent, one-seeded fruit with a woody or leathery pericarp developing from an inferior compound ovary; generally partially or wholly encased in an involucre or husk.

Nutlet. A small nut.

Obovoid. Inversely egg-shaped; ovoid with the broad end toward the apex.

Ovary. In angiosperms, the basal portion of a pistil that bears the ovules.

Ovoid. Egg-shaped with the broad end toward the point of attachment.

Ovule. A rounded outgrowth of the ovary in seed plants that develops into a seed usually after fertilization.

Panicle. A compound raceme.

Parthenogenesis. Reproduction from an unfertilized ovule.

Parthenocarpy. The development of fruit without viable seeds.

Pedicel. The stalk of a single flower within a flower cluster.

Peduncle. A stalk that bears a flower or a flower cluster.

Pericarp. The wall of a ripened ovary, homogeneous in some genera and composed of distinct layers in others.

Phenology. The study of the relations between seasonal climatic changes and periodic biological phenomena, such as the flowering and fruiting of plants.

Pistil. The ovule-bearing organ of an angiosperm composed of stigma, style, and ovary.

Pistillate. Having pistils but no stamens.

Plumule. The primary bud of a plant embryo situated at the apex of the hypocotyl.

Polygamodioecious. Having a mixture of single sex and perfect flowers on different plants of the same species; applied to species that are funtionally dioecious, but have a few bisexual flowers.

Polygamomonoecious. Having a mixture of single sex and perfect flowers on the sample plant; applied to species that are functionally monecious, but have a few bisexual flowers.

Polygamous. Having both bisexual and unisexual flowers on the same plant.

Pome. The many-seeded fruit of the rose family, consisting of an enlarged fleshy receptacle surrounding a pericarp; that is papery and fleshy or hard and stony.

Prechilling. A pregerminative treatment of seeds to induce germination. Seeds are place in cold-moist conditions for a period of time to break dormancy. Synonym **Cool-moist stratification**.

Propagule. Any part of the plant that can be used to propagate.

Pyrene. The stone of a small drupe.

Pyriform. Pear-shaped.

Raceme. An elongated inflorescence with flower on stalks of equal lengths arising from a main axis.

Rachis. The axis of an inflorescence.

Radicle. The portion of the axis of an embryo from which the root develops.

Ramet. The independent member of a clone.

Raphe. The external ridge on a seed developed from an inverted ovule and formed by part of the funiculus adnate to the ovule.

Receptacle. The end of the flower stalk on which the floral organs are borne.

Samara. A dry, indehiscent, winged fruit, 1 or 2 seeded.

Scarification. A pregerminative treatment to make seed coats permeable to water and/or gases; accomplished by mechanical abrasion, or by soaking in strong chemical solutions such as acids.

Seed. The mature ovule containing an embryo and nutritive tissue enclosed in layers of protective tissue; capable under suitable conditions of developing into a plant.

Seedcoat. The protective layer on a seed derived from an integument.

Seedcoat dormancy. Dormancy imposed on seeds by their seed coats through impermeability to water or gas exchange or by mechanical restriction of seedling growth.

Serotinous. Used of cones that fail to open but remain on the tree for several seasons after maturity and are therefore late in dispersing seeds.

Simple fruit. Fruit formed by a single ovary and sometimes including other flower parts. Compare **Aggregate fruit** and **Multiple fruit**.

Spike. An elongated inflorescence with sessile flowers on the main axis.

Stamen. The pollen-bearing organ of flowers of angiosperms, consisting of a filament and an anther.

Staminate. Having stamens but no pistils.

Stigma. The part of the pistil that receives the pollen.

Stone. A part of the drupe; consisting of a seed enclosed in a hard, bony endocarp.

Stratification. A pregerminative treatment to break dormancy in seeds; accomplished by exposing seeds for a specific time to moisture in cold or warm conditions. There has been a movement by seed physiologists to replace stratification with prechilling, as we have done in this volume. Prechilling is self-explanatory for cold-moist stratification, but no equally good term has been developed for warm-moist stratification.

Striate. Marked with parallel grooves.

Strobile. A spiky, pistillate inflorescence or the resulting fruit, not a true strobilus. Compare **Catkin**.

strobilus. A male or female fruiting body of gymnosperms.

Tegmen. The inner seed coat.

Testa. The outer seed coat.

Umbel. A inflorescence with flower stalks arising from the apex of the main floral axis and reaching approximately equal lengths.

Utricle. A bladdery, one-seeded, usually indehiscent fruit consisting of an achene surrounded by bracts.

Viability. The capacity of a seed to germinate.

LITERATURE CITED

Abbott, H. G. 1974. Some characteristics of fruitfulness and seed germination in red maple (*Acer rubrum*). *Tree Planter's Notes* 25:25–27.

Abseitov, S. Y. 1983. Installation for extracting seeds from juniper berries. *Lesnoe Khozyaistov* 64–66.

Abseitov, S. Y., and & Y. S. Osipov. 1985. Basis of the process of extracting seeds from juniper berries. *Lesnoi Zhurnal* 26–30.

Adams, J. C., & B. A. Thieges. 1978. Seed treatment for optimum pecan germination. *Tree Planter's Notes* 29:12–13.

Adkins, C. R., L. E. Hinesly, & F. A. Blazich. 1984. Role of stratification and light in Fraser fir germination. *Canadian Jour. Forest Res.* 14:61–62.

Agiar, I. V., & S. Bisarro. 1978. Behaviour of *Eucalyptus* species under arboretum conditions. *Cientifica* 6:341–348.

Agiar, I. V., & J. T. Nakane. 1983. Size of seeds of *Eucalyptus citriodora* influence germination and vigor. *Brazil Forestal.* 13:25–28.

Agpoa, A. 1980. Germination of seed in growth of Benguet Pine (*Pinus insularis*) seedlings and root production of cuttings on mine tailings and different soil media. *Sylvatrop* 5:173–179.

Agpoa, A., & E. Pulmano. 1978. Seed treatment of *Pinus kesiya* for germination. *Philippine Forest Research Inst.* 3:241:242.

Ahmad, D. 1983. The effect of sowing media on the germination of *Calamus manan* and *Calamus caeius*. *Malaysian Forestry* 46:77–80.

Ahmed, F. U., D. Sharmila, & M. A. Hossain. 1986. Effect of seed treatment on the germination of Rakta Kambal seeds. *Bano Biggya Patrika* (Bangladesh) 12:62–65.

Aissa, D. 1983. Study on the germination of seeds of holly oak. 1. Effect of the producer tree and size of seeds. *Review Cytology Biol. Vegetation Botany* (Paris) 6:5–14.

Alam, M. 1983. The quality of seeds of some representatives of the genus *Quercus* and pattern of acorn germination in the greenhouse. *Gorskostopanska Nauka* 20:10–19.

———. 1984. Seedling emergence and the growth of seedlings of some species of the genus *Quercus*. *Gorskostopanska Nauka* 21:15–23.

Alaniz, M. C., & J. H. Everitt. 1980. Germination of anaua seeds. *Jour. Rio Grande Valley Hort. Soc.* 34:75–80.

Alariz, M. C., & S. C. Martin. 1974. *Cereus.* In: *Seeds of Woody Plants in the United States.* Forest Service, USDA, Washington, DC. 313–314.

Alcorn, S. M., & S. C. Martin. 1974. *Cereus.* In: *Seeds of Woody Plants in the United States.* Forest Service, USDA, Washington, DC. 313–314.

Aleksandrou, A. K. 1985. Variation in seed weight and germination of *Picea abies* in Bulgaria. *Gorskostopanska Nauka* 22:3–9.

Alexander, R. R. 1974. *Ginkgo.* In: *Seeds of Woody Plants in the United States.* Forest Service, USDA, Washington, DC. 429–430.

Alexander, R. R., K. Jorgensen, & A. P. Plummer. 1974. *Cowania.* In: *Seeds of Woody Plants in the United States.* Forest Service, USDA, Washington, DC. 353–355.

Alexander, R. R., & F. W. Pond. 1974. *Yucca.* In: *Seeds of Woody Plants in the United States.* Forest Service, USDA, Washington, DC. 857–858.

Alexander, R. R., & W. D. Sheppard. 1974. *Ceratonia.* In: *Seeds of Woody Plants in the United States.* Forest Service, USDA, Washington, DC. 303–304.

Alfjorden, G., & G. Remrod. 1975. The year's pine and spruce seed crop. A new method for predicting viability. *Skogen* 62:634–635.

Alhaya, C. P. 1985. Ecological studies of some forest tree seeds. II. Seed storage and viablity. *Indian Jour. Forestry* 8:137–140.

Ali, M. N., A. A. Khan, & S. H. Zaidi. 1979. Germination of *Ephedra nebrodensis*. *Pakistan Jour. Forestry* 29:118–119.

Al-Kinany, A. 1981. Effect of some pre-treatments on seed germination and subsequent development of *Acacia longifolia* seedlings. *Pakistan Jour. Forestry* 31:818.

Allen, P. S., S. E. Meyer, & T. E. Davis. 1987. Determining seed quality of winterfat. *Jour. Seed Tech.* 11:7–14.

Allen, R., & R. E. Farmer. 1977. Germination characteristics of bear oak. *Southern Jour. Applied Forestry* 1:19–20.

Allue-Andrade, J. L. 1983. Morphology, types, attributes, difficulties and treatments in production and germination of seeds of *Colutea arborescens*. *Annual—Inst. Nac. Invest. Agr. Ser. Forest.* El Institute. 1987:129–154.

Aminunddin, M., & I. Zakaria. 1980. Grading of *Gmelina arborea* fruits by color. *Malaysian Forester* 43:337–339.

Anderson, D. R. 1983. A simple method for temporary cone storage. *Tree Planter's Notes* 34:28–30.

Anderson, L. C. 1986. An overview of the genus *Chrysothamnus*. In: *Proc. Symp. Biol.* Artemisia *and* Chrysothamnus. Forest Service, USDA, Ogden, UT. 20–29.

Anoak, J. 1973. Storage of oak seed. In: *Proc. Inter. Symp. Seed Processing and Seed Problems.* Inter. Union Forestry Res. Org. 1:26.

Ansely, R. J., & R. H. Abernethy. 1983. Overcoming seed dormancy in Gardner saltbush [*Atriplex gardneri* (Moq.) D. Dieti.]. In: *Proc. Symp. Biol.* Atriplex *and Related Chenopods.* Gen. Tech. Rpt. 172. Forest Service, USDA, Ogden, UT. 152–157.

Appleton, B. L. & C. E. Whitcomb. 1983. Effect of seed germination position. Oklahoma Agr. Exp. Sta., Stillwater, OK.

Appleton, B. W., C. E. Whitcomb, & S. W. Akers. 1986. Effect of seed handling, pregermination and planting position on tree seedling root development. *Jour. Environmental Hort.* 4:69–72.

Aquila, M. E. A., & A. G. Ferreira. 1984. Germination of scarified seeds of *Araucaria angustifolia* in soil. *Ciencia Cultura* 36:1583–1589.

Aragua, R. G. M., J. F. Alves, R. Barros, & F. M. E. Scuza. 1980. Influence of presoaking in gibberellic acid on jojoba seeds. *Cleno Agron. Fortaleza* 10:1–4.

Arai, K. 1982. Storage and germination of *Quercus crispula* acorns. *Jour. Japanese Forestry Soc.* 64:23–34.

———. 1983. Germination of subalpine coniferous seeds. *Bulletin Forestry Products Res. Inst. Japan.* 133–145.

Arrehhini, R. I. n.d. Pre-sowing treatments of seed of *Prosopis caldenia*. Univ. Nacional de Cuyo, Argentina.

Asanuma, S., K. Hayashi, & S. Ohba. 1984. Effect of herbicide on the germination of *Fagus crenata* seed. Bulletin Forestry and Forestry Products Res. Inst. Japan. 157–185.

ASOA (Association of Official Seed Analysis). 1985 Rules for testing seeds. *Jour. Seed Tech.* 6:1–118.

Auchmoody, L. R. 1979. Nitrogen fertilization stimulates germination of dormant pin cherry seed. *Canadian Jour. Forest Res.* 9:514–516.

Auld, T. D. 1986a. Dormancy and viability in *Acacia suaeolens*. *Australian Jour. Botany* 34:463–472.

――――. 1986b. Population dynamics of the shrub *Acacia suaveolens* (Sm.) Willd. Fire and the transition of seedlings. *Australian Jour. Ecology* 11:373–385.

Ayaz, M. 1980. Anatomy of juniper (*Juniperus excelsa*) seed. *Pakistan Jour. Forestry* 30:99–101.

Babeley, G. S., S. P. Guatam, & A. K. Kandya. 1986. Pretreatment of *Albizza lebbek* Benth. seeds to obtain better germination and vigor. *Jour. Tropical Forestry* 2:105–115.

Babeley, G. G., & A. K. Kandya. 1985. Effect of various pretreatments on germination and vigor of *Leucaena leucocephala* (Lam.) de Wit. seeds. *Jour. Tropical Forestry* 1:85–90.

Bachelard, E. P. 1985. Effects of soil moisture stress on the growth of seedlings of three Eucalypt species. I. Seed germination. *Australian Forest Res.* 15:103–114.

Bagchi, S., & C. J. S. K. Emmanuel. 1983. Germination studies in *Tectona grandis* L. *Myforest* 19:209–213.

Baghugana, V. K., & M. M. S. Rauat. 1986. Seed germination behavior of *Rhus parviflora*. *Van Vigyan* 24:113–115.

Bailey, L. H. 1951. *Manual of Cultivated Plants*. MacMillan Publ. Co., New York.

Baker, F. A., & D. W. French. 1986. Dispersal of *Arceuthobium pusillom* seeds. *Canadian Jour. Forest Res.* 16:1–5.

Balboa, Z. 1982. Bound gibberellins in *Prunus avium* cultivar Mericier seeds. *Turrialba* 32:340–343.

Ball, J., & R. Kisor. 1985. Acid scarification requirements of Kentucky coffee tree seeds from south central Minnesota. *Tree Planter's Notes* 36:23.

Banik, R. L. 1977. Studies on grading of teak fruits—one. Fruit size is a factor in germination of teak seeds. *Bano Biggyan Patrika* 6:1–7.

Barbour, M. G. 1968. Germination requirements of the desert shrub *Larrea divaricata*. *Ecology* 50:679–685.

Barnett, J. P. 1976. Sterilizing southern pine seeds with hydrogen peroxide. *Tree Planter's Notes* 27:17–19.

――――. 1977a. The effect of burial by squirrels on germination and survival of oak (*Quercus alba*) and hickory nuts (*Carya glabra*). *American Midland Naturalist* 98:319–300.

――――. 1977b. Effects of soil wetting agent concentration on southern pine seed germination. *Southern Jour. Applied Forestry* 1:14–15.

――――. 1979. Storing spruce pine seeds. *Tree Planter's Notes* 30:20–21.

Barnett, J. P., & O. Hall. 1977. Subfreezing conditions after seeding can reduce southern pine seed germination. *Tree Planter's Notes* 28:3–4.

Barnett, J. P., & J. M. McGilvary. 1976. Storing sand pine (*Pinus clausa*) seeds and cones. *Tree Planter's Notes* 27:10, 13.

Barnett, J. P., & B. F. McLemore. 1984. Germination speed as a predictor of nursery seedling performance. *Southern Jour. Applied Forestry* 8:157–162.

Barnett, J. P., & J. A. Vozzo. 1985. Viability and vigor of slash and short leaf pine seeds after 50 years of storage. *Forest Sci.* 31:316:320.

Barnett, P. E., and R. E. Farmer. 1978. Altitudinal variation in germination characteristics of yellow-poplar in the southern Appalachians. *Silvae Genetica* 27:101–104.

Barnhill, M. A., M. Cunningham, & R. E. Farmer. 1982. Germination characteristics of *Paulownia tomentosa*. *Seed Sci. Tech.* 10:217–221.

Baron, F. J. 1978. Moisture and temperature in relation to seed structure and germination of sugar pine. *American Jour. Botany* 65:804–810.

Barrett, S. C. H., & J. D. Thomson. 1982. Spatial pattern, floral sex ratio and fecundity in dioecous *Aralia nudicaule* L. *Canadian Jour. Botany* 60:1662–1670.

Barton, I. L. 1978. Temperature and its effect on the germination and initial growth of kauri (*Agathis australis*). *New Zealand Jour. Forest Sci.* 8:327–331.

Bartow, L. V. 1961. Experimental seed physiology at Boyce Thompson Institute for plant research; Yonkers, NY, 1924–1961. *Proc. ASOA* 26:561–596.

Basu, S. K., & D. P. Mukenjee. 1972. Studies on the germination of palm seeds. *Principes* 16:136–137.

Baua, K. S., C. R. Keegan, & R. H. Voss. 1982. Sexual diamorphism in *Aralia nudicaulis* L. *Evolution* 36:371–378.

Bazzaz, F. A. 1973. Seed germination in relation to salt concentration in three populations of *Prosopis farcta*. *Oecologia* 13:73–80.

Beadle, N. C. W. 1951. The germination of seed and establishment of seedlings of five species of *Atriplex* in Australia. *Ecology* 33:49–62.

――――. 1952. Studies in halophytes. The germination of seed and establishment of seedling of five species of *Atriplex* in Australia. *Ecology* 33:49–62.

Bebawi, F. F., & S. M. Mohamed. 1985. The pretreatment of six Sudanese acacias to improve their germination response. *Seed Sci. Tech.* 13:111–119.

Becker, P., & M. Wong. 1985. Seed dispersal, seed predation, and juvenile mortality of *Aglaia* spp. in lowland dipterocarp rainforest. *Biotropica* 17:230–237.

Beckjord, P. P. 1982. Containerized and nursery production of *Paulownia tomentosa*. 33:29–34.

Beckman, K. M., & L. F. Roth. 1968. The influence of temperature on longevity and germination of seeds of western dwarf mistletoe. *Phytopathology* 58:147–150.

Becwar, M. R., P. C. Stanwood, & K. W. Leonhardt. 1983. Dehydration effects on freezing characteristics and survial in liquid nitrogen of desiccation-tolerant and desiccation-sensitive seeds. *Jour. American Hort. Sci.* 108:613–618.

Becwar, M. R., S. A. Verhagon, & S. R. Wann. 1987. The frequency of plant regeneration from Norway spruce somatic embryos. *Proc. Southern Forest Tree Improvement Conf.* 92–100.

Beetle, A. A. 1960. A Study of Sagebrush; the Section *Tridentatae* of *Artemisia*. Univ. of Wyoming Agr. Exp. Sta., Laramie.

Belcher, E. W., Jr. 1974. Influence of substrate moisture level on germination of four *Picea* species. *Proc. ASOA* 63:129–130.

――――. 1982. Storing stratified seed for extended periods. *Tree Planter's Notes* 33:23–25.

Belcher, E. W., G. N. Leach, & H. H. Gresham. 1984. Sizing slash pine seeds as a nursery procedure. *Tree Planter's Notes* 35:5–10.

Belcher, E. W., & L. Miller. 1975. Influence of substrate moisture level on the germination of sweetgum and sand pine seed. *Proc. ASOA* 65:88–101.

Belcher, E. W., & B. Perkins. 1985. Effect of substrate moisture on germination of Scots pine. *Tree Planter's Notes* 36:24–26.

Bell, D. T., S. Vlahos, & L. E. Watson. 1987. Stimulation of seed germination of understory species of the northern jarrah forest of Western Australia. *Australian Jour. Botany* 35:593–599.

Berg, A. R. 1974. *Arctostaphylos*. In: *Seeds of Woody Plants in the United States*. Forest Service, USDA, Washington DC. 228–231.

Beri, R. M., P. C. Dobhal, & K. S. Ayyar. 1982. Chemical examination of the seeds of *Prosopis cineraria*. *Indian Forester* 108:669–672.

Berry, A. M., & J. G. Torrey. 1985. Seed germination, seedling inoculation and establishment of *Alnus* spp. in containers in greenhouse trials. *Plant and Soil* 87:161–173.

Bevington, J. 1986. Geographical differences in the seed germination of paper birch. *American Jour. Botany* 73:546–573.

Bhaskar, V. 1983. Effect of forest soils on the germination of soobabul and agase seeds. *Myforest* 19:72–73.

———. 1984. Seed germination of agarwood tree under Bangalore conditions. *Myforest* 20:2–3.

Bhatnagar, H. P. 1980. Preliminary studies on the effect of gibberellic acid on seed germination of *Pinus caribaea* and *Pinus patula*. *Indian Jour. Forestry* 3:156–158.

Bhumibamon, S., & L. Atipanumpai. 1980. Investigations on the seed quality of *Pinus merkusii* by using X-ray techniques. Res. Note 34, Faculty of Forestry, Kasetsart Univ. 8.

Bhumibamon, S., & W. Kanchanabhum. 1980. Studies on the content of fat and protein in some tropical tree seeds. Faculty of Forestry, Kasetart Univ. 1–32.

Bhumibamon, S., N. Komasatit, K. Jenkarnying, & L. Atipanumpai. 1980. Cone and seed studies in *Pinus merkusii*. Res. Note 37, Faculty of Forestry, Kasetsari Univ. 6.

Binet, P. 1965. Etude de quelques aspects physiologiques de la germination chez *Atriplex tronabeni* Tin. Bul. Soc. Bot. Nord, France 18:40–55.

———. 1966. Proprietes physiologiques fundamentales des semences d'*Atriplex babingtonii* Wood. Bul. Soc. Bot. Nord, France 19:121–137.

Bjorkbom, J. C. 1971. Production and germination of paper birch seed and its dispersal into a forest opening. Forest Service, USDA, Washington, DC.

Bjorkbom, J. C., & D. A. Marquis. 1965. The variablity of paper birch seed production, dispersal, and germination. Forest Service, USDA, Washington, DC.

Blanche, C. A., W. U. Elae, & J. D. Hodges. 1980. Some aspects of acorn maturation in water oak. *Jour. Seed Tech.* 5:42–51.

Blazich, F. A., & L. E. Hinesley. 1984. Low temperature germination of Fraser fir seed. *Canadian Jour. Forest Res.* 14:948–949.

Blum, B. M. 1974. *Aralia*. In: *Seeds of Woody Plants in the United States*. Forest Service, USDA, Washington, DC. 220–222.

Blum, B. M., & A. Krochmal. 1974. *Epigaea*. In: *Seeds of Woody Plants in the United Statess*. Forest Service, USDA, Washington, DC. 380–381.

Boado, E. L. 1976. Germination of Bargras using different soil media. *Philippine Forest Res. Jour.* 1:34–37.

Boado, E. L., & V. T. Lasmarias. 1976. Extraction of seeds of green and green-brown Mindro Pine (*Pinus merkusii*) cones soaked in laquer thinner. *Philippine Forest Res. Jour.* 1:15–20.

Bobrinev, V. P. 1978. Times for harvesting pine cones in the eastern Trans-Baikalia. *Lesovedenie* 73–75.

Boe, K. N. 1974a. *Sequoia*. In: *Seeds of Woody Plants in the United States*. Forest Service, USDA, Washington, DC. 764–766.

———. 1974b. *Sequoiadendron*. In: *Seeds of Woody Plants in the Unites States*. Forest Service, USDA, Washington, DC. 767–768.

Bogdanov, B. 1978. Sources of good quality planting stock of Weymouth pine. *Gorsko Stopanstvo* 34:16–21.

———. 1979. Vegetative propagation of *Pinus strobus* for production of high quality seed. Nauchni Trudove, Vissh Lesoteknicheski Institut, Sofiya. *Gorsko Stopanstvo* 24:91–97.

Bogoroditskii, I. I., & L. V. Sholokhov. 1974. Mositure regime of seeds of *Robinia pseudoacacia* prepared for sowing by vacuum water saturation method and by scalding in boiling water. *Referativnyi Zhurnal* 16:115–118.

Bolland, D. J. 1984. *Forest Trees of Australia*. Thomas Nelson, Melbourne.

Bolland, D. J., M. I. H. Brecker, & J. W. Turnbull. 1976. *Eucalyptus Seed*. Division of Forest Res. CISRO, Canberra.

Bolte, M. L., W. D. Crow, N. Takahashi, A. Sakurai, M. Uji-ie, &

S. Yoshida. 1985. Structure activity relationships of grandinol, a germination inhibitor in *Eucalyptus*. *Agr. and Biol. Chemistry* 49:761–768.

Bonker, E. J. 1976. Germinating palm seeds. *Plant Propagation* 25:373–378.

Bonner, F. T. 1974a. *Campsis*. In: *Seeds of Woody Plants in the United States*. Forest Service, USDA, Washington, DC. 260–261.

———. 1974b. *Celtis*. In: *Seeds of Woody Plants in the United States*. Forest Service, USDA, Washington, DC. 298–300.

———. 1974c. *Cephalanthus*. In: *Seeds of Woody Plants in the United States*. Forest Service, USDA, Washington, DC. 301–302.

———. 1974d. *Fraxinus*. In: *Seeds of Woody Plants in the United States*. Forest Service, USDA, Washington, DC. 411–416.

———. 1974e. *Ilex*. In: *Seeds of Woody Plants in the United States*. Forest Service, USDA, Washington, DC. 450–453.

———. 1974f. *Liquidambar*. In: *Seeds of Woody Plants in the United States*. Forest Service, USDA, Washington, DC. 505–507.

———. 1974g. Maturation of acorns of cherry bark, water, and willow oaks. *Forest Sci.* 21:238–242.

———. 1974h. *Nyssa*. In: *Seeds of Woody Plants in the United States*. Forest Service, USDA, Washington, DC. 554–557.

———. 1974i. *Platanus*. In: *Seeds of Woody Plants in the United States*. Forest Service, USDA, Washington, DC. 641–644.

———. 1974j. *Sapium*. In: *Seeds of Woody Plants in the United States*. Forest Service, USDA, Washington, DC. p. 760.

———. 1974k. *Taxodium*. In: *Seeds of Woody Plants in the United States*. Forest Service, USDA, Washington, DC. 796–798.

———. 1974l. Testing the vigor in cherry bark oak acorns. *Proc. ASOA* 64:109–114.

———. 1974m. *Zanthoxylum*. In: *Seeds of Woody Plants in the United States*. Forest Service, USDA, Washington, DC. 859–861.

———. 1975. Germination temperatures and prechill treatments for white ash. *Proc. ASOA* 65:60–65.

———. 1976. Storage and stratification recommendations for pecan and shagbark hickory. *Tree Planter's Notes* 27:3–5.

———. 1977. Effect of gibberellin on germination of forest tree seeds with shallow dormancy. In: *Symp.: Physiology of Seed Germination*. Govt. Forest Exp. Sta., Tokyo.

———. 1979. Collection and care of sycamore seeds. *Southern Jour. Applied Forestry* 3:23–25.

———. 1983. Germination response of loblolly pine to temperature differentials on a two-way thermal gradient plate. *Jour. Seed Tech.* 8:6–14.

———. 1984. Testing for seed quality in southern oaks. Forest Service, USDA, New Orleans.

———. 1986. Measurement of seed vigor for loblolly pine and slash pine. *Forest Sci.* 32:170–178.

———. 1987a. Collection and care of sweetgum seeds. *New Forest* 1:207–214.

———. 1987b. Cone storage and seed quality in longleaf pine. Res. Note. Forest Service, USDA, New Orleans.

Bonner, F. T., and J. D. Burton, 1974. *Paulownia*. In: *Seeds of Woody Plants in the United States*. Forest Service, USDA, Washington, DC. 572–573.

Bonner, F. T., and J. D. Burton, & H. C. Grigsby. 1974. *Gleditsia*. In: *Seeds of Woody Plants in the United States*. Forest Service, USDA, Washington, DC. 431–433.

Bonner, F. T., & J. A. Crossley. 1974. *Vitis*. In: *Seeds of Woody Plants in the United States*. Forest Service, USDA, Washington, DC. 853–854.

Bonner, F. T., & E. R. Ferguson. 1974. *Maclura*. In: *Seeds of Woody Plants in the United States*. Forest Service, USDA, Washington, DC. 525–526.

Bonner, F. T., & D. L. Graney. 1974. *Catalpa*. In: *Seeds of Woody*

Plants in the United States. Forest Service, USDA, Washington, DC. 281–283.

Bonner, F. T., & C. X. Grano. 1974. *Melia.* In: *Seeds of Woody Plants in the United States.* Forest Service, USDA, Washington, DC. 535–536.

Bonner, F. T., & L. K. Halls. 1974a. *Asimina.* In: *Seeds of Woody Plants in the United States.* Forest Service, USDA, Washington, DC. 238–239.

_____. 1974b. *Gaylussacia.* In: *Seeds of Woody Plants in the United States.* Forest Service, USDA, Washington, DC. 427–428.

Bonner, F. T., & L. C. Maisenhelder. 1974a. *Carya.* In: *Seeds of Woody Plants in the United States.* Forest Service, USDA, Washington, DC. 269–272.

_____. 1974b. *Sassafras.* In: *Seeds of Woody Plants in the United States.* Forest Service, USDA, Washington, DC. 761–762.

Bonner, F. T., & A. L. Mignery. 1974. *Halesia.* In: *Seeds of Woody Plants in the United States.* Forest Service, USDA, Washington, DC. 441–442.

Bonner, F. T., & P. O. Rudolf. 1974. *Ziziphus.* In: *Seeds of Woody Plants in the United States.* Forest Service, USDA, Washington, DC. 862–863.

Bonner, F. T., & T. E. Russell. 1974. *Liriodendron.* In: *Seeds of Woody Plants in the United States.* Forest Service, USDA, Washington, DC. 508–511.

Bonner, F. T., & R. S. Schmidtling. 1974. *Bumelia langinosa.* In: *Seeds of Woody Plants in the United States.* Forest Service, USDA, Washington, DC. 258–259.

Bonner, F. T., & B. J. Turner. 1980. Rapid measurement of the moisture content of large seeds. *Tree Planter's Notes* 31:9–10.

Bonnet-Masimbert, M. 1975a. Effect of the temperature of extraction on the germination and storage of seeds of *Pinus pinaster* seeds. 1. Evidence of the role of light. *Annales des Sciences Forestieres* 32:93–112.

_____. 1975b. Germination of *Pinus pinaster* seeds. 1. Evidence of the role of light. *Annales des Sciences Forestieres* 32:93–112.

Bonnet-Masimbert, M. & C. Muller. 1973. Storage of beech mast and acorns; research and prospects. CNRF Office National des Forets, Nancy, France.

_____. 1974. The use of H_2O_2 for breaking dormancy in Douglas fir seeds can only be considered as an emergency measure. *Revue Forestiere Francaise* 26:135–138.

_____. 1975. Storage of beechnuts is feasible. *Revue Forestiere Francaise* 27:129–138.

_____. 1976. A rapid germination test for *Fagus sylvatica. Canadian Jour. Forest Res.* 6:281–286.

Book, J. H., & C. E. Book. 1985. Patterns of reproduction in Wright's sycamore. Forest Service, USDA, Fort Collins, CO.

Booth, D. T. 1984. Threshing damage to radicle apex affects geotropic reaponse of winterfat. *Jour. Range Manage.* 37:222–225.

Borges, E. E., R. D. G. Borges, J. F. Candids, & J. M. Gomes. 1982. Comparison of dormancy breaking methods in capaiba seed germination. *Brazilian Seed Jour.* 4:9–12.

Borges, E. E., A. J. Regazzi, G. R. Carvalo, & P. C. Correa. 1980. Effects of air flow rate and temperature during fruit drying on germination of *Eucalyptus* seed. *Brazilian Seed Jour.* 2:97–106.

Borghetti, M., G. G. Vendramin, A. Venesiano, & R. Giannini. 1986. Influence of stratification on germination of *Pinus leucodermis. Canadian Jour. Forest Res.* 16:867–869.

Borland, J. 1986. *Fraxinus anomala.* American Nurseryman 164:198. Bormann, B. T. 1983. Ecological implications of phytochrome mediation seed germination in red alder (*Alnus rubra*). *Forest Sci.* 29:734–738.

Borno, C., & I. E. P. Taylor. 1975. The effect of high concentrations of ethylene on seed germination of Douglas fir. *Canadian Jour. Forest Res.* 5:419–423.

Borowicz, V. A. 1986. The effect of fruit conditions on germination success of two dogwood species. *Michigan Botany* 25:11–15.

Borr, D. A., & W. J. Atkinson. 1970. Stabilization of coastal sand dunes after mining. *Jour. Soil Conservation Service* (New South Wales) 67:89–105.

Bouvier-Durand, M., M. Dawidowicz-Grezgorzeuski, C. Thevenof, & D. Come. 1984. Dormancy of apple embryos: Are starch and reserve protein changes related to dormancy breaking? *Canadian Jour. Botany* 62:2308–2315.

Bowen, M. R. 1980a. *Eucalyptus deglupta.* A note on seed collection, handling and storage techniques. FAO/UNDP-MAL78/009. Sabah, Malaysia.

_____. 1980b. *Gmelina arborea.* A note on fruit collection, handling and seed storage techniques with data on nuts and seeds, FAO/UNDP-MAL, Sabah, Malaysia.

Bowen, M.R, & T. V. Eusebio. 1983. Seed handling practices: four fast growing trees for humid tropical plantations in the eighties. *Malay Forest* 45:534–547.

Boyer, J. N., D. B. Smith, C. Muller, H. Vanderveer, W. Chapman, & W. Rayfield. 1985. Speed of germination affects diameter of lifting of nursery loblolly pine seedlings. *Southern Jour. Applied Forestry* 9:243–247.

Bramlett, D. L. n.d. Seed and aborted ovules from cones of *Pinus echinata.* Georgia Forest Res. Paper No. 71. Forest Service, USDA, Blacksburg, VA.

Bramlett, D.L, E. W. Belcher, Jr., G. L. DeBarr, G. D. Hertel, R. P. Karrfalt, C. W. Lantz, T. Miller, K. D. Ware, & H. O. Yates III. 1977. Cone analysis of southern pines—a guidebook. Forest Service, USDA, Altanta, GA.

Bramlett, D.L, T. R. Dell, & W. D. Pepper. 1983. Genetic and maternal influences of Virgina pine seed germination. *Silvae Genetica* 32:1–4.

Brandbeer, J. W. 1968. Studies in seed dormancy. IV. The role of endogenous inhibitors and gibberellin in the dormancy and germination of *Corylus avellana. Planta* 78:266–276.

Brandbeer, J.W. I. E. Arias, & I. S. Nirmala. 1978. The role of chilling in breaking the seed dormancy of *Corylus avellana. Pesticide Sci.* 9:184–186.

Brandbeer, J.W, & N. J. Pinfield. 1967. Studies in seed dormancy. III. The effects of gibberellin on dormant seeds of *Corylus avellana. New Phytologist* 66:515–523.

Brandis, D. 1970. *Indian Trees.* Archibald Constable & Co., London.

Brice-Bruce, A. P. 1976/1977. Some observations on raising eucalypt and black nursery stock. Rpt. Wattle Res. Inst. South Africa. :76–86.

Brinar, M. 1976. Viability of beech seed and growth of seedlings in relation to some characteristics of provenance sites. *Zbornik Gozdartva in Lesarstva Ljbljana* 13:61–79.

Brinkman, K. A. 1974a. *Amelanchier.* In: *Seeds of Woody Plants in the United States.* Forest Service, USDA, Washington, DC. 212–215.

_____. 1974b. *Amorpha.* In: *Seeds of Woody Plants in the United States.* Forest Service, USDA, Washington, DC. 216–219.

_____. 1974c. *Betula.* In: *Seeds of Woody Plants in the United States.* Forest Service, USDA, Washington, DC. 252–257.

_____. 1974d. *Cornus.* In. *Seeds of Woody Plants in the United States.* Forest Service, USDA, Washington, DC. 336–342.

_____. 1974e. *Corylus.* In: *Seeds of Woody Plants in the United States.* Forest Service, USDA, Washington, DC. 343–345.

_____. 1974f. *Crataegus.* In: *Seeds of Woody Plants in the United States.* Forest Service, USDA, Washington, DC. 356–360.

_____. 1974g. *Hamamelis.* In: *Seeds of Woody Plants in the United States.* Forest Service, USDA, Washington, DC. 443–444.

———. 1974h. *Juglans*. In: *Seeds of Woody Plants in the United States.* Forest Service, USDA, Washington, DC. 454–459.

———. 1974i. *Lonicera*. In: *Seeds of Woody Plants in the United States.* Forest Service, USDA, Washington, DC. 515–519.

———. 1974j. *Rhus*. In. *Seeds of Woody Plants in the United States.* Forest Service, USDA, Washington, DC. 715–719.

———. 1974k. *Rubus*. In. *Seeds of Woody Plants in the United States.* Forest Service, USDA, Washington, DC. 738–743.

———. 1974l. *Salix*. In: *Seeds of Woody Plants in the United States.* Forest Service, USDA, Washington, DC. 746–750.

———. 1974m. *Sambucus*. In. *Seeds of Woody Plants in the United States.* Forest Service, USDA, Washington, DC. 754–757.

———. 1974n. *Tilia*. In. *Seeds of Woody Plants in the United States.* Forest Service, USDA, Washington, DC. 810–812.

———. 1974o. *Ulmus*. In. *Seeds of Woody Plants in the United States.* Forest Service, USDA, Washington, DC. 829–834.

Brinkman, K. A., & G. G. Erdmann. 1974. *Mitchella*. In: *Seeds of Woody Plants in the United States.* Forest Service, USDA, Washington, DC. 543.

Brinkman, K. A., & H. M. Phipps. 1974a. *Lindera*. In: *Seeds of Woody Plants in the United States.* Forest Service, USDA, Washington, DC. 503–504.

———. 1974b. *Menispermum*. In: *Seeds of Woody Plants in the United States.* Forest Service, USDA, Washington, DC. 537–538.

Brinkman, K. A., & R. C. Schlesinger. 1974. *Ptelea*. In: *Seeds of Woody Plants in the United States.* Forest Service, USDA, Washington, DC. 684–685.

Brooke, B. M., R. E. McDiarmid, & W. Majak. 1985. The cyanide potential in two varieties of *Amelanchier alnifolia. Canadian Jour. Plant Sci.* 68:543–547.

Brown, F. M., & E. Belcher. 1979. Improved techniques for processing *Prosopia* seed. *Tree Planter's Notes* 30:19.

Brown, K. E. 1976. Ecological studies of the cabbage palm, *Sabal palmetto.* 3. Seed germination and seedling establishment. *Principes* 20:98–115.

Brown, S. C., B. G. Coombe, C. L. Tolley, & G. B. Gotty. 1984. Investigations of germination and benching of sweet orange. *Plant Propagation* 33:145–152.

Browse, P. D. A. 1979. *Hardy, Woody Plants from Seed.* Grower Books, London.

Browse, P. M. 1987. Dormancy control in *Magnolia* seed germination. *Plant Propagation* 36:116–120.

Burns, R. M., & C. W. Coggus. 1969. Sweet orange germination and growth aided by water and gibberellin seed soak. *California Agr.* 4:18–19.

Burton, P. J. 1982. The effect of temperature and light on *Metrosideros polymorpha* seed germination. *Pacific Science* 36:229–240.

Burton, P. J., & D. Mueller-Dombois. 1984. Response of *Metrosideros polymorpha* seedlings to experimental canopy openings. *Ecology* 65:779–791.

Burton, P. W., W. F. Campbell, A. S. Bittner, & A. J. Johnson. 1984. Cell wall composition and enzymatic degradation of *Atriplex gardeneri* (Moq.) Dieti. seed bracteola. *Jour. Seed Tech.* 9:68–78.

Buta, J. G., & W. R. Lushy. 1986. Catechins as germination and growth inhibitors in *Lespedeza* seeds. *Phytochemistry* 25:93–95.

Calamansi, R. 1982. Effects of light and temperature on the germination of seed of some provenances of *Pinus halepensis* and *Pinus brutia. L'Italia Forestale e Montana* 37:174–187.

Calamansi, R., M. Falusi, & A. Tocci. 1984. The effects of germination temperature and stratification on the germination of *Pinus halepensis* seed. *Silvae Genetica* 33:133–139.

Campbell, T. E. 1982a. The effects of presoaking longleaf pine seeds in sterilants on direct seeding. *Tree Planter's Notes.* 33:8–11.

———. 1982b. Imbibition, desiccation, and reimbibition effects on light requirements for germinating southern pine seeds. Forest Service, USDA, Pineville, LA.

Campbell, P. L., D. J. Erasmus, & J. Van Staden. 1988. Enhancing seed germination of sand blackberry. *Hort. Sci.* 23:560–561.

Campbell, R. C., & G. C. Martin. 1976. Determination of moisture in walnut (*Juglans regia*) seeds by near infrared spectrophotometry. *Hort. Sci.* 11:494–496.

Campbell, R. K., & S. M. Ritland. 1982. Regulation of seed germination timing by moist chilling in western hemlock. *New Phytologist* 92:173–182.

Campbell, R. K., & F. C. Sorenson. 1979. A new basis for characterizing germination. *Jour. Seed Tech.* 4:24–34.

Campbell, T. E. 1982a. The effects of presoaking longleaf pine seeds in sterilants on direct seeding. *Tree Planter's Notes* 33:8–11.

———. 1982b. Imbibition, desiccation, and reimbibition on light requirements for germinating southern pine seeds. *Forest Science* 28:539–543.

Capon, B., G. L. Maxwell, and P. H. Smith. 1978. Germination response to temperature pretreatment of seed from 10 populations of *Salvia columbaria* in the San Gabriel Mountains and Mojave Desert, California. *Aliso* 9:365–373.

Carl, C. M., Jr. 1983. Stratification of sugar maple seeds. *Tree Planter's Notes* 34:25–27.

Carl, C. M., Jr, & H. W. Yauney. 1977. Fall versus spring sowing of sugar maple seeds in a nursery. *Tree Planter's Notes* 28:24–26.

Carlson, J. R., & W. C. Sharp. 1975. Germination of high elevation manzanitas. *Tree Planter's Notes* 26:10–11, 25.

Carpenter, S. B., T. R. Cunningham & N. D. Smith. 1980. Germination of *Paulownia* seed after stratification, dry storage, and pretreatment with gibberellic acid. *Proc. Inter. Symp. Forest Tree Seed Storage.* Mexico City, Mexico.

Carpenter, S. B., & N. D. Smith. 1979. Germination of *Paulownia* seed after stratification and dry storage. *Tree Planter's Notes* 30:4–6.

———. 1981. Germination of paulownia seeds in the presence and absence of light. *Tree Planter's Notes* 32:27–29.

Carpenter, W. J. 1987. Temperature and imbibition effects on seed germination of *Sabal palmetto* and *Serenca repens. Hort. Sci.* 22:660.

Carpita, N. C., A. S. Karic, J. P. Barnett, & J. R. Dunlap. 1983. Cold stratification and growth of radicles of loblolly pine embryos. *Physiologia Plantarium* 59:601–606.

Carpon, B., & P. E. Brecht. 1970. Variation in seed germination and morphology among populations of *Salvia columbariae* in southern California. *Aliso* 7:207–216.

Carr, J. D. 1976. *The South African Acacias.* Conservation Press, Johannesburg.

Carrillo, S. A., F. Patino Valera, & I. Talavera Armas. 1980. Moisture content of stored seed of 7 *Pinus* species and *Abies regligiosa* and its relation to percentage germination. *Ciencia Forestal* 5:39–48.

Carvalho, N. M., S. P. Dematte, & T. T. Graziano. 1980. Germination of seeds of Brazilian plants. *Brazilian Seed Jour.* 2:81–87.

Cavanaugh, T. 1987. Germination of hard seeded species. In: *Germination of Australian Native Plants.* Inkata Press, Melbourne. 58–70.

Cecich, R. A., & E. O. Bauer. 1987. The acceleration of jack pine seed development. *Canadian Jour. Forest Res.* 17:1408–1415.

Cecich, R. A., & R. A. Rudolph. 1982. Time of jack pine seed maturity in Lake States provenances. *Canadian Jour. Forest Res.* 12:368–373.

Chandra, J. P., & P. S. Chauhan. 1977. Note on germination of spruce seeds with gibberellic acid. *Indian Forester* 102:721–725.

Chandra, J. P., & A. Ram. 1980. Studies on depth of sowing deodar seeds. *Indian Forester* 12:852–855.

Chaney, W. R., & T. T. Kozlowski. 1974. Effect of antitranspirants on seed germination, growth, and survival of tree seedlings. *Plant and Soil* 40:225–229.

Chang, S., & D. J. Wenner. 1984. Relationships of seed germination and respiration during stratification with chilling requirements in peach. *Jour. American Soc. Hort. Sci.* 109:42–45.

Chappelka, A. H., & B. I. Chevone. 1986. White ash seedling growth response to ozone and simulated acid rain. *Canadian Jour. Forest Res.* 16:786–790.

Chauhan, P. S., & V. Raina. 1980. Effect of seed weight on germination and growth of chir pine. *Indian Forester* 106:53–57.

Chem, J. D. 1985. A preliminary study on increasing the seed yield of *Cunninghamia lanceolata* seed orchards. *Forest Science and Tech.* 1985:4–7.

Chen, R. Z., & J. R. Fu. 1984. Physiological studies on the seed dormancy and germination of *Erythrophleum fordii*. *Scientia Silvae Sinicae* 20:35–41.

Chen, Y. D. 1981. The relationship between seed quality and seedling quality of Korean pine. *Jour. North-Eastern Forestry Inst.* (China) 79–84.

Chen, Y. S., & O. Sziklai. 1985. Preliminary study on the germination of *Toona sinensis* seed from eleven Chinese provenances. *Forest Ecology and Manage.* 10:269–281.

Chernik, V. V. 1983. Endosperm in the seed of Ulmaceae. Byulletin Glavnogo Botanicheskogo Sada 1983:62–66.

Chin, H. F., M. Azia, B. B. Ang, & S. Hamzah. 1981. The effect of moisture and temperature on the ultrastructure and viability of seeds of *Hevea brasiliensis*. *Seed Sci. Tech.* 9:441–442.

Ching, T. M., & M. C. Parker. 1985. Hydrogen peroxide for rapid viability tests of some coniferous tree seeds. *Forest Sci.* 4:128–134.

Chou, Y., S. Y.Ho, & H. M. Zhu. 1982. Determination of seed vigour of *Pinus thunbergii*. *Jour. Nanjing Tech. College Forest Products* 1982:200–203.

Christensen, P. E. S., & C. J. Schuster. 1979. *Some Factors Affecting the Germination of Karri* (Eucalyptus diversicolor) *Seed.* Forest Dept. Western Australia.

Citharel, M. S. 1979. Variation de la composition azotee de *Pinus canariensis* au cour de la germination. *Phytochemistry* 18:47–50.

Clatenbuck, W. K. 1985. Utilization of food reserves in *Quercus* seed during storage. *Seed Sci. Tech.* 13:121–128.

Clay, K. 1983. Myrmecology in the trailing arbutus. *Bul. Torrey Bot. Club* 110:166–169.

Cochran, P. H. 1984. Lodgepole pine-the species and its management. In: *Lodgepole Pine Symposium.* Forest Service, USDA, Portland, OR. 89–93.

Coffman, M. S. 1978. Eastern hemlock germination influenced by light, germination media, and moisture content. *Mich. Botany* 17:99–103.

Cohen, A. 1956. Studies on the viability of citrus seeds and certain properties of their coats. *Bul. Res. Council of Israel* 50:200–209.

Come, D., C. Perion, & J. Ralambosoa. 1985. Oxygen sensitivity of apple embryos in relation to dormancy. *Israel Jour. Botany* 34:17–23.

Conkle, M. T. 1971. Isozyme specificity during germination and early growth of Kucheene Pine. *Forest Sci.* 17:494–498.

Conn, J. S., & E. K. Snyder-Conn. 1981. Relationship of the rock outcropping microhabitat to germination, water relations, and phenology of *Erythrina flabelliformis* in southern Arizona. *Southwestern Naturalist* 25:443–451.

Conner, K., P. P. Feret, & R. E. Adams. 1976. Variation in *Quercus* mast production. *Virginia Jour. Sci.* 27:54.

Connett, J. W. 1985. Germination of *Washingtonia filitera* seeds eaten by coyotes. *Inter. Palm Soc.* 29:19.

Corbineau, F., S. Defresne, & D. Come. 1985. Some characteristics of seed germination and growth of seedlings of *Cedrela odorata*. *Bois et Forets des Tropiques* 207:17–22.

Corbineau, F., Kante, & D. Come. 1986. Seed germination and seedling development in the mango. *Tree Physiology* 1:151–160.

Corner, E. J. H. 1952. *Wayside Trees of Malaya*. Govt. Printing House, Singapore.

Coughenour, M. B., & J. K. Detling. 1986. *Acacia tortilis* seed germination response to water potential and nutrients. *African Jour. Ecology* 24:203–205.

Cram, W. H. 1982. Seed germination of elder and honeysuckle. *Hort. Sci.* 17:618–619.

_____. 1984. Presowing treatments and storage of green ash seeds. *Tree Planter's Notes* 35:20–21.

Cremer, K. W., & S. B. Mucha. 1985. Effects of freezing temperatures on mortality of air-dry, imbibed, germinating seeds of eucalypts and radiata pine. *Australian Forest Res.* 15:243–251.

Crisosta, C., & E. G. Sutter. 1985. Role of the endocarp in 'Manzanillo' olive seed germination. *Jour. American Soc. Hort. Sci.* 110:50–52.

Critchfield, W. B. 1988. Hybridization of the California firs. *Forest Sci.* 34:139–145.

Crocker, W., & L. V. Barton. 1931. After-ripening, germination, and storage of certain rosaceous seeds. Contrib. Boyce Thompson Inst. 3:385–404.

Crossley, J. A. 1974a. *Malus.* In: *Seeds of Woody Plants in the United States.* Forest Service, USDA, Washington, DC. 531–534.

_____. 1974b. *Solanum.* In: *Seeds of Woody Plants in the United States.* Forest Service, USDA, Washington, DC. 777–778.

_____. 1974c. *Vaccinium.* In: *Seeds of Woody Plants in the United States.* Forest Service, USDA, Washington, DC. 840–843.

Cruise, J. E. 1964. Studies of natural hybrids in *Amelanchier*. *Canadian Jour. Botany* 46:651–633.

Cubbererly, B., & E. R. Hasselkus. 1987. Amelanchiers: Trees and shrubs with year-round enchantment. *American Nurseryman* 165:111–112, 116–117.

Cunningham, T. R., & S. B. Carpenter. 1980. The effect of diamonium phosphate fertilizer on the germination of *Paulownia tomentosa* seed. *Tree Planter's Notes* 31:6–8.

Czernik, A. 1983. Geometrical features of seeds of Scot pine, Norway spruce, European larch. *Sylvan* 127:31–40.

Dabral, S. L. 1976. Extraction of teak seeds from fruits, their storage and germination. *Indian Forester.* 102:650–658.

Dafni, A., & M. Negbi. 1978. Variability in *Prosopis farcta* in Israel : seed germination as affected by temperature and salinity. *Israel Jour. Botany* 27:147–159.

Dahab, A. M. A., Y. Shafiq, & A. Al-Kinay. 1975. Effects of gibberellic acid. B-nine and scarification on germination of seeds of *Pistacia khinjuk*. *Mesoptamia Jour. Agr.* 10:13–19.

Dakin, A. J., & B. R. McClure. 1975. Aspects of Kauri propagation by seed. *Plant Propagation* 25:338–343.

Dale, A., & B. C. Jarvis. 1983. Studies on germination in raspberry. *Crop Res.* 23:73–81.

Dale, J. W., & J. A. Schenk. 1978. Cone production and insect-caused seed losses of ponderosa pine in Idaho and adjacent Washington and Montana. Bulletin No. 24, Forest, Wildlife, and Range Exp. Sta., Univ. of Idaho, Moscow.

Dallimore, W., & A. B. Jackson. 1967. *A Handbook of Coniferae and Ginkgoaceae.* St. Martins Press, New York.

Dalmacio, M. V. 1976. Coating ipil-ipil (*Leucaena leucocephala*) seeds with Arasan–75. *Sylvatrop* 1:148–149.

Daniels, F. W., & H. A. Sijde. 1975. Cold stratification of *Pinus elliotti*, *Pinus taeda* and *Pinus patula* seed. *Forestry in South Africa* 63–68.

Daniels R. A. n.d. Phenotypic and geographic variation in seed characteristics of white ash. Forest Service, USDA, Starkville, MS.

Danielson, H. R. n.d. Air drying does not adversely affect the quality of stratified ponderosa pine and Douglas fir seeds. Seed Lab. Oregon State Univ., Corvallis.

Danielson, H. R., & Y. Tanaka. 1978. Drying and storing stratified ponderosa pine and Douglas fir seeds. *Forest Sci.* 24:11–16.

Danielson, R. 1986. Stratification and germination of western white pine seeds. Forest Service, USDA, Fort Collins, CO.

Dan'shin, I. I., S. A. Kazadaev, & F. G. Ponomarev. 1975. Highly effective stimulants of the growth of *Pinus sylvestris* seedlings. *Referativnyi Zhurnal* 90–94.

Davidson, E. D., & M. G. Barbour. 1977. Germination, establishment, and demography of coastal bush lupine at Bodega Head, California. *Ecology* 58:592–600.

Dealy, J. E. 1978. Autecology of curlleaf mountain mahogany. Proc. First Inter. Rangeland Congres:398–400.

DeCarli, M. E., B. Baldan, P. Mariana, & N. Rascoi. 1987. Subcellular and physiological changes in *Picea excelsa* seeds during germination. *Cytobios* 50:29–39.

DeHayes, D. H., & C. E. Waite. 1982. The influence of spring sowing on black walnut germination in northern Vermont. *Tree Planter's Notes* 33:16–18.

Deitschman, G. H. 1974. *Artemisia.* In: *Seeds of Woody Plants in the United States.* Forest Service, USDA, Washington, DC. 225–237.

Deitschman, G. H., K. R. Jorgensen, and A. P. Plummer. 1974a. *Cercocarpus.* In: *Seeds of Woody Plants in the United States.* Forest Service, USDA, Washington, DC. 309–312.

———. 1974b. *Chrysothamnus.* In: *Seeds of Woody Plants in the United States.* Forest Service, USDA, Washington, DC. 326–328.

———. 1974c. *Fallugia.* In: *Seeds of Woody Plants in the United States.* Forest Service, USDA, Washington, DC. 406–408.

———. 1974d. *Purshia.* In: *Seeds of Woody Plants in the United States.* Forest Service, USDA, Washington, DC. 686–688.

Delkov, N. 1975. Study on the seed quality of *Platanus orientalis. Gorsko Stopanstvo* 20:23–28.

Deltredion, P. 1981. *Magnolia virginiana* in Massachusetts. *Arnoldia* 41:36–49.

DenHeyer, J., & N. Seymoir. 1978. Aspen and balsam popular seed collection and storage. *Tree Planter's Notes* 29:35.

Desch, H. E. 1941 Manual of Malayan Timbers. Malayan Forest Res., Kuala Lumpur.

Dettori, M. L., J. F. Balliette, J. A. Young, & R. A. Evans. 1984. Temperature profiles for germination of two species of winterfat. *Jour. Range Manage.* 37:218–222.

Dharmaratne, H. R. W., S. Sotheesuaran, & S. Balasubran. 1984. Triterpenes and neoflavenoides of *Calophyllum lankaensis* and *C. thuaitesii. Phytochemistry* 23:260–263.

Diagana, D. 1985. Treatment to accelerate germination of *Acacia mangium*, *Albizia*, *Calliandra*, and *Leucaena. Nitrogen Fixing Tree Rpt.* 3:2–3.

Dickie, J. B., & J. T. Bower. 1985. Estimation of provisional seed viability constants for apple. *Annals Botany* 56:271–274.

Dietz, D. R., & P. E. Slabaugh. 1974. *Caragana.* In: *Seeds of Woody Plants in the United States.* Forest Service, USDA, Washington, DC. 262–263.

Dimalla, G. G., & J. J. Van Staden 1977. Pecan nut germination—a review of the nursery industry. *Scientia Hort.* 8:1–9.

Dimock, E. J., W. F. Johnston, & W. I. Stein. 1974. *Gaultheria.* In: *Seeds of Woody Plants in the United States.* Forest Service, USDA, Washington, DC. 422–426.

Dimock, E. J., & W. I. Stein. 1974. *Osmaronia.* In: *Seeds of Woody Plants in the United States.* Forest Service, USDA, Washington, DC. 561–563.

Dirr, M. A. 1983. *Manual of Woody Landscape Plants.* Stipes Publ. Co., Champaign, IL.

———. 1987. Native amelanchiers: a sampler of northeastern species. *American Nurseryman* 166:66–74, 76–78, 82–83.

Dirr, M. A., & B. Brinson. 1985. *Magnolia grandflora*: a propagation guide. *American Nurseryman* 162:8–49.

Dirr, M. A., & C. W. Heuser, Jr. 1987. *The Reference Manual of Woody Plant Propagation.* Varsity Press, Athens, GA.

Djavanshir, K., & G. H. Fechner. 1976. Epicoptyl and hypocotyl germination of eastern red cedar (*Juniperus virginiana*) and Rocky Mountain juniper (*J. scopulorum*). *Forest Sci.* 22:261–266.

Djavanshir, K., & C. P. P. Reid. 1975. Effect of moisture stress on germination and radicle development of *Pinus eldarica* and *Pinus ponderosa. Canadian Jour. Forest Res.* 5:80–83.

Dnyansager, V. R., & V. S. Kothekar. 1982. Problem of teak seed germination. *Indian Jour. Forestry* 5:94–98.

Dobrin, V. A., I. L. Kameshkov, A. L. Klebanov, & V. I. Kryuk. 1983. Effects of laser radiation on the viability of seeds of *Larix sukaczewii. Lesnoi Zhurnal* 16–19.

Dodd, M. C., & J. Van Staden. 1982. Cytokinins in *Podocarpus henkelii* seeds. 2. Transport and metablism of C^{14} zeaton during germination. *Plant Physiology* 108:401–407.

Dolgolikov, V. I. 1977. Size of cones and seeds of *Picea abies* in clonal seed orchards. *Lesnoe Khozyaistvo* 46–47.

Dominx, S. W. J., & J. E. Wood. 1986. Shelter spot seeding trials with jackpine, black spruce and white spruce in northern Ontario. *Forestry Chronicle* 62:446–450.

Donald, D. G. M., 1981. Dormancy control in *Pinus patula* seed. *South African Forestry Jour.* 14–19.

———. 1987a. The effect of long term stratification on the germination of *Pinus pinaster. South African Forestry Jour.* 25–29.

———. 1987b. Studies on the viability of *Paulounia* seeds. *Sveriges Lantbruksuniversitet* 76–83.

Dong, L., D. X. Yoa, & D. Han. 1987. Breaking dormancy of *Pinus bungensis* with microorganisms. Swedish Univ. of Agr. Sci., Dept. Forest Genetics & Plant Physiology 7:159–164.

Dong, L., X. Zhang, & H. Zhang. 1987. Breaking dormancy of *Pinus bungeana* Zucc. with trichoderma–4030 inoculations. *New Forest* 3:248–249.

Doniushkina, E. A., & V. M. Novikova. 1984. Morphological features of clematis seed. *Gosudarstvennyi Nikitskii* 92:111–118.

Doran, J. C., J. W. Turnbull, D. J. Boland, & B. V. Gund. 1983. *Handbook of Seeds of Dry Zone Acacias.* FAO, Rome.

Doran, J. C., J. W. Turnbull, & E. M. Kariuki. 1987. Effects of storage conditions on germination of five tropical tree species. Swedish Univ. of Agr. Sci., Dept. Forest Genetics and Physiology 84–94.

Drake, D. W. 1975. Seed abortion in some species and interspecific hybrids of *Eucalyptus. Australian Jour. Botany* 23:991–995.

Dreimanis, A. 1975. Variation in the fall of female strobili in *Pinus sylvestris. Zinatine* 89–93.

———. 1978. Genetic characteristics of cones and seeds of Scots pine in a seed orchard. Bulletin No. 143, Elgava, Latvain. 3–8.

Dunlap, J. R., & J. P. Barnett. 1983a. Influence of seed size on germination and early seedling development of loblolly pine germinants. *Canadian Jour. Forest Res.* 13:40–44.

———. 1983b. Stress induced distortions of some seed germination patterns in *Pinus taeda.* In: *Proc. Soc. American Foresters.* Washington, DC. 392–397.

———. 1984. Manipulating loblolly pine seed germination with simulated moisture and temperature stress. *Forest Sci.* 14:61–74.

Durzan, D. J., A. J. Mia, & B. S. P. Uaus. 1971. Effect of tritiated water on the metabolism and germination of jack pine seeds. *Canadian Jour. Botany* 49:2139–2149.

Eakle, T. W., & A. S. Garica. 1977. Hastening the germination of lumbang seeds. *Sylvatrop* 2:291–295.

Edgar, J. G. 1977. Effects of moisture stress on germination of *Eucalyptus camaldulensis* and *E. regnans. Australian Forest Res.* 7:241–245.

Edwards, D. G. W. 1982. Improving seed germination in *Abies. Plant Propagation* 31:69–78.

———. 1985. Collection, processing, testing and storage of true fir in the Pacific Northwest. Univ. of Washington. 113–137.

Edwards, D. G. W., & O. E. Olson. 1973. A photoperiod response in germination of western hemlock seeds. *Canadian Jour. Forest Res.* 3:146–18.

Edwards, D. G. W., & J. R. Sutherland. 1979. Hydrogen peroxide treatment of *Abies amabilis* and *A. grandis* seeds. *Bi-Mon. Res. Notes* 31:3–4.

Edwards, J. L., & J. L. McConnell. 1982. Forest tree seed harvesting system for loblolly pine. American Society of Agricultural Engineers. 10.

Eijsackers, H., & D. Ham. 1984. Germiniation of *Prunus serotina* under laboratory and field conditions. *Nederlands Bosbouwtijdschrift* 56:178–185.

Eldridge, K. G. 1982. Genetic improvements from a radiata pine seed orchard. *New Zealand Jour. of Forestry Sci.* 12:404–411.

Elkinauy, M. 1982. Physiological significance of indoleacetic acid and factors determining its level in cotyledons of *lupinus albus durus* germination and growth. *Physiologia Plantarium* 54:302–308.

Elliott, D. M., & I. E. P. Taylor. 1981. Germination of red alder (*Alnus rubra)* seed from several locations in its natural range. *Canadian Jour. Forestry* 11:517–521.

Ellis, R. H., T. D. Hong, & E. H. Roberts. 1983. A note on the development seed of grape. *Vitis* 22:211–219.

———. 1985. *Handbook of Seed Technology for Genebanks.* Vol. 2, *Compendium of Specific Germination and Test Recommendations.* International Board for Plant Genetic Resources.

Emery, D. 1964. Seed propagation of native California plants. Leaflets of the Santa Barbara Botanical Garden 1:81–96.

Engler, J. M., H. Louarn, & F. Tacon. n.d. Influence of birds and small rodents on beech nuts disappearing during winter. INRA-CNRF, Seichamps, France.

Enu-Kwesi, L., & E. B. Dumbroff. 1980. Changes in phenolic inhibitors in seeds of *Acer saccharum* during stratification. *Jour. Experimental Botany* 31:425–436.

Etejere, E. O., M. O. Fawole, & A. Sani. 1982. Studies on the germination of *Parkia clapertoniana. Turrialba* 32:181–185.

Evans, K. E. 1974. *Symphoricarpos.* In: *Seeds of Woody Plants in the United States.* Forest Service, USDA, Washington, DC. 787–790.

Evans, R. A., H. H. Biswell, & D. E. Palmquist. 1987. Seed dispersal in *Ceanothus cuneatus* and *C. leucodermis* in an oak woodland-savanna. *Madrono* 28:283–293.

Everett, P. C. 1964. The culture of manzanitas at Rancho Santa Ana Botanical Garden, Claremont, Calif. *Jour. Calif. Hort. Soc.* 25:37–52.

Everett, R. L., & R. O. Meeuwig. 1975. Hydrogen peroxide and thiourea treatment of bitterbrush seed. Forest Service, USDA, Ogden, UT.

Everitt, J. H. 1983a. Seed germination characteristics of three woody plant species from south Texas (blackbrush), *Acacia rigidula,* (guajillo), *Acacia berlandieri,* and (guayacau), *Portieria angustifolia. Jour. Range Manage.* 36:246–249.

———. 1983b. Seed germination characteristics of two woody legumes (retana and twisted acacia) from south Texas. *Jour. Range Manage.* 36:411–414.

———. 1984. Germination of Texas persimmon seed. *Jour. Range Manage.* 37:189–192.

Falusi, M. 1982. Dormancy and seed germination of some provenances of *Fagus sylvatica. L'Italia Forestale e Montana* 37:313–333.

Farmer, R. E. 1974. Germination of northern red oak: effect of provenance, chilling, gibberellic acid. In: *Proc. 8th Central Forest Tree Improvement Conf.* 16–19.

———. 1981. Variation in seed yield of white oak. *Forest Sci.* 27:377–380.

Farmer, R. E., P. Charrette, J. E. Searle, & D. P. Tarjan. 1984. Interaction of light, temperature and chilling in the germination of black spruce. *Canadian Jour. Forest Res.* 14:131–133.

Farmer, R. E., & J. C. Goelz. 1984. Germination characteristics of red maple in northwestern Ontario. *Forest Sci.* 30:670–672.

Farmer, R. E., G. C. Luckley, & M. Cunningham. 1983. Germination patterns of the sumacs, *Rhus glabra,* and *R. copallina. Seed Sci. Tech.* 10:223–231.

Farmer, R. E. Jr., M. L. Maley, M. U. Stoehr, & F. Schenburger. 1985. Reproduction characteristics of green alder in northwestern Ontario. *Canadian Jour. Botany* 12:2243–2247.

Farmer, R. E., & R. W. Reinholt. 1986. Seed quality and germination characteristics of tamarack in northwestern Ontario. *Canadian Jour. Forest Res.* 16:680–683.

Farrant, J. M., P. Berjak, & N. W. Pammenter. 1984. The effect of drying rate on viability retention of recalcitrant propagules of *Avicennia marina. South African Jour. Botany* 42:431–438.

Fechner, G. H., K. E. Burr, & J. F. Myers. 1981. Effects of storage, temperature, and moisture stress on seed germination and early seedling development of trembling aspen. *Canadian Jour. Forest Res.* 11:718–722.

Felker, P., P. R. Clark, J. F. Osborn, & G. H. Cannell. 1984. *Prosopis* pod production-comparison of North American, South American, Hawaiian, and African germplasm in young plantations. *Economic Botany* 38:36–51.

Fernald, M. L. 1950. *Gray's Manual of Botany.* American Book Co. New York.

Fernandes, P. S. 1977. Induction of flowering and fruiting. Publ. Inst. Florestal, Sao Paulo, Brazil.

Ferreira, M., R. M. Brandi, & G. Schneider. 1977. Flowering and fruiting in *Eucalyptus grandis* of hybrid origin at Vicosa, Minas Gerais. *Revista Ceres* 24:341–344.

Ffolliott, P. F., & J. L. Thames. 1983. Collection, handling, storage and pretreatment of *Prosopis* seeds in Latin American. FAO, Rome, Italy.

Fins, L. 1981. Seed germination of giant sequoia. *Tree Planter's Notes.* 32:3–8.

Fisher, R. F. 1979. Possible allelopathic effects of reindeer-moss on jackpine and white spruce. *Forest Sci.* 25:256–260.

Flanagan, L. B., & W. Moser. 1985. Flowering phenology, floral display and reproductive succession in dioecious *Aralia nudicaulis* L. *Oecologia* 68:23–28.

Fleming, R. L., & S. A. Lister. 1984. Stimulation of black

spruce germination by osmotic priming: laboratory studies. Great Lakes Forest Res. Centre, Sault Ste. Marie, MI.

Flores, E. M., & B. Mora. 1984. Germination and seedling growth of *Pethecellobium arboreum. Turrialba* 34:485–488.

Flowerdew, J. R., & G. Gardner. 1978. Small rodent population and food supply in a Derbyshire ashwood. *Jour. Animal Ecology* 47:725–740.

Foiles, M. W. 1974. *Atriplex.* In: *Seeds of Woody Plants in the United States.* Forest Service, USDA, Washington, DC. 240–243.

Fordham, A. F. 1969. *Acer griseum* and its propagation. *Plant Propagation* 19:346–348.

Fosket, E. B., & W. R. Briggs. 1970. Photosensitive seed germination in *Catalpa speciosa. Botanical Gazette* 131:167–172.

Fowler, C. J., & D. K. Fowler. 1987. Stratification and temperature requirements for germination of autumn olive. *Tree Planter's Notes* 38:14–17.

Franco, J. D. 1950. *Abetos.* An. Inst. Super. Agron., Lisbon, Portugal.

Franklin, J. F. 1968. Cone production by upper-slope conifers. Forest Service, USDA, Portland, OR.

_____ . 1974. *Abies.* In: *Seeds of Woody Plants in the United States.* Forest Service, USDA, Washington, DC. 168–183.

Fraser, J. W. 1971. Cardinal temperatures for germination of six provenances of white spruce seeds. Canadian Dept. Agr.

Fraser, J. W., & M. J. Adams. 1980. The effect of pelleting and encapsulation on germination of some conifer seeds native to Ontario. Great Lakes Forest Res. Centre, Sault Ste. Marie, MI.

Frett, J. J. 1989. Germination requirements of *Hovenia dulcis* seeds. *Hort. Sci.* 24:152.

Frutostomas, D. 1981. State of seed moisture and effectiveness of gibberellic acid in the germination of *Juglans regia. Recursos Naturales Inst. Nac. de Invest. Agrarias.* 14.

Fuhrer, E., & M. Pall. 1984. On the selection of beech seed stands. Possibilities of increasing seed production by the use of fertilizer. *Centralbatt fur das Gesamte Forstwesen* 101:33–48.

Fulbright, T. E., K. S. Flennika, & Waggerman. 1986. Enhancing germination of spiny hackberry seeds. *Jour. Range Manage.* 39:552–554.

Fung, M. Y. P. 1984. Silverberry seed pretreatment and germination techniques. *Tree Planter's Notes* 35:32–33

Gabriel, W. J. 1978. Genetic variation in seed and fruit characteristics in sugar maple. Forest Service, USDA, Portland, OR.

Galaaen, R., & K. Venn. 1981. Effects of benomyl on the germination of seeds of *Picea abies. Meddelelser fra Norsk Inst. for Skogforskning* 34:229–236.

Galeev, L. V., & P. I. Chikizov. 1976. Collection and processing of cones. *Lesnoe Khozyaistvo* 14–16.

Galil, & L. Meri. 1981. Druplet germination in *Ficus regligiosa. Israel Jour. Botany* 30:41–47.

Galli, M. G., & P. Miracca. 1979. Interaction between abscissic acid and fusicoccin during germination and post germination growth in *Haplopappus gracilis. Plant Sci. Letters* 14:105–111.

Galli, M. G., E. Sparvoli, & M. Caroi. 1975. Comparative effects of fusicoccic and gebberellic acid on the promotion of germination and DNA synthesis initiation in *Hapopappus gracilis. Plant Sci. Letters* 5:351–357.

Gardner, G. 1977. The reproduction capacity of *Fraxinus excelsior* on the Derbyshire limestone. *Jour. Ecology* 65:107–118.

Gaussen, H. 1964. Les gymospermes actuelles et fossiles. In: *Genres Pinus (suite), Cedrus, et Abies.* Toulouse, France. 273–480.

Gaut, P. C., & J. N. Roberts. 1984. *Hamamelis* seed germination. *Plant Propagation* 34:334–342.

Geary, T. F., G. A. Cortes, & H. H. Hadley. 1971. Germination and growth of *Pinus caribaea* directly sown into containers as influenced by shade, phosphate and captan. *Turrialba* 21:336–342.

Geary, T. F., & W. F. Miller. 1982. Pelleting small volumes of *Eucalyptus* seeds for precision sowing. *South African Forestry Jour.* 79–83.

Gendel, S. M., D. E. Fosket, & J. P. Miksche. 1977. Increasing white ash seed germination by embryo dissection. Forest Service, USDA, Blacksburg, VA.

Genebank Handbook. 1985. See Ellis et al. 1985.

Generalo, M. L. 1977. Effects of pre-treatment media on the germination of palasan and limuran seeds in Pagbilao, Quezon. *Sylvatrop* 2:215–218.

Georgieva, I. 1978. Effect of irradiation of pine seeds with beta-rays. *Gorsko Stopanstvo* 34:20–24.

Georgievskii, A. B. 1974. Regeneration of *Haloxylon ammodendron* associations in the Repetek Valley. *Botanicheskii Zhurnal* 59:1033–1045.

Georgievskii, A. B., & A. Khodzamkuliev. 1977. Ecology of reforestation of black saxaul. *Ekologiya* 21–26.

Ghai, S. K., D. N. L. Roa, & L. Batra. 1985. Effect of salinity and alkalinity on seed germination of three tree type sesbanias. Nitrogen Fixing Tree Res. Rpt. 3:10–12.

Ghosh, R. C., & A. Kumar. 1981. Thermal sensitivity of chir pine seed for germination. Inst. Nac. de Invest. Forest, Mexico City.

Ghosh, R. C., B. Singh, & K. K. Sharma. 1974a. Standardization of nursery techniques of tropical pines. I. Hastening germination of *Pinus caribaea* and *Pinus patula. Indian Forestry* 100:407–421.

_____ . 1974b. Standardization of nursery techniques for tropical pines. 2. Germination medium for *Pinus caribaea* var. *hondurensis* and *Pinus patula. Indian Forestry* 100:491–496.

_____ . 1976. Effect of seed grading by size on germination and growth of pine seedlings. *Indian Forestry* 102:850–858.

Giannini, R., G. S. Mugnozza, & R. Bellarosa. 1983. Effect of year of seed ripening on some characteristics of seeds and seedlings of stone and maritime pine. *L'Italia Forestale e Montana* 38:173–183.

Gibson, A., & E. P. Bachelard. 1986. Germination of *Eucalyptus gieberi* seeds. I. Response to substrate and atmospheric moisture. *Tree Physiology* 1:57–65.

_____ . 1987. Provenance variation in germination response to water stress of seeds of some eucalypt species. *Australian Forest Res.* 17:49–58.

Gill, J. D., & F. L. Pogge. 1974a. *Aronia.* In: *Seeds of Woody Plants in the United States.* Forest Service, USDA, Washington, DC. 232–234.

_____ . 1974b. *Chionanthus.* In: *Seeds of Woody Plants in the United States.* Forest Service, USDA, Washington, DC. 323–325.

_____ . 1974c. *Cytisus:* In: *Seeds of Woody Plants in the United States.* Forest Service, USDA, Washington, DC. 370–371.

_____ . 1974d. *Parthenocissus.* In: *Seeds of Woody Plants in the United States.* Forest Service, USDA, Washington, DC. 568–571.

_____ . 1974e. *Physocarpus.* In: *Seeds of Woody Plants in the United States.* Forest Service, USDA, Washington, DC. 584–586.

_____ . 1974f. *Pyrus.* In: *Seeds of Woody Plants in the United States.* Forest Service, USDA, Washington, DC. 689–691.

_____ . 1974g. *Rosa.* In: *Seeds of Woody Plants in the United States.* Forest Service, USDA, Washington, DC. 732–737.

_____ . 1974h. *Viburnum.* In: *Seeds of Woody Plants in the*

United States. Forest Service, USDA, Washington, DC. 844–850.

Gilmore, A. R. 1985. Allelopathic effects of giant foxtail on germination and radicle elongation of loblolly pine seed. *Jour. Chemical Ecology* 11:583–592.

Girgidov, D. Y., & S. P. Gusev. 1976. Size grading of pine and spruce. *Lesnoe Khozyaistvo* 47–50.

Gladysz, A. 1983. Quality of aspen seed produced in greenhouse. *Sylvan* 127:9–20.

Gladysz, A., & M. Ochlewska. 1983. Effects of the embryo size of aspen seed on seedling growth and development. *Sylvan* 127:31–37.

Godman, R. M., & G. A. Mattson. 1980. Low temperature optimum for field germination of northern red oak. *Tree Planter's Notes* 31:32–34.

Golyadkin, A. I., I. M. Demidenko, & V. P. Shalamov. 1972. The effect of electromagnetic field on the sowing quality of conifer seeds. *Referativnyi Zhurnal* 78–81.

Gomes, J. M., R. M. Brandi, L. Couto, & J. G. Lelles. 1978. Effect of previous treatment of the soil with methyl bromide on the growth of seedlings of *Pinus caribaea* var. *hondurensis* in the nursery. *Brasil Florestal* 9:18–23.

Goo, M. I., I. Ishikawa, & H. Ikoda. 1979. Longevity of *Acacia mearnsii* De Wild. seeds. 1. Results of 17 years storage. *Jour. Jap. Forestry Soc.* 61:53–57.

Goor, A. Y., & C. U. Barney. 1976. *Forest Tree Planting in Arid Zones*. The Ronald Press, New York.

Gordon, A. G., D. C. Wakeman, & H. Ghazal. 1979. Trials of the fluid drilling technique on conifer seeds. Forestry Comm., Farnham, Surrey, UK.

Gorobets, A. M. 1978. Experimental-ecological study of seed germination. 2. Germination of *Salix phylicifolia* and *Salix caprea* seed. *Vestnik. Biologia Leningradski Univ.* 57–62.

Gorobov, D. L. 1985. Fruit ripening asynchrony is related to variable seed number in *Amelanchier* and *Vaccinium*. *American Jour. Botany* 72:1939–1943.

Gottried, G. T., & L. J. Heidmann. 1986. Effect of cold stratification treatment on pinyon germination. Forest Service, USDA, Ogden, UT.

Graaff, J. L., & J. V. Staden. 1984. The germination characteristics of two *Sesbania* species. *South African Jour. Botany* 3:59–62.

Granstron, A., & C. Fries. 1985. Depletion of viable seeds of *Betula pubescens* and *B. verrucosa* on some northern Swedish forest soils. *Canadian Jour. Forest Res.* 6:1176–1180.

Gregorius, H. R., J. Kraukausen, & G. Muller-Stark. 1986. Spatial and temporal genetic differentiation among the seed in stands of *Fagus sylvatica*. *Heredity* 57:255–262.

Grisez, T. J. 1974. *Prunus*. In: *Seeds of Woody Plants in the United States*. Forest Service, USDA, Washington, DC. 658–673.

Groot, A. 1985. Application of germination inhibitors in organic solvents to conifer seeds. Great Lakes Forest Res. Centre, Sault Ste. Marie, MI.

Grubisic, D., M. Noskovic, & R. Konjevic. 1985. Changes in light sensitivity of *Paulownia tomentosa* and *Paulownia fordunei* seeds. *Plant Sci.* 39:13–16.

Grzeskowiak, H., & B. Suszka. 1983. Storage of partially afterripened and dried mazzard seeds. *Arbor. Kornickie* 28:261–281.

Grzywacz, A. P., & J. Rosochacka. 1980. The colour of *Pinus sylvestris* seeds and their susceptibility to damping -off. I. The colour and quality of seeds and fatty acid content of the seed coat. *European Jour. Forest Pathology* 10:138–144.

Guariglia, R. D., & B. E. Thompson. 1985. The effect of sowing depth and mulch on germination and 1+0 growth of Douglas-fir seedlings. Forest Service, USDA, Ogden, UT.

Gulati, N. K., & B. K. Sharma. 1983. Propagation of *Calamus*

tenuis. *Indian Forestry* 109:541–546.

Gul'binene, N., & R. Murkaite. 1978. Effect of ultrasonic treatment on the germination of spruce seeds. *Lesnoe Khozyaistvo* 36–37.

Gunatilake, A. A., A. M. Y. Silva, S. Sotheeswaran, & S. Balasubramania. 1984. Terpenoid and biflavonoid constituents of *Calophyllum calaba* and *Garoinia specata* from Sri Lanka. *Phytochemistry* 23:233–238.

Gunn, C. R. 1984. *Fruits and Seeds in the Subfamily Mimosoideae (Fabaceae)*. ARS, USDA, Washington, DC.

Guo, W. M., L. C. Zhang, & Y. S. Cheng. 1981. Preliminary research on isoperoxidases in Korean pine (*Pinus koraiense*) seeds. *Jour. North-Eastern Forestry Inst.* (China) 4:51–57.

Gupta, B. N., & A. Kumar. 1976. Estimation of potential germinability of teak fruits from twenty-three Indian sources by cutting test. *Indian Forester* 102:808–813.

Gupta, B. N., P. S. Pathak, & R. D. Roy. 1983. Seedling growth of *Leucaena leucocephala* (Lam.) de Wit. 2. Effect of seed size. *Indian Jour. Forestry* 6:202–204.

Gupta, B. N., & P. G. Pattanath. 1975. Factors affecting germination behaviour of teak seeds of eighteen Indian sources. *Indian Forester* 101:584–588.

Haase, S. 1986. Effect of prescribed burning on soil moisture and germination of southwestern ponderosa pine seed on basaltic soils. Forest Service, USDA, Fort Collins, CO.

Haavisto, V. F., & D. A. Winston. 1974. Germination of black spruce and jackpine seed at 0.5°C. *Forest Chronicle* 50:240.

_____. 1977. Germination of black spruce and jack pine seed on soil and germination paper media following paraquat spraying. *Bi-Mon. Res. Notes* 33:1–2.

Hager, M. 1985. Pregerminated seeds. *Sveriges Skogsvardsforbunds Tidskrift* 57–63.

Haines, R. J., & R. J. Gould. 1983. Separation of the seeds of *Auraucaria cunninghamii* by flotation. *Australian Forest Res.* 13:299–304.

Halevy, G. 1974. Effects of gazelle and seed beetles (Bruchidae) on germination and establishment of *Acacia* species. *Israel Jour. Botany* 23:120–126.

Hall, H. M., & P. E. Clements. 1923. *The Phylogenetic Method in Taxonomy; The North American Species of* Artemisia, Chrysothamnus, *and* Atriplex. Carnegie Inst. f Wash., Washington, DC.

Hall, I. V., & G. E. Biel. 1970. Seed germination, pollination, and growth of *Vaccinium vitis-idaea* var. *minus*. *Canadian Jour. Plant Sci.* 50:731–734.

Hall, ŁJ. P. 1981. Yield of seed in *Larix laricina* in Newfoundland. *Canadian Forestry Service Res. Notes* 1:1–2.

_____. 1985. Variation in seed quantity and quality in two grafted clones of European larch. *Silvae Genetica* 34:51–56.

Hall, J. P., & I. R. Brown. 1977. Embryo development and yield of seed in *Larix*. *Silvae Genetica* 26:77–84.

Hall, O. 1984. The effects of dry heat on tree seed germination: lodgepole pine, slash pine and Douglas fir. *Newsletter AOSA* 58:117–121.

_____. 1987. Methods for stimulating eastern white pine seed germination. *Newsletter ASOA* 58:114–116.

Hall, O., & E. Olson. 1986. Effect of stratification, drying and cold storage on Noble fir and Pacific silver fir. *Jour. Seed Tech.* 10:58–61.

Hamilton, D. F. 1972. Ethrel for breaking dormancy in seeds of some woody plants. *Plant Propagation* 22:368–373.

_____. 1976. Regulation of seed dormancy in *Elaeagnus umbellata* by endogenous growth substances. *Canadian Jour. Botany* 54:1068–1073.

Hamilton, D. F., & P. L. Carpenter. 1975. Regulation of seed dormancy in *Elaeagnus umbellata* by endogenous growth substances. *Canadian Jour. Botany* 53:2303–2311.

Hams, A. S., & W. I. Stein. 1974. *Sorbus.* In: *Seeds of Woody Plants in the United States.* Forest Service, USDA, Washington, DC. 789–784.

Hanna, P. J. 1984. Anatomical features of the seed coat of *Acacia kempeana* Mueller which relate to increased germination rate induced by heat treatment. *New Phytologist* 96:23–29.

Hardin, E. D. 1984. Variation in seed weight, number per capsule and germination in *Populus deltoides* trees in southeastern Ohio. *American Midland Naturalist* 112:29–34.

Hare, R. C. 1981. Nitric acid promotes pine seed germination. Forest Service, USDA, Gulfport, MS.

Harif, I. 1977. Influence of weight of *Quercus calliprimos* acorns on establishment of seedlings in the Judean Hills. *Israel Jour. Botany* 26:43–44.

Halou, W. M., & E. S. Harrar. 1958. *Textbook of Dendrology.* McGraw-Hill Book Co., New York.

Harniss, R. O., & W. T. McDonough. 1976. Yearly variation in the germination of three subspecies of big sagebrush. *Jour. Range Manage.* 29:167–168.

Harrar, E. S., & J. George. 1962. *Guide to Southern Trees.* Dover Publ. Inc., New York.

Harre, J. 1987. Propagation of South African Proteaceae by seed. *Plant Propagation* 36:470–476.

Harrington, M. G. 1987. Phytotoxic potential of gambel oak on ponderosa pine seed germination and initial growth. Forest Service, USDA, Fort Collins, CO.

Harris, A. S. 1974. *Chamaecyparis.* In: *Seeds of Woody Plants in the United States.* Forest Service, USDA, Washington, DC. 316–320.

Harris, A. S., & W. I. Stein. 1974. *Sorbus.* In: *Seeds of Woody Plants of the United States.* Forest Service, USDA, Washington, DC. 780–784.

Harvey, A. M. 1978. Effect of seedbed cover on germination of patual pine seed. Queensland Dept. Forestry, Queensland.

Hashizume, H. 1979. Changes in chemical constituents during the development of Japanese beechnut. *Jour. Japanese Forestry Soc.* 61:342–345.

Hashizume, H., & H. Fukutomi. 1978. Development and maturation of fruits and seeds of *Fagus crenata. Jour. Japanese Forestry Soc.* 60:163–168.

Haverboke, D. F., & M. R. Barnhart. 1978. A laboratory technique for depulping Juniper cones. *Tree Planter's Notes* 29:33–34.

Hawksworth, F. G. 1965. Life tables for two species of dwarf mistletoes. 1. Seed dispersal, interception, and movement. *Forest Sci.* 11:142–151.

Hawksworth, F. G., & D. Wiens. 1984. Biology and classification of *Arceuthobium:* an update. Forest Service, USDA, Fort Collins, CO.

Hedderwick, G. W. 1970. Prolonged drying of stratified Douglas-fir seed affects laboratory germination. *Commonwealth Forest Rev.* 49:130–131.

Hedlin, A. F. 1966. Cone and seed insects of grand fir, *Abies grandis* (Dougl.) Lindl. *Bi-Mon. Res. Notes* 22:3.

Hedlin, A. F., & D. S. Ruth. 1978. Examination of Douglas-fir clones for differences in susceptibility to damage by cone seed insects. *Jour. Entomological Soc.* (British Columbia) 75:33–34.

Heidmann, L. J. 1986. Acetone is unreliable as a solvent for introducing growth regulators into seeds of southwestern ponderosa pine. Forest Service, USDA, Ogden, UT.

Heit, C. E. 1955. The excised embryo method for testing germination quality in dormant seeds. *Proc. ASOA* 45:108–117.

———. 1967a. Fall planting of fruit and hardwood seeds.

American Nurseryman 126:12–13, 85–90.

———. 1967b. Laboratory germination and suggested testing methods for ten less common exotic pine species. *Proc. ASOA* 57:161–169.

———. 1967c. Propagation from seed. Part 6: Hardseededness—a critical factor. *American Nurseryman* 125:10–12, 88–96.

———. 1967d. Propagation from seed. Part 2. Storage of deciduous tree and shrub seeds. *American Nurseryman* 126:12–13, 86–94.

———. 1968a. Propagation from seed. Part 15: Fall planting of shrub seeds for successful seedling production. *American Nurseryman* 128:8–10, 70–81.

———. 1968b. Thirty-five years' testing of tree and shrub seeds. *Jour. Forestry* 66:632–634.

———. 1970a. Germination characteristics and optimum testing methods for twelve western shrub species. *Proc. ASOA* 60:197–205.

———. 1970b. Laboratory germination of barrel and saguaro cactus seed. *Newsletter ASOA* 44:11–15.

———. 1971. Propagation from seed. Part 22: Testing and growing western desert and mountain shrub seeds. New York Agr. Exp. Sta. Bulletin.

———. 1973a. Germination testing of four tropical pine species from Mexico and Central America. *Newsletter ASOA* 47:46–48.

———. 1973b. Optimum germination testing conditions for cacti mixtures. *Newsletter ASOA* 47:55–56.

———. 1974. Laboratory germination and testing method for *Syringa villosa* seed. *Newsletter ASOA* 48:26–28.

———. 1976a. Laboratory germination and recommended testing methods for 15 less common and exotic *Abies* species. *Newsletter ASOA* 50:19–24.

———. 1976b. Laboratory germination testing of *Hippophae rhamnoides. Newsletter ASOA* 50:44–46.

———. 1976c. Seed studies and laboratory germination testing of *Pinus brutia* and *Pinus eldarica. Newsletter ASOA* 50:50–55.

Heit, C. E., & J. J. Natti. 1969. Accurate germination of *Abies balsamae* and *Abies fraser* in laboratory tests by control of *Phizootonia. Proc. ASOA* 59:148–153.

Hellum, A. K. 1973. Seed storage and germination of balsam popal. *Canadian Jour. Plant Sci.* 53:227–228.

Hellum, A. K., & I. Dymock. 1986. Cold stratification for lodge pole pine seeds. In: *Proc. Symp. Biol.* Atriplex *and Related Chenopods.* Gen. Tech. Rpt. 172. Forest Service, USDA, Ogden, UT. 107–111.

Hellum, A. K., & L. Hackett. 1988. Variable dormancy in seed of *Pinus contorta. Scandinavian Jour. Forestry Res.* 3:137–146.

Hellum, A. K., A. Loken, & S. Gjelsvik. 1983. Seed and cone maturity in *Pinus contorta* from Alberta. Agr. and Forestry Bul., Univ. Alberta 6:24–30.

Helms, J. A. 1987. Invasion of *Pinus contorta* var. *murrayana* into mountain meadows of Yosemite National Park, California. *Madrono* 34:91–97.

Herrera, C. M., & P. Jordano. 1981. *Prunus mahaleb* and birds: the high-efficiency seed dispersal system of a temperature fruiting tree. *Ecol. Monog.* 51:203–218.

Hess, C. W. M. Jr. 1973. Seedling propagation of difficult species. *Plant Propagation* 23:284–285.

Hickey, J. E., A. J. Blakesley, & B. Turner. 1983. Seedfall and germination of *Nothofagus cunninghamii* (Hook.) Oerst., *Eucryphia lucida* (Labill.) Baill and *Atherosperma moschatum* Labill. Implications for regeneration practice. *Australian Forest Res.* 13:21–28.

Hill, J. A. 1976. Viability of several species of conifer seeds after long term storage. *Tree Planter's Notes* 27:2–3.

Hodgson, L. M. 1977a. In situ grafting in *Eucalyptus grandis* seed orchards at J. D. M. Keet Forest Research Station. *South African Forestry Jour.* 58–60.

———. 1977b. Methods of seed orchard management for seed production and ease of reaping in *Eucalyptus grandis*. *South African Forestry Jour.* 38–42.

Hodgson, T. J. 1977. Effects of organic flotation media on germination of *Pinus elliotii, P. patula*, and *P. taeda* seed. *South African Forestry Jour.* 17–21.

Hoff, R. J. 1981. Cone production of western white pine seedlings and grafts. Forest Service, USDA, Ogden, UT.

———. 1986. Effect of stratification time and seed treatment on germination of western white pine seed. Forest Service, USDA, Ogden, UT.

Hoff, R. J., & R. J. Steinhoff. 1986. Cutting stratified seed of western white pine to determine viability or to increase germination. *Tree Planter's Notes* 37:25–26.

Hollis, C. A., J. E. Smith, & R. F. Fisher. 1982. Allelophathic effects of common understory species on germination and growth of southern pines. *Forest Sci.*. 28:509–515.

Holmes, G. D., & G. Buszeuicz. 1958. The storage of seeds of temperature forest species. *Forest Abstracts* 19:313–322.

Hopper, G. M., D. Parrish, & D. W. Smith. 1985. Adenylate energy charge during stratification and germination of *Quercus rubra*. *Canadian Jour. Forest Res.* 15:829–832.

Hopper, G. M., D. W. Smith, & D. J. Parrish. 1985. Germination and seedling growth of northern red oak: effects of stratification and pericarp removal. *Forest Sci.* 31:31–39..

Horn, P. E., & G. D. Hill. 1974. Chemical scarification of seeds of *Lupinus cogentini Guss. Jour. Australian Agr. Sci.* 40:84–87.

Houston, D. B. 1976. Determining the quality of white ash seed lots by x-ray analysis. *Tree Planter's Notes* 27:8,23.

Howcroft, N. H. S. 1974. Observations on the development of ovulate strobili and maturation and ripening of cones of *Pinus merkusii*. *Tropical Forestry Res. Note,* Papua New Guinea. 12.

Hu, T., T. P. Lin, Y. L. Chung, & W. Yang. 1978. The germination study of the seeds of *Chamaecyparis formosensis*. Bul. Taiwan Forest Res. Inst. 315:11.

Hubbard, R. L. 1974a. *Castanopsis*. In: *Seeds of Woody Plants in the United States*. Forest Service, USDA, Washington, DC. 276–277.

———. 1974b. *Rhamnus*. In: *Seeds of Woody Plants in the United States*. Forest Service, USDA, Washington, DC. 704–708.

Huluta, C., & A. Tomescu. 1973. The simulation of seed germination and seedling emergence, survival and growth in *Pinus nigra, P. sylvestris, Fraxinus excelsior*, and *F. tomentosa* by ionizing radiation and folcysteine Oeriu 'P'. *Silviculturel* 29:171–213.

Hunt, D. R. 1971. *Eurya japonica. Curtis's Bot. Mag.* (May): 3.

Hylton, L. O. 1974. *Penstemon*. In: *Seeds of Woody Plants in the United States*. Forest Service, USDA, Washington, DC. 574–575.

Hyun, S. K. n.d. The conservation of forest plant and animal genetic resources in Korea. Inst. Forest Gen., Suweon, Korea.

Ignaciuk, R., & J. A. Lee. 1980. The germination of four strand-line species. *New Phytologist* 84:581–591.

Il'ina, N. A. 1982. Effect of low temperature on the germination of sea buckthorn. Byulleten Vsesoivzny Inst. Rastenievodstva 1982:51–53.

Illick, J. S., & E. F. Brouse. 1926. *The Ailanthus Tree in Pennsylvania*. Penn. Dept. Forest and Water.

Inyushin, V. M., V. M. Fedorova, & N. N. Lazorenko. 1983. Effect of a laser beam on the germination of Scots pine seeds. *Lesnoe Khozyaistvo* 1983:31–33.

Ishihata, K. 1974. Studies on the morphology and cultivation of palms. Bul. Faculty of Agr. Kagoshima Univ. 24:11–23.

Ishii, Y. 1979. Effect of light and temperature on the germina-tion of *Zelkova serrata* seeds with various prechilling times. *Jour. Japanese Forestry Soc.* 61:362–366.

Isik, K. 1986. Altitudinal variation in *Pinus brutia*; seed and seedling characteristics. *Silvae Genetica* 35:58–67.

Istanbouli, A., M. Arban, & A. Kasbi. 1974. Rapid reproduction of olive trees from seeds. *Olivae* 4:30–35.

Istanbouli, A., & P. Neville. 1974. "Dormancy" of the seeds of *Olea europaea*. Evidence for an inhibition caused by the albumen. *Comptes Rendus Hebdomadaires des Seances de l'Academie des Sciences* (France) 279:1441–1442.

———. 1977. Study of dormancy in seeds of *Olea europaea*. Favourable effect of light in the presence of obstacles to germination. *Comptes Rendus Hebdomadaires des Seances de l'Academie des Sciences* (France) 285:41–44.

Iwegbulam, E. R. 1983. Effect of pre-sowing treatments on germination of seeds of *Gmelina arborea. Forestry Abstracts* 44:606.

Jacobson, T. L., & B. L. Welch. 1987. Planting depth of 'Hobble Creek' mountain big sagebrush. *Great Basin Naturalist* 47:497–499.

Janerette, C. A. 1977. *The Physiology of Water Uptake in Seeds of Acer saccharum Marsh*. Ph. D. Thesis, North Carolina State, Durham, N.C.

———. 1978. A method stimulating the germination of sugar maple seeds. *Tree Planter's Notes* 29:7–8.

———. 1979a. Cold soaking reduces the stratification requirement of sugar maple seeds. *Tree Planter's Notes* 30:3.

———. 1979b. The effects of water soaking on the germination of sugar maple seeds. *Seed Sci. Tech.* 7:341–346.

Janson, L. 1979. Storage of acorns for more than one year. Prace Inst. Badawezego Lesnictwa (Poland) 46–65.

Jarvis, B. C. 1966. *Nucleotide Metabolism in Dormant and Non-Dormant Seeds*. Ph. D. Thesis, University of Wales.

———. 1975. The role of seed parts in the induction of dormancy of hazel. *New Phytologist* 75:491–494.

Jarvis, B. C., & D. A. Wilson. 1978. Factors influencing growth of embryonic axes from dormant seeds of hazel. *Planta* 138:189–191.

Jaynes, R. A. 1971. Seed germination of six *Kalmia* species. *Hort. Sci.* 96:668–672.

———. 1982. Germination of *Kalmia* seed after storage of up to 20 years. *Hort. Sci.* 17:203.

Jeavons, R. A., & B. C. Jarvis. 1984. The breaking of dormancy in hazel seed by pretreating with ethanol and mercuric chloride. *New Phytologist* 96:551–554.

Jeffers, R. M. 1985. Seed quality in five jack pine stands in northern Wisconsin. Res. Note No. 328. Forest Service, USDA, North Central Exp. Sta.

Jeffery, D. J., P. M. Holmes, & A. G. Robelo. 1988. Effects of dry heat on seed germination in selected indigenous and alien legume species in South Africa. *South African Assoc. Botantists* 54:28–34.

Jo, D. G., H. M. Kwon, S. G. Choi, & S. U. Han. 1983. Self-pollination effects in a *Pinus densiflora* seed orchard on seed production, germination, and survival of seedlings. Res. Rpt. Korean For. Gen. 19:66–72.

Jobidon, R. 1986. Allelopathic potential of conifer species to old-field weeds in eastern Quebec. *Forest Sci.* 32:112–118.

Johnsen, T. N. Jr., & R. A. Alexander. 1974. *Juniperus*. In: *Seeds of Woody Plants in the United States*. Forest Service, USDA, Washington, DC. 460–469.

Johnson, C. D. 1983. *Handbook of Seed Insects of Prosopis Species*. FAO, Rome.

Johnson, G. R. 1983. Removing cull acorns from a water-willow oak acorn lot using salt solutions. *Tree Planter's Notes* 34:10–12.

Johnson, G. R. Jr., & R. C. Kellison. 1984. Sycamore seedlings from the nursery—not the same genetic composition as the collected seedlot. *Tree Planter's Notes.* 35:34–35.

Johnson, L. C., 1974a. *Cupressus.* In: *Seeds of Woody Plants in the United States.* Forest Service, USDA, Washington, DC. 63–369.

———. 1974b. *Metasequoia.* In: *Seeds of Woody Plants in the United States.* Forest Service, USDA, Washington, DC. 540–542.

Johnson, R. L. 1979. A new method of storing Nuttal oak acorns over winter. *Tree Planter's Notes* 30:6–8.

Johnson, R. L., & R. M. Krinard. 1985. Oak seeding on an adverse field site. Forest Service, USDA, New Orleans.

Jones, L. 1962. Recommendations for successful storage of tree seeds. *Tree Planter's Notes* 55:9–20.

Jones, R. J., G. Villamizar, & S. J. Cook. 1983. The effect of seed treatments and nitrogen fertilizer on seeding growth of *Leucaena.* Leucaena Res. Rpts. 4:4–5.

Julin-Tegelman, A., & N. Pinfield. 1982. Changes in the level of endogenous cytokinin-like substances in *Acer pseudoplatanus* embryos during stratification and germination. *Physiologia Plantarum* 54:318–322.

Junttila, O. 1974. Seed quality and seed production of woody ornamentals in Scandinavia. *Meldinger fra Norges Landbrukshogskole* 53:41.

———. 1976. Seed germination and viability in five *Salix* species. *Astarte* 9:19–24.

Kamenick, A., & M. Rypak. 1985. Effect of growth regulators upon the growth of embryos of the American ash in vitro before the onset of dormancy. *Slovenska Akademia Vied* 40:859–865.

Kamra, S. K. 1982a. Factors effecting seed storage. In: *Proc. Inter. Symp. Forest Tree Seed Storage.* Canadian Forestry Service, Chalk River, Ontario. 136–220.

———. 1982b. Seed biology of lodgepole pine. Rpt. Inst. for Skoglig Genetik och Vaxtfysiologi, Sverges Lantbuksuniversitet, Umea, Sweden. 51.

Kandya, A. K. 1978. Relationship among seed weight and various growth factors in *Pinus oocarpa* seedlings. *Indian Forester* 104:561–567.

Kandya, A. K., & K. Ogino. 1987. Vigor of excised embryos of four weight groups of *Pinus thunbergii, P. densiflora,* and *P. pentaphylla. Indian Jour. Forestry* 10:279–282.

Kardell, L. 1974. Stratification of Scots pine seeds. Inst. for Skogsskotsel, Sweden.

Karrfalt, R. P. 1983. Fungus-damaged seeds can be removed from slash pine seed lots. *Tree Planter's Notes* 34:38–40..

Karrfalt, R. P., & R. E. Helmuth. 1984. Preliminary trials with upgrading *Platanus occidentalis* with the Helmuth electroastatic seed separator. Forest Service, USDA, Ogden, UT.

Karschow, R. 1975. Seedling morphology and seed germination of *Arbutus andrachne* L. Yaaron For. Israel Forest Assoc. 25:31–32.

Katsuta, M., M. Saito, C. Yamamoto, T. Kaneko, & M. Itoo. 1981. effect of gibberellins on the promotion of strobilus production in *Larix leptolepis* Gord. and *Abies homolepis* Sieb. et Zucc. Bul. Forestry and Forest Products Res. Inst., Japan, No. 313:37–45.

Kaul, R. B. 1986. Evolution and reproductive biology of inflorescences in *Lithocarpus, Castanopsis, Castanea,* and *Quercus. Annals Missouri Bot. Gardens* 73:284–296.

Kawecki, Z. 1973. Studies concerning the physiology of stratified walnut seeds. *Zezyty Naukowe Akademii Rolniczo-Technicznej w Olsytynie, Rolnictwo* 41–85.

Kay, B. L., W. L. Graves, & J. A. Young. 1988. *Long-term Storage of Desert Shrub Seeds.* Agronomy and Range Science, Univ. California, Davis.

Kay, B. L., W. L. Ross, W. L. Graves, C. R. Brown, & J. A. Young. 1977. *Mojave Revegetation Notes.* Agronomy and Range Science, Univ. California, Davis, CA.

Kazadaev, S. A. 1985. Propagation of *Fraxinus pensylvanica* by winter cuttings. *Lesnoe Khozyaistvo* 12:32–33.

Ke, S., R. M. Skirvin, A. G. McPheeters, & G. Galletta. 1985. In vitro germination and growth of *Rubus* seeds and embryos. *Hort. Sci.* 20:1047–1049.

Keeley, J. E. 1968. Seed germination patterns of *Salvia mellifera* in fire-prone environments. *Oecologia* 71:1–5.

Keeley, J. E., & D. A. Tufenkian. 1984. Garden comparison of germination and seedling growth of *Yucca whipplei. Madrono* 31:14–29.

Keen, F. P. 1968. Cone and seed insects of western forests. Forest Service, USDA, Washington, DC.

Keiding, H. 1985. Teak, *Tectonia grandis* L. Seed Leaflet DANIDA. Forest Seed Center, Denmark. 21.

Keiding, H., & F. Knudsen. 1974. Germination of teak seed in relation to international provenance testing. *Forest Tree Improvement* 19–29.

Kellman, M. 1985. Forest seedling establishmen in Neotropical savannas: transplant experiments with *Xylopia frutescens* and *Calophyllum brasiliense. Jour. Biog.* 12:373–379.

Kenady, R. M. 1978. Regeneration of red alder. In: *Utilization and Management of Alder.* Forest Service, USDA, Portland, OR. 183–191.

Keng, H. 1977. Theaceae. In: *Tree Flora of Malaya.* Forest Res. Inst., Kepong. 275–296.

Khan, D., S. Shaukat, & M. Faheemuddin. 1984. Germination studies of certain desert plants. *Pakistan Jour. Botany* 16:231–254.

Khan, M. A., N. Sankhla, D. J. Weber, & E. D. McArthur. 1987. Seed germination characteristics of *Chrysothamnus nauseosus* ssp. *viriduls. Great Basin Naturalist* 47:220–226.

Khan, M. A., and I. A. Ungar. 1984a. The effect of salinity and temperature on the germination of polymorphic seeds and growth of *Atriplex triangularis* Willd. *American Jour. Botany* 71:481–489.

———. 1984b. Seed polymorphism and germination responses to salinity stress in *Atriplex triangularis* Willd. *Botanical Gazette* 145:487–494.

———. 1985. The role of hormones in regulating the germination of polymorphic seeds and early seedling growth of *Atriplex triangular* Willd. *under saline conditions. Physiologia Plantarum* 63:109–113.

———. 1986. Inhibition of germination in *Atriplex triangular* seeds by application of phenols and reversal of inhibition by growth regulators. *Botanical Gazette* 147:148–151.

Khan, M. I. 1982. Allelopathic potential of dry fruits of *Washingtonia filifera*: Inhibition of seed germination. *Physiologia Plantarum* 54:323–328.

Khanal, B. K. 1975. Nursery techniques for raising teak in Nepal. Forest Dept., Nepal. 4.

Khutortsov, I. I., & N. B. Anikeeko. 1981. The food value of Fagaceae and their destruction by animals. *Legovedenie* 19:83–85.

Kim, Y. S., & W. S. Kim. 1983. Influence of several factors on seed germination of *Ziziphus jojoba.* Res. Rpt. Rural Develop. Hort., Republic Korea 25:125–130.

King, J. P., & S. L. Krugman. 1980. Test of 36 *Eucalyptus* species in northern California. Forest Service, USDA, Berkeley, CA.

Kiprianov, A. I., T. I. Prokhorchuk, T. V. Sokolova, & A. P. Evdokimov. 1984. The duration of effect of stimulants on growth of spruce and pine seedlings in the open. *Lesnoi Zhurnal* 30–33.

Kiprianov, A. I., T. I. Prokhorchuk, T. V. Sokolova, M. M. Librofanova, E. N. Kibasova, V. I., Sedykh, E. V. Trambitskaya, & N. I. Petrov. 1982. Effect of the organic materials in sulphate liquors on the germination of spruce and pine seeds. *Lesnoi Zhurnal* 14–17.

Kiprianov, A. I., T. I. Prokhorchuk, T. V. Sokolova, M. M. Librofanova, & N. I. Petrov. 1982. Stimulation of the growth of spruce and pine seedlings in the greenhouse. *Lesnoi Zhurnal* 23–27.

Knapp, A. K., & J. E. Andersen. 1980. Effect of heat and germination of seeds from serotinous lodgepole pine cones. *American Midland Naturalist* 104:370–372.

Knutson, D. M. 1971. Dwarf mistletoe seed storage best at low temperatures and high relative humidity. Forest Service, USDA, Portland, OR.

———. 1984. Seed development, germination behavior, and infection characteristics of several species *Arceuthobium*. Forest Service, USDA, Fort Collins, CO.

Knypl, J. S., D. S. Lethau, & L. M. S. Palni. 1985. Cytokinins in maturing and germinating *Lupinus luteus* seeds. *Biology Plant Praha Czechoslovak Acad. Sci.* 27:188–194.

Koa, C., J. H. Tsens, & Z. T. Tsai. 1980. Biochemical changes in seeds of Taiwan white pine and Taiwan acacia during germination. *Quarterly Jour. Chinese Forestry* 13:25–38.

Kochkar, N. T. 1977. Determining the ripeness of seeds of *Picea abies*. *Lesnoe Khozyaistvo* 59–60.

———. 1978. Evaluating the yield of seeds of *Robinia pseudoacacia*. *Lesnoe Khozyaistvo* 35–36.

———. 1983. Fruiting of *Populus nigra* and variation in seed quality with age. *Lesnoe Khozyaistvo* 41–42.

———. 1984. Features of the fruiting of *Populus nigra* in a natural stand and in an avenue planting. *Lesovedenie* 88–91.

Kofman, P. D., & C. Workhoven. 1977. Machine harvesting of tree seeds. *Nederlands Bosbouw Tijdso Arift* 49:264–273.

Koller, D. 1957. Germination mechanisms in some desert seeds. I. *Atriplex dimorphostegia*. *Ecology* 38:1–13.

Komarova, T. A. 1986. Role of forest fires in germination of seeds dormant in the soil. *Soviet Jour. Ecology* 311–315.

Kormanik, P. P. 1983. Third-year seed production in out planted sweet gum related to nursery root colonization by endomycorrhizal fungi. *Proc. Southern Forest Tree Improvement Conf.* 39:49–54.

Koshioka, M., J. Kanazaua, & Y. Murakami. 1968. Identification of gibberellins Al, A9, A17, and A20 in immature seeds of *Pithecellobium microcarpus*. *Agr. Biol. Chem.* (Tokyo) 50:1899–1901.

Kosnikov, B. I., & G. A. Nikulin. 1985. Method of calculating the expected yield and seed damage in *Betula pendula* in production stands. *Lesnoe Khozyaistvo* 12:28–29.

Kowalski, R. M., & Z. Kowecki. 1982. Physiology of breaking hazel seed dormancy. I. Germination of stratified seeds in the year of harvest and after storage for one year. *Roczniski Nauk Rolniczych* 105:179–190.

Kral, R. 1960. A revision of *Asimina* and *Deeringothamnus* (Annonaceae). *Brittonia* 12:233–278.

Kral'ova, J. 1974. The effects of some disinfectants and their concentration in the protection of *Picea abies* and *Pinus sylvestris* seed. *Lesnicky Casopis* 20:97–109.

Krass, N., & K. H. Kohler. 1985. Stratification and germination of ash seeds. *Flora Moiphol Gecht Oekophysiol.* 177:91–105.

Krochmal, A. 1974. *Myrica.* In: *Seeds of Woody Plants in the United States.* Forest Service, USDA, Washington, DC. 548–550.

Krugman, S. L. 1974a. *Eucalyptus.* In: *Seeds of Woody Plants in the United States.* Forest Service, USDA, Washington, DC. 384–392.

———. 1974b. *Menodora.* In: *Seeds of Woody Plants in the United States.* Forest Service, USDA, Washington, DC. 539.

———. 1974c. *Olea.* In: *Seeds of Woody Plants in the United States.* Forest Service, USDA, Washington, DC. 558–559.

———. 1974d. *Olyneya.* In: *Seeds of Woody Plants in the United States.* Forest Service, USDA, Washington, DC. 560.

———. 1974e. *Washingtonia.* In: *Seeds of Woody Plants in the United States.* Forest Service, USDA, Washington, DC. 855–856.

Krugman, S. L., & J. L. Jenkinson. 1974. *Pinus.* In: *Seeds of Woody Plants in the United States.* Forest Service, USDA, Washington, DC. 598–638.

Kruse, W. H. 1970. Temperature and moisture stress affect germination of *Gutierrezia sarothrae*. *Jour. Range Manage.* 23:143–144.

Krusi, B. O., & M. Debussche. 1988. The fate of flowers and fruit of *Cornus sanguinea* in three contrasting Mediterranean habitats. *Oecologia* 74:592–599.

Kudashova, F. N., & F. N. Osetrova. 1976. Cell-physiological and biochemical changes in the seeds of some conifers by space flight. *Institut Botanik Litovskoi* 202–210.

Kulygin, A. A. 1974. Production of seed of *Robinia pseudoacacia*. *Referativnyi Zhurnal* 16:18–25.

———. 1977. The effect of drought on the quality of seeds of *Robinia pseudoacacia*. *Lesnoi Zhurnal* 146–148.

Kumar, A. 1979. Effect of fruit size and source on germination of teak seeds. *Sri Lanka Forester* 14:58–63.

Kumar, A., & R. Sharma. 1982. Determination of viability of ipil-ipil (*Leucaena leucocephala*) (Lam) de Wit seed by tetrazolium. *Van Vigyan* 20:26–29.

Kung, F. S. 1976. Study on the effect of stratification on coniferous tree seed. *Bul. of Taiwan Forestry Res. Inst.* Taipei 282:18.

Kunhikrishnan Nambiar, K. 1946. A novel method of improving the germination of *Prosopis juliflora* seeds. *Indian Forester* 72:193–195.

Kuo, S. R. 1983. Studies on the factors affecting survival and early performance of Luchu pine. National Taiwan Univ., Taipei, Taiwan.

Kushalapa, K. A. 1981. *Leucaena leucocephala* provenances in Philippines-varietal mix-up galore. *Indian Forester* 107:635–637.

Kuznetsova, V. 1978. Storage of seeds of *Larix gmelinii* in snow. *Lesnoe Khozyaistvo* 37–38.

Lagarda, A., G. Martin, & D. E. Kester. 1983. Influence of environment, seed tissue, and seed maturity on 'Manzanillo' olive seed germination. *Hort. Sci.* 18:868–869.

Lagarda, A., G. Martin, & V. S. Polito. 1983. Anatomical and morphological development of 'Manzanil' olive seed in relation to germination. *Jour. American Soc. Hort. Sci.* 108:741–743.

Laidlaw, T. F. 1987. Drastic temperature fluctuations-the key to efficient germination of pin cherry. *Tree Planter's Notes* 38:30–32.

Lajtha, K., J. Weishampel, & W. H. Schlesinger. 1987. Phosphorus and pH tolerance in the germination of the desert shrub *Larrea tridentata*. *Madrono* 34:63–68.

Lalman, M., & A. Misra. 1980a. Effect of growth regulators on seed germination and seedling growth of *Diospyros*. *Kalikasan Laguna* (Kalikasan) 9:49–54.

———. 1980b. Endogenous inhibitors of germination of seeds. *Indian Jour. Experimental Biology* 18:292–295.

Lamond, M. 1978. The pericarp and speed of germination of acorns of pedunculate oak. *Annales des Sciences Forestieres* 35:203–212.

Lamond, M., & J. Levert. 1980. Effects of the pericarp on the water absorption of pedunculate oak acorns. *Annals Botany* 37:73–83.

Lamprey, H. F., G. Halevy, & S. Makacha. 1974. Interaction between *Acacia* seed, bruchid beetles and large herbivores. *East African Wildlife Jour.* 12:81–85.

Langkamp, P., ed. 1980. *Germination of Australian Native Plant Seed.* Inkata Press, Melbourne.

Larsen, J. A. 1922. Some characteristics of seed of coniferous trees from the Pacific northwest. *National Nurseryman* 30:246–248.

Larson, M. M., & M. Davault. 1974. Pine seeds that withstand severe drying before germination may have after germination drought tolerance reduced. *Tree Planter's Notes* 26:22–23.

Laura, M. P. 1978. Flowering of pine clones in seed orchards. *Lesovedenie* 56–62.

Laurence, G. H. M. 1951. *Taxonomy of Vascular Plants*. MacMillan Co., New York.

Laurent, N., & S. A. O. Chamshama. 1987. Studies on the germination of *Erythrina abyssinica* and *Juniperus procera*. *Inter. Tree Crop Jour.* 4:291–298.

Layton, P. A., & R. Goddard. 1983. Low level in breeding effects on germination, survival, and early height growth of slash pine. *Proc. Southern Forest Tree Improvement Conf.* 39,106,116.

Leadem, C. L. 1986. Seed dormancy in three *Pinus* species of the inland mountain west. Forest Service, USDA, Ogden, UT.

Lebeck, F., & G. Skofitsch. 1984. Distribution of serotonin in *Juglan resia* seed during ontogenetic development and germination. *Jour. Plant Physiology* 114:349–353.

Ledgard, N. J., & P. W. Cath. 1983. Seed of New Zealand *Nothofagus* species: studies of seed weight, viability, shape and the effect of varying stratification periods. *New Zealand Jour. Forestry* 28:150–162.

Lehtiniemi, T. 1977. Factors affecting gamma radiation sensitivity of Scots pine and Norway spruce seeds. *Silvia Fennica* 11:69–80.

LePage-Degiury, M. T. 1973. Influence de la acide absoissique sur le developement des embryos de *Taxus baccata* cultives in vitro. *Z. Pflanzeuphysiol.* 70:406–413.

LePage-Degiury, M. T., & G. Garello. 1973. La dormance embryonnaire chez *Taxus baccata*: influence de la composition du milieu liquide sur l'induction de la germination. *Physiologia Plantarum* 29:204–207.

Levert, J. and M. Lamond. 1979. Temperature and germination of English oak. *Acad. d'agriculture de France* 65:1006–1017.

Lill, R. E., & J. H. McWha. 1979. The influence of volatile substances from the litter of *Pinus radiata* on seed germination. *Annals Botany* 43:81–85.

Lin, T. P., T. W. Hu, & J. J. Leu. 1979. The effect of alternating temperatures on the germination of the seed of *Phellodendron* spp. Taiwan Forestry Res. Inst., Taipei.

Lincoln, W. C. 1980. Laboratory germination of *Epigaea repens*. *Newsletter ASOA* 54:72–73.

Lindgren, K., I. Ekberg, & G. Eriksson. 1977. External factors influencing female flowering in *Picea abies*. *Studia Forestalia Suecica* 1977:1–53.

Lineberger, R. D. 1983. Shoot proliferation, rooting, and transplant survival of tissue cultured Hally Jolivette cherry. *Hort. Sci* 18:182–185.

Ling, S. Y., & Y. D. Dong. 1983. Researches on the physiology of seed dormancy of Manchurian ash. *Scientia Silvae Sinicae* 19:349–359.

Lisbao, L., & W. S. Filho. n.d. Preservation of seeds of *Eucalyptus saligna* in several levels of relative humidity. Luiz de Queiroz Univ., Sao Paulo, Brazil.

Little, E. L., Jr. 1953. Checklist of Native and Naturalized Trees of the United States (including Alaska). Forest Service, USDA, Washington, DC.

Little, S. 1974. *Ailanthus*. In: *Seeds of Woody Plants in the United States*. Forest Service, USDA, Washington, DC. 201–202.

Livingston, W. H., & R. A. Blanchette. 1986. Eastern dwarf mistletoe seed storage, germination and inoculation of spruce seedlings. *Forest Sci.* 32:92–96.

Lobov, A. I. 1973. Storing seeds of *Pinus sylvestris* in snow in the Amur region. *Referativnyi Zhurnal* 1975:53–55.

———. 1985. Phenological indicators of the time to start collecting *Larix sibirica* cones. *Lesnoe Khozyaistvo* 12:31–32.

Lockley, G. C. 1980. Germination of chokecherry seed. *Seed Sci. Tech.* 8:237–244.

Lodhi, M. A. K., & E. I. Rice. 1971. Allelopathic effects of *Celtis laevigata*. Bul. Torrey Bot. Club 98:83–89.

Loffler, J. 1976. Experience of seed production in seed orchards. Mitteilungen, Verein fur Forstliche Standortskunde und Forstpflanzenzuchtung 25:53–58.

Logan, K. T., & D. F. W. Polland. 1981. Effect of seed weight and germination rate on the initial growth of Japanese larch. *Bi-Mon. Res. Rpt.* 5:28–29.

Loken, A. 1977. Germination and pre-treatment of conifer seed in clear water. *Landbruksdepartmentet Skogavdelingen* 68–78.

Loneragan, O. W. 1979. Karri (*Eucalyptus diversicolor*). Phenological studies in relation to reforestation. Forest Dept., Western Australia.

Loomis, H. F. 1958. The preparation and germination of palm seeds. *Principes* 2:98–102.

———. 1961. Culture of the palms. Preparation and germination of palm seeds. *American Hort. Mag.* 40:128–130.

Lui, T. S. 1971. *A Monograph of the Genus* Abies. Dept. Forestry Taiwan Univ., Taipei.

Luk'yanets, V. B. 1978. Contents of amino acids in acorns of various species and climatypes of oaks. *Lesnoi Zhurnal* 29–32.

Lulandala, L. L.L. 1981. Seed viability, germination and pretreatment of *Leucaena leucocephala*. Leucaena *Res. Rpts.* 2:59.

Lumley, M. L., & I. B. Oliver. 1987. The cultivation of *Rhus erosa*. *Veld and Flora* 72:112–113.

Lundergan, C. A., & J. A. Carlisi. 1984. Acceleration of the reproduction cycle of the cultivated blackberry. *Hort. Sci.* 19:102–103.

Lush, W. M., P. E. Kaye, & R. H. Groves 1984. Germination of *Clematis Microphylla* seeds following weathering and other treatments. *Australian Jour. Botany* 32:121–129.

Lyuhich, E. S. 1985. Quality of the seed resources ways of improving it. *Lesnoe Khozyaistvo* 23–26.

Ma, C. G., & D. Y. Liu. 1986. Effect of experimental soaking of the seeds of 14 tree species. *Forest Sci. and Tech.* 1986:10–13.

McArthur, E. D., S. E. Meyer, & D. J. Weber. 1987. Germination rate at low temperatures; rubber rabbitbrush population differences. Jour. Range Manage. 40:530–533.

McArthur, E. D., C. L. Pope, & D. C. Freeman. 1981. Chromosomal studies of subgenus Tridentatae: evidence of autopolyploidy. *American Jour. Botany* 68:589–605.

McArthur, E. D., & S. C. Sanderson. 1983. Distribution systematics and evolution of Chenopodiaceae. In: *Proc. Symp. Biol.* Atriplex *and Related Chenopods*. Gen. Tech. Rpt. 172. Forest Service, USDA, Ogden, UT. 14–24.

MacArthur, J. D., & J. W. Fraser. 1963. Low temperature germination of some eastern Canadian tree seeds. *Forestry Chronicle* 39:478–497.

McBride, J. R., & J. R. Dickson. 1972. Gibberellin, citric acid and stratification enhance white ash germination. *Tree Planter's Notes* 23:1–2.

McCarter, P. S., & C. E. Hughes. 1984. *Liquidambar styrachiflua* L.—a species of potential for the tropics. *Commonwealth Forestry Review* 63:207–216.

McClaran, M. D. 1987. Yearly variation of blueoak emergence in northern California. Forest Service, USDA, Berkeley, CA.

McCleary, J. A., & K. A. Wagner. 1973. Comparative germination and early growth studies of six species of the genus *Yucca*. *American Midland Naturalist* 90:503–508.

McCutchen, C. W. 1977. The spinning rotation of ash and tulip tree samaras. *Science* 197:691–692.

McDonald, P. M., D. R. Vogler, & D. Mayhew. 1988. Unusual decline in tanoak sprouts. Forest Service, USDA, Berkeley, CA.

McDonough, W. T. 1979. Quaking aspen-seed germination and early seedling growth. Forest Service, USDA, Ogden, UT.

———. Sexual reproduction, seeds, and aspen seedlings. Forest Service, USDA, Fort Collins, CO.

McDonough, W. T., & R. O. Harniss. 1974a. Effects of temperature on the germination of the seeds of three subspecies of big sagebrush. *Jour. Range Manage.* 27:204–205.

———. 1974b. Seed dormancy in *Artemisia tridentata*. Nutt. subspecies *vaseyana* Rydb. *Northwest Sci.* 48:17–20.

McDowell, C. R., & E. J. Moll. 1981. Studies of seed germination and seedling competion in *Virgilia oroboides* (Berg.) Salter, and *Albizia lophantha* (Willd.) Benth., and *Acacia longifolia* (Andr.) Willd. *Jour. South African Botany* 47:653–685.

McGill, D., & C. E. Whitcomb. 1977. Using scarification to promote seed germination of black gum. Res. Rpt. Oklahoma Agr. Exp. Sta. 760:87–88.

McIntosh, M., & G. L. Rolfe. 1977. Effect of beef cattle waste on germination and survival of loblolly and short leaf pine. Illinois Forest Res. Rpt.

McLean, A. 1967. Germination of forest and range species from southern British Columbia. *Jour. Range Manage.* 25:321–322.

McLemore, B. F. 1976. Viable seed from a shortleaf pine 13 months old. *Silvae Genetica* 25:25–26.

———. 1977. Strobilli and conelet losses in four species of southern pines. Forest Service, USDA, Pineville, LA.

McMinn, H. E., & E. Maino. 1959. *An Illustrated Manual of Pacific Coast Trees*. Univ. Calif. Press, Berkeley.

Macoboy, S. 1982. *Trees*. Landsowe Press, Sydney.

McTavish, B. 1986. Seed propagation of some native plants is surprisingly successful. *American Nurseryman* 164:55–56, 60, 62–63.

Madden, G. D., & H. W. Tisdale. 1975. Effect of chilling and storage on the germination of northern and southern pecan cultivars. *Annual Rpt. Northern Nut Growers Assoc.* 66:30–32.

Madsen, E. 1975. Determination of moisture content in fruits of teak. *Statsfrokontrollen, Beretning* 103–108.

Madhuaraja, K. A. 1982. Effects of presowing stratification of seed of *Pinus caribaea* var. *hondurensis* on germination. *Indian Forester* 108:60–61.

Maeda, J. A., & L. A. F. Matthes. 1984. Conservation of ipa seeds. *Bragantia* 43:51–61.

Magill, A. W. 1974a. *Chamaebatia*. In: *Seeds of Woody Plants in the United States*. Forest Service, USDA, Washington, DC. 315.

———. 1974b. *Chilopsis*. In: *Seeds of Woody Plants in the United States*. Forest Service, USDA, Washington, DC. 321–322.

———. 1974c. *Photinia*. In: *Seeds of Woody Plants in the United States*. Forest Service, USDA, Washington, DC. 582–583.

Magini, E., & N. P. Tulstrup. 1955. *Tree Seed Notes*. FAO, Rome.

Mahmoud, A., & A. M. El-Sheikh. 1978. Germination of *Prosopis chilensis*. *Egyptian Jour. Botany* 21:69–74.

———. 1981. Germination of *Parkinsonia aculeata*. *Jour. College of Science Univ. Riyad* 12:53–64.

Maithani, G. P., V. K. Bahuguna, M. M. S. Rawat, & O. P. Sood. 1986. Potential of artifical heat in extraction from the cones of *Pinus roxburghii*. *Jour. Tropical Forestry* 2:211–215.

Majer, A. 1982. Periodicity of seed production in beech *Erdo* 31:388–392.

Malone, P. L. 1985. Germination of western white pine seed. *Forest Service*, USDA, Ogden, UT.

Mamiopia, N. G., L. J. Empia, & B. P. Welgas. 1983. Physiological changes in ipil-ipil seed during storage. *Leucaena Res. Rpts.* 4:67–68.

Mamonov, N. I. 1976a. Optimum technology for processing Scots pine cones. *Lesnoe Khozyaistvo* 57–58.

———. 1976b. Respiration of acorns in various storage conditions. *Referativnyi Zhurnal* 9.

Mamonov, N. I., & M. V. Smurova. 1978. Treatment of pine seeds before sowing. *Lesnoe Khozyaistvo* 65–66.

Mann, W. F. 1979. Relationships of seed size, number of cotyledons, and initial growth of southern pines. *Tree Planter's Notes* 30:22–23.

Manokaran, N. 1978. Germination of fresh seeds of Malaysian rattans. *Malaysian Forester* 41:319–324.

———. 1979. Germination of Malaysian palms. *Malaysian Forester* 42:50–52.

Matheson, A. C., & K. W. Willcocks. 1976. Seed yield in a radiata pine seed orchard following pollarding. *New Zealand Jour. Forest Science* 6:14–18.

March, S. G. 1976. *Eurya japonica* Thunb. *Hort. Sci.* 11:269.

Marinov, I. 1977a. Effect of molybdenum on the germination of ash seeds. *Gorsko Stopanstvo* 33:15–19.

———. 1977b. Growth and development of embryos of seeds of *Fraxinus excelsior* and *F. oxycarpa* during ripening and stratification. *Gorskostopanska Nauka* 14:13–20.

———. 1977c. Sowing seeds of *Fraxinus excelsior* and *F. oxycarpa* of different ripeness. *Gorsko Stopanstvo* 33:14–19.

———. 1977d. Stratification periods for seeds of *Fraxinus oxycarpa*. *Gorsko Stopanstvo* 33:9–13.

———. 1978. Germination of Caucasian ash seed after different periods of storage. *Gorsko Stopanstvo* 34:21–24.

Marjal, Z. 1972. The yield of *Robinia pseudoacacia* seed from the soil and possibilities of collecting it. *Erdeszeti Kutatasok* 68:87–100.

Marks, P. L. 1979. Apparent fire stimulated germination of *Rhus typhina* seeds. *Bul. Torrey Bot. Club* 106:41–42.

Marshall, P. E. 1981. Methods for stimulating green ash seed germination. *Tree Planter's Notes* 32:9–11.

Marshall, P. E., & T. T. Kozlowski. 1976. Importance of endosperm for nutrition of *Fraxinus pensylvanica* seedlings. *Jour. Experimental Botany* 27:572–574.

Martin, C. E., R. L. Miller, & C. T. Cushwa. 1975. Germination response of legume seeds subject to moist and dry heat. *Ecology* 56:1441–1445.

Martin, S. C. 1974. *Larrea*. In: *Seeds of Woody Plants in the United States*. Forest Service, USDA, Washington, DC. 486–487.

Martin, S. C., & R. R. Alexander. 1974. *Prosopis*. In: *Seeds of Woody Plants in the United States*. Forest Service, USDA, Washington, DC. 656–657.

Martynchuk, A. I. 1984. Growing *Ephedra equisetina*. *Lesnoe Khozyaistvo* 31–33.

Marusov, A. A. 1976. Seed production of *Picea obovata* in the Urals. *Lesnoe Khozyaistvo* 58–59.

Mathiasen, R. L. 1986. Infection of young Douglas-fir and spruce by dwarf mistletoes in the southwest. *Great Basin Naturalist* 46:528–534.

Mathur, R. S., K. K. Sharma, & M. S. Rawat. 1984. Germination behavior of various provences of *Acacia* locations in India. *Indian Forester* 110:435–449.

Matsudaik, K., & J. R. McBride. 1987. Germination and shoot development of seven California oaks at different elevations. Forest Service, USDA, Berkeley, CA.

Maury-Lechon, G., A. M. Hassan, & D. R. Bravo. 1981. Seed storage of *Shorea parvifoli* and *Dipterocarpus humeratus*. *Malaysian Forester* 44:267–280.

Mayeux, H. S., & L. Leotta. 1981. Germination of broom snake weed and thread leaf snakeweed seed. *Weed Sci.* 29:530–534.

Mehanna, H. T., & G. C. Martin. 1985. Effect of seed coat on peach seed germination. *Scientia Horticulturae* 25:247–254.

Mehanna, H. T., G. C. Martin, & C. Nishijima. 1985. Effect of temperature, chemical treatments and endogenous hormone content on peach seed germination and subsequent growth. *Scientia Horticulturae* 27:63–73.

Menninger, E. A. 1962. *Flowering Trees of the World.* Hearthside Press, New York.

Merwin, M. L. 1987. Current research on *Eucalyptus* and *Casuarina* in California. Forest Service, USDA, Berkeley, CA. 436–438.

Meyer, S. E. 1989. Warm pretreatment effects on antelope bitterbrush germination response to chilling. *Northwest Sci.* 63:146–153.

Meyers, J. F., & G. H. Fechner. 1980. Seed hairs and seed germination in *Populus. Tree Planter's Notes* 31:3–4.

Meyers, S. E., S. L. Kitchen, G. R. Wilson, & R. Stevens. 1989. *Cowania mexicana. Newsletter ASOA* 63:24–25.

Michalska, S. 1983. Embryonal dormancy and induction of secondary dormancy in seeds of mazzard cherry. *Arbor. Kornickie* 27:311–322.

Migliaccio, E., G. Gasparrini, & S. G. Albonetti. n.d. Experimental trials on direct seeding in Italy. EUTECO Co., Rome.

Miller, G. E. 1983. When is controlling cone and seed insects in Douglas-fir seed orchards justified? *Forestry Chronicle* 56:304–307.

Min, K. H., K. S. Kim, & C. Y. An. 1974. The study on stimulating seed production in *Pinus koraiensis* and *Larix leptolepis.* Res. Rpt. 21, Forest Res. Inst., Korea. 215–220.

Minina, E. G., & A. I. Iroshnikov. 1977. Provenance trials and seed orchards of conifers in Siberia. Novosibirsk, USSR. 168.

Minko, G. 1975. Effects of soil physical properties, irrigation and fertilization on *Pinus radiata* seedling development in the Benalla nursery. *Forestry Tech. Papers,* Forest Comm., Victoria 22:19–24.

Minore, D. 1984. Germination and growth of Douglas fir and incense cedar seedlings on two southwestern Oregon soils. *Tree Planter's Notes* 36:3–6.

———. 1986a. Effect of madrone, chinkapin, and tanoak sprouts on light intensity, soil moisture, and soil temperature. *Canadian Jour. Forest Res.* 16:654–658.

———. 1986b. Germination, survival and early growth of conifer seedlings in two habitat types. Forest Service, USDA, Portland, OR.

Mirov, N. T., & C. J. Kraebel. 1939. *Collecting and Handling Seeds of Wild Plants.* Civilian Conservation Corp.

Mishnev, V. G. 1984. Fruiting of beech in the reserve forest of the Crimea. *Lesovedenie* 54–58.

Misra, C. M., & S. L. Singh. 1985. Salt stress studies upon germination of *Sesbania grandiflora.* Nitrogen Fixing Tree Res. Rpt. 3:22.

Mittal, R. K., B. S. Wang, & and D. Harmsworth. 1982. Effect of extended prechilling on laboratory germination and fungal infection in seeds of white spruce and eastern white pine. *Tree Planter's Notes* 38:6–9.

Mohan, E., N. Mitchell, & P. Lovell. 1984. Environmental factors controlling germination of *Leptospermum scoparium. New Zealand Jour. Botany* 22:95–101.

———. 1984b. Seasonal variation in seedfall and germination of *Leptospermum scoparium. New Zealand Jour. Botany* 22:103–108.

Mohamad, A., & F. S. P. Nig. 1982. Influence of light on germination of *Pinus caribaea, Gmelina arborea, Sapium baccatum,* and *Vitex pinnata. Malaysian Forester* 45:62–68.

Mohn, C. A., J. W. Hanover, H. Kong, & R. A. Strive. 1988. Survival and growth of tamarack seed sources in ten NC 99 test. Univ. Wisconsin, Madison.

Monov, N. E. 1975. Respiration of acorns in various storage conditions. *Referativnyi Zhurnal* 81–89.

Moore, M. B., & F. A. Kidd. 1982. Seed source variation in induced moisture stress germination of ponderosa pine. *Tree Planter's Notes* 33:12–14.

Morgan, D. L., & T. P. Brohaker. 1986. Dehydration effect on germination of liveoak seeds. *Jour. Environmental Hort.* 4:95–96.

Morgenson, G. 1986. Seed stratification treatments for two hardy cherry species. *Tree Planter's Notes* 37:35–38.

Mori, T. 1979. Effect of light on adenosine triphosphate levels in *Pinus thunbergii* seeds during imbibition. *Jour. Japanese Forestry Soc.* 61:399–404.

Morita, E. 1979. Seed crops of hinoki (*Chamaecyparis obtusa*) in the recent three decades in Kyushu and the forecast for the future. *Jour. Japanese Forestry Soc.* 61:242–248.

Morrison, D. A. 1987. A review of the biology of *Acacia suaveolens* (Smith) Willd. *Proc. Linnean Soc. New South Wales* 109:271–292.

Moura, V. P. G. 1982. Effect of temperature on the germination of seeds of *Eucalyptus urophylla.* EMBRAPA, Planaltina, Brazil.

Moyer, D. T., & R. L. Lang. 1976. Variable germination response to temperature for different sources of winter fat. *Jour. Range Manage.* 29:320–321.

Mozingo, H. N. 1987. *Shrubs of the Great Basin.* Nevada Press, Reno.

Mpri, T., Z. H. Rahman, & C. H. Tan. 1980. Germination and storage of ratan manau seeds. *Malaysian Forester* 43:44–45.

Msaga, H. P., & J. A. Maghembe. 1986. Effect of hot water and chemical treatments on the germination of *Albizia schimperana* seed. *Forest Ecology and Manage.* 17:137–146.

Muhle, O., & A. J. Hewicker. 1976. Trials with pelleted forest seeds and seeds mounted on paper strips. *Allgemeine Frost- und Jagdzeitung* 147:10–16.

Mukai, Y., & T. Yokoyama. 1985. Effect of stratification on seed germination of *Phellodendron amurense. Jour. Japanese Forestry Soc.* 67:103–104.

Muller, C., & M. Bonnet-Masimbert. 1983. Improvement of germination of beech by pretreatment in the presence of polyethylene glycol. *Annales des Sciences Forestieres* 40:157–164.

———. 1985. Breaking of dormancy of beechnuts before storage: preliminary results. *Annales des Sciences Forestieres* 42:385–395.

Muller, C., & E. T. Cross. 1982. Storage of *Populus nigra* seed for five years. *Annales des Sciences Forestieres* 39:179–185.

Mullin, R. E. 1980. Water dipping and frozen over winter storage of red and white pine. *Tree Planter's Notes* 31:25.

Munson, R. H. 1984. Germination of western soapberry as affected by scarification and stratification. *Hort. Sci.* 19:712–713.

Muradyan, V. M. 1982. Propagation of juniper in Armenia. *Lesnoe Khozyaistvo* 62–63.

Murphy, J. B. 1985. Acetone production during the germination of fatty seeds. *Physiologia Plantarum* 63:231–234.

Murphy, J. B., & R. G. Stanley. 1975. Increased germination rates of baldcypress and pondcypress seeds following treatments affecting the seed coat. *Physiologia Plantarum.* 35:135–139.

Murugi, K. E. 1987. Effects of presowing treatments on seed germination of four important tree species in Kenya. Swedish Univ. of Agr. Sci., Dept. Forest Genetics and Plant Physiology 143–153.

Muselem, S. 1975. Some correlations between cone characteristics and the viability and germination of seeds of *Pinus caribaea* var. *hondurensis* obtained from plantations. *Tecnica de Bosques* 1:23–27.

Muttiah, S. 1975. Some data on teak seed and further pregermination treatment trials. *Sri Lanka Forester* 12:25–36.

Myer, R. E., M. G. Merkle, & C. R. Thowood. 1970. Texas persimmon fruit inhibition of seedling growth. *The Station* (Texas Agr. Exp. Sta.). 74–76.

Nagata, H., & M. Black. 1977. Phytochrome-chilling interaction in the control of seed dormancy of *Betula maximowicziana* Reg. *Jour. Japanese Forestry Soc.* 59:368–371.

Nagata, H., & Y. Tsuda. 1975. Action of far-red and blue lights on the germination of Japanese white birch seed. *Jour. Japanese Forestry Soc.* 57:160–163.

Nagaveni, H. C., & R. A. Srimathi. 1980. Studies on the germination of sandal seeds. *Indian Forester* 106:792–799.

_____. 1981. Studies on the germination of sandal. Pretreatment of sandal seeds. *Indian Forester* 107:348–354.

Nagy, M., & A. Keri. 1984. Role of the embryo in the cytolysis of the endosperm cell during the germination of the seeds of *Tilia platyphyllos*. *Biochemie und Physiologie de Pflanzen* 179:145–148.

Naqui, H. H., & G. P. Hanson. 1982. Germination and growth inhibitors in guayule chaff and their possible influence in seed dormancy. *American Jour. Botany* 69:985–989.

Nather, J., & W. Krissl. 1983. Cotyledon number of Austrian provenances of Norway spruce. *Allgemeine Forstzeitung* 94:208–210.

Neal, D. L. 1974a. *Carpenteria*. In: *Seeds of Woody Plants in the United States*. Forest Service, USDA, Washington, DC. 265.

_____. 1974b. *Dendromecon*. In: *Seeds of Woody Plants in the United States*. Forest Service, USDA, Washington, DC. 372.

Neal, M. C. 1965. *In Gardens of Hawaii*. Bishop Museum Press, Honolulu, HI.

Nekrasov, V. I., Y. Podgornyi, and N. G. Smirnoa. 1974. Variation in seed quality of *Pinus nigra* var. *caramania*. *Lesovedenie* 3:51–55.

Nesme, Y. 1985. Respective effects, testa and endosperm, and embryo on germination of raspberry seeds. *Canadian Jour. Plant Sci.* 65:125–130.

Ng, F. S. P. 1975. Germination of fresh seeds of Malaysian tree. II. *Malaysian Forester* 40:160–163.

_____. 1977a. Germination of fresh seeds of Malaysian trees. 3. *Malaysian Forester* 40:160–163.

_____. 1977b. Strategies of establishment in Malayan forest trees. In: *Tropical Trees As Living Systems*. Cambridge Univ. Press, Cambridge. 128–162.

_____. 1980. Germination ecology of Malaysian woody plants. *Malaysian Forester* 43:406–437.

Ng, F. S. P., & M. A. Sanah. 1979. Germination of fresh seeds of Malaysian trees. 4. *Malaysian Forester* 42:221–224.

Nicholson, E., & M. E. Demeritt. 1978. Cleaning *Populus* seeds with a blender. *Silvae Genetica* 27:216.

Nicholson, R. 1984. Propagation notes on *Cedrus deodara* 'Shalimar' and *Cedrus decurrens*. *Plant Propagator* 30:5–6.

Niembro, A., M. A. Musalem, & H. Ramires. n.d. Effect of size and color of *Pinus hartwegii* seeds on germination. Dep. Bosques Esc. Nal. Agr., Chapingo, Mexico.

Nikolaeva, M. G., & N. S. Vorob'eva. 1978. Seed biology of different provenances of *Fraxinus excelsior*. *Botanicheskii Zhurnal* 63:115–1167.

Nikolova, M. 1984. Changes in the free amino acids in acorns of *Quercus cerris* during dry and moist storage. *Gorskostopanska Nauka* 21:31–37.

Noble, D. L., C. B. Edminster, & W. P. Shepperd. 1978. Effect of watering treatment on germination, survival, and growth of Rocky Mountain Douglas fir. Forest service, USDA, Fort Collins, CO.

Noel, A. R. A., & J. Staden. 1976. Seed coat structure and germination in *Podocarpus henkelii*. *Zeitschrift fur* Pflanzenphysiologie 77:174–186.

Nogaev, V. M. 1977. Seed production and morphological features of larch. *Lesnoe Khozyaistvo* 5:40–41.

Nord, E. C. 1974. *Fremontodendron*. In: *Seeds of Woody Plants in the United States*. Forest Service, USDA, Washington, DC. 417–419.

Nord, E. C., & L. E. Gunter. 1974. *Salvia*. In: *Seeds of Woody Plants in the United States*. Forest Service, USDA, Washington, DC. 751–753.

Nord, E. C., & A. Kadish. 1974. *Simmondsia*. In: *Seeds of Woody Plants in the United States*. Forest Service, USDA, Washington, DC. 774–776.

Nord, E. C., & A. T. Leiser. 1974. *Nama*. In: *Seeds of Woody Plants in the United States*. Forest Service, USDA, Washington, DC. pp 551–552.

Norman, E. M., & D. Clayton. 1986. Reproductive biology of two Florida pawpaws: *Asimina ovata* and *A. pygmaea*. Bul. Torrey Bot. Club. 113:16–22.

Norton, C. R. 1985. The use of gibberellic acid, thaphon and cold treatments to promote germinatiom of *Rhus typhina* seeds. *Scientia Hort.* 27:163–169.

_____. 1986. Low temperature and gibberellic acid stimulation of germination in *Sambucus caerulea*. *Scientia Hort.* 28:323–329.

Ntima, O. O. 1968. *Fast Growing Timber Trees of the Lowland Tropics—The Araucarias*. Commonwealth Forest Inst., Oxford, UK.

Nussbau, E. S., & H. B. Lagerstedt. 1983. The effect of stratification and GA on germination of filbert seed as influenced by dry storage. *Pacific Nut Growers Assoc.* 86–88, 90, 92.

Olovyannikova, I. N. 1981. Growth of *Ulmus pumilia* var. *arborea* on dark soils in the semi-desert near the Caspian Sea. *Lesovedenie* 66–74.

Olson, D F. 1974a. *Baccharis*. In: *Seeds of Woody Plants in the United States*. Forest Service, USDA, Washington, DC. 224–226.

_____. 1974b. *Elaeagnus*. In: *Seeds of Woody Plants in the United States*. Forest Service, USDA, Washington, DC. 376–389.

_____. 1974c. *Quercus*. In: *Seeds of Woody Plants in the United States*. Forest Service, USDA, Washington, DC. 692–703.

_____. 1974d. *Rhododendron*. In: *Seeds of Woody Plants in the United States*. Forest Service, USDA, Washington, DC. 709–712.

_____. 1974e. *Robinia*. In: *Seeds of Woody Plants in the United States*. Forest Service, USDA, Washington, DC. 728–731.

Olson, D F., & R. L. Barnes. 1974a. *Cladrastis*. In: *Seeds of Woody Plants in the United States*. Forest Service, USDA, Washinton, DC. 329–330.

_____. 1974b. *Diospyros*. In: *Seeds of Woody Plants in the United States*. Forest Service, USDA, Washington, DC. 373–375.

_____. 1974c. *Kalmia*. In: *Seeds of Woody Plants in the United States*. Forest Service, USDA, Washington, DC. 470–471.

_____. 1974d. *Oxydendrum*. In: *Seeds of Woody Plants in the United States*. Forest Service, USDA, Washington, DC. 566–567.

_____. 1974e. *Sabal*. In: *Seeds of Woody Plants in the United States*. Forest Service, USDA, Washington, DC. 744–745.

_____. 1974f. *Serenoa*. In: *Seeds of Woody Plants in the United States*. Forest Service, USDA, Washington, DC. 769–770.

Olson, D. F., R. L. Barnes, & L. Jones. 1974. *Magnolia*. In: *Seeds of Woody Plants in the United States*. Forest Service, USDA, Washington, DC. 527–530.

Olson, D. F., & W. J. Gabriel. 1974. *Acer*. In: *Seeds of Woody Plants in the United States*. Forest Service, USDA, Washington, DC. 187–194.

Olson, D. F., & E. Q. Pettys. 1974. *Casuarina*. In: *Seeds of Woody Plants in the United States*. Forest Service, USDA, Washington, DC. 278–280.

Olson, D. L., & R. R. Silver. 1975. Influence of date of cone collection on Douglas-fir seed processing and germination: case history. Forest Service, USDA, Portland, OR.

Olvera, E., & S. H. West. 1986a. The influence of IAA and GA hormones on the germination and early development of *Leucaena. Leucaena Newsletter* 1:53.

———. 1986b. *Leucaena* seed scarification. I. Leucaena *Newsletter* 1:52.

Olvera, E., & S. H. West, & W. G. Blue. 1982. Aspects of *Leucaena* germination. Leucaena *Res. Rpts.* 3:86–87.

Opler, P. A. 1975. Reproductive biology of *Cordia. Biotropics* 7:234–247.

Osipov, Y. S., S. Y. Abseitov, A. B. Stenyukov, & L. A. Yozlov. 1985. The MIS-O, 2 machine for extracting seeds from the arils of juniper. *Lesnoe Khozyaistvo* 54–55.

Ostroshenko, V. 1977. Treatment of conifer seed before sowing. *Lesnoe Khozyaistvo* 78.

Owston, P. W., & W. I. Stein. 1974. *Pseudotsuga.* In: *Seeds of Woody Plants in the United States.* Forest Service, USDA, Washington, DC. 674–683.

Paal, H. 1984a. Norway spruce seed harvest in Estonia in 1978/79. *Metsanduslikud Uurimused Estonian SSR* 19:28–35.

Pal, R. N., R. Singh, V. K. Vu, & J. N. Sharma. 1976. Effect of gibberellins, GA$_{31}$, GA$_{4+7}$, and GA$_{13}$ on germination and subsequent seedling growth in Early Muscat grape. Vitis 14:265–268.

Pall, H. 1984. Potential seed yield of Norway spruce in Estonia. *Metsanduslikud Uurimused Estonian SSR* 19:18–27.

Palmer, E., & N. Pitman. 1972. *Trees of South Africa.* A. H. Balkewa, Capetown.

Panetta, F. D. 1972. The effects of vegetation development upon achene production in the woody weed groundsel bush. *Australian Jour. Agr. Res.* 30:1053–1065.

Pardos, J. A., & L. Ayerbe. 1978. Growth regulators in *Arctostaphylos* fruits. *Phyton* 36:31–39.

———. 1980. Gibberellin-like substances in the germination of *Arctostaphylos uva-ursil* seeds. *Phyton* 38:1–7.

Pardos, J. A., & G. Lazaro. 1983. Germination aspects of juniper oxycedrus. *Annals Inst. Nac. Agr. Madrid* 7:155–163.

Pathak, P. S., 1985. Seed germination and seedling growth of *Hardwickia binata* Roxb. Nitrogen Fixing Tree Res. Rpt. 3:25–26.

Pathak, P. S., & S. K. Gupta. 1984. Seed germination and seedling growth of *Hardwickia binata* Roxb. Nitrogen Fixing Tree Res. Rpt. 2:7–8.

Pathak, P. S., & C. M. Misra. 1978. Studies on the seeds and their germination in *Robinia pseudoacacia. Annals Arid Zone* 17:363–369.

Pathak, P. S., M. P. Pal, & R. Debroy. 1981. Seed weight affecting early seedling growth attributes in *Leucaena leucocephala* (Lam.) de Wit. *Van Vigyan* 19:97–101.

Pathak, P. S., & B. D. Patil. 1985. Seed weight affecting early seedling growth of *Butes monosperma* (Lam.) Taub. Nitrogen Fixing Tree Res. Rpt. 3:24–24.

Pathak, P. S., B. D. Patil, & R. Debroy. 1980. Studies of seed polymorphism, germination, and seedling growth of *Acacia tortilis* Hayne. *Indian Jour. Forestry* 3:64–67.

Pathak, P. S., K. A. Shankararayan, P. Rai, & R. Debroy. 1976. Effect of reduced moisture levels on seed germination of fodder trees. *Forage Res.* 2:179–182.

Patil, B. D., & P. S. Pathak. 1977. Energy plantations and silvipastoral systems for rural areas. *Invention Intelligence* 12:78–87.

Patil, S. G., & V. P. Patil. 1984. Seed variation in grape cultivars. *Biovisyanan* 10:115–119.

Paves, H. 1979. Seed yield and quality from larch seed orchards in 1976. *Metsanduslikud Uurimused, Estonian SSR* 15:73–80.

Payandeh, B., & V. F. Haavisto. 1982. Predicting equations for black spruce seed production and dispersal in northern Ontario. *Forestry Chronicle* 58:96–99.

Pederick, L. A. n.d. An analysis of seed production of 15 clones in a *Pinus radiata* seed orchard. *Res. Branch Forestry Comm. Victoria, Melbourne.* 4.

Pegeckiewe, A. 1972. The bearberry in Lithuanian SSR: germination. *Akad. Nauk. Litou.* 58:49–59.

Penafiel, S. R. 1982. Effects of three pasture plant extracts on germination of benguet pine. *Sri Lanka Forester* 15:154–156.

Penafiel, S. R., & B. F. Noble. 1978. Germination of benguet pine (*Pinus kesiya*) seeds gathered from different crown exposures. *Philippine Forest Res. Jour.* 3:37–40.

Penafiel, S. R., B. F. Noble, & L. P. Ngales. 1982. Growth of benguet pine seedlings planted under alnus stands. *Sylvatrop* 7:45–48.

Pereira, J. C. D., & M. A. Garrido. 1975. Effect of seed size on the germination and initial growth of seedlings of *Eucalyptus grandis. Silvicultura en Sao Paulo* 9:117–124.

Perrin, R., C. Muller, & M. Bonnet-Masimbert. n.d. Attempted improvement of storage method for beechnuts. INRA-CNRF, Seichamps, France.

Perry, M. S. 1956. *Tree Planting Practices for Tropical Africa.* FAO, Forest Development, Rome.

Peterson, J. K. 1983. Mechanisms involved in delayed germination of *Quercus nigra* seeds. *Annals Botany* 52:81–92.

Petrov, M., & K. Genov. 1984. Some features of the fruiting biology of *Quercus suber* in Bulgaria. *Gorskostopanska Nauka* 21:13–17.

Petrukzzelli, L. 1984. The effect of ethanol on germination of Aleppo pine seeds. *Biologia Plantatrum* 26:235–238.

Pettys, E. Q. P. 1974. *Tristania.* In: *Seeds of Woody Plants in the United States.* Forest Service, USDA, Washington, DC. 817–818.

Pfister, R. D. 1974. *Ribes.* In: *Seeds of Woody Plants in the United States.* Forest Service, USDA, Washington, DC. 720–727.

Pinfield, N. J., & A. K. Stobart. 1982. Hormonal regulation of germination and early seedling growth of *Acer pseudoplatanus. Planta* 104:134–145.

Pintel, J. A., & W. M. Chliak. 1986. Enzyme activities during inhibition and germination of seeds of tamrack. *Physiologia Plantarum* 67:562–563.

Pintel, J. A., & W. M. Chliak, & B. S. P. Wang. 1984. Changes in isoenzyme patterns during imbibition and germination of lodgepole pine. *Canadian Jour. Forest Res.* 14:743–746.

Pintel, J. A., & B. S. P. Wang. 1980. A preliminary study of dormancy in *Pinus albicaulis* seeds. *Bi-Mon. Res Rpt.* 36:4–5.

———. 1985. Physical and chemical treatment to improve laboratory germination of western white pine seeds. *Canadian Jour. Forest Res.* 15:1187–1190.

Pishchik, A. A. 1978. The seed yield in permanent seed stands of Scots pine. *Lesnoe Khozyaistvo* 32–34.

Platonova, R. N., U. P. Parfenov, U. P. Ol'khovenko, N. I. Karpova, & M. E. Pichugov. 1977. Germination of pine seeds in weightlessness on board "Kosmos-782" satellite. *Biol. Bul. Acad. Sci.* 4:628–634.

Plumb, T. R. 1982. Factors affecting germination of southern California oaks. Forest Service, USDA, Berkeley, CA.

Plummer, M. 1983. Considerations in selecting chenopod species for range seedings. In: *Proc. Symp. Biol.* Atriplex *and Related Chenopods.* Gen. Tech. Rpt. 172. Forest Service, USDA, Ogden, UT.

Poggenpool, P. V. 1978. Collection and pretreatment of seed from black wattle (*Acacia mearsii*) seed orchards. Rpt. Wattle Res. Inst. South Africa. 1978:85–90.

Polupannev, Y. I., T. S. Smogunova, & B. I. Fabrichnyi. 1979.

Effective methods of sowing Scots pine seed in nurseries. *Lesnoe Khozyaistvo* 32–34.

Ponomarenko, P. V., & V. S. Petrovskii. 1977. Some physical and mechanical properties of seeds of *Pinus sylvestris, Picea obovata,* and *Larix sibirica. Lesnoi Zhurnal* 31–34.

Pontailler, J. Y. n.d. Beech-nut production and consumption in a natural beech stand in Fontainbleau forest. Univ. Paris, Orsay Cedex, France.

Popov, P. P. 1980. The sowing qualities of *Picea obovata* seeds. *Lesnoe Khozyaistvo* 64–65.

———. 1982. Influence of ecological conditions on the germination of *Picea obovata* seeds. *Lesnoe Khozyaistvo* 19–22.

———. 1986. Effect of stratification on the germination capacity of spruce seeds. *Lesnoe Khozyaistvo* 65–66.

Posey, C. E., & J. F. Goggans. 1967. Observations on spores of cypress indigenous to the United States. Auburn Univ. Agr. Exp. Sta. Cir. 153.

Pozdova, L. M. 1985. Changes of cytokinin activity during breaking of dormancy in seeds. *Soviet Plant Physiology* 32:286–374.

Pratt, C. R. 1986. Environmental factors affecting seed germination of gray birch collected from abandoned anthracite coal spoils in northeast Pennsylvania. *Annals Applied Biology* 108:649–658.

Prieston, D. S. 1979. Stump sprouts of swamp and water tupelo produce viable seeds. *Southern Jour. Applied Forestry* 3:149–151.

Prinsen, J. H. 1986. Potential of *Albizia lebbek* as a tropical fodder tree: a literature review. *Tropical Grasslands* 20:78–83.

Pritchard, H. W., & F. G. Prendergast. 1986. Effects of desiccation and cryopreservation on the vitro viability of embryos of the recalcitrant seed species *Araucaria hunsteinii. Jour. Experimental Botany* 37:1388–1397.

Pukittayacamee, P., & A. K. Hellum. 1988. Seed germination in *Acacia auriculiformis:* developmental aspects. *Canadian Jour. Botany* 66:388–393.

Pulliainen, E., L. H. J. Lajunen. 1984. Chemical composition of *Picea abies* and *Pinus sylvestris* seeds under subarctic conditions. *Canadian Jour. Forest Res.* 14:214–217.

Quinlivan, B. J. 1970. The interpretation of germination test on seeds of *Lupinus* species which develop in permeability. *Inter. Seed Testing Assoc. Proc.* 35:349–359.

Radwin, M. A., & G. L. Crouch. 1977. Seed germination and seedling establishment of red stem ceanothus. *Jour. Wildlife Manage.* 41:760–766.

Radwin, M. A., & D. S. DeBell. 1981. Germination of red alder seeds. Forest Service, USDA, Portland, OR.

Raevskikh, V. M. 1979. Seed quality of *Larix gmelinii* in the Magadan region. *Lesnoe Khozyaistro* 39.

Rai, S. N. 1978. Pretreatment of seeds of *Albizzia falcata, A. chinensis,* and *A. richardiana. My Forest* 241–245.

Rajasekaran, K., J. Vine, & M. G. Mullins. 1982. Dormancy in somatic embryos and seeds of *Vitis:* changes in endogenous abscissic acid during embryogeny and germination. *Planta* 154:139–144.

Ramaiah, P. K., D. J. Durzanand, & A. J. Mia. 1971. Amino acid, soluble proteins, and isoenzyme patterns of peroxidase during the germination of jack pine. *Canadian Jour. Botany* 49:215–216.

Ramirez, C. M., M. Romero, & M. Henriquez. 1980. Germination of seeds of Chilean Myrtaceae. *Bosque* 3:106–114.

Rao, P. B. 1988. Effect of environmental factors on germination and seedling growth of *Quercus floribunda* and *Cupressus torulosa,* tree species of the central Himalaya. *Annals Botany* 61:531–540.

Rao-Vasudeva, P. H. V., & E. R.R. Ivensar. 1982. Studies in seed morphology and germination of jojoba. *Current*

Science 51:516–518.

Rasmussen, G. A., & H. A. Wright. 1988. Germination requirements of flameleaf sumac. *Jour. Range Manage.* 41:48–52.

Ratliff, R. D. 1974a. *Eriogonum.* In: *Seeds of Woody Plants in the United States.* Forest Service, USDA, Washington, DC. 382–383.

———. 1974b. *Haplopappus.* In: *Seeds of Woody Plants in the United States.* Forest Service, USDA, Washington, DC. 445–447.

———. 1974c. *Lupinus.* In: *Seeds of Woody Plants in the United States.* Forest Service, USDA, Washington, DC. 520–521.

Ray, P. K., & S. B. Sharma. 1985. Viability of *Litchi chinensis* seeds when stored in air and water. *Jour. Agr. Sci.* 104:247–248.

Read, R. A. 1974a. *Phellodendron.* In: *Seeds of Woody Plants in the United States.* Forest Service, USDA, Washington, DC. 578–579.

———. 1974b. *Sapindus.* In: *Seeds of Woody Plants in the United States.* Forest Service, USDA, Washington, DC. 758–759.

Read, R. A., & R. L. Barnes. 1974. *Morus.* In: *Seeds of Woody Plants in the United States.* Forest Service, USDA, Washington, DC. 544–547.

Record, S. J., & R. W. Hess 1943. *Timbers of the New World.* Yale Univ. Press, New Haven, CT.

Reed, M. J. 1974. *Ceanothus.* In: *Seeds of Woody Plants in the United States.* Forest Service, USDA, Washington, DC. 284–290.

Rehder, A. 1940. *Manual of Cultivated Trees and Shrubs.* The MacMillian Co., New York.

Reid, W. H. 1972. Germination of *Pinus aristata. Great Basin Naturalist* 32:235–237.

Reukema, D. L. 1982. Seedfall in a young-growth Douglas-fir stand: 1950–1978. *Canadian Jour. Forest Res.* 12:249–254.

Reynolds, H. G., & R. R. Alexander. 1974a. *Garrya.* In: *Seeds of Woody Plants in the United States.* Forest Service, USDA, Washington, DC. 420–421.

———. 1974b. *Tamarix.* In: *Seeds of Woody Plants in the United States.* Forest Service, USDA, Washington, DC. 794–795.

Richardson, D. M., B. W. Van Wilgen, & D. T. Mitchell. 1987. Aspects of the reproductive ecology of four Australian *Hakea* species in South Africa. *Oecologia* 71:345–359.

Rietveld, W. J. 1983. Allelopathic effects of julone on germination and growth of several herbaceous and woody species. *Jour. Chemical Ecology* 9:295–308.

Rim, Y. D., & T. Shidei. 1974. Studies on the seed production of Japanese red and black pine. Bulletin Kyoto Univ. Forest 46:75–84.

Rimando, E. F., & M. V. Dalmacio. 1978. Direct-seeding of ipil-ipil (*Leucaena leucocephala*). *Sylvatrop* 3:171–175.

Rink, G., T. R. Dell, G. Sartzer, & F. T. Bonner. 1979. Use of the Werbell function to quantify sweetgum germination sites. *Silvae Genetica* 28:9–12.

Rink, G., & R. D. Williams. 1984. Storage technique affects white oak acorn viability. *Tree Planter's Notes* 35:3–5.

Roa, P. B. 1988. Effect of environmental factors on germination and seedling growth in *Quercus floribunda* and *Cupressus torulosa* tree species of the central Himalaya. *Annals Botany* 61:531–540.

Roa, P. B., & S. P. Singh. 1985. Response breadths on environmental gradients of germination and seedling growth in two dominate forest species of central Himalaya. *Annals Botany* 56:783–794.

Robbins, T. A. 1983. *Pinus oocharpa.* Danida Forest Seed Center, Denmark.

Roberts, T. A., & C. H. Smith. 1982. Growth and survial of black oak seedlings under different germination, watering, and planting regimes. *Tree Planter's Notes* 33:10–12.

Robinson, W. A. 1986. Effect of fruit ingestion on *Amelanchier* seed germination. Bul. Torrey Bot. Club 113:131–134.

Rock, J. F. 1920. *The Leguminous Plants of Hawaii*. Hawaii Sugar Planters Assoc., Honolulu, HI.

Ronis, E. Y., & B. Z. Kodola. 1977. Stimulation of flowering of Scots pine. *Lesnoe Khozyaistvo* 58–59.

Rosochacka, J., & A. P. Grzywacz. 1980. The colour of *Pinus sylvestris* L. seeds and their chemical composition. II. Colour of seed coats and their chemical composition. *European Jour. Forest Pathology* 10:193–201.

Ross, J. D., & J. W. Brandbeer. 1971. Studies in seed dormancy. V. The content of endogenous gibberellin in seeds of *Corylus avellana*. *Plants* 100:288–302.

Rostovtsev, S. A., E. S. Lyubich, & A. A. Solomonova. 1975. The seasonal variation in the germination of Scots pine seeds. *Lesnoe Khozyaistvo* 57–60.

Rouskas, D. 1983. Seed germination of the peach cultivar INGR GF 305: effects of benzyl aminopurine and gibberellins on breaking dormancy and on the absence of foliar abnormalities in plants grown from nonstratified seeds. Arerud Publ., New Delhi.

Rowan, S. J. 1982. Effects of rate and kind of seedbed mulch and sowing depth on germination of southern pine seeds. *Tree Planter's Notes* 33:19–21.

Rowan, S. J., & G. L. DeBarr. 1974. Moldy seed and poor germination linked to seed bag damage in slash pine. *Tree Planter's Notes* 25:25–27.

Roy, D. F. 1974a. *Arbutus*. In: *Seeds of Woody Plants in the United States*. Forest Service, USDA, Washington, DC. 226–227.

———. 1974b. *Cercis*. In: *Seeds of Woody Plants in the United States*. Forest Service, USDA, Washington, DC. 305–308.

———. 1974c. *Lithocarpus*. In: *Seeds of Woody Plants in the United States*. Forest Service, USDA, Washington, DC. 512–514.

———. 1974d. *Torreya*. In: *Seeds of Woody Plants in the United States*. Forest Service, USDA, Washington, DC. 815–816.

Rudolf, P. O. 1974a. *Aesculus*. In: *Seeds of Woody Plants in the United States*. Forest Service, USDA, Washington, DC. 195–200.

———. 1974b. *Berberis*. In: *Seeds of Woody Plants of the United States*. Forest Service, USDA, Washington, DC. 247–251.

———. 1974c. *Cedrus*. In: *Seeds of Woody Plants of the United States*. Forest Service, USDA, Washington, DC. 291–294.

———. 1974d. *Clematis*. In: *Seeds of Woody Plants in the United States*. Forest Service, USDA, Washington, DC. 331–334.

———. 1974e. *Colutea*. In: *Seeds of Woody Plants in the United States*. Forest Service, USDA, Washington, DC. 335.

———. 1974f. *Cotinus*. In: *Seeds of Woody Plants in the United States*. Forest Service, USDA, Washington, DC. 346–348.

———. 1974g. *Euonymus*. In: *Seeds of Woody Plants in the United States*. Forest Service, USDA, Washington, DC. 393–397.

———. 1974h. *Kalopanux*. In: *Seeds of Woody Plants in the United States*. Forest Service, USDA, Washington, DC. 472–473.

———. 1974i. *Koelreuteria*. In: *Seeds of Woody Plants in the United States*. Forest Service, USDA, Washington, DC. 474–475.

———. 1974j. *Laburnum*. In: *Seeds of Woody Plants of the United States*. Forest Service, USDA, Washington, DC. 476–477.

———. 1974k. *Larix*. In: *Seeds of Woody Plants in the United States*. Forest Service, USDA, Washington, DC. 478–485.

———. 1974l. *Ligustrum*. In: *Seeds of Woody Plants in the United States*. Forest Service, USDA, Washington, DC. 500–502.

———. 1974m. *Lycium*. In: *Seeds of Woody Plants of the United States*. Forest Service, USDA, Washington, DC. 522–524.

———. 1974n. *Rhodotypos*. In: *Seeds of Woody Plants in the United States*. Forest Service, USDA, Washington, DC. 713–714.

———. 1974o. *Sciadopitys*. In: *Seeds of Woody Plants in the United States*. Forest Service, USDA, Washington, DC. 763.

———. 1974p. *Sorbaria*. In: *Seeds of Woody Plants in the United States*. Forest Service, USDA, Washington, DC. 779.

———. 1974q. *Taxus*. In: *Seeds of Woody Plants in the United States*. Forest Service, USDA, Washington, DC. 799–802.

———. 1974r. *Ulex*. In: *Seeds of Woody Plants in the United States*. Forest Service, USDA, Washington, DC. 828.

Rudolf, P. O., & W. B. Leak. 1974. *Fagus*. In: *Seeds of Woody Plants of the United States*. Forest Service, USDA, Washington, DC. 401–405.

Rudolf, P. O., & H. Phipps. 1974. *Carpinus*. In: *Seeds of Woody Plants in the United States*. Forest Service, USDA, Washington, DC. 266–268.

Rudolf, P. O., & P. E. Slabaugh. 1974. *Syringa*. In: *Seeds of Woody Plants in the United States*. Forest Service, USDA, Washington, DC. 791–793.

Rudolph, T. D., & R. A. Cecich. 1979. Provenance and plantation location variation in jack pine seed yield and quality. In: *Proc. First North Central Tree Improvement Conf.* Forest Service, USDA, Rhinelander, WI. 10–20.

Russo, V. M. 1978. Development of *Pinus* seedlings grown from seeds subjected to drying and wetting cycles. *Forest Sci.* 24:537–541.

Ruth, R. H. 1974. *Tsuga*. In: *Seeds of Woody Plants in the United States*. Forest Service, USDA, Washington, DC. 819–827.

Rypak, M. 1979. Some morphological and biochemical changes in *Fraxinus excelsior* seeds during stratification. *Acta Dendrobiologica* 293, 295–334.

Ryynanen, L. 1980. Storage of Scots pine seed and seed ageing. *Inst. Forestale Fenniae* No. 428. Kokari, Finland. 11.

Sabeev, A. G., & V. A. Olisaev. 1983. The seed resource of beech and fir in the northern Caucasus. *Lesnoe Khozyaistov* 42–43.

Sabiiti, E. N., & R. W. Wein. 1987. Fire and acacia seeds: a hypothesis of colonizing species. *Jour. Ecology* 75:937–946.

Sadjad, S., & S. Hadi. 1976. The time for the first and the second counting in standard seed germination test method for *Pinus merkusii*. Center for Tropical Biology, Bogor, Indonesia.

Safford, L. E. 1974. *Picea*. In: *Seeds of Woody Plants in the United States*. Forest Service, USDA, Washington, DC. 587–597.

Sagwal, S. S. 1984. Studies on production from individual cones of chir pine. *Indian Jour. Forestry* 7:4–6.

———. 1986. Pre-sowing treatment of puna seed. *Indian Forestry* 112:261–263.

St. Johns, S. 1982. Acid treatment of seeds of *Crataegus monogyna* and other *Crataegus* species. *Plant Propagator* 32:203–204.

Salmia, M. A., S. A. Nyman, & J. J. Mikola. 1978. Characterization of the proteinases present in germinating seeds of Scots pine. *Physiologia Plantarum* 42:252–256.

Salomonson, M. G. 1978. Adaptations for annual dispersal of one seed juniper seeds. *Oecologia* 32:333–339.

Samoshkin, E. N. 1977. Reaction of pine to the action of aqueous solutions of N-nitrosomethyurea. *Lesnoi Zhurnal* 27–31.

Samsonova, A. E. 1974. Gibberellin and the growth of of *Picea abies* seedlings. *Referativnyi Zhurnal* 127–131.

Sandberg, G. 1988a. Application of growth regulators in aqueous media and organic solvents to seeds of *Picea abies* and *Pinus sylvestris*. *Scandinavian Jour. Forest Res.* 3:97–105.

———. 1988b. Effect of growth regulators on germination of *Picea abies* and *Pinus sylvestris* seeds. *Scandinavian Jour. Forest Res.* 3:83–95.

Sandberg, G., & A. Ernstsen. 1987. Dynamics of indo-3-acetic acid during germination of *Picea abies* seeds. *Tree Physiology* 3:185–192.

Sander, I. L. 1974a. *Castanea.* In: *Seeds of Woody Plants in the United States.* Forest Service, USDA, Washington, DC. 273–275.

———. 1974b. *Gymnocladus.* In: *Seeds of Woody Plants in the United States.* Forest Service, USDA, Washington, DC. 439–440.

Sankary, M. N., & M. G. Barbour. 1972. Autecology of *Atriplex polycarpa* from California. *Ecology* 53:1155–1162.

Saplaco, S. R., & A. Revilla. 1973. Comparative seed germination and seedling height growth of cashew. *Philippine Lumberman* 19:16,18.

Saralidze, G. M., & B. G. Saralidze. 1976. A modified machine for extracting seeds from larch cones. *Lesnoe Khozyaistvo* 44.

Sargent, C. S. 1965. *Manual of Trees of North America.* Dover Publ. Inc., New York.

Sasaki, S. 1980. Storage and germination of dipterocarp seeds. *Malaysian Forester* 43:290–308.

Sashi, M., & J. Citharel. 1979. Variation in the nitrogen composition of *Pinus canariensis* during germination. *Phytochemistry* 18:47–50.

Sato, M., & C. Yamamoto. 1977. Effective and ineffective cone scales of *Pinus densiflora* et Zucc. and *P. thunbergii* Parl. for viable seed production. Bulletin Govt. Forest Exp. Sta., Japan. 89–103.

Saxena, S. K., & W. A. Khan. 1974. A quick method for obtaining clean seed of *Prosopis juliflora*. *Annals Arid Zone* 13:269–272.

Schalin, I. 1967. Germination analysis of *Alnus incana* (L.) Moench. and *A. glutinosa* (L.) Gaertn. seeds. *Acta Oecol. Scand.* 18:253–260.

Scharpf, R. F. 1970. Seed viability germination and radicle growth of dwarf mistletoe in California. *Forest Service, USDA,* Berkeley, CA.

Scharpf, R. F., & R. Parmeter. 1962. The collection, storage, and germination of seeds of dwarf mistletoe. *Jour. Forestry* 60:551–552.

Scherbatskay, T., R. M. Klein, & G. J. Badger. 1987. Germination response of forest tree seed to acidity and metal ions. *Environmental and Experimental Botany* 27:157–164.

Schopmeyer, C. S. 1974a. *Alnus.* In: *Seeds of Woody Plants in the United States.* Forest Service, USDA, Washington, DC. 204–211.

———. 1974b. *Nemopanthus.* In: *Seeds of Woody Plants in the United States.* Forest Service, USDA, Washington, DC. 553.

———. 1974c. *Thuja.* In: *Seeds of Woody Plants in the United States.* Forest Service, USDA, Washington, DC. 805–809.

———. 1974d. *Vitex.* In: *Seeds of Woody Plants in the United States.* Forest Service, USDA, Washington, DC. 851–852.

Schopmeyer, C. S., & W. B. Leak. 1974. *Ostrya.* In: *Seeds of Woody Plants in the United States.* Forest Service, USDA, Washington, DC. 564–565.

Schreiner, E. J. 1974. *Populus.* In: *Seeds of Woody Plants in the United States.* Forest Service, USDA, Washington, DC. 645–655.

Schroeder, W. R., & D. S. Walker. 1987. Effects of moisture content and storage temperature on germination of *Quercus macrocarpa* acorns. *Jour. Environmental Hort.* 5:22–24.

Schubert, T. H. 1974. *Tectona.* In: *Seeds of Woody Plants in the United States.* Forest Service, USDA, Washington, DC. 803–804.

Schwitter, R. 1984. Germination of *Commiphora* species. *Tech. Centra de formation Prof. Forestiere de Morondava* Madagascar:7.

Scott, L. 1983. Germinating eucalypt seeds. *Plant Propagation* 32:357–362.

Scowcroft, P. G. 1978. Germination of *Sophora chrysophylla* increased by presowing treatment. Forest Service, USDA, Berkeley, CA.

———. 1981. Regeneration of mamane: effects of seed coat treatment and sowing depth. *Forest Sci.* 27:771–779.

———. 1982. Distribution and germination of mamane seeds. Forest Service, USDA, Berkeley, CA.

———. 1988. Germinability of cork pine (*Araucaria columnaris*) seeds under different storage conditions. *Forest Service Summer* 39:17–25.

Scoz, A., & P. Grossoni. 1985. Germination of seed of some *Pinus cimbra* provenances. *Dendronatura* 6:19–33.

Seiffert, N. F. 1982. Practical method for *Leucaena* seed scarification using NAOH. *Leucaena Res. Rpts.* 3:5–6.

Seitner, P. G. 1981. Factors in the storage of *Magnolia macrophylla* seeds. *Magnolia* 17:13–20.

Sequelquist, C. A. 1971. Moistening and heating improves germination of two legume species. *Jour. Range Manage.* 24:393–394.

Sento, T. 1972. Studies of germination of palms. IV. *Jour. Japanese Hort. Soc.* 40:255–261.

Shafiq, Y. 1978. Studies on the cones and seeds of *Pinus brutia* Ten. *Mesopotamia Jour. Agr.* 13:79–84.

———. 1979. Some effects of light and temperature on the germination of *Pinus brutia, Nothofagus obliqua* and *Nothofagus procera* seeds. *Seed Sci. Tech.* 7:189–193.

———. 1981. Effect of gibberellic acid and prechilling on germination percent of *Nothofagus obliqua* and *N. procera* seeds. *Turrialba* 31:365–368.

Shannon, P. R. M., R. A. Jeavos, & B. C. Jarvis. 1983. Light sensitivity of hazel seeds with respect to the breaking of dormancy. *Plant and Cell Physiology* 24:933–936.

Sharma, K. E., K. C. Naithani, & P. M. Sangal. 1985. Effect of high temperatures of seeds-results of a study with *Leucaena leucocephala*. *Indian Forester* 11:182–194.

Sharma, M. L. 1976. Interaction of water potential and temperature effects on germination of three semi-arid plant species. *Agron. Jour.* 68:390–394.

Sharma, M. M., & A. N. Purohit. 1980. Seedling survial and seed germination under natural and laboratory conditions in *Shorea robusta*. *Seed Sci. Tech.* 8:283–287.

Sharma, S. K. 1977. A further contribution to the study of nursery behaviour of *Diospyros marmorata*. *Indian Forester* 103:542–549.

Shea, G. M., & P. A. Armstrong. 1978. Factors affecting survival in op-root plants of Caribbean pine-coastal Queensland, Dept. Forestry.

Shearer, R. C. 1977. Maturation of western larch cones and seeds. Forest Service, USDA, Ogden, UT.

———. 1984. Influence of insects on Douglas fir and western larch cone and seed production. Forest Service, USDA, Ogden, UT.

Sheikh, M. I., 1979. Tree seeds respond to acid scarification. *Pakistan Jour. Forestry* 29:253–254.

———. 1980. Effect of different treatments to hasten tree germination. *Pakistan Jour. Forestry* 30:176–180.

———. 1983. Germination trials of juniper seeds. *Pakistan Jour. Forestry* 33:41–43.

Shepperd, W. D., & D. L. Noble. 1976. Germination, survival, and growth of lodgepole pine under simulated precipation regimes, a greenhouse study. Forest Service, USDA, Fort Collins, CO.

Shibakusa, R. 1980. Effect of GA, IAA, ethrel and BA on seed germination of *Picea glehnii*. *Jour. Japanese Forestry Soc.* 62:440–443.

Shih, C. Y., E. B. Dumbroff, & C. A. Peterson. 1985. Developmental studies of the stratification-germination process in sugar maple embryos. *Jour. Botany* 63:903–908.

Sholokhov, L. V., I. I. Bogoroditskii, O. F. Bogoroditskaya, & B. G. Naumenko. 1978. Accelerated preparation of *Juglans regia* seeds for sowing. *Lesnoe Khozyaistvo* 49–52.

Shultz, L. M. 1984. Taxonomic and geographic limits of *Artemisia* subgenus *Tridentatae* (Beetle) McArthur. In: *Proc. Sym. Biol.* Artemisia *and* Chrysothamus. Forest Service, USDA, Ogden, UT. 22–28.

Siddiqui, K. M., & M. Parvez. 1981. Seed storage and germination studies of blue pine. *Pakistan Jour. Forestry* 31:51–60.

Sigaffoos, R. S. 1977. Relations among surfical materials, light intensity and sycamore seed germination along the Potomac River near Washington, DC., *Jour. Res. Geologic Survey* 4:733–736.

Silvia, L. L. 1979. Effects of shading and underlying materials on the germination and survival of *Eucalyptus grandis*. *Brasil Florestal* 9:15–18.

Simak, M. 1973. Storage experiments on *Pinus sylvestris* seed never exposed to light. In: *Proc. Inter. Symp. Forest Tree Seed Storage*. Canadian Forestry Service, Chalk River, Ontario. 5.

_____ . 1980. Germination and storage of *Saliz caprea* and *Populus tremula* seeds. In: *Proc. Inter. Symp. Forest Tree Seed Storage*. Canadian Forestry Service, Chalk River, Ontario. 142–160.

Simeonov, A. 1974. Germination of seeds of *Fraxinus excelsior* treated with an aqueous solution of molybdenum. *Gorsko Stopanstvo* 30:51–52.

_____ . 1975. Stimulating the germination of *Fraxinus excelsior* seeds by treatment in an aqueous solution of molybdenum. *Gorsko Stopanstov* 31:45–47.

Simpson, J. D., & G. R. Powell. 1981. Some factors influencing cone production on young black spruce in New Brunswick. *Forestry Chronicle* 57:267–269.

Sims, H. P. 1970. Germination and survival of jackpine on three prepared cutover sites. Canadian Dept. Agr.

Singh, R., & R. Bawa. 1982. Effect of leaf leachates from *Eucalyptus globulus* Labill. and *Aesculus indica* Colebr. on germination of *Glaucium flavum* Grantz. *Indian Jour. Ecology* 9:21–28.

Singh, R. V., J. P. Chandra, & S. N. Sharma. 1973. Effect of depth of sowing on germination of kail seed. *Indian Forestry* 99:367–371.

Singh, R. V., & V. Singh. 1983. Germination of *Populus ciliata* seed as influenced by moisture stress. *Indian Forester* 109:357–358.

Singh, S. S., & G. S. Palical. 1986. Influence of ethyl-hydrogen propylphosphate (Niagard) on seed germination and early growth of *Leucaena leucocephala*. *Leuczensa Res Reports* 1986(7):70–71.

Singh, V., & S. R. Arya. 1987. *Populus ciliata*-effect of parts of catkins on seed germination. *Van Visyan* 25:26–28.

Singh, V., O. Singh, & H. D. Sharma. 1986. Germination of silver fir (*Abies pindrow*) seed as affected by moisture stress. *Indian Jour. Forestry* 9:293–295.

Sinha, M. M., R. G. Dal, & D. S. Koranga. 1977. Studies in the seed germination of walnut. *Prog. Hort.* 8:69–74.

Sinska, I., & R. Gladen. 1984. Ethylene and removal of embryonal apple seed dormancy. *Hort. Sci.* 19:73–75.

Skrynuikov, B. M. 1985. Cleaning tree seed in a flow of air. *Lesnoi Zhurnal* 26–30.

Slabaugh, P. E. 1974a. *Cotoneaster*. In: *Seeds of Woody Plants in the United States*. Forest Service, USDA, Washington, DC. 349–352.

_____ . 1974b. *Hippophae*. In: *Seeds of Woody Plants in the United States*. Forest Service, USDA, Washington, DC. 446–447.

Small, J. G. C., J. E. McNaughton, & J. H. Greeff. 1977. Physiological studies of the germination of *Erythrina caffra* seeds. *Jour. South African Botany* 43:213–222.

Smirnov, I. A. 1979. Growth and resistance of seedlings of introduced conifers raised from seed at various stages of development. *Byulleten Glavnogo Botanicheskogo Sada* 1979. 14–20.

Smith, C. C. 1968. The adaptive nature of social organization in the genus of tree squirrels, *Tamiasciurus*. *Ecol. Monogr.* 38:31–63.

Smith, J. G. 1974a. *Grayia*. In: *Seeds of Woody Plants in the United States*. Forest Service, USDA, Washington, DC. 434–436.

_____ . 1974b. *Peraphyllum*. In: *Seeds of Woody Plants in the United States*. Forest Service, USDA, Washington, DC. 576–577.

Smith, M. N. 1985. Seeding the California oaks: old problems, new tricks. *Plant Propagation* 31:14–15.

Smurova, M. V. 1975. Long-term storage of seeds of *Haloxylon aphyllum*. *Lesnoe Khozyaistvo* 63–65.

Sniezko, R. A., & D. P. Guaze. 1987. Effect of seed treatments on germination of *Acacia albida* Del. Swedish Univ. of Agr. Sci., Dept. of Forest Genetics and Plant Physiology 7:325–333.

Song, X. Z., Q. D. Chen, D. F. Wang, & J. Yang. 1983. A study ultrastructural changes in radicle-tip cells and seed vigor of *Hopea* and *Vatica* in losing water process. *Scientia Silvae Sinicae* 19:121–125.

_____ . 1984. A study on the principal storage conditions of *Hopea hainanensis* seeds. *Scientia Silvae Sinicae* 20:225–236.

Sopp, D. F., S. S. Salac, & R. K. Sutton. 1977. Germination of Gambel oak seed. *Tree Planter's Notes* 28:4–5.

Sorensen, F. C., & R. K. Campbell. 1981. Germination rate of Douglas-fir seeds affected by their orientation. *Annals Botany.* 47:467–471.

_____ . 1985. Effect of seed weight on height growth of Douglas-fir seedlings in a nursery. *Canadian Jour. Forest Res.* 15:1109–1115.

Sork, V. L. 1984. Observation on fruiting and dispensers of *cecropia obtusifolia* at Los Tuxtlas, Mexico. *Biotropica* 16:315–318.

_____ . 1985. Germination response in a large-seeded neotropical tree species, *Gustavia superba*. *Biotropica* 17:130–136.

Speer, E. R., & H. A. Wright. 1981. Germination requirements of loto bush in Texas, Arizona, New Mexico, and northern Mexico. *Jour. Range Manage.* 34:365–368.

Speers, C. F. 1962. Fraser fir seed collection, stratification and germination. *Tree Planter's Notes* 53:7–8.

_____ . 1967. Insect infestation distorts Fraser fir seed text. *Tree Planter's Notes* 18:19–21.

_____ . 1968. Balsam fir chalcied causes loss of Fraser fir seed. *Tree Planter's Notes* 19:18–20.

Sprackling, J. A. 1976. Germination of *Pinus gerardiana* seeds following storage and stratification. *Tree Planter's Notes* 27:5–6, 22.

Springfield, H. W. 1974. *Eurotia lanata*. In: *Seeds of Woody Plants in the United States*. Forest Service, USDA, Washington, DC. 398–400.

Staafilt, H., M. Junsson, & L. G. Olsen. 1987. Buried germinative seeds in native beech forest with different herbaceous vegetation and soil types. *Holarctic Ecology* 10:268–277.

Staden, J. Van, D. P. Webb, & P. F. Waring. 1972. The effect of stratification on endogenous cytokinin levels in seeds of *Acer saccharum*. *Planta* 104:110–114.

Standnitskii, G. V. 1985. Requirements for permanent seed stands of Scot pine. *Lesnoe Khozyaistvo* 43–45.

State seed orchards in Lot. 1975. *Revue Forestiere Francaise* 27:351:356.

Stein, W. I. 1951. Germination of noble and silver fir seed on snow. *Jour. Forestry* 49:448–449.

———. 1974a. *Libocedrus decurrens*. In: *Seeds of Woody Plants in the United States*. Forest Service, USDA, Washington, DC. 494–499.

———. 1974b. *Umbellularia*. In: *Seeds of Woody Plants in the United States*. Forest Service, USDA, Washington, DC. 835–839.

Steven, D. P. 1982. Seed production and seed mortality in a temperature forest shrub. *Jour. Ecology* 79:437–443.

Stevens, P. 1974. A revision of *Calophyllu* L. in Papuasia. *Australian Jour. Botany* 22:349–412.

Stevens, R., K. R. Jorgenson, J. N. Davis, & S. B. Monson. 1986. Seed pappus and placement influences on white rabbitbrush establishment. In: *Proc. Symp. Biol.* Artemisia and Chrysothamnus. Forest Service, USDA, Ogden, UT. 353–357.

Stickney, P. F. 1974a. *Holodiscus*. In: *Seeds of Woody Plants in the United States*. Forest Service, USDA, Washington, DC. 448–449.

———. 1974b. *Philadephus*. In: *Seeds of Woody Plants in the United States*. Forest Service, USDA, Washington, DC. 580–581.

———. 1974c. *Spiraea*. In: *Seeds of Woody Plants in the United States*. Forest Service, USDA, Washington, DC. 785–786.

Stimart, D. P. 1981. Factors regulating germination of trifoliate maple seeds. *Hort. Sci.* 16:341–343.

Stinemetz, C. L.,, & B. R. Roberts. 1984. An analysis of the gibberellic and abscissic acid content of white ash seeds. *Jour. Arboriculture* 10:283–285.

Stoehr, M. V., & R. E. Farmer. 1986. Genetic and environmental variance in cone size, seed yield, and germination properties of black spruce cones. *Canadian Jour. Forest Res.* 16:1149–1151.

Street, R. J. 1962. *Exotic Forest Trees in the British Commonwealth*. Clarendon Press, Oxford.

Struck, D., & C. E. Whitcomb. 1977. Effects of nutrition on germination and growth of *Cedrus deodara* seedlings. Oklahoma Agr. Exp. Sta. Res. Rpt. 760:32–34.

Struve, D. K., & R. D. Lineberger. 1985. Field transplant survival of *Amelanchier* liners produced by tissue culture. *Plant Propagation* 31:11–13.

Stutz, H. C. 1983. *Atriplex* hybridization in western North America. In: *Proc. Symp. Biol.* of *Atriplex and Related Chenopods*. Gen. Tech. Rpt. 172. Forest Service, USDA, Ogden, UT. 25–26.

Subbaiah, C. C. 1982/83. Effect of gibberellic acid on seed germination and seedling growth of cashew. *Scientia Hort.* 18:137–142.

Suiter, W., & L. Lishao. 1973. Influence of the relative humidity on the characteristics of the seeds of *Eucalyptus saligna*. In: *Proc. Inter. Symp. Seed Processing and Seed Problems*. Inter. Union Forestry Res. Org. 2:15.

Suleimanov, B. 1980. The seeding of *Haloxylon aphyllum* in south Kazakbstan. *Lesnoe Khozyaistvo* 40–41.

Sulli, A. Z. 1975. Results of an international trial on improved methods of testing the germination of beech seeds. *L'Italia Forestale e Montana* 30:18–25.

Sullivan, T. P. 1979. The use of alternative foods to reduce conifer predation by deer mice. *Jour. Applied Ecology* 16:475–495.

Sullivan, T. P., & D. S. Sullivan. 1982. The use of alternative foods to reduce lodgepole pine seed predation by small mammals. *Jour. Applied Ecology* 19:33–45.

Suo, Q. S. 1982. Studies on the fruiting role in natural stands of *Larix gmelinii* var. *olgenisis* Henry in Zhambei Mountain Range. *Scientia Silvae Sinicae* 18:347–356.

Surber, E., I. Katin, A. Simonett, & E. Frehner. 1973. Freeze drying of forest tree seeds, especially of spruce, for long storage. In: *Proc. Inter. Symp. Seed Processing and Seed Problems*. Inter. Union Forestry Org. Res.

Suresh, K. K., & R. S. V. Rai. 1987. Studies on the allelopathic effects of some agroforestry tree crops. *Inter. Tree Crop Jour.* 4:109–115.

Suszka, B. 1968. Conditions for breaking of dormancy and germination of hornbeam seeds. *Arbor. Kornickie* 13:147–172.

———. 1975. Cold storage of already after-ripened beech seeds. *Arbor. Kornickie* 20:299–315.

———. n.d. How to achieve simultaneous germination of after-ripened hardwood seeds? Inst. Dendrology, Kornik, Poland.

Suszka, B., & A. Kluczynska. 1980. Seedling emergence of stored beech seed chilled without medium at a controlled hydration level and pregerminated in cold-moist conditions. *Arbor. Kornickie* 25:231–255.

Suszka, B., & T. Tylkowski. 1981. Storage of acorns of the northern red oak over 1–5 winters. *Arbor. Kornickie* 26:253–306.

Suszka, B., & L. Zieta. 1976. Further studies on the germination of beech seed stored in an already afterripened condition. *Arbor. Kornickie* 21:279–296.

———. 1977. A new presowing treatment for cold-stored beech seed. *Arbor. Kornickie* 22:237–255.

Sutherland, J. R. 1981. Time, temperature and moisture effects on incidence of seed infected by *Caloscypha fulgens* in Sitka spruce cones. *Canadian Jour. Forest Res.* 11:727–730.

Sutherland, J. R., & D. G. W. Edwards. 1976. Pigments for use on conifer seeds sown in forest nurseries. Pacific Forest Res. Centre, Victoria, B.C., Canada.

Sutherland, J. R., T. A. Woods, W. Lock, & D. A. Gaudet. 1978. Evaluation of surface sterilants for isolation of the fungus *Geniculodendron pyriforme* from Sitka spruce seeds. *Bi-Mon. Res. Notes* 34:20–21.

Sutherland, J. R., & T. A. D. Woods. 1978. *Geniculodendron pyriforme* in stored Sitka spruce seeds: effects of seed extraction and cone collection methods on disease incidence. Phytopathology 68:747–750.

Szalia, I., & M. Nagy. 1974. Dormancy in fruits of *Tilia platyphyllos* Scop. II. The inhibitor-substance content of dormant fruits. *Acta Botanica Acad. Sci.* Hungarice 20:389–394.

Szczotka, Z. 1974. Amylolytic activity in acorns of *Quercus borealis* during storage under controlled conditions. *Arbor. Kornickie* 19:129–134.

———. 1975. Changes in the intensity of protein synthesis in the embryo axes of northern red oak and English oak during storage under controlled conditions. *Arbor. Kornickie* 20:291–297.

———. 1978. Intensity of respiration in the embryo axes of *Quercus borealis* and *Q. rubur* acorns during storage and aging under controlled conditions. *Arbor. Kornickie* 23:145–151.

Tamari, C. 1976. *Phenology and Seed Storage Trials of Dipterocarpus*. Forest Research Inst., Kepong.

Tanaka, Y., & D. G. W. Edwards. 1985. An improved and more versatile method for prechilling *Abies procera* Rehd. seed. *Seed Sci. Tech.* 14:457–464.

Tanako, Y., N. J. Kleyn, & L. M. Harper. 1986. Seed stratification of Engelmann spruce and lodge pole pine. The effect of stratification duration and timing of surface drying. *Forestry Chronicle* 62:147–151.

Tang, H. T., & C. Tamari. 1973. Seed description and storage test of some dipterocarps. *Malaysian Forester* 36:38–53.

Tappeiner, J. C., P. M. McDonald, & T. F. Hughes. 1986. Survival of tanoak and Pacific madrone seedlings in forests of southern Oregon. *New Forest* 1:43–55.

Targonskii, P. N., & G. A. Bogdanova. 1984. Seed and vegetative reproduction of *Vaccinium vitis-idaea*. *Rastit-Resur. Lennigrad* 20:29–35.

Tauer, C. G. 1979. Seed tree, vacuum, and temperature effects on western cottonwood seed viability during extended storage. *Forest Sci.* 25:112–114.

Taylor, J. S., & P. F. Waring. 1979a. The effect of light on the endogenous levels of cytokinins and gibberellins in seed of Sitka spruce. *Plant Cell and Environment* 2:173–179.

———. 1979b. The effect of stratification on the endogenous levels of gibberellins and cytokinins in seeds of Douglas-fir and sugar pine. *Plant Cell and Environment* 2:165–171.

Taylor, R. J., & D. C. Shaw. 1983. Allelopathic effects of Engelmann spruce bark stilbenes and tannin-stilbene combinations on seed germination and seedling growth of selected conifers. *Canadian Jour. Botany* 61:279–289.

Teclaw, R. M., & J. G. Isebrands. 1986. Collection procedures affect germination of northern red oak acorns. *Tree Planter's Notes* 37:8–12.

Thalourian, P. 1976. Stimulation of germination in *Pinus halepensis* by successive applications of mercury and chloride ions: research on the mechanisms responsible. *Comptes Rendus Hebdomadaires des Seances de l'Academie des Sciences* (France) 282:1857–1860.

Thaplyal, R. C., & B. N. Gupta. 1980. Effect of seed source and stratification on the germination of deodar seeds. *Seed Sci. Tech.* 8:145–150.

Therios, I. N. 1981. Effect of temperature, moisture stress and pH on the germination of seeds of almond. *Seed Sci. Tech.* 10:585–594.

Thevenot, C., C. Perino, & D. Comes. 1983. Influence of temperature on breaking of dormancy germination sensu stricto and growth of apple embryo: thermal optimum of the phenomena. *Israel Jour. Botany* 32:139–145.

Thilenius, J. F., K. E. Evans, & E. C. Garrett. 1974. *Shepherdia*. In: *Seeds of Woody Plants in the United States*. Forest Service, USDA, Washington, DC. 771–773.

Thomas, P. A., & R. W. Wein. 1985. Water availability and the comparative emergence of four conifer species. *Canadian Jour. Botany* 63:1740–1746.

Thomson, J. D., W. P. Maddison, & R. C. Plowright. 1982. Behavior of bumble bee pollinators of *Aralia hispida* Vent. *Oecologia* 54:326–336.

Tilberg, E., & N. J. Pinfield. 1981. The dynamics of indole-3-acetic acid in *Acer platanoides* seeds during stratification and germination. *Physiologia Plantarum* 53:34–38.

———. 1982. Changes in abscissic acid levels during after-ripening and germination of *Acer platanoides* L. seeds. *New Phytologist* 92:167–172.

Timmis, R., & J. Worrall. 1975. Environmental control of cold acclimation in Douglas fir during germination, active growth and rest. *Canadian Jour. Forest Res.* 5:464–477.

Timofeev, V. V. 1984. Shipbuilding forests of *Larix sukaczewii*-the basis for modern seed production in the Ivanov region. *Lesnoi Zhurnal* 126–129.

Tinnin, R. O., & L. A. Kirkpatrick. 1985. The allelopathic influence of broadleaf trees and shrubs on seedlings of Douglas-fir. *Forest Sci.* 31:945–952.

Tinus, R. W. 1987. Modification of seed covering material yields more and larger seedlings. *Tree Planter's Notes* 38:11–13.

Tipton, J. L. 1984. Evaluation of three growth curve models for germination data analysis. *Jour. American Hort. Sci.* 1109:451–454.

Tokarz, Z. 1974. Trials in storing seed of silver fir, beech, and oak over one growing season. *Sylvan* 118:53–56.

Tompsett, R. B. 1982. The effect of desiccation on the longevity of seed of *Araucaria hunsteinii* and *A. cunninghamii*. *Annals Botany* 50:693–704.

———. 1983. The influence of gaseous environment on the storage life of *Araucaria hunsteinii* seed. *Annals Botany* 52:299–237.

———. 1984. Desiccation studies in relation to the storage of *Araucaria* seed. *Annals Applied Biology* 105:581–586.

———. 1985. The influence of moisture content and storage temperature on the viability of *Shorea almon*, *S. robusta*, and *S. roxburghii* seed. *Forest Res.* 15:1074–1079.

———. 1986. The effect of temperature and moisture content on the longevity of seed of *Ulmus carpinifolia* and *Terminalia brassii*. *Annals Botany* 57:875–883.

Toth, J. 1980. Cedar II. Seed dissemination, extraction, quality, germination, and pregermination. *La Foret Privee* Jan Feb:78–84.

Tran, V. N. 1979. Effect of microwave energy on the strophiole, seed coat and germination of *Acacia* seeds. *Australian Jour. Plant Physiol.* 6:277–287.

———. 1981. Germinating acacia seeds by distribing the aril. *Australian Plants* 11:66–67.

Trenin, V. V., & N. N., Chernobrovkina. 1984. Embryogenesis and quality of the seeds in *Larix sibirica* plantations. *Lesovedenie* 84–88.

Truus, M. 1984. Stimulation of seed production in Scots pine with gibberellic acid. *Metsanduslikud Uurumused, Estonian* 19:7–17.

Turnpull, J., & J. Doran. 1987a. Germination in the Myrtaceae: Eucalyptus. In: *Germination of Australian Native Plants*. Intata Press, Melbourne. 186–193.

———. 1987b. Preliminary germination test data for Myrtaceae other than *Eucalyputus*. In: *Germination of Australian Native Plants*. Inkata Press, Melbourne. 194–199.

Tuskan, G. A., & C. A. Blanche. 1980. Mechanical shaking improves Shumard oak acorn germination. *Tree Planter's Notes* 31:3–6.

Tylkowski, T. 1982. Thermal conditions for the presowing treatment of European elder and red elder. *Arbor. Kornickie* 27:347–355.

Tyystjarvi, P. 1978. The need for development of harvesting techniques in pine (*Pinus sylvestris*) seed orchards in Finland. *Metsanjalostus Saatio*. 5.

Uchiyama, Y. 1981. Studies of germination of salt bushes. I. The relationship between temperature and germination of *Atriplex nummularia*. *Japanese Jour. Tropical Agr.* 25:62–67.

Urgenc, S. 1973. Cold storage for 10 years of *Pinus nigra* and *P. brutiaten*. In: *Proc. Inter. Symp Seed Processing and Seed Problems*. Inter. Union Forestry Res. Org. 8.

Ustin, S. L., R. A. Woodard, M. G. Barbour, and J. L. Hatfield. 1984. Relationship between subfleck dynamics and red fir seedling distribution. *Ecology* 65:1420–1428.

Valahos, S., & D. T. Bell. 1986. Soil seed-bank components of the northern jarrah forest of Western Australia. *Australian Jour. Forestry* 11:171–179.

Valase, I, A. Mihalache, S. Radulescu, & L. Voinescu. 1973. Influence of the duration and method of storage on the viability of seeds of *Tillia tomentosa* and *Fraxinus excelsior*. *Studii s Cerctari Inst. de Cercetare, Proiectare si Documentare Silvica* 29:141–170.

Vander-Kloet, S. P. 1983. Seed and seedling characteristics in *Vaccinium-Myrtillus*. *Naturalisle Canada* 110:285–292.

Vander-Kloet, S. P., P. J. Austin-Smith. 1986. Energetics, patterns, and timing of seed dispersal in *Vaccinium* section Cyanococcus. *American Midland Naturalist* 115:386–396.

Vander-Kloet, S. P., & P. Cabilio. 1984. Annual variation in seed production in a population of *Vaccinium corymbosum*. *Bul. Torrey Bot. Club* 111:483–488.

Vanesse, R. 1974a. Changes in the degree of dormancy of Douglas fir seeds. *Bul. de la Societe Royale Forestiere de Belgique* 81:149–158.

_____. 1974b. Measuring the active germination of *Pseudotsuga menziesii*. Bul. des Res. Agrononiques de Gembloux 9:131–138.

_____. 1975. The breaking of dormancy of Douglas fir seed. Bul. de la Societe Royale Forestiere de Belgique 82:157–164.

Van Haverbeke, D. F., & C. W. Comer. 1985. Effective treatment and seed sources for germination of eastern red cedar seed. Forest Service, USDA, Fort Collins, CO.

Vanselow. 1981. Experiments concerning the effect of seed size and seed provenance on the development of spruce. Translation Environmental Canada TR–2047. 13.

Van Staden, J., B. N. Wolstenholme, & G. G. Dimalla. 1976. Effect of temperature on pecan seed germination. *Hort. Sci.* 11:261–262.

Vanstone, D. E. 1978. Basswood seed germination. *Plant Propagation* 27:566–569.

Vanstone, D. E., & L. J. LaCroix. 1975. Embryo immaturity and dormancy of black ash. *Jour. American Soc. Hort. Sci.* 100:630–632.

Vanstone, D. E., & W. G. Ronald. 1982. Seed germination of American Basswood in relation to seed maturity. *Canadian Jour. Plant Science* 62:709–713.

Vazquez-Yanes, C. 1979. Notes on the ecophysiology of germination of *Cecropia obtusifolia*. *Turrialba* 29:147–149.

_____. 1980. Light quality and germination in *Cecropia obtusifolia* and *Piper auritum* from a tropical rain forest in Mexico. *Phyton* 38:33–35.

_____. 1981. Germination of two species of Tiliaceae from secondary tropical vegetation: *Belotia cambellii* and *Heliocarpus donnell-smithii*. Turrialba 33:81–83.

Vedenyapina, N. S., & A. P. Badanov. 1974. Use of *Azotobacter* in growing Scots pine seedlings. *Lesnoi Zhurnal* 1974:19–22.

Velkov, D., & S. Popov. 1976. Dynamics of moisture absorption and germination of seeds of *Juglans regia* when treated before sowing. *Gorskostopanska Nauka* 13:3–12.

_____. 1978. Moisture absorption and pretreatment of walnut fruits. *Gorsko Stopanstvo* 34:15–20.

Venator, C. R. 1973. The relationship between seedling height and date of germination in *Pinus caribaea* var. *hondurensis*. *Turrialba* 23:473–474.

Verma, A., & P. Tandan. 1983. Growth resultor treatment of *Pinus kesiya* seed for enhancement of germination and early seedling growth. *Advancing Frontiers Plant Science* 290–291.

_____. 1984a. Seed germination and seedling growth of *Pinus kesiya* Royal. I. Influence of inhibition, substrate pH and moisture level. *Proc. Indian National Sci. Acad.* 50:326–331.

_____. 1984b. Seed germination and seedling growth of *Pinus kesiya* Royal. II. Light and temperature requirements. *Proc. Indian National Sci. Acad.* 50:512–518.

Verzar-Petri, & M. Then. 1974. Biosynthesis of volatile oils in *Salvia sclarea* in the course of germination. *Plant Med.* 25:366–372.

Villagomez A. Y., & M. C. Garcia. 1979. Effects of stratification on the germination of seeds of three *Pinus* species. *Ciencia Forestal* 4:31–55.

Vines, R. A. 1960. *Trees, Shrubs, and Woody Vines of the Southwest*. Univ. Texas Press, Austin.

Vogel, W. G. 1974. *Lespedeza*. In: *Seeds of Woody Plants in the United States*. Forest Service, USDA, Washington, DC. 488–490.

Vogt, A. R. 1970. Effect of gibberellic acid on germination and initial seedling growth of northern red oak. *Forest Sci.* 16:453–459.

_____. 1974a. Physiological importance of changes in endogenous hormones during red oak acorn stratification. *Forest Sci.* 20:187–191.

_____. 1974b. Possible methods for improvement of yellow poplar seed germination. In: *Forestry Research Review—1974*. ARS-USDA, Wooster, OH. 4–5.

Voichal, P. I., & A. I. Barabin. 1983. Possibility of determining spruce seed yields by calculation. *Lesnoi Zhurnal* 14–19.

Vongkaluang, I. 1984. A preliminary study of germination and some ecological aspects of *Calamus peregrinus* in Thailand. In: *Proc. Rattan Seminar, Kuala Kepong*, Malaya. 41–46.

Vora, R. S. 1989. Seed germination characteristics of selected native plants of the lower Rio Grande Valley, Texas. *Jour. Range Manage.* 42:36–40.

Vorchleva, N. S. 1981. Some features of embryo growth in two species of *Fraxinus* in relation to seed dormancy. *Botanicheskii Zhurnal* 66:1763–1769.

Voyiatzis, D. G., & I. C. Porlingis. 1987. Temperature requirements for the germination of olive seeds. *Hort. Sci.* 62:405–411.

Vozzo, A. 1973. Germination analysis of excised embryo cylinders and whole acorns of water oak. Forest Service, USDA, New Orleans, LA.

_____. 1978. Carbohydrates, lipids, and proteins in ungerminated and germinated *Quercus alba* embryos. *Forest Sci.* 24:486–493.

_____. 1985. Pericarp changes observed during *Quercus nigra* germination. *Seed Sci. Tech.* 13:1–9.

Wadia, K. R.R., C. Manocharachary, & K. Janakr. 1984. Fruit rot of Indian gooseberry and survival of its pathogen on *Calophyllum inophyllum* fruits. *Phytopathological Notes* 37:565–566.

Walters, G. A. 1974a. *Araucaria*. In: *Seeds of Woody Plants in the United States*. Forest Service, USDA, Washington, DC. 223–225.

_____. 1974b. *Cryptomeria*. In: *Seeds of Woody Plants in the United States*. Forest Service, USDA, Washington, DC. 361–362.

_____. 1974c. *Toona*. In: *Seeds of Woody Plants in the United States*. Forest Service, USDA, Washington, DC. 813–814.

Walters, G. A., F. T. Bonner, & E. Q. P. Petteys. 1974. *Pithecellobium*. In: *Seeds of Woody Plants in the United States*. Forest Service, USDA, Washington, DC. 639–640.

Wang, B. S. P. 1973. Laboratory germination criteria for red pine seed. *Proc. ASOA* 63:94–101.

_____. 1974. Testing and treatment of Canadian white spruce seed to overcome dormancy. *Proc. ASOA* 64:8.

Wang, B. S. P. & B. D. Hadden. 1978. Germination of red maple seed. *Seed Sci. Tech.* 6:785–790.

Wang, T. T., & F. W. Hung. 1979. A study of seed germination and seedling growth of the genus *Paulownia* in Taiwan. Yen Chiu Pao Kao Tech. Bul. 123. 41–71.

Ward, J. M. 1967. Studies in ecology of a shell barrier beach. III. Chemical factors of the environment. *Vegetato* 15:77–112.

Warr, H. J., D. R. Savory, & A. K. Bal. 1979. Germination studies of bake apple seeds. *Canadian Jour. Plant Sci.* 59:69–74.

Warren, D. C., & B. L. Kay. 1983. Pericarp inhibition of germination of *Atriplex confertifolia* (Torr. and Frem.). In: *Proc. Symp. Biol. Atriplex and Related Chenopods*. Gen. Tech. Rpt. 172. Forest Service, USDA, Ogden, UT. 168–174.

Washitani, I. 1988. Effects of high temperature on the permeability and germinability of the hard seeds of *Rhus javanica*. *Annals Botany* 62:13–16.

Washitani, I., & T. Saeki. 1986. Germination response of *Pinus densiflora* seeds to temperature, light and interrupted inhibition. *Jour. Experimental Botany* 37:1376–1387.

Wassermann, V. D., & H. C. J. Agenbas. 1978. Effect of harvesting at different ripening stages on seed quality and germination capacity of the narrow-leaped lupinus. *South African Crop Production* 7:65–68.

Wcislinska, B. 1977. Studies of the effects of temperature and other factors on the germination of ash. *Sylva* 121:65–77.

Webb, D. P., & E. B. Dombroff. 1969. Factors influencing the stratification process in seeds of *Acer saccharum*. *Canadian Jour. Botany* 47:1555–1563.

Webb, D. P., and P. F. Waring. 1972a. Seed dormancy in *Acer*: Endogenous germination inhibitors and dormancy in *Acer pseudoplatanus* L. *Planta* 104:115–125.

_____. 1972b. Seed dormancy in *Acer*: the role of covering structures in the dormancy of *Acer platanoides*. *Jour. Experimental Botany* 23:813–829.

Weber, G. P., L. E. Wiesner, & R. E. Lund. 1982. Improving germination of Skunkbush sumac and serviceberry seed in the western United States. *Jour. Seed Tech.* 7:60–71.

Weerakeen, W. L., & J. V. Levett. 1986. Studies of *Salvia reflexa* Hornew. 3. Factors controlling germination. *Weed Res.* 28:269–276.

Weissen, F. 1980. Germination of beech trees preserved at low temperature. *Bul. de la Societe Royale Forestiere de Belgique* 87:81–88.

_____. n.d. Ten years' observations on the regeneration of beech forest in the Ardennes. IRSIA, Gemblous, Belgium.

Wendel, G. W. 1974. *Celastrus scandens*. In: *Seeds of Woody Plants in the United States*. Forest Service, USDA, Washington, DC. 295–297.

_____. 1977. Longevity of black cherry, wild grape, and sassafras seed in the forest floor. Forest Service, USDA, Hamden.

Wheeler, N. C., C. J. Masters, S. C. Cade, S. D. Ross, & L. Y. Hsin. 1985. Girdling: an effective and practical treatment for enhancing seed yields in Douglas-fir seed orchards. *Canadian Jour. Forest Res.* 15:505–510.

White, J., & P. H. Lovell. 1984. Anatomical changes which occur in cuttings of *Agathis australis* (D.Don) Lindl. 2. The initiation of root primordia and early root development. *Annals Botany* 54:633–645.

White, P. S. 1984. The architecture of devil's walking stick, *Aralia spinosa*. *Arboretum* 65:403–418.

Whitesell, C. D. 1974a. *Acacia*. In: *Seeds of Woody Plants in the United States*. Forest Service, USDA, Washington, DC. 184–186.

_____. 1974b. *Lucaena*. In: *Seeds of Woody Plants in the United States*. Forest Service, USDA, Washington, DC. 491–493.

Whitmore, T. C. 1972. *Tree Flora of Malaya*. Longman, London.

Wick, H. L. 1974. *Flindersia*. In: *Seeds of Woody Plants in the United States*. Forest Service, USDA, Washington, DC. 409–410.

Wick, H. L., & G. A. Walters. 1974. *Albizia*. In: *Seeds of Woody Plants in the United States*. Forest Service, USDA, Washington, DC. 203–205.

Widmoyer, F. B., & A. Moore. 1968. The effect of storage period temperature and moisture on the germination of *Aesculus hippocastanum* seed. *Plant Propagation* 14:14–15.

Wilcox, M. D., & A. Firth. 1980. Artificial ripening of green *Pinus radiata* cones does not reduce seed germination or seedling vigor. *New Zealand Jour. Forestry* 10:363–366.

Willan, R. F. 1985. *A Guide to Forest Seed Handling*. DANDA-FAO, Rome.

Williams, L. 1963. Lactiferous plants of economic importance. 4. Jelutong (*Dyera* spp.). *Economic Botany* 17:110–126.

Williams, P. M., and I. Arias. 1978. Physio-ecological studies of plant species from the arid and semi-arid regions of Venezuela. I. The role of endogenous inhibitors in the germination of the seeds of *Cereus griseus*. *Acta Cientifica Venezolana* 29:93–97.

Williams, R. D. 1980. Period in stratificaion hastens germination of black walnut seed. *Proc. Indiana Acad. Sci.* 1980:80–94.

Williams, R. D., & J. E. Winstead. 1972. Population variation in seed germination and stratification of *Acer negundo* L. *Kentucky Acad. Sci. Trans.* 33:43–48.

Willson, M. F., & D. W. Schemske. 1980. Pollinators' limitations, fruit production and floral display in paupau (*Asimina triloba*). *Bul. Torrey Bot. Club* 107:401–408.

Winer, N. 1983. Germination of pretreated seed of mesquite under arid conditions in northern Sudan. *Forest Ecology and Manage.* 5:307–312.

Win Wink, M. 1983. Inhibition of seed germination by quinolizdme alkalides, aspects of allelopathy in *Lupinus albus*. *Planta* 158:365–368.

Winstead, J. E. 1972. Population differences in seed germination and stratification requirements of sweetgum. *Forest Sci.* 17:34–36.

Winston, D. A., & B. D. Hadden. 1981. Effect of early cone collection and artificial ripening on white spruce and red pine germination. *Canadian Jour. Forest Res.* 11:817–826.

Wirges, G., & J. Yeiser. 1984. Stratification and germination of Arkansas oak acorns. *Tree Planter's Notes* 35:36–38.

Wisenant, S. G., & D. N. Ueckert. 1982. Germination responses of *Eysenhardtia texana* and *Leucaena*. *Jour. Range Manage.* 35:748–750.

Wolf, C. B., & W. W. Wagener. 1948. *The New World Cypress*. Rancho Santa Ana Botanical Garden, El Aliso, CA.

Wong, W. H. C., Jr. 1974. *Grevillea*. In: *Seeds of Woody Plants in the United States*. Forest Service, USDA, Washington, DC. 437–438.

Wood, B. W. 1984. Free and bound abscissic acid and free gibberellin-like substances in pecan kernel tissue during seed development. *Jour. American Soc. Hort. Sci.* 109:626–629.

Wood, M. K., R. W. Knight, & J. A. Young. 1976. Spiny hopsage germination. *Jour. Range Manage.* 29:53–56.

Woodward, P. M. 1983. Germination success of *Pinus contorta* and *Picea engelmannii* on burned seedbeds. *Forest Ecology and Manage.* 5:301–306.

Woodward, P. M., & G. Cummins. 1987. Engelmann spruce, lodgepole pine, and subalpine fir germination success on ash bed conditions. *Northwest Sci.* 61:233–238.

Woodward, R. A., & H. Land. 1984. Suppression of sugar pine by Douglas fir in a northern California plantation. *Tree Planter's Notes* 35:11–12.

Workman, J. P., & N. E. West. 1969. Ecotypic variation of *Eurotia lanata* populations in Utah. *Botanical Gazette* 130:26–35.

Wynens, J. C., & T. L. Brooks. 1979. Seed collection from loblolly pine. Georgia Forest Res. Paper No. 3.

Yamamori, N., & H. Oyama. 1974. Studies on the optimum temperature of germination of Ryukyu-matsu (*Pinus luchuensis* Mayr) seeds. *Sci. Bul. College Agr., Univ. Ryukyus* 21:603–607.

Yamamoto, N., & A. Masuda. 1979. Germinability of pine seeds during maturation stages. *Bul. Utsunomiya Forest* 75–79.

Yanagisawa, T. 1965. Effect of cone maturity on the viability and longevity of coniferous seed. *Jap. Govt. Forest Exp. Sta. Bul.* 172:45–49.

Yang, J. C. 1976. Studies on viability test of Taiwan red pine seeds by TTC staining and X-ray contrast methods. *Quarterly Jour. Chinese Forestry* 9:77–90.

Yao, N. Y. N. 1979. Introduction of tropical and subtropical pine species for testing in Taiwan. Four year-nine-month results. Memoirs of the College of Agriculture, National Taiwan University 19:29–51.

Yap, S. K. 1980. Jelutong: phenology, fruit and seed biology. *Malaysian Forester* 43:309–316.

———. 1981. Collection, germination and storage of dipterocarp seeds. *Malaysian Forester* 44:281–300.

Yap, S. K., & S. M. Wong. 1983. Seed biology of *Acacia mangium, Albizia falcataria, Eucalyptus* spp., *Gmelina arborea, Maesopsis eminii, Pinus caribaea* and *Tecona grandis*. *Malay-Forest* 46:26–45.

Yeiser, J. L. 1983. Germinative pretreatment and seed coat impermeability for the Kentucky coffeetree. *Tree Planter's Notes* 34:33–35.

Ying, C. C., J. C. Murphy, & S. Anderson. 1985. Cone production and seed yield of lodgepole pine grafts. *Forestry Chronicle* 61:223–228.

Young, J. A., & R. A. Evans. 1976. Stratification of bitter-brush. *Jour. Range Manage.* 29:421–425.

———. 1980. Identification of seeds of big sagebrush taxa. *Newsletter ASOA* 52:26–27.

———. 1981. Germination of seeds of antelope and desert bitterbrush and cliffrose. Agr. Res. Ser., USDA, Oakland, CA.

Young, J. A., & R. A. Evans, J. D. Budy, & D. E. Palmquist. 1988. Stratification of seeds of western and Utah juniper. *Forest Sci.* 34:1058–1060.

Young, J. A., & R. A. Evans, & B. L. Kay. 1977. *Ephedra* seed germination. *Agronomy Jour.* 69:209–211.

Young, J. A., & R. A. Evans, & D. L. Neal. 1978. Treatment of curlleaf cercocarpus seeds to enhance germination. *Jour. Wildlife Manage.* 42:614–620.

Young, J. A., & R. A. Evans, R. Stevens, & R. L. Everett. 1981. Germination of *Kochia prostrata* seeds. *Agronomy Jour.* 73:957–961.

Young, J. A., D. W. Hedrick, & R. F. Kenston. 1967. Forest cover and logging-herbage and browse production in the mixed coniferous forest of northeastern Oregon. *Jour. Forestry* 65:807–813.

Young, J. A., B. L. Kay, H. George, & R. A. Evans. 1980. Germination of three species of *Atriplex*. *Agronomy Jour.* 72:705–709.

Young, J. A., & C. G. Young. 1985. *Collecting, Processing and Germinating Seeds of Wildland Plants*. Timber Press, Portland, OR.

Zasada, J. C. 1983. Effect of cone storage method and collection date on Alaskan white spruce seed quality. In: *Proc.*

Symp. Seed Processing and Seed Problems. Inter. Union Forestry Res. Org. 5.

Zasada, J. C., & R. Densmore. 1980. Alaskan willow and balsam poplar seed viability after 3 years storage. *Tree Planter's Notes* 31:9–10.

Zavarin, E., K. Saiberk, & P. Senter. 1976. Analysis of terpenoides from seed coats as a means of identifying seed origin. *Forest Sci.* 25:20–24.

Zelenskii, M. A., & V. M. Sidorova. 1973. Growing Scots pine seedlings with the use of trace elements. *Lesnoi Zhurnal* 16:26–28.

Zensen, F. 1980. Improved techniques for western larch. *Tree Planter's Notes* 31:23–25.

Zentsch, W., Y. Diaz. 1977. Investigations on the germination of *Calophyllum brasiliense*. *Beitrage fur die Forstwirtschaft* 11:73–74.

Zhang, L. C., W. M. Guo, & Y. S. Chen. 1981a. Experimental investigations of the physiology of afterripening of Korean pine seed. *Jour. North-Eastern Forest Inst.* (China) 4:43–50.

———. 1981b. Studies on the relationship between peroxide activity and germination of *Pinus koraiensis* seeds and on metabolic inhibitors in the seeds. *Acta Botanica Sinica* 25:53–61.

Zhoa, H. Z. 1983. The effect of hormones on dormancy and germination of *Fraxinus mandshurica* seeds. *Jour. North-Eastern Forest Inst.* (China) 11:7–12.

Zhou, Y. X. 1987. Characteristics of dormancy and germination of *Pinus taeda* seeds. *Plant Physiology Communications* 1987:22–26.

Zietsman, P. C., & F. C. Botha. 1987. Seed germination of *Ziziphus mucronata* subsp. *mucronata*. *South African Jour. Botany* 53:341–344.

Zimmerman, G. A. 1960. Hybrids of American paupau. *Jour. Heredity* 32:82–92.

Zobel, D. B. 1979. Seed production in forest of *Chamaecyparis lawsoniana*. *Canadian Jour. Forest Res.* 9:327–335.

Zohar, Y., Y. Waiseland, & R. Karschon. 1975. Effect of light, temperature, and osmotic stress on seed germination of *Eucalyptus occidentalis*. *Australian Jour. Botany* 23:391–397.

Zviedre, A. A., A. Y. Dzintare, & G. Igaunis. 1984. Storing a reserve supply of seed of Norway spruce. *Lesnoe Khozyaistvo* 1984:333–338.

Zwaaw, J. D. 1978. The effects of hot water treatment and stratification on germination of blackwood (*Acacia melanoxylon*) seed. *South African Forestry Jour.* 105:40–42.